Applied Signal and Image Processing:
Multidisciplinary Advancements

Rami Qahwaji
University of Bradford, UK

Roger Green
University of Warwick, UK

Evor Hines
School of Engineering, University of Warwick, UK

T0338751

Senior Editorial Director:	Kristin Klinger
Director of Book Publications:	Julia Mosemann
Editorial Director:	Lindsay Johnston
Acquisitions Editor:	Erika Carter
Development Editor:	Joel Gamon
Production Coordinator:	Jamie Snavely
Typesetters:	Keith Glazewski & Natalie Pronio
Cover Design:	Nick Newcomer

Published in the United States of America by
Information Science Reference (an imprint of IGI Global)
701 E. Chocolate Avenue
Hershey PA 17033
Tel: 717-533-8845
Fax: 717-533-8661
E-mail: cust@igi-global.com
Web site: http://www.igi-global.com

Library of Congress Cataloging-in-Publication Data

Applied signal and image processing : multidisciplinary advancements / Rami
Qahwaji, Roger Green, and Evor Hines, editors.
      p. cm.
   Includes bibliographical references and index.
   Summary: "This book highlights the growing multidisciplinary nature of
signal and image processing by focusing on emerging applications and recent
advances in well-established fields, covering state-or-the-art applications in
both signal and image processing, which include optical communication and
sensing, wireless communication management, face recognition and facial
imaging, solar imaging and feature detection, fractal analysis, and video
processing"-- Provided by publisher.
   ISBN 978-1-60960-477-6 (hardcover) -- ISBN 978-1-60960-478-3 (ebook) 1.
Image processing. 2. Signal processing. 3. Image analysis. 4. Wireless
communication systems. I. Qahwaji, Rami, 1972- II. Green, Roger, 1951- III.
Hines, Evor, 1957-
   TA1637.A69 2011
   621.382'2--dc22
                                2011003448

British Cataloguing in Publication Data
A Cataloguing in Publication record for this book is available from the British Library.

List of Reviewers

Abbas Mohammed, *Blekinge Institute of Technology, Sweden*
Al-Omari Mohammad, *Applied Science University, Jordan*
Banat Moahammad, *Jordan University of Science and Technology, Jordan*
Colak Tufan, *University of Bradford, UK*
David Perez-Suarez, *Trinity College Dublin, Ireland*
Dobrescu Radu, *Politehnica University, Romania*
Evor Hines, *University of Warwick, UK*
Fu Zhang, *University of Warwick, UK*
Grecos Christos, *University of the West of Scotland, UK*
Higgins Paul, *Trinity College Dublin, Ireland*
Hui Fang, *Swansea University, UK*
Ipson Stanley, *University of Bradford, UK*
Mark Leeson, *University of Warwick, UK*
Moi Hoon Yap, *University of Bradford, UK*
Popescu Dan, *Politehnica University, Romania*
Qawhaji Rami, *University of Bradford, UK*
Reza Ghaffari, *University of Warwick, UK*
Roger Green, University of Warwick, UK
Simant Prakoonwit, *University of Reading, UK*
Thierry Dudok de Witt, *University of Orleans, France*
Waleed Al-Nuaimy, *University of Liverpool, UK*

Table of Contents

Section 2
Multidisciplinary Advancements in Image Processing

Detailed Table of Contents

Section 1
Multidisciplinary Advancements in Signal Processing

Roger J. Green, University of Warwick, UK
Matthew Higgins, University of Warwick, UK
Harita Joshi, University of Warwick, UK

This chapter covers a broad area within the domain of optical communications, and, specifically, optical wireless communications. There are several challenges within the field, concerned with distribution of the optical field, signal bandwidth, and noise. The authors examine three specific areas of the technology which address these issues, namely modulation methods for optical wireless, genetic algorithm-based methods for optimisation of the optical fields indoors for power and bandwidth uniformity, and then receiver-amplifier techniques for bandwidth and sensitivity maximisation.

Abbas Mohammed, Blekinge Institute of Technology, Sweden
David Last, University of Bangor, UK

Skywave interference commonly affects the performance of LOng RAnge Navigation (Loran) receivers. Traditional skywave rejection methods that use fixed, worst-case, sample timing are far from optimal. This chapter reports on novel signal processing techniques for measuring in real-time the delay and strength of the varying skywave components of a Loran signal relative to the groundwave pulse. The merits and limitations of these techniques will be discussed. Their effectiveness will be assessed by theoretical analysis, computer simulations under a range of realistic conditions, and by testing using off-air signals. A prototype Loran system employing the proposed techniques is also presented.

Chapter 3

Application of Space-Time Signal Processing and Active Control Algorithms for the
Suppression of Electromagnetic Fields.. 45

Tommy Hult, Lund University, Sweden
Abbas Mohammed, Blekinge Institute of Technology, Sweden

Several studies have been conducted on the effects of radiation on the human body. This has been especially important in the case of radiation from hand held mobile phones. The amount of radiation emitted from most mobile phones is very small, but given the close proximity of the phone to the head it might be possible for the radiation to cause harm. In this chapter adaptive active control algorithms and a full space-time processing system setup (i.e. multiple antennas at both the transmitter and receiver side or MIMO) are implemented to reduce the possibly harmful electromagnetic radiation emitted by hand held mobile phones. Simulation results show the possibility of using the adaptive control algorithms and MIMO antenna system to attenuate the electromagnetic field power density.

Chapter 4

Data Broadcast Management in Wireless Communication: An Emerging Research Area................... 61

Seema Verma, Bansathali University, India
Rakhee Kulshrestha, Birla Institute of Technology and Science, India
Savita Kumari, University of Seventh April, Libya

In this chapter, recent advances in the field of Wireless Communication Technology (WCT) within the context of mobile information services are dicussed. Various technologies and the advent of new gadgets of communication have made many mobile communication applications a reality. How to disseminate data effectively to a large number of users with minimum consumption of time and physical resources of client in WCT environment is a challenge to system. The WCT is hindered by factors like low battery power, frequent disconnection, asymmetric and heterogeneous broadcast and scalability etc. To overcome these difficulties various data management strategies along with broadcast mode of dissemination are implemented.

Chapter 5

Blind Equalization for Broadband Access using the Constant Modulus Algorithm.......................... 76

Mark S. Leeson, University of Warwick, UK
Eugene Iwu, DHL Supply Chain, UK & Ireland Consumer Division, Solstice House, 251

The cost of laying optical fiber to the home means that digital transmission using copper twisted pairs is still widely used to provide broadband Internet access via Digital Subscriber Line (DSL) techniques. However, copper transmission systems were optimally designed for voice transmission and cause distortion of high bandwidth digital information signals. Thus blind equalization, such as the Constant Modulus Algorithm (CMA,) is needed to ameliorate the effects of the distortion. This algorithm is investigated in more details in this chapter.

Chapter 6

Fu Zhang, School of Engineering, University of Warwick, UK
Reza Ghaffari, School of Engineering, University of Warwick, UK
Daciana Iliescu, School of Engineering, University of Warwick, UK
Evor Hines, School of Engineering, University of Warwick, UK
Mark Leeson, School of Engineering, University of Warwick, UK
Richard Napier, Warwick HRI, University of Warwick, UK

This chapter presents the initial studies on the detection of two common diseases and pests, the powdery mildew and spider mites, on greenhouse tomato plants by measuring the chemical volatiles emitted from the tomato plants as the disease develops using a Field Asymmetric Ion Mobility Spectrometry (FAIMS) device. The processing on the collected FAIMS measurements using Principal component analysis (PCA) shows that clear increment patterns can be observed on all the experimental plants representing the gradual development of the diseases. Optimisation on the number of dispersion voltages to be used in the FAIMS device shows that reducing the number of dispersion voltages by a factor up to 10, preserves the key development patterns perfectly, though the amplitudes of the new patterns are reduced significantly.

Chapter 7

Reza Ghaffari, School of Engineering, University of Warwick, UK
Fu Zhang, School of Engineering, University of Warwick, UK
Daciana Iliescu, School of Engineering, University of Warwick, UK
Evor Hines, School of Engineering, University of Warwick, UK
Mark Leeson, School of Engineering, University of Warwick, UK
Richard Napier, Warwick HRI, University of Warwick, UK

This chapter introduces the principles of some of the most widely used supervised and unsupervised Pattern Recognition (PR) techniques and assesses behaviour and performances. A dataset acquired from a set of experiments conducted at University of Warwick is employed to construct a case study in which the techniques will be applied. We will also evaluate the integration of PR methods with an Electronic Nose (EN) device to develop and implement a plant diagnosis tool based on discriminating the Organic Volatile Compounds (VOC) released by plants when attacked by pest. The chapter concludes with a performance comparison and a brief discussion of how an appropriate PR technique can be coupled with an EN to produce a greenhouse plant pest and disease diagnosis system for day-to-day utilisation. Some consideration of further work is also presented

Chapter 8

Ali Al-Ataby, University of Liverpool, UK
Waleed Al-Nuaimy, University of Liverpool, UK

Non-destructive testing (NDT) is commonly used to monitor the quantitative safety critical aspects of manufactured components and forms one part of quality assurance (QA) procedures. The strategic trend in the development of NDT has changed towards the issue of safety in the broadest sense, to the protection of the population and the environment against man-made and natural disasters. This chapter describes some recent advances in signal processing as applied to NDT problems. This is an area that has made progress for over twenty years and its importance is gaining attention gradually especially since the new advanced techniques in signal processing and pattern recognition.

Chapter 9

Mark J. Bentum, ASTRON, The Netherlands & University of Twente, The Netherlands
André W. Gunst, ASTRON, The Netherlands
Albert Jan Boonstra, ASTRON, The Netherlands

The Low Frequency Array (LOFAR) is a large radio telescope based on phased array principles, distributed over several European countries with its central core in the Northern part of the Netherlands. LOFAR detects the incoming radio signals by using an array of simple omni-directional antennas. The antennas are grouped in so called stations mainly to reduce the amount of data generated. More than forty stations will be built, mainly within a circle of 150 kilometres in diameter. The signals of all the stations are transported to the central processor facility, where all the station signals are correlated with each other, prior to imaging. In this chapter the signal processing aspects on system level will be presented. Methods to image the sky will be given and the mapping of these concepts to the LOFAR phase array radio telescope will be presented. Challenges will be addressed and potentials for further research will be presented.

Section 2
Multidisciplinary Advancements in Image Processing

Chapter 10

Hui Fang, Swansea University, UK
Nicolas Costen, Manchester Metropolitan University, UK
Phil Grant, Swansea University, UK
Min Chen, Swansea University, UK

In this chapter, approaches to extracting features via the motion subspace for improving face recognition from moving face sequences, are discussed. Although the identity subspace analysis has achieved reasonable recognition performance in static face images, more recently, there has been an interest in motion-based face recognition. This chapter reviews several state-of-the-art techniques to exploit the motion information for recognition and investigates the permuted distinctive motion similarity in the motion subspace. The motion features extracted from the motion subspaces are used to test the performance based on a verification experimental framework.

Chapter 11

Moi Hoon Yap, University of Bradford, UK
Hassan Ugail, University of Bradford, UK

The application of computer vision in face processing remains an important research field. This chapter reviews and demonstrates the computer vision techniques applied to facial image processing and analysis. In addition, recent advances in facial image processing in computer vision are discussed.

Chapter 12

Thierry Dudok de Wit, LPC2E, CNRS and University of Orléans, France

The emergence of a new discipline called space weather, which aims at understanding and predicting the impact of solar activity on the terrestrial environment and on technological systems, has led to a growing need for analysing solar images in real time. The rapidly growing volume of solar images, however, makes it increasingly impractical to process them for scientific purposes. This situation has prompted the development of novel processing techniques for doing feature recognition, image tracking, knowledge extraction, etc. In this chapter the concepts and applications of Blind Source Separation (BSS) and multiscale (multiresolution, or wavelet) analysis are investigated for solar images.

Chapter 13

David Pérez-Suárez, Trinity College Dublin, Ireland
Paul A. Higgins, Trinity College Dublin, Ireland
D. Shaun Bloomfield, Trinity College Dublin, Ireland
R.T. James McAteer, Trinity College Dublin, Ireland
Larisza D. Krista, Trinity College Dublin, Ireland
Jason P. Byrne, Trinity College Dublin, Ireland
Peter. T. Gallagher, Trinity College Dublin, Ireland

The solar surface and atmosphere are highly dynamic plasma environments, which evolve over a wide range of temporal and spatial scales. Large-scale eruptions, such as coronal mass ejections, can be accelerated to millions of kilometers per hour in a matter of minutes, making their automated detection and characterisation challenging. Additionally, there are numerous faint solar features, such as coronal holes and coronal dimmings, which are important for space weather monitoring and forecasting, but their low intensity and sometimes transient nature makes them problematic to detect using traditional image processing techniques. These difficulties are compounded by advances in ground- and space-based instrumentation, which have increased the volume of data. These issues are tackled in this chapter.

Texture analysis research attempts to solve two important problems: texture segmentation and texture classification. In this chapter, two classes of features are proposed in the decision theoretic recognition problem for textured image classification. Practical experiments are conducted to test the efficiency of the proposed methods.

In this chapter, a multitask primary processing pipeline configuration is introduced to combine speed and flexibility of an optimum hardware/software configuration. This structure, which is an interface between the sensing element (camera) and the main processing system, achieves real time video signal preprocessing, during the image acquisition time. A case study is presented and fully discussed.

In this chapter, recent advances in corneal imaging, especially in the fields of image enhancement and registration are discussed. The challenges facing this emerging area of research are discussed along with future research trends.

This chapter presents a generalised framework for multi-objective optimisation of video CODECs for use in off-line, on-demand applications. In particular, an optimization scheme is proposed to determine the optimum coding parameters for a H.264 video standard in a memory and bandwidth constrained environment, which minimises codec complexity and video distortion.

A rapid 3D reconstruction of bones and other structures during an operation is an important issue. However, most of existing technologies are not feasible to be implemented in an intraoperative environment. Normally, a 3D reconstruction has to be done by a Computed tomography (CT) or a Magnetic resonance imaging (MRI) pre operation or post operation. Due to some physical constraints, it is not feasible to utilise such machine intraoperatively. In this chapter a special type of MRI has been developed to overcome the problem. This chapter discusses a possible method to use a small number, e.g. 5, of conventional 2D X-ray images to reconstruct 3D bone and other structures intraoperatively.

Arabic text recognition is receiving more attentions from both Arabic and non-Arabic-speaking researchers. This chapter provides a general overview of the state-of-the-art in Arabic Optical Character Recognition (OCR) and the associated text recognition technology. It also investigates the characteristics of the Arabic language with respect to OCR and discusses related research on the different phases of text recognition including: pre-processing and text segmentation, common feature extraction techniques, classification methods and post-processing techniques. Moreover, the chapter discusses the available databases for Arabic OCR research and lists the available commercial Software. Finally, it explores the challenges related to Arabic OCR and discusses possible future trends.

Preface

Image and signal processing techniques are receiving increasing interest because of their different applications in our daily life. We live in a world that is becoming more and more data rich. Today's data is available in different forms (e.g. textual, visual, audio, etc) and in different wavelengths (e.g., x-rays, infra-red, radio, microwave, optical, etc) and even in different dimensions (e.g., 1-D textual and audio, 2-D images and videos, 3-D Graphics, etc.). This creates the need for novel multidisciplinary solutions for automated data processing and analysis, data associations, knowledge extraction, representation of spatial relations, time-based analysis, et cetera.

This inspired us to edit a book that highlights the growing multi-disciplinary nature of signal and image processing by focusing on some of the emerging applications and some of the recent advances in well-established fields. This book covers different state-of-the-art applications in both signal and image processing. The signal processing applications include recent advances in: optical communication and sensing, minimising interference signals in long range navigation receivers, suppression of electromagnetic fields, wireless communication management, broadband access equalisation, electronic nose, spectrometry, non-destructive testing, and application of the new low frequency array. The image processing applications include recent trends in: face recognition and facial imaging, solar imaging and feature detection, fractal analysis, real-time imaging, corneal imaging, video processing, 3-D reconstruction and Arabic optical character recognition. This book will also cover the state-of-the-art in the investigated topics and will identify open research problems that need to be resolved along with some future research directions.

We believe nothing represents this book better than the visionary words of Norbert Wiener (1894-1964)

"The most fruitful areas for growth of the sciences are those between established fields. Science has been increasingly the task of specialists, in fields which show a tendency to grow progressively narrower. Important work is delayed by the unavailability in one field of results that may have already become classical in the next field. It is these boundary regions of science that offer the richest opportunities to the qualified investigator."

Finally, the editors would like to thank all the chapter authors and the reviewers. The sincere efforts and hard work of these people, made this book a reality.

Rami Qahwaji, Roger Green, and Evor Hines

21/06/2010 – United Kingdom

Section 1
Multidisciplinary Advancements in Signal Processing

Chapter 1
Signal Processing for Optical Wireless Communications and Sensing

Roger J. Green
University of Warwick, UK

Matthew Higgins
University of Warwick, UK

Harita Joshi
University of Warwick, UK

ABSTRACT

This chapter covers a broad area within the domain of optical communications, and, specifically, optical wireless communications, as it shares particular features with both RF wireless and optical fiber communications. There are several challenges within the field, concerned with distribution of the optical field, signal bandwidth, and receiver SNR. This chapter examines three specific areas of the technology which address these issues, namely, modulation methods for optical wireless using OFDM, genetic algorithm-based methods for optimisation of the optical fields indoors for power and bandwidth uniformity, and receiver-amplifier techniques for bandwidth and sensitivity maximisation.

INTRODUCTION

Optical wireless is a subject which more recently has attracted much attention, as it is able to offer high bandwidth communications amidst a growing clamouring for space in the electromagnetic

DOI: 10.4018/978-1-60960-477-6.ch001

spectrum. It offers direct competition to terahertz technologies, as the devices used for optical wireless are, in most respects, very similar to those used in optical fibre communications, and so are well developed, cheap, and versatile. The topic is not without certain specific problems, in spite of its theoretical ease of use. The main issues are associated with the background illumination aspects, as solar radiation and even room lighting can be quite strong contenders for the usually weak infrared signature which comprises the signal to be detected. Associated with the ambient illumination is also noise. On the one hand, therefore, there is the availability of bandwidth which is compatible with the many Gigahertz available on fiber systems, and then the other aspect is that the shot noise associated with the front end receiver amplifier is powerfully related to bandwidth by a cube law, above the 3 dB point, if equalisation is applied. Equalisation can be used to flatten frequency response, as, unlike the situation for an optical fiber with a 9μm diameter core, an optical wireless receiver requires a large area device to collect the signal, resulting in a commensurate adverse capacitance. Here are ways around this, and work at Warwick University on optical antennas for example has found a way around this problem using an Optical Antenna (Green & Ramirez-Iniguez, 2000). Additionally, optimised use of the receiver amplifier as well can give significant improvements overall in terms of the SNR versus bandwidth trade-off. OFDM is an effective method for making the most of system bandwidth once this has been optimised, and it poses some interesting challenges in the optical wireless scenario, which can be evaluated, and, to some extent, mitigated against. Infrared sources tend to distribute their output over an area, and this inevitably results in an optical field which is anything but uniform. Fortunately, systems can be devised which use the genetic algorithm (GA) approach to regulation of the optical field, resulting in, not only equalisation of the SNR over the areas of interest, but also the available bandwidth over the same area. We shall firstly discuss OFDM for modulation of the optical carrier, and then examine the GA approach to regulate the optical field, and finally examine standard- and new optical receiver-amplifiers.

OFDM FOR THE INDOOR WIRELESS OPTICAL CHANNEL

Background

Orthogonal Frequency division multiplexing (OFDM) is now becoming the most extensively employed modulation technique for emerging broadband wired and wireless communication systems due to its robustness against intersymbol interference (ISI) arising out of multipath dispersion within a communication channel. At high data rates and increased channel dispersion, OFDM imparts less degree of complexity at the receiver end due to employment of frequency domain equalisation when compared with conventional serial modulation techniques employing time domain equalisation such as quadrature amplitude modulation (QAM). Performance of conventional serial modulation techniques is also significantly impacted by precise design of analogue filters while OFDM on the other hand, transfers the complexity of transceivers to the digital domain so that any phase variation with frequency can be corrected in the digital parts of the receiver with relative ease and negligible cost.

Simplistically explained, the basic concept of OFDM involves transmitting data in parallel over a number of different frequencies so that the symbol period is much longer when compared with a serial system involving a single carrier frequency. This results in reduced symbol rate for the same total bit rate. Thus in the worst case, ISI affects at the most one symbol which implies that the OFDM signal experiences little distortion due to multipath dispersion, and equalisation at the receiver end is simplified significantly. The residual ISI can be easily removed by using a cyclic prefix as a guard interval.

It has to be noted that there is a fundamental parallel among OFDM, frequency division multiplexing (FDM) used in conventional wireless systems, and wavelength division multiplexing

Table 1. Key features of OFDM and FDM/WDM

Key Features	OFDM	FDM/WDM
Transmission Format	Parallel - Data are transmitted simultaneously over different frequencies	Parallel - Data are transmitted simultaneously over different frequencies/wavelengths
Subcarrier Frequencies	Mutually orthogonal over one symbol period	Not orthogonal
Modulation/Demodulation	Digital – IFFT/FFT	Analog – demodulation by filtering at the receiver end
Subcarrier Spectra	Overlapping	Non-overlapping – Separated by frequency guard bands
Bandwidth Efficiency	Higher due to overlapping subcarriers	Lower due to frequency guard bands between individual subcarriers
Frequency offset and phase noise sensitivity	Highly sensitive as sidelobes of each subcarrier extend well into other subcarriers	Almost negligible as each subcarrier is spectrally well separated from other subcarriers

(WDM) used in optical systems, as all three modulation schemes involve simultaneously transmitting information in parallel over different frequencies. However, in OFDM these frequencies are chosen so that the transmitted signals are mutually orthogonal over the duration of one symbol period. This property of orthogonality ensures, for a linear channel, that the received subcarriers can be demodulated without interference and without actually separating them through analogue filtering. At the transmitter end, modulation and multiplexing is implemented through an Inverse Fast Fourier Transform (IFFT), generating precise orthogonal subcarriers in a computationally efficient way, whilst at the receiver end, the process of demodulation and demultiplexing is implemented through a Fast Fourier Transform (FFT). The major concern for OFDM is its high sensitivity to frequency offset and phase

noise. This arises from the fact that the spectrum of an individual OFDM subcarrier is of the form of $\left| \sin(x) / x \right|^2$, which implies that the sidelobes of each OFDM subcarrier extend well into the spectra of other OFDM subcarriers. Table 1 provides comparative summary of key features of OFDM and FDM/WDM.

The spectral difference between FDM/WDM and OFDM is shown below.

OFDM for Optical Wireless

The performance of OFDM over an indoor optical wireless channel was investigated in (Gonzalez, *et al*, 2005), where it was shown that this scheme is highly efficient in mitigating the effects of quality fluctuations of the optical wireless channel induced due to variations in spatial distribution

Figure 1. Spectral difference between FDM/WDM and OFDM

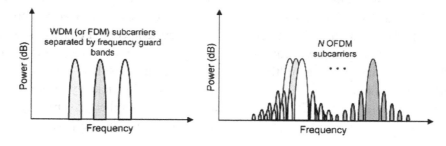

Figure 2. Transmitter block diagram for multiple sub-carrier modulation system

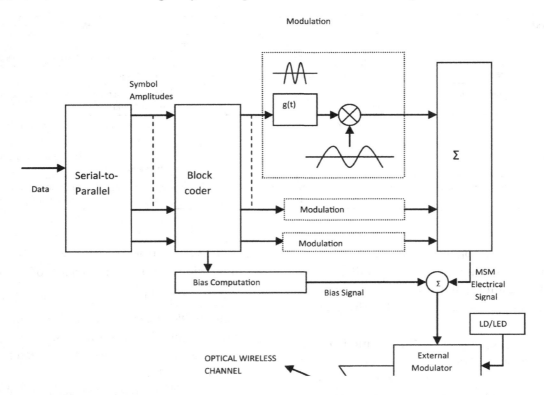

Power Penalty

of emitters and receivers. In order to increase the system throughput over noisy indoor environment, a simple adaptive OFDM scheme was proposed which takes into account the fact that in a wireless optical channel, different subcarriers undergo different gains. Therefore, the number of transmitted bits assigned to each subcarrier has to be adaptively tailored to the channel characteristics by estimating the instantaneous channel response at the receiver. Although this scheme causes an additional amount of transmission overhead due to the feedback from receiver, this does not affect the performance significantly due to the slow time-varying nature of a low mobility indoor wireless environment. The schematic below gives an outline of an optical wireless communication link using OFDM as a special case of a multiple subcarrier modulation technique.

Though OFDM signals avoid the ISI penalty experienced due to high data rate transmission through a multipath optical wireless channel, they do incur a significant power penalty of several dB due to the requirement of a dc bias to be added to the transmitter drive signal in the case of an intensity-modulated system. Basically, the contest between single carrier modulation and multi-carrier modulation is essentially a trade-off between bandwidth efficiency and power efficiency of the transmitted signal. For comparison purpose, figure below shows power penalty incurred by OFDM systems compared with a single sub-carrier system for identical level of modulation. For an uncompensated OFDM system, without utilising any power efficiency improvement scheme, a reasonable number of sub-carriers that can be used for an acceptable performance is four, as

Figure 3. Power penalty incurred by uncompensated multiple sub-carrier systems compared with a single sub-carrier system for identical level of modulation

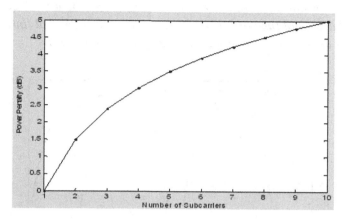

the power penalty in this case is 3dB, i.e. power requirement doubles compared with the single sub-carrier system in this case to maintain the same BER performance.

PAPR Reduction Techniques for OFDM

The main disadvantage of OFDM in a power-limited channel such as the optical wireless channel is the large Peak-to-Average Power Ratio (PAPR). When the overlapping subcarriers are added with the same phase, they produce a peak power which is equal to number of subcarriers times the average power. For a large number of subcarriers, this may cause non-linearity concerns for the external intensity modulators in the optical transmitter and subsequently, result in performance degradation due to optical signal distortion. It can be shown that a higher PAPR, resulting in higher modulation amplitude, worsens the link performance by causing the external intensity modulator to be driven into the nonlinear region of operation.

The simplest and most effective approach to reducing the PAPR appears to be straightforward clipping and filtering (Ochai, 2004), and, if the bit rate per subcarrier is low, the clipping distortion can be tackled with powerful channel coding.

Unfortunately, it is not the same case for a higher data rate system with a large constellation size, where clipping distortion results in a significant performance degradation.

However, a technique proposed in (Akhtman, *et al*, 2003) employs the frequency domain guard band of an OFDM spectrum to accommodate the spectrum of a clipping signal which assists in reducing the PAPR, and is designed such that the in-band clipping distortion is minimised. This approach of using redundant-subcarriers for PAPR reduction is similar to the block coding schemes which incorporate redundant information bits in the time domain. This particular scheme is spectrally efficient, has a low implementation complexity and is suitable for a low-power, optical wireless channel.

Furthermore, a few recent approaches for reducing the PAPR of an OFDM system through block coding are discussed below, which can also be applied to the optical wireless channel. The common idea of these approaches is to introduce redundancy in OFDM symbols in some way so that the OFDM symbol having the smallest PAPR can be chosen for transmission.

An iterative algorithm to reduce the PAPR for a power-limited OFDM system is proposed in (Huang, *et al*, 2007) It works on the known result

Figure 4. BER performance for 16-QAM combined with OFDM modulation for a semiconductor Mach-Zehnder modulator with ideal transfer function with modulating amplitude of (a) 100mV, (b) 110mV, (c) 200mV and (d) 350mV [Transmission data rate was 25 Mbps at a subcarrier frequency of 3.5GHz with 32 OFDM carriers and a symbol time of 5.12µs for a point-to-point optical link with perfectly aligned transmitter and receiver as the simplest case to provide proof of concept]

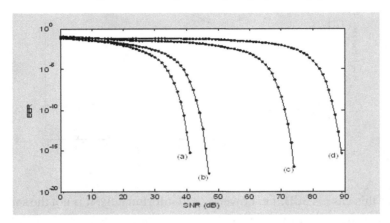

that, given only the frequency-domain magnitude of an unknown time signal, it is possible to recover the frequency domain phase, provided the unknown signal meets a mild set of conditions (Thomas, *et al*, 1981) The algorithm transmits information through the frequency magnitude of the OFDM symbol only, and the phase information is recovered from a low PAPR time signal with desired frequency magnitudes. At the transmitter, this operation causes a loss in the data rate as the information bits to be transmitted through the frequency-phase channel are now sent through the frequency-amplitude channel. At high SNR, this loss of data rate is prohibitive, but at low SNR, as in case of a power-limited noisy optical wireless channel, the rate loss is non-existent, as the frequency-phase channel has zero capacity. The so-called phase synthesis algorithm achieves at least 3dB reduction in PAPR, with a zero rate penalty for low SNR channels.

Another interesting approach to PAPR reduction is proposed in (Ochai *et al*, 2004)) which is based on recursive minimisation of the autocorrelation sidelobes of an OFDM data sequence and which flattens the resulting band-limited OFDM signals. The Viterbi algorithm is applied to devise a metric design to search for a valid code word which minimises the autocorrelation function of the complex OFDM data sequence. The performance of this scheme depends upon signal constellation mapping and dynamic range as well as average power, or the dynamic range only, can be reduced depending upon the respective choice of mapping schemes.

A signal scrambling method, based on combined symbol rotation and inversion (CSRI), is proposed in (Tan, *et al*, 2003) for PAPR reduction of OFDM signals, which divides OFDM sequences into sub-blocks, and performs symbol rotation and inversion in each sub-block so that high degrees of freedom are available to offset the possibility of encountering poor symbol sequences with large PAPR. Symbol inversion by a phase change of π is added to symbol rotation, so that the possibility of having strings of same symbols in an OFDM block (which can be responsible for a high PAPR) is avoided.

INTELLIGENT OPTICAL POWER DISTRIBUTION FOR OPTICAL WIRELESS

The Design Challenge

Indoor optical wireless (OW) is, inherently, a cellular system (Bellon, *et al*, 1999, and Parand, *et al*, 2003). Each room, or section of a room, will have within it a base station linked to the backbone network (Barry, *et al*, 1991). The advantages of OW, such as high speed communications, may be beneficial to scenarios in the workplace, but each workplace will have a different topology of furniture and seating arrangements. The low cost appeal of OW will allow systems to be deployed in residential environments, where, for example in living rooms and bedrooms, the type of furnishing and flooring is also unique. The immunity to RF interference may appeal to heavy industrial users, where there is no furniture but sparsely open-plan factories or warehouses with surfaces such as concrete and steel for example, that are not normally found elsewhere. From a channel perspective, it is therefore reasonable to assume that no two system deployment environments are identical.

These channel variability challenges are complicated further, due the fact that one cannot even assume the environment to be a static quantity. An OW system is about connecting users in a flexible way, such that simple human behaviour, like rearranging the furniture or substituting carpet for wooden flooring, will cause the channel to change slowly over time. The users themselves require the freedom of movement which, through their actions, are going to cause instantaneous (or at least very fast) changes to the channel. Furthermore, it needs to be assumed that there are multiple end users, each one unique, and each one of these users has upon them a battery powered portable device such a mobile phone or laptop. Historically, with work from standards bodies such as the IrDA, each of the devices is most likely to be very similar, or

conform to some minimum technical specification of operational requirement. Moreover, from a marketing point of view, the standards agencies or the parties with financial interest in the devices, try and make them cheap, and attempt to put them in as many devices as possible. This leads to somewhat challenging system design requirements, of how to make a single, mass-producible and cheap receiver design operate in an essentially infinite number of channel scenarios, whilst still being able to market the device as user-friendly.

Many authors have proposed, with great success, various receiver designs and modulation schemes that can be used to mitigate the limitations of the OW channel (Green *et al*, 2008). Quasi-diffuse configurations employing multisport diffusion (MSD) and diversity receivers (O'Brien, *et al*, 2005) improve the bandwidth and ambient noise rejection through the use of an array of photodetectors coupled to either a single imaging lens (Djahani, *et al*, 2000) or several optical concentrators (Ramirez-Iniguez & Green, 2005). Implementation of automatic gain control can compensate for the variations in received power at different positions within the room (Zhang, *et al*, 2008). Modulation techniques, such as trellis coded pulse position modulation (Lee, *et al*, 1997) and amplitude shift key digital demodulation (Uno, *et al*, 1997) are capable of mitigating the effects of ISI and noise from fluorescent lights, respectively. Each of the aforementioned receiver and modulation techniques proposed have their respective merits in mitigating certain channel factors, such as bandwidth or received power limitations. However, given the cellular nature of OW, one has to weigh up the added complexity, cost and physical size of each receiver within the cell against the benefits they can provide. There is an argument with economies of scale, such that the larger the uptake, the lower the cost of each unit. The argument, however, can also be made the other way, if one considers that the cost and complexity overhead of the entire system will be influenced more by the number of users or receivers present in a cell than by the single base station. Therefore, why should one not shift the complexity of the system into the transmitter?

The transmitter should be cost-effective, but it does not have to be disposably cheap as does the receiver. The transmitter should be power-efficient, but not necessarily using low power, as it is connected to the backbone network, and so is potentially near a power source. Finally, the transmitter should be unobtrusive, but does not have to be miniaturized to the point where it can fit into a mobile device. The transmitter is therefore one possible avenue to explore that could lead to a solution to the system design challenge. Furthermore, this available freedom in the design of the transmitter allows for the possibility of incorporating some intelligent methods as a potential solution to the problems being faced.

Intelligent Methods

Intelligent methods, such as genetic algorithms (GA), neural networks and fuzzy logic have found numerous applications within telecommunications applications. Whether they have been applied to optical fibre systems (Fetterolf & Anandalingam, 1991, Yener & Boult, 1994, and Bannister, *et al*, 1990), free space optics (Desai & Milner, 2005), or RF applications such as antenna design and receiver array control (Telzhensky & Leviatan, 2006, Chen *et al*, 2007, Allard *et al*, 2003, and Haupt, 1995), an argument has always been found to use them from their beneficial properties. However, there has been very little application of the techniques to indoor optical wireless communications.

Very recently, work was published (Wong *et al*, 2005) based upon the use of optimising, through simulated annealing (Kirkpatrick *et al*, 1983) the multispot diffusion pattern generated by holographic diffusers. It was shown that the simulated annealing method could be employed to find the design of a holographic diffuser required to produce either a low power deviation, lower multipath dispersion, or a combination of both.

By changing the holographic diffuser design, an improvement of up to 85% could be achieved in reducing the standard deviation of the received power throughout the room. The work also covers the application in a second environment with lesser improvements. The work was later extended (Wen *et al*, 2005) to include the use of a modified GA that incorporates elements of the original simulated annealing process.

In (Dickenson & Ghassemlooy, 2003) a complex system was demonstrated that makes great improvements in overcoming the channel-induced multipath distortion resulting from the multiple reflections around the diffuse environment. By sending the received signal through a feature extraction algorithm based upon wavelet analysis, an artificial neural network pattern recognition post processor was employed to rebuild the original signal. The results were successful in that, for the system to run at a BER of 10^{-5}, required 2 dB, 8 dB and 17 dB, less SNR than systems with maximum likelihood detection, filtered or unfiltered receivers, respectively.

In (Hou & O'Brien, 2005, and Hou & O'Brien, 2006), a fuzzy logic-based handover decision algorithm was developed for use in systems where OW and RF networks coexist. In these types of networks, handover from one system to another is a problem, due to different interruption schemes and traffic modes. The strength of the fuzzy logic approach to uncertainties and conflicting decision metrics, gives the handover algorithm an improved QoS though lower packet transfer delay. One of the most important aspects to this work is that the authors highlighted the dynamics of the mobile channel. Each handover decision is different, and previous structured algorithms are only effective under certain conditions, making their approach very beneficial.

One other solution proposed recently is based upon the use of a GA-controlled MSD transmitter, where, instead of pairing it with a traditional diversity receiver, a simpler single element receiver was implemented (Higgins *et al*, 2008,

2009, 2009a). The method has shown several advantages applicable to the mitigating the channel variability of OW system deployment. Firstly, it has been shown to be capable of adapting the received power deviation around the room, independent of its characteristics. Secondly, it has been shown to work with user movement and random user receiver alignment. Thirdly is it possible to implement the technique on transmitter of different source technology, such as RCLED, VCSEL or individual LED based transmitters, and finally the optimisation of the received power distribution is achieved with negligible RMS delay spread and bandwidth penalties.

Within the remaining section, a brief overview of the GA controlled MSD method will be presented, finishing with some new results not previously published.

BASIC PRINCIPLES

Considering Figure 5, to form a MSD system, a transmitter emits I fine beams that form I diffusion spots S_i upon the ceiling. The diffusion spots reflect the incident radiation off the ceiling and through an optical geometrical path back out into the room, where the radiation can be incident upon the J single element receivers, R_j. For this kind of non-directed IR channel employing IM/DD, if each source S_i reflects an identical instantaneous optical power $X_i(t)$, it will induce an instantaneous

Figure 5. Multispot diffusion geometry, (a) Source to ceiling; (b) Ceiling to receiver including reflections

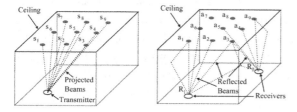

photocurrent $Y_j(t)$ at the receiver R_j with photodiode responsivity r_j in the presence of additive, white Gaussian shot noise $N_j(t)$ (Carruthers & Kahn, 1997) given by:

$$Y_j\left(t\right) = \sum_{i=1}^{I}\left(r_j X_i\left(t\right) * h\left(t; S_i, R_j\right)\right) + N_j\left(t\right)$$

(1)

where $h(t; S_i, R_j)$, is the impulse response between S_i and R_j. The determination of $h(t; S_i, R_j)$ can be achieved through multiple methods, each with their own advantages and disadvantages.

Closed form approximations (Gfeller & Bapst, 1979) do exist, and provide the potential to investigate basic configurations and conduct basic analysis on factors such as material reflectivity and the source intensity profiles, but are too complex when considering multiple reflections. Experimental characterisation can also be conducted (Hashemi *et al*, 1994, Pakravan & Kavehrad, 2001, Kahn *et al*, 1995), but is an expensive and lengthy task that has to be done on a channel by channel basis (Lomba *et al*, 2000).

A more practical, general simulation method, first proposed by Barry (Barry *et al*, 1993) incorporating techniques found in the ray tracing community (Goral *et al*, 1994) provides the ability, with relative ease of implementation, to determine the impulse response for a system where the signal undergoes any number of reflections for any configuration of source and receiver inside an arbitrary empty rectangular room. However, due to the algorithm being recursive, the computational time is exponential, and the memory requirements were impractical beyond three reflections. To solve this issue, Barry's method was refined by Carruthers in (Carruthers *et al*, 2003, Carruthers & Kannan, 2002) allowing the algorithm to be applied iteratively, reducing the computational time to be proportional to the square of the number reflections required. The refinement also allowed the possibility of simulating scenarios with multiple transmitters and receivers without considerable time penalties.

Alternatively, one may wish to determine the impulse response, based upon a mixed deterministic Monte Carlo ray tracing algorithm (López-Hernández *et al*, 1998, 1998, 2000). Using this method, by sending out rays from the source in a pseudo random nature and literally tracing the rays through the reflections, the room does not need to be partitioned into elements, as was the case with the Barry method, so the level of accuracy is determined not by the spatial segmentation, but by the choice in the number of rays. The authors still base their impulse response calculations on the original Barry formulations, but the use of rays allows the unnecessary calculations from rays that would never be incident upon the receiver to be omitted. The method presented is potentially beneficial in terms of computational time, albeit through the choice of a reduction in the number of rays, and would be convenient to implement on vectorised graphics processors. However, as the rays are generated statistically, each run of the simulation provided different results, and it has been shown that simulation of the same system configuration 100 times provides a relative error of between 5% and 20% between each run (González *et al*, 2002).

The original work on the GA-controlled MSD transmitter (Higgins *et al*, 2009) did determine the impulse response via the method of Barry (Barry *et al*, 1993) including the refinements of Carruthers (Carruthers *et al*, 2003, Carruthers & Kannan, 2002) but any of the aforementioned techniques is suitable, depending upon the users requirements. Once $h(t; S_i, R_j)$ has been determined, the GA control method assumes that each diffusion spot power is controlled by some scaling factor $a_i \forall i \in \{1, 2, ..., I\}$, such that the $Y_j(t)$ in (1) becomes:

$$Y_j\left(t\right) = \sum_{i=1}^{I}\left(r_j a_i X_i\left(t\right) * h\left(t; S_i, R_j\right)\right) + N_j\left(t\right)$$

$$(2)$$

which, furthermore, as one is only concerned with a single receiver design, the photodiode responsivity r_j, is constant for each receiver or receiver location, such that there may exist a set of I scaling factors a_i that can be applied to the I identical signal waveforms $X_i(t)$, that will allow for the J receivers, to attain the same or very similar instantaneous photocurrents

$$Y_1\left(t\right) \approx Y_2\left(t\right) \approx ... \approx Y_J\left(t\right)$$

$$(3)$$

Knowing that the IM/DD channel is linear (Carruthers & Kahn, 1997), (2) can be written as

$$Y_j\left(t\right) = \sum_{i=1}^{I}\left(r_j X_i\left(t\right) * a_i h\left(t; S_i, R_j\right)\right) + N_j\left(t\right)$$

$$(4)$$

Such that to solve (3), one needs to solve

$$\sum_{i=1}^{I} a_i h\left(t; S_i, R_1\right) \approx \sum_{i=1}^{I} a_i h\left(t; S_i, R_2\right) \approx ... \approx \sum_{i=1}^{I} a_i h\left(t; S_i, R_J\right)$$

$$(5)$$

By inspection of (2) through (5), it can be seen that a solution may require some scaling factors of ≤ 1, lowering the total received power, compared to if all sources emitted full power. This problem can be solved by drawing on the experience of previous work with rate-adaptive transmission in the IR domain, (Garcia-Zambrana & Puerta-Notario, 2003) whereby, for example, the pulse characteristic can be adjusted to increase or decrease the received power and make the power distributions similar. This then allows for the same optimal receiver design to be used in different environments, albeit under the compromise of variable data rates in the same manner as most other variable data rate systems, such as IEEE802.11 {WiFi} (Haratcherev *et al*, 2005).

A final problem simplification that was made to the previous work (Higgins *et al*, 2009) is based upon the fact that the impulse response $h(t; S_i, R_j)$, is in effect, a finite train of scaled delta functions of width Δt. This implies that for a room of width and depth of 6m and a height of 3m, the maximum time of flight for radiation undergoing 3 reflections is $t = (4(6^2 + 6^2 + 4^2)^{0.5})c^{-1} \approx 120\,\text{ns}$, when it undergoes a path reflecting off the opposite corners of the room. Using an impulse response bin width of 0.1ns would produce 1200 samples for each impulse response train, for every combination of I sources and J receivers in (5). Proposing a GA that can solve (5) for the possibly infinite number of source and transmitter combinations would be far too unwieldy. By replacing the need to evaluate each bin of the impulse response train, with the need to find only the scaling factor solutions for the time integral, or DC value of the frequency response

$$H\left(0; S_i, R_j\right) = \int_{-\infty}^{\infty} h\left(t; S_i, R_j\right) dt$$ allows (5) to reduce to:-

$$\sum_{i=1}^{I} a_i H\left(0; S_i, R_1\right) \approx \sum_{i=1}^{I} a_i H\left(0; S_i, R_2\right) \approx \ldots \approx \sum_{i=1}^{I} a_i H\left(0; S_i, R_J\right)$$

(6)

such that, solving (6), allows a solution close to the original aim in (3).

The Genetic Algorithm

Genetic Algorithms are not to be considered as ready to use algorithms, but rather a general framework that must be tailored to a specific problem (Bäck *et al*, 1997). In this section a brief overview, as a full description can be found in (Higgins *et al*, 2009, 2009a), is provided upon how the representation, fitness function, selection, recombination and mutation sub-routines were adapted from a typical GA framework. Throughout the description below, we refer the reader to Figure 6, which provides a very simplified example of how the GA would perform under reduced conditions.

In GA terminology, the genotype represents all the information stored in the chromosome and allows for an explicit description of an individual solution at the level of the genes. To solve (6), a set of scaling factors $a_i \forall i \in \{1,2,\ldots,I\}$ is defined, that can take on the value in the set $\{0,0.01,\ldots,1\}$, such that the search space $\Phi_g = \{0,0.01,\ldots,1\}^I$, will provide a possible $|\Phi_g| = 101^I$ solutions [46]. From this initialisation, a population of solutions $\Psi(t)$ at time t, of μ chromosomes $\hat{a}_\nu = \left(a_1, \ldots a_I\right) \in \Phi_g, \forall \nu \in \left\{1, \ldots \mu\right\}$, is defined as the initial starting conditions for the GA. Figure 6 illustrates this initial population as an example of when $I=9$ and $\mu=6$.

At a time t, each solution $\hat{a}_\nu$ is evaluated by a fitness function F, which, for the problem here, that of optimising the received power distribution, is given by

$$F\left(\hat{a}_\nu\right) = 100 - \left(100\left(\frac{\max H\left(0; \hat{a}_\nu\right) - \min H\left(0; \hat{a}_\nu\right)}{\max H\left(0; \hat{a}_\nu\right)}\right)\right)$$

(7)

where $\max H\left(0; \hat{a}_\nu\right)$ and $\min H\left(0; \hat{a}_\nu\right)$, are the maximum and minimum received DC frequency responses of all the receiver locations within the environment post application of the GA. As per Figure 6, we have illustrated the hypothetical fitness evaluations of our 6 initial solutions whereby it can be seen, solution (or chromosome) $\hat{a}_1$ is the fittest and $\hat{a}_2$ is the weakest.

The primary objective of the selection operator is to emphasize the fitter solutions such that they are passed on to the next generation (Rothlauf, 2002). Results are shortly to be presented based upon the use of two selection routines that were found to be particularly effective (Higgins *et al*, 2009, 2009a), namely, stochastic uniform sampling (SUS), and tournament selection. The SUS selection schemes assign a probability of selection proportional to an individual's relative fitness

Figure 6. Representation of the GA based upon stochastic uniform sampling with I=9 and μ=6

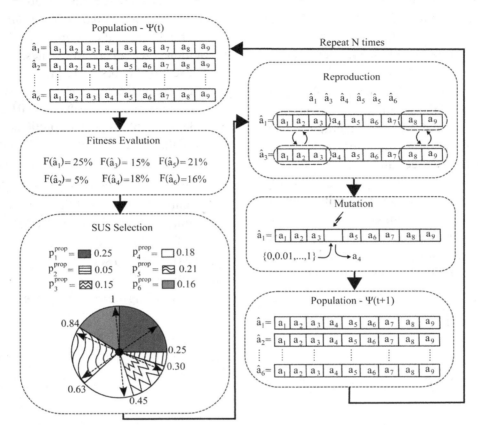

within the population, such that the individual's probability of selection, p_ν^{prop}, is given by

$$p_\nu^{prop} = \frac{F\left(\hat{a}_\nu\right)}{\sum_{\nu=1}^{\mu} F\left(\hat{a}_\nu\right)} \quad (8)$$

The probabilities are then contiguously mapped onto a wheel, such that $\sum_{\nu=1}^{\mu} p_\nu^{prop} = 1$. Next, μ uniformly spaced numbers in the range {0,1} are generated, and then circularly shifted by a pseudo-randomly generated offset value. Each member of the population whose cumulative probability intersects each of the μ offset lines is then chosen for reproduction. Figure 6 illustrates this whereby the SUS routine has discarded $\hat{a}_2$ as statisti-

cally it is the weakest, selected $\hat{a}_5$ twice, and the remainder only once. One may ask why $\hat{a}_1$ was not chosen twice, as technically it is the fittest, but one must remember this is a statistical process, and to make an analogy with natural selection, the fittest is not always chosen the most. In the example, μ=6, and for the results presented shortly using the SUS algorithm, μ=200.

Tournament selection is carried out by first ranking all members of the population $\Psi\left(t\right) = \left\{\hat{a}_1, \ldots, \hat{a}_\mu\right\}$ by their absolute fitness in the population $F\left(\hat{a}_\nu\right)$, where $\hat{a}_1$ is the fittest and $\hat{a}_\mu$ is the weakest. Then, by randomly selecting q members of the ranked members, the fittest is simply chosen. The probability of a member $\hat{a}_\nu$ being selected is given by (Bäck, 1996):-

$$p_{\nu}^{torn} = \frac{1}{\mu^q}\left[\left(\mu - \nu + 1\right)^q - \left(\mu - \nu\right)^q\right] \qquad (9)$$

Increasing the size of the tournament q increases the selective pressure, giving fitter members of the population a higher probability of selection. For example, it is known, (Poli, 2005) that for $q=3$ and $\mu>50$, approximately 50% of the genetic material will be lost through the selection process alone, which therefore gives rise to a very exploitative algorithm, but it loses genetic diversity and risks finding non-optimal solutions. For simplicity, this GA sub-routine is not depicted in Figure 6, but forms a direct replacement to the selection block. For the results presented here, $q=3$, and is therefore going to be known as tournament 3 selection (T3), where we set $\mu=100$.

Crossover imitates the principles of natural reproduction, and is applied here with a probability $\rho_c=0.7$ to random individuals chosen by the selection scheme. From (Higgins *et al*, 2009, 2009a) that the most effective choice of reproduction was when the algorithms apply a double point crossover. This is carried out by generating two unique numbers in the range $\{1,2,...I-1\}$, that are subsequently sorted into ascending order, followed by simply exchanging the substrings between successive crossover points. For the example in Figure 6, we illustrate the effect on randomly selected chromosomes $\hat{a}_1$ and $\hat{a}_3$ undergoing crossover at locations 3 and 8.

Mutation was originally developed as a background operator (Bäck, 1996) able to introduce new genetic material into the search space, such that the probability of evaluating a string in Φ_g will never be zero. Mutation is performed on each individual scaling factor $a_i \in \hat{a}_i \forall v \in \{1,...,\mu\}$ after crossover with a probability of $\rho_m=0.05$ and $\rho_m=0.1$ for the SUS and T3 based GAs respectively. As shown in Figure 6, if a given scaling factor a_i is chosen for mutation, in this case, a_4 from $\hat{a}_1$, it is simply replaced with another ran-

domly-generated number in the set $\{0,0.01,...,1\}$, and the original value discarded.

This process then finishes one generation of the GA, and the new generation $\Psi(t+1)$ is then used to start the process again for N times. For the results presented here, $N=5000$.

Results

Considering an empty environment, hereon in known as environment 1, with a width and depth of 6m, and a height of 3m, whereby the ceiling and the walls have a reflectivity of 0.7, and the floor has a reflectivity of 0.3. Upon the ceiling, 25 uniformly-distributed diffusion spots are generated by the transmitter. In the centre of the room, at a height of 1m, a single element receiver with an active photodiode area of 1cm² and a FOV=55° is initially vertically orientated, i.e towards the ceiling, that is then subsequently rotated through ±90° in both its x and y axes. The resultant received power can be seen in Figure 7 where crucially, for this example, the received power can be seen varying between 18.1μW to 57.2μW, equating to a 39.1μW, or 68% power deviation from the peak, purely from alignment effects. It should be stressed that this result is

Figure 7. The received power for a single receiver within an empty environment 1, as a function of its orientation

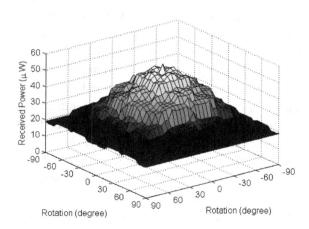

Figure 8. Received power distribution at 1024 locations in an empty environment 1 when (a) Non-optimised; (b) Optimised with the SUS based GA

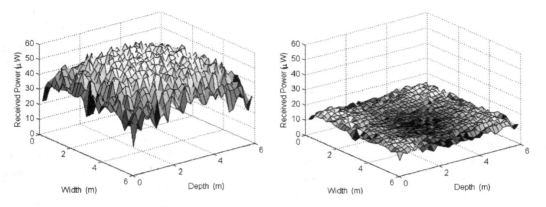

based purely upon one receiver, at one location in one environment. Receivers in different positions within this or any other environment will have their own rotationally-dependent received power distribution. Therefore, taking into account the presence of many receivers that can be independently orientated, the problem of user alignment to an OW system designer becomes apparent.

As can be seen from the variability of the received power in Figure 7, some assumption has to be made as to the degree of freedom that the user can have with their receiver. As in previous work (Higgins *et al*, 2009a) this constraint is set to assume the user rotates their receiver in each of the x and y axes as a normal distribution with mean of 0, (no rotation) and a standard deviation 11.7. This therefore means the user will align, with a probability of 0.8 and 0.99, the receiver within ±15° and ±30° of the un-rotated case respectively. It is next assumed that, in this environment, 1024 independent receivers are uniformly distributed at a height of 1m around the room. Figure 8(a) provides the received power distribution at each location, where it can be seen that the received power varies between 18.3μW and 55.6μW, a deviation of 37.3μW or 67.1% from the peak received power in the room. Upon application of the SUS-based GA, this received power distribution, as in Figure 8(b), can now be

seen varying between 9.8μW and 17.1μW, a deviation of 7.3μW or 42.6% from the peak received power in the room. This equates to a GA optimisation gain of 24.5%.

As stated earlier, the primary objective and benefit of this GA optimisation method is to be able to compensate for multiple dynamic users in multiple environments. Therefore, in addition to environment 1, a second environment of the same dimensions, but where the ceilings reflectivity is increased to 0.75, the north wall reflectivity is reduced to 0.5, and the south and west walls are increased to 0.8, is considered. Furthermore, consider two users, each having a reflectivity of 0.3, a shoulder to shoulder width of 0.7m, a chest to back depth of 0.4m and a height of 1.8m. In each environment, each pair of users undergoes

Figure 9. User movement scenarios for: (a) Environment 1; (b) Environment 2

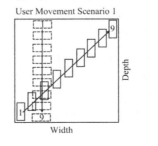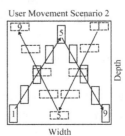

different movement paths, consisting of 9 discrete locations per user, which, for environment 1 and environment 2, are illustrated in Figures 9(a) and 9(b) respectively.

For environment 1, as shown in Figure 10, when empty as discussed previously, the non-optimised power deviation is 67.1%, which, upon application of the SUS-based GA is reduced to 42.6%, which is an optimisation gain of 24.5%. Alternatively, upon application of the T3-based GA, the power deviation is reduced to 48.5%, which is an optimisation gain of 18.6%. As the users move within the environment, they perturb the power deviation by up to 6.6%, as it now varies between 60.5% and 69.8%. However, upon application of the SUS-based GA, this variation is reduced to vary between 43.9% and 50.6%. This is a slightly increased perturbation of up to 8% from the optimised empty room case, but an overall gain of up to 22% from the non-optimised case at a given location. Under application of the T3 based GA, the variation in received power is reduced to vary between 47.3% and 52.2%, This is a reduction in the perturbation to within 2.9%

of the optimised empty case, and an overall gain of up to 18.1% from the non-optimised case at a given location.

For environment 2, as shown in Figure 11, when empty, the non-optimised power deviation is 66.1%, which upon application of the SUS- and T3-based GAs is reduced to 44.6% and 48.1%, a gain of 21.5% and 18.0% respectively. Upon instigation of user movement, the users perturb the received power by up to 4.9% as it varies between 64.0% and 71.0%. Upon application of the SUS-based GA, the variation is reduced to between 41.2% and 50.8%. Similar in performance to the SUS-based GA within environment 1, this is an increase in the perturbation by up to 6.2% from the optimised empty case, but an overall gain of 24.9% from the non-optimised case at a given location. Upon application of the T3-based GA, the deviation is reduced to vary between 44.5% and 52.8%. This results in a negligible difference to the user perturbation of the received power, compared to the optimised empty case, but provides a gain of up to 20.6% from the non-optimised case at a given location.

Figure 10. Deviation from peak power for environment 1, movement scenario 1. —◊— non-optimised empty room. - - + - - non-optimized with movement. —□— T3 optimised empty room. —○— SUS optimised empty room. - -- - T3 optimised with movement. - -▷- - SUS optimized with movement.*

Figure 11. Deviation from peak power for environment 2, movement scenario 2. —◊— non-optimised empty room. - - + - - non-optimized with movement. —□— T3 optimised empty room. —○— SUS optimised empty room. - -- - T3 optimised with movement. - -▷- - SUS optimized with movement.*

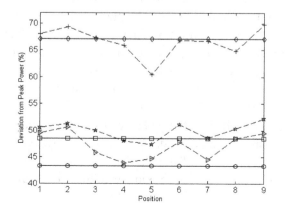

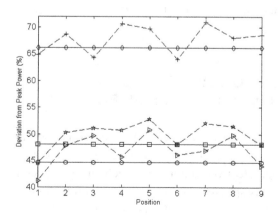

Figure 12. Basic detector interface

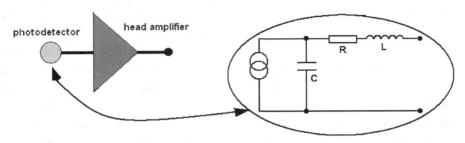

Conclusions about GA Power Balancing

In conclusion, this section of the chapter has illustrated the concept of using a GA-controlled MSD transmitter to compensate for the variability in the OW channel in multiple dynamic environments. The method has been shown capable of optimising the received power deviation by up to 24.5% when empty, and by up to 20.6% when a user movement and random receiver alignment is considered. As in previous work (Higgins *et al*, 2009, 2009a) the SUS-based GA outperforms the T3-based GA by a few percent, but the T3-based GA would be slightly easier to implement in hardware, as it does not require proportional fitness assignment, and requires a lower population, thus lowering the processing and memory requirements, respectively. In applications such as OW, where user-friendliness and mobility is paramount, and standards bodies push for a single design of receiver, this method could prove to be very beneficial.

OPTICAL WIRELESS RECEIVER-AMPLIFIERS

The purpose of this section is to demonstrate particular aspects of signal processing when it comes to optoelectronic interfaces. Such interfaces occur in many spheres of operation, for example: imaging, optical communications, optical measurements,

and sensing. To facilitate understanding, and to demonstrate the commonality in the different application areas, consider firstly the issue of the basic optical detector interface: (See Figure 12)

For the majority of applications and devices used, the equivalent circuit of the detector may be considered as a current source with associated capacitance C, followed by resistance R and inductance L, as shown in the figure above. The detector is then connected directly to a head amplifier. The exact mechanisms within the device which lead to this relatively simple equivalent circuit include depletion capacitance, bulk resistance, and lead inductance for a semiconductor photodetector. Vacuum tube detectors, which were once very popular and now are used only in special applications, such as the vidicon and plumbicon, surprisingly, can be similarly regarded in terms of their small-signal equivalent circuit, even though physically quite different in operation. In the majority of situations the issue is a matter of dealing with the conflicting requirements of bandwidth and gain. Another very important consideration is the noise performance, which can interact with the previous two requirements adversely. Consider what happens when such a signal source is connected to a high impedance amplifier, as shown in Figure 13.

Z_{in} is the input impedance, usually considered as purely real i.e a resistance. In a simplistic view, the impedance of the amplifier is set as high as possible in order to maximise gain, as, excluding

Figure 13. High impedance front end amplifier interface

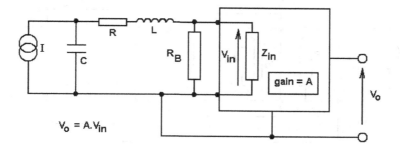

$$V_o = A.V_{in}$$

the effect of C, R and L at low frequencies, the output of the amplifier is given by:

$$V_o = A.V_{in} = I.Z_{in} \approx I.R_x$$

where R_x is the real component of Z_{in} in parallel with the bias resistor for the device, R_B.

Consider what happens when the actual equivalent circuit is considered. If R and L are neglected for frequencies where their impedances are small compared to the value of R_x, then the equivalent circuit can be simplified to the following, (see Figure 14)

In this case, the shortcomings of the technique become very clear: the voltage V_{in} becomes frequency dependent, and the output reduces to:

$$V_o = A. V_{in} = I.R_x /(1 + j\omega CR_x)$$

with the standard half power frequency at $f_{3db} = 1/2\pi CR_x$. Therefore, to maximise the bandwidth, R_x must be as small as possible and C as small as possible – the latter being not usually adjustable except by using as high a bias voltage as is safe for the device in the case of a PN or PIN diode. However, to maximise the gain, R_x must be as large as possible. One solution to this dilemma is to have R_x small and follow the head amplifier by an equalisation stage, which is a technique well used in the video industry: (see Figure 15).

A2, the gain characteristic of the equalisation amplifier, must therefore have a transfer characteristic of the form:

$$A2 = G.(1 + j\omega\tau) \text{ where } \tau = CR_x$$

Thus $V_{o2} = G.I.R_x$

The consequences of this are desirable in terms of overall signal characteristic, but at the

Figure 14. Approximated high impedance front end

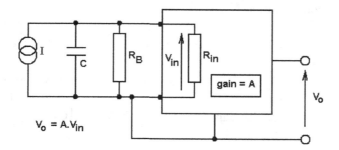

$$V_o = A.V_{in}$$

Figure 15. High impedance amplifier followed by an equalisation stage

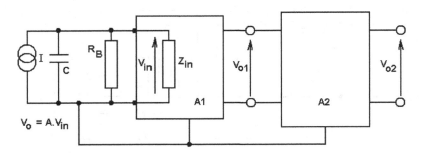

expense of the noise spectrum in terms of the noise originating from the front end amplifier. In an optical fibre communication system, if the system is not limited in performance by the photodiode shot noise, then that shot noise, plus any Johnson noise in the front end (the head amplifier), is exacerbated due to its previously white spectrum being converted to triangular above the 3 dB point $(f_0 = 1/2\pi\tau)$. This results in the integrated noise power, over a typical signal bandwidth **B**, greater than the original 3 dB bandwidth, being a function of $\mathbf{B}^3$ – a very undesirable feature. In turn, this produces increased jitter and Bit Error Rate (BER) in digital systems. In multichannel fibre analogue video using FM/FDM, this also produces severe picture degradation near threshold conditions, and in VSB/FDM video systems a gradual increase in visible granularity.

When the detector device is fixed in terms of its operating parameters, an improvement in performance can be achieved by the transimpedance amplifier approach: (see Figure 16)

I is the device current, $\mathbf{C}_d$ the device capacitance, and **A** the open loop gain of the amplifier. The gain response of the configuration is given by the standard formula:

$$V_o = - (I.R_f)/(1 + j\omega C_d R_f /A)$$

which thus reduces the effect of device capacitance by a factor of **A**, subject to the stability of

the amplifier at high frequencies. Clearly, though, the effect of noise must be taken into account. The majority of the noise in both the transimpedance amplifier and the high impedance amplifier is due to the effective load resistance. In the case of the high impedance amplifier, then this is $\mathbf{R}_x$ as defined earlier involving the bias resistor for the device in combination with the input resistance of the amplifier itself. In the case of the transimpedance amplifier, then the effective load is the feedback resistor $\mathbf{R}_f$ in Figure 16. It can be shown that for optical fibre communications applications, which are not generally shot noise limited (i.e. where the noise originates from the photodiode detection process), then the transimpedance amplifier has no particular noise advantage compared to the high impedance amplifier – in fact, it is arguably detrimental from this point of view. However, in many sensing and optical wireless applications, then the detector shot noise dominates and so there is benefit to be gained in terms of the much im-

Figure 16. Basic transimpedance amplifier model

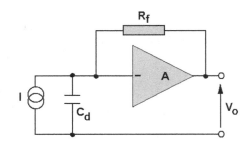

Figure 17. FET transimpedance amplifier

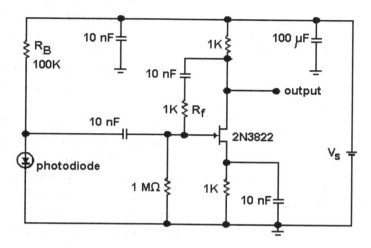

proved bandwidth without the amplifier incurring a significant noise penalty to the overall system performance. A practical circuit for a transimpedance amplifier implemented using a RF JFET is shown in Figure 17.

The gain of the above circuit is around 51dB at a bandwidth of 20 MHz in a load resistance of 1K (i.e. in R_f). Yet another approach to the problem of interfacing effectively and maintaining bandwidth at low noise penalty is the bootstrap amplifier, which achieves bandwidth enhancement without requiring the amplifier to have high voltage gain. Instead, the overall system gain is achieved by being able to increase the value of the effective load that the photodetector device "sees" without reducing the bandwidth. A typical configuration is shown in Figure 18.

In this case, the value of A is set to unity in theory, or nearly unity in practice, using circuits such as source followers, emitter followers, or operational amplifiers of sufficiently high bandwidth, but set so that $A \approx 1$. In theory, when $A = 1$, then the bandwidth is infinite and the system would oscillate. Note that R_B serves only to provide a point such that positive feedback can be applied. C_{bp} is chosen to provide a low impedance path to ground at signal frequencies of interest.

The gain as a function of frequency can thus be shown as:

$$Vo = I.R_L/(1 + j\omega C_d.R_L(1/A - 1))$$

The bandwidth is therefore strongly related to the value of A. As in the case of the two previous amplifiers, the effective load resistance is the principal factor governing noise performance. In this example, it is essentially R_L in parallel with the input resistance of the amplifier, in fact, very similar to the high impedance amplifier situation. This configuration therefore can, in principle,

Figure 18. Bootstrap amplifier basic configuration

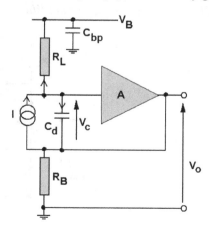

Figure 19. FET implementation of a bootstrap amplifier

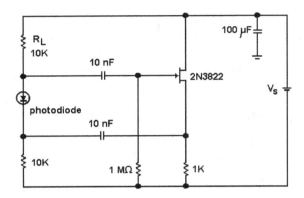

offer a better noise performance in terms of its noise spectral density than the transimpedance amplifier, and offer a larger bandwidth than the high impedance amplifier configuration.

The above circuit does not take full advantage of the possibilities for bandwidth improvement. An improved version of the circuit is given in Figure 20.

This offers better performance because:

a. High frequency low noise bipolar transistors are used. These offer lower output impedances in common collector mode than FET common drain amplifiers, and the gain of the stage is nearer unity.

b. Boostrapping is applied to the first transistor by the second, which reduces the effect of the base-collector capacitance for the first transistor.

c. The load resistance is differently defined in comparison to the circuit of Figure 19.

As mentioned earlier, usually bandwidth is at a premium because of the effect of device capacitance. A method of extending this has been developed, based on the idea of the "Percival" peaking coil from the earlier days of video communications, where the relatively large capacitance of vacuum tubes and the subsequent amplifying devices required compensation to enable workable video bandwidths. In the case of a modern optical communications receiver-amplifier, shown in Figure 21, the input capacitance of the first transistor, T1, is resonated with an inductance L1 at a just out-of-band frequency, as is the second bootstrapping device, T2, with inductance L2: (see Figure 21).

Thus the bootstrapping is doubly compensated, which, in the example shown above extends the bandwidth from an original value of 236 MHz to a final value of 1.7 GHz, using the same transistors and effective load resistance (Green *et al*, 2008).

The analysis of the noise performance is an interesting challenge, and depends quite critically

Figure 20. Improved bootstrap amplifier using bipolar devices

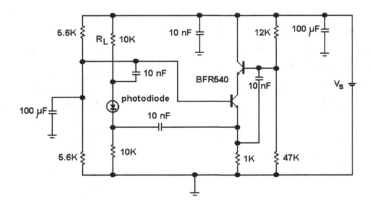

Figure 21. Frequency-extended bootstrap amplifier

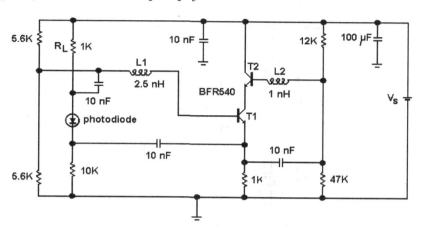

on the application. In optical fibre communications, the system tends to be Johnson noise-limited, where the primary source of such noise is the load resistance, which is just under 1K in the example above. Clearly, gain may be exchanged for both noise performance and/or bandwidth in this example, as in the case of the transimpedance amplifier. In the case of optical wireless, as earlier, shot noise in the detector is frequently large compared to the Johnson noise, except under very small signal conditions. The bootstrap amplifier and the transimpedance amplifier can be combined, and an example was described in (Green & McNeill, 1989).

There are ways to make better use of detector devices if the requirements for bandwidth are not too stringent. A concept called photoparametric amplification may be employed (Leeson *et al*, 2008) which takes advantage of previous work undertaken by, amongst many others, D.P. Howson and R.B. Smith (Howson & Smith, 1970) on parametric amplifiers for microwave applications. It is well known that a semiconductor junction can act as a varicap diode. When such a diode is pumped by a high frequency source, it can therefore be regarded as a time-varying capacitance. In other words, its parameters are time-varying. When a small signal at a (usually) lower frequency is introduced to the device, then,

because the CV characteristic of the diode is a nonlinear function, then parametric mixing takes place which produces a spectrum of frequencies. The benefit of this arrangement, compared to a resistive mixer such as a diode or transistor type, is that it is performed in a reactance, which theoretically offers a very low noise figure – shot noise and Johnson noise do not exist in the ideal case. A typical value for such an amplifier depends on the ratio of the pump frequency, f_p, to the signal frequency, f_s. For example, for $f_p / f_s = 50$, a value of around 0.85 dB is obtained, whereas for $f_p / f_s = 100$, then the noise figure has reduced to 0.45 dB, for a varactor figure-of-merit of $\chi = 0.2$ – the latter being defined in terms of the maximum and minimum capacitances of the device between zero bias and breakdown voltages.

In the concept of photoparametric amplification (PPA), instead of the signal being introduced electrically, the varactor is, instead, a photodiode biased into a region where its nonlinearity of CV characteristic is optimum for mixing (Green *et al*, 2005). This even includes zero bias for some configurations of devices. The benefit of this arrangement is that the problem of signal interfacing is avoided, which used to lead to some conflicts and spurious RF by-product generation in standard parametric amplifiers. A typical PPA configuration is shown in Figure 22.

Figure 22. Basic photoparametric amplifier configuration

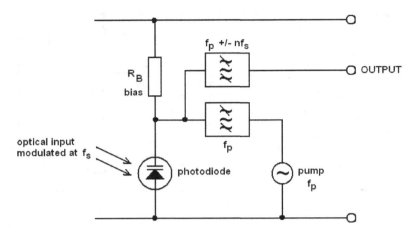

Analysis has shown that the gain of such a configuration is, in an analogous way to the standard parametric amplifier, proportional to $\{f_p / f_s\}$ according to the expression derived in (Alhagagi *et al*, 2010) and expanded below:

If $E = \dfrac{V_P}{(V_b + V_0)}$ where V_p is the pump voltage, V_0 is the DC bias, and V_b is the inbuilt barrier potential, then it can be shown that the gain of the system is given by:

$$A_v = (\omega_i/\omega_s).\beta_x$$

where the gain is defined as the output voltage for pumping divided by the output voltage without pumping (i.e. using the photodetector as a direct detector). ω_i is the idler frequency, being equal to the sum of the signal- and pump frequencies, ω_s is the signal frequency, and β_x is approximately equal to *E/4*. Thus the gain is dependent on device fabrication features (through β_x because of V_b), and external conditions via the other parameters. Note that the standard PPA operation is as an upconverter: there is no direct equivalent to the electrical parametric amplifier in this respect. In this case, a simple detector at ω_i is sufficient to utilise the upconverted signal to full advantage.

However, in many applications it is clearly desirable to recover the signal at baseband as originally transmitted from the other end. For this, a double mixing approach has been proposed [69]. This employs a PPA front end, whilst using a more standard mixer subsequently. As might be expected, this more sophisticated arrangement leads to certain matching requirements, and yet a beneficial overall signal gain of typically 10 dB at low noise penalty, can be achieved with a passive mixer. The configuration appears as in Figure 23.

The input as designated in Figure 23 is an optical signal modulated at some baseband frequency, then upconverted when detected in the PPA approach, and then passband filtered. The output of the filter is passed along to the mixer, using the same local oscillator as did the PPA for its pump source, and then the output is channelled through a low pass filter from which the baseband modulation can be recovered intact.

FUTURE DIRECTIONS

The subject of optical wireless (sometimes called FSO – Free Space Optics) is an interesting one because it bridges the gap between RF and optical communications within a fiber environment.

Figure 23. Double PPA-mixer configuration

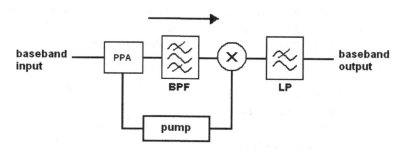

Future directions of the ongoing research into the aspects described within this chapter will consider ways to ensure optical fields are controlled in a wide variety of environments, and the challenge clearly remains that, when those environments are subject to change, then the parameters have to change accordingly so as to optimise the Quality Of Service (QOS). As it happens, GA techniques are well suited to this challenge, as they naturally include optimisation methods which are not rigid, but are flexible. Additionally, the multichannel approach using OFDM can be extended to a range of other scenarios using optical wireless carriers, and the optimisation of the parameters for different optical environments, such as within other optical media, remains an interesting area to consider. Integration of Optical Wireless with RF systems is a valuable area to consider, as it can therefore lead to the "holy grail" of a seamless communications infrastructure.

CONCLUSION

The chapter has given a detailed look at three important aspects of the Optical Wireless (OW) channel. The techniques which have been discussed serve to provide a sample of what is actually a large international effort in many research directions, which has continued for several decades. Although OW has, ironically, been somewhat overshadowed by the RF communications lobby,

it is now emerging as a competitive technology to terahertz systems, which will develop but over some considerable time. In the meantime, OW can perform very well and offer immunity from electromagnetic interference, at a time when the RF environment is steadily getting more polluted, a susceptibility which will be shared with terahertz systems, even when they reach maturity.

REFERENCES

Akhtman, J., Bobrovsky, B. Z., & Hanzo, L. (2003). Peak-to-average power ratio reduction for OFDM modems. In *Proceedings of IEEE GLOBECOM, 2003*, 1188–1192.

Alhaghagi, H., Green, R. J., & Hines, E. L. (2010). Double heterodyne photoparametric amplification techniques for optical wireless communications and sensing applications. *Proceedings of ICTON 10*, session We.D3.2, Munich.

Allard, R. J., Werner, D. H., & Werner, P. L. (2003). Radiation pattern synthesis for arrays of conformal antennas mounted on arbitrarily-shaped three-dimensional platforms using genetic algorithms. *IEEE Transactions on Antennas and Propagation, 51*(5), 1054–1062. doi:10.1109/TAP.2003.811510

Bäck, T. (1996). *Evolutionary algorithms in theory and practice: Evolution strategies, evolutionary programming, genetic algorithms*. Oxford University Press.

Bäck, T., Hammel, U., & Schwefel, H. P. (1997). Evolutionary computation: Comments on the history and current state. *IEEE Transactions on Evolutionary Computation*, *1*(1), 3–17. doi:10.1109/4235.585888

Bannister, J. A., Fratta, L., & Gerla, M. (1990). *Topological design of the wavelength-division optical network*. INFOCOM '90. Ninth Annual Joint Conference of the IEEE Computer and Communication Societies, (pp. 1005–1013).

Barry, J. R., Kahn, J. M., Krause, W. J., Lee, E. A., & Messerschmitt, D. G. (1993). Simulation of multipath impulse response for indoor wireless optical channels. *IEEE Journal on Selected Areas in Communications*, *11*(3), 367–379. doi:10.1109/49.219552

Barry, J. R., Kahn, J. M., Lee, E. A., & Messerschmitt, D. G. (1991). High-speed nondirective optical communication for wireless networks. *IEEE Network*, *5*(6), 44–54. doi:10.1109/65.103810

Bellon, J., Sibley, M. J. N., Wisely, D. R., & Greaves, S. D. (1999). Hub architecture for infrared wireless networks in office environments. *IEEE Proceedings in Optoelectronics*, *146*(2), 78–82. doi:10.1049/ip-opt:19990313

Carruthers, J. B., Carroll, S. M., & Kannan, P. (2003). Propagation modelling for indoor optical wireless communications using fast multireceiver channel estimation. *IEE Proceedings. Optoelectronics*, *150*(5), 473–481. doi:10.1049/ip-opt:20030527

Carruthers, J. B., & Kahn, J. M. (1997). Modeling of nondirected wireless infrared channels. *IEEE Transactions on Communications*, *45*(10), 1260–1268. doi:10.1109/26.634690

Carruthers, J. B., & Kannan, P. (2002). Iterative site-based modelling for wireless infrared channels. *IEEE Transactions on Antennas and Propagation*, *50*(5), 759–765. doi:10.1109/TAP.2002.1011244

Chen, K., Yun, X., He, Z., & Han, C. (2007). Synthesis of sparse planar arrays using modified real genetic algorithm. *IEEE Transactions on Antennas and Propagation*, *55*(4), 1067–1073. doi:10.1109/TAP.2007.893375

Desai, A., & Milner, S. (2005). Autonomous reconfiguration in free-space optical sensor networks. *IEEE Journal on Selected Areas in Communications*, *23*(8), 1556–1563. doi:10.1109/JSAC.2005.852183

Dickenson, R. J., & Ghassemlooy, Z. (2003). A feature extraction and pattern recognition receiver employing wavelet analysis and artificial intelligence for signal detection in diffuse optical wireless communications. *IEEE Transactions on Wireless Communications*, *10*(2), 64–72. doi:10.1109/MWC.2003.1196404

Djahani, P., & Kahn, J. M. (2000). Analysis of infrared wireless links employing multibeam transmitters and imaging diversity receivers. *IEEE Transactions on Communications*, *48*(12), 2077–2088. doi:10.1109/26.891218

Fetterolf, P. C., & Anandalingam, G. (1991). Optimizing interconnections of local area networks: An approach using simulated annealing. *ORSA Journal on Computing*, *3*(4), 275–287.

Garcia-Zambrana, A., & Puerta-Notario, A. (2003). Novel approach for increasing the peak-to-average optical power ratio in rate-adaptive optical wireless communication systems. *IEEE Proceedings in Optoelectronics*, *150*(5), 439–444. doi:10.1049/ip-opt:20030526

Gfeller, F. R., & Bapst, U. (1979). Wireless in-house data communication via diffuse infrared radiation. *Proceedings of the IEEE*, *67*(11), 1474–1486. doi:10.1109/PROC.1979.11508

González, O., Militello, C., Rodriguez, S., Pérez-Jiménez, R., & Ayala, A. (2002). Error estimation of the impulse response on diffuse wireless infrared indoor channels using a Monte Carlo ray-tracing algorithm. *IEEE Proceedings in Optoelectronics*, *149*(5-6), 222–227. doi:10.1049/ip-opt:20020545

Gonzalez, R. (2005). OFDM over indoor wireless optical channel. *IEEE Proceedings in Optoelectronics, 152,* 199–204. doi:10.1049/ip-opt:20045065

Goral, C. M., Torrance, K. E., Greenberg, D. P., & Battaile, B. (1984). Modeling the interaction of light between diffuse surfaces. *SIGGRAPH '84: Proceedings of the 11th annual conference on Computer graphics and interactive techniques.* (pp. 213–222). New York: ACM Press.

Green, R. J., Joshi, H., Higgins, M. D., & Leeson, M. S. (2008). Recent developments in indoor optical wireless systems. *IET Communications, 2*(1), 3–10. doi:10.1049/iet-com:20060475

Green, R. J., Joshi, H., Higgins, M. D., & Leeson, M. S. (2008). Bandwidth extension in optical wireless receiver-amplifiers. In *International Conference on International Transparent Optical Networks, ICTON 2008.* (pp. 201–204.)

Green, R. J., & McNeill, M. G. (1989). The bootstrap transimpedance amplifier-a new configuration. *IEE Proceedings on Circuits & Systems, 136*(2), 57–61. doi:10.1049/ip-g-2.1989.0009

Green, R. J., Sweet, C., & Idrus, S. (2005). Optical wireless links with enhanced linearity and selectivity. *Journal of Optical Networking, 4*(10), 671–684. doi:10.1364/JON.4.000671

Haratcherev, I., Taal, J., Langendoen, K., Lagendijk, R., & Sips, H. (2005). Automatic IEEE802.11 rate control for streaming applications. *Wireless Communications in Mobile Computing, 5*(4), 421–437. doi:10.1002/wcm.301

Hashemi, H., Yun, G., Kavehrad, M., Behbahani, F., & Galko, P. A. (1994). Indoor propagation measurements at infrared frequencies for wireless local area networks applications. *IEEE Transactions on Vehicular Technology, 43*(3), 562–576. doi:10.1109/25.312790

Haupt, R. L. (1995). An introduction to genetic algorithms for electromagnetics. *IEEE Antennas and Propagation Magazine, 37*(2), 7–15. doi:10.1109/74.382334

Higgins, M. D., Green, R. J., & Leeson, M. S. (2008). *Genetic algorithm channel control for indoor optical wireless communications.* In *International Conference on International Transparent Optical Networks, ICTON 2008.* (pp. 189–192).

Higgins, M. D., Green, R. J., & Leeson, M. S. (2009). A genetic algorithm method for optical wireless channel control. *Journal of Lightwave Technology, 27*(6), 760–772. doi:10.1109/JLT.2008.928395

Higgins, M. D., Green, R. J., & Leeson, M. S. (2009a). Receiver alignment dependence of a GA controlled optical wireless transmitter. *Journal of Optics. A, Pure and Applied Optics, 11*(7), 375–403. doi:10.1088/1464-4258/11/7/075403

Hou, J., & O'Brien, D. C. (2005). Adaptive inter-system handover for heterogeneous rf and ir networks. In *Proceedings of 19th IEEE International Parallel and Distributed Processing Symposium, 2005.* (pp. 125a–125a).

Hou, J., & O'Brien, D. C. (2006). Vertical handover-decision-making algorithm using fuzzy logic for the integrated radio-and-ow system. *IEEE Transactions on Wireless Communications, 5*(1), 176–185. doi:10.1109/TWC.2006.1576541

Howson, D. P., & Smith, R. B. (1970). *Parametric amplifiers.* McGraw-Hill.

Huang, E. W., & Wornell, G. W. (2007). Peak-to-average power reduction for low-power OFDM systems. In. *Proceedings of, ICC,* 2924–2929.

Kahn, J. M., Krause, W. J., & Carruthers, J. B. (1995). Experimental characterization of non-directed indoor infrared channels. *IEEE Transactions on Communications, 43*(234), 1613–1623. doi:10.1109/26.380210

Kirkpatrick, S., Gelatt, C. D., & Vecchi, M. P. (1983). Optimization by simulated annealing. *Science, 220*(4598), 671–680. doi:10.1126/science.220.4598.671

Lee, D. C. M., Kahn, J. M., & Audeh, M. D. (1997). Trellis-coded pulse-position modulation for indoor wireless infrared communications. *IEEE Transactions on Communications, 45*(9), 1080–1087. doi:10.1109/26.623072

Leeson, M. S., Green, R. J., & Higgins, M. D. (2008). Photoparametric amplifier frequency converters. *Proceedings of ICTON, 08*, 197–200.

Lomba, C. R., Valadas, R. T., & de Oliveira Duarte, A. M. (2000). Efficient simulation of the impulse response of the indoor wireless optical channel. *International Journal of Communication Systems, 13*(7-8), 537–549. doi:10.1002/1099-1131(200011/12)13:7/8<537::AID-DAC455>3.0.CO;2-6

López-Hernández, F. J., Pérez-Jiménez, R., & Santamará, A. (1998). Monte Carlo calculation of impulse response on diffuse ir wireless indoor channels. *Electronics Letters, 34*(12), 1260–1262. doi:10.1049/el:19980825

López-Hernández, F. J., Pérez-Jiménez, R., & Santamará, A. (1998). Modified Monte Carlo scheme for high-efficiency simulation of the impulse response on diffuse IR wireless indoor channels. *Electronics Letters, 34*(19), 1819–1820. doi:10.1049/el:19981173

López-Hernández, F. J., Pérez-Jiménez, R., & Santamará, A. (2000). Ray-tracing algorithms for fast calculation of the channel impulse response on diffuse IR wireless indoor channels. *Journal of Optical Engineering, 39*(10), 2775–2780. doi:10.1117/1.1287397

O'Brien, D. C., Katz, M., Wang, P., Kalliojarvi, K., Arnon, S., & Matsumoto, M. (2005). *Short range optical wireless communications.* Wireless World Research Forum.

Ochiai, H. (2004). A novel Trellis-shaping design with both peak and average power reduction for OFDM systems. *IEEE Transactions on Communications, 52*, 1916–1926. doi:10.1109/TCOMM.2004.836593

Ochiai, H., & Imai, H. (2002). Performance analysis of deliberately clipped OFDM signals. *IEEE Transactions on Communications, 50*, 89–101. doi:10.1109/26.975762

Pakravan, M. R., & Kavehrad, M. (2001). Indoor wireless infrared channel characterization by measurements. *IEEE Transactions on Vehicular Technology, 50*(4), 1053–1073. doi:10.1109/25.938580

Parand, F., Faulkner, G. E., & O'Brien, D. C. (2003). Cellular tracked optical wireless demonstration link. *IEEE Proceedings in Optoelectronics, 150*(5), 490–496. doi:10.1049/ip-opt:20030961

Poli, R. (2005). Tournament selection, iterated coupon-collection problem, and backward-chaining evolutionary algorithms. In *Foundations of Genetic Algorithms*, (pp. 132–155).

Ramirez-Iniguez, R., & Green, R. J. (2000). *Optical antennae for infrared mobile communications.* (Patent GB01/03812).

Ramirez-Iniguez, R., & Green, R. J. (2005). Optical antenna design for indoor optical wireless communication systems. *International Journal of Communication Systems, 18*(3), 229–245. doi:10.1002/dac.701

Rothlauf, F. (2002). *Representations for genetic and evolutionary algorithms.* Physica-Verlag.

Tan, M., & Bar-Ness, Y. (2003). OFDM peak-to-average power ratio reduction by combined symbol rotation and inversion with limited complexity. In. *Proceedings of IEEE GLOBECOM, 2003*, 605–610.

Telzhensky, N., & Leviatan, Y. (2006). Novel method of UWB antenna optimization for specified input signal forms by means of genetic algorithm. *IEEE Transactions on Antennas and Propagation*, *54*(8), 2216–2225. doi:10.1109/TAP.2006.879201

Thomas, J., Quatieri, F., & Oppenheim, A. V. (1981). Iterative techniques for minimum phase signal reconstruction from phase magnitude. *IEEE Transactions on Acoustics, Speech, and Signal Processing*, *29*, 1187–1193. doi:10.1109/TASSP.1981.1163714

Uno, H., Kumatani, K., Okuhata, H., Shirakawa, I., & Chiba, T. (1997). ASK digital demodulation scheme for noise immune infrared data communication. *Wireless Networks*, *3*(2), 121–129. doi:10.1023/A:1019188729979

Wen, M., Yao, J., Wong, D. W. K., & Chen, G. C. K. (2005). Holographic diffuser design using a modified genetic algorithm. *Optical Engineering (Redondo Beach, Calif.)*, *44*(8), 85801–85808. doi:10.1117/1.2031268

Wong, D. W. K., Chen, G., & Yao, J. (2005). Optimization of spot pattern in indoor diffuse optical wireless local area networks. *Optics Express*, *13*(8), 3000–3014. doi:10.1364/OPEX.13.003000

Yang, H., & Lu, C. (2000). Infrared wireless LAN using multiple optical sources. *IEE Proceedings. Optoelectronics*, *147*(4), 301–307. doi:10.1049/ip-opt:20000610

Yener, B., & Boult, T. E. (1994). A study of upper and lower bounds for minimum congestion routing in lightwave networks. In *13th Proceedings IEEE INFOCOM '94: Networking for Global Communications*, (pp. 138–147).

Zhang, Y., Joyner, V., Yun, R., & Sonkusale, S. (2008). A 700mbit/s cmos capacitive feedback front-end amplifier with automatic gain control for broadband optical wireless links. In *IEEE International Symposium on Circuits and Systems, 2008. ISCAS 2008.* (pp. 185–188).

Chapter 2

Application of Novel Signal Processing Algorithms for the Detection and Minimization of Skywave Interfering Signals in Loran Receivers

Abbas Mohammed
Blekinge Institute of Technology, Sweden

David Last
University of Bangor, UK

ABSTRACT

Skywave interference commonly affects Loran receivers' performance. Traditional skywave rejection methods that use fixed, worst-case, sample timing are far from optimal. This chapter reports on novel signal processing techniques for measuring in real-time the delay and strength of the varying skywave components of a Loran signal relative to the groundwave pulse. The merits and limitations of these techniques will be discussed. Their effectiveness will be assessed by theoretical analysis, computer simulations under a range of realistic conditions, and by testing using off-air signals. A prototype Loran system employing the proposed techniques is also presented. This work establishes a basis on which to design a Loran receiver capable of adjusting its sampling point adaptively to the optimal value in a constantly-changing skywave environment. Such receivers promise to improve significantly the accuracy and reliability of positioning under adverse operational conditions.

DOI: 10.4018/978-1-60960-477-6.ch002

INTRODUCTION

LOng RAnge Navigation (or Loran) is a pulsed, low-frequency (100 kHz) radio-navigation system for position fixing by reference to terrestrial transmitting stations. Loran-C was the third generation of Loran developed since World War II and it became one of the world's most widely used terrestrial radio-navigation systems. The current system provides continuous, reliable and cost-effective navigation, location and timing services for both civil and military air, land and marine users. Although a long-established and well-known system, Loran has been studied intensively in recent years as a potential complement to the Global Navigation Satellite Systems (GNSS), especially the US Global Positioning System (GPS) [e.g., Federal Aviation Administration (FAA), Loran Accuracy Performance Panel (LORAPP), Loran Integrity Performance Panel (LORIPP) programmes]. The US Volpe Report (Volpe, 2001) and many other studies have demonstrated the vulnerability of GPS to accidental and intentional interference and identified Loran as a most promising complement, since it shares almost no vulnerabilities with GPS.

The combination of Loran and GPS has been explored in Europe in the form of Eurofix (van Willigen, 1989). This employs a data channel added to the Loran-C transmissions to broadcast differential GPS (DGPS) corrections. Eurofix went on-air in April 2001 at the Bø, Vaerlandet, Sylt and Lessay stations of the Northwest European Loran-C System (NELS). This integrated Loran-C/DGPS transmission provides high accuracy position fixing over long ranges. In addition, its introduction demonstrated the concept of using integrated Loran-C and DGPS to provide an uninterrupted navigation service even if one of the basic systems had failed, in accordance with the rational view of not relying on a sole-means system. This principle has driven the development of a system that provides true radio navigation redundancy and safety for a wide range of positioning applications, plus the precise timing of telecommunications infrastructures, all of which currently rely on GPS. This system is known as Enhanced Loran or eLoran (ILA, 2007).

Enhanced Loran is an internationally standardized positioning, navigation, and timing (PNT) service for use by many modes of transport and in other applications. It takes full advantage of 21st century technology, including the use of advanced digital signal processing (DSP) techniques. These have produced dramatic improvements in the performance of receivers. By employing a data channel such as Eurofix to pass differential Loran data and integrity messages, an eLoran accuracy of 10-20m has been achieved in place of the 460m of traditional Loran-C. As a consequence, eLoran meets the accuracy, availability, integrity, and continuity performance requirements for aviation non-precision instrument approaches, maritime harbor entrance and approach maneuvers, land-mobile vehicle navigation, and location-based services. It is also a precise source of time and frequency for telecommunications systems.

Among the properties of Loran-C that underwent intensive evaluation in the development of eLoran was its integrity and, specifically, the confidence with which cycles of the 100 kHz signal could be identified within the pulses. This chapter sets out a contribution to that discussion by addressing the question of skywave contamination. In it, the term "Loran" will be used to encompass both traditional Loran-C and the new eLoran. Loran employs the groundwave components of the transmitted signals for position determination, since their propagation velocities are normally exceptionally stable in time. However, noise affects the received signals. Also, various propagation effects and the front-end filters of receivers alter the shapes of the received pulses. These factors all lead to inaccuracies in the measured positions. So, too, do unwanted skywave signal components received via ionospheric paths.

Historically, the ability of Loran-C receivers to resist this skywave contamination was its major advantage over earlier continuous-wave low-frequency navigation aids. As a consequence of the techniques employed, a single chain of Loran-C transmitters could provide coverage of a large geographical area.

It has conventionally been assumed that Loran receivers avoid skywave contamination by processing only samples taken prior to the arrival of the first skywave component, typically 35-60μs after the groundwave. This technique has significant limitations when implemented in receivers of finite bandwidth, since the receivers' filters increase the rise times of the Loran pulses and substantially reduce the amplitudes of the groundwave signals at the sampling point (Last & Farnworth et al., 1992). As a result, receivers have had to be designed to take samples later in the pulse and consequently suffer skywave errors. An attractive solution to this skywave problem is a receiver that adaptively adjusts the sampling point (for each station) to the optimal value as the skywave delay varies (Mohammed & Last, 1994; Mohammed & Last, 1995; Yi & Last, 1992). This technique enables the receiver to minimize the errors due to skywave interference while maximizing the signal-to-noise and signal-to-interference ratios. The question of skywave contamination, and receivers' ability to deal with it, are of special interest to those studying the question of the early skywaves experienced exceptionally at higher latitudes (Lo, 2005).

A receiver designed according to this "skywave-adaptive" concept (Mohammed & Last, 1994; Mohammed & Last, 1995; Yi & Last, 1992) must analyze the incoming signal to determine the skywave delay. We will demonstrate that the problem of estimating the arrival times of the groundwave and skywave components of a Loran signal is analogous to that of isolating the components of a composite signal in the frequency domain. That has let us take advantage of advances in frequency-domain signal-processing estimation techniques. This chapter investigates novel application of signal processing techniques

that employ both classical Fourier analysis and modern high-resolution algorithms for identifying skywaves and measuring their delays in Loran receivers.

For these tests, the chapter will employ a mathematical model of a Loran signal that describes it in the time and frequency domains. The classical Fourier analysis and modern high-resolution skywave detection techniques will be reviewed and their principal limitations discussed. The performance of the various algorithms will be assessed by theoretical analysis, by computer simulations under variety of adverse conditions and by testing off-air Loran signals. A prototype Loran system that employs the proposed techniques and practical real-time results will be also presented.

1 REJECTION OF SKYWAVE SIGNALS IN LORAN RECEIVERS

This section will describe how conventional Loran receivers distinguish between groundwave and skywave components. The technique will be shown to have significant limitations when implemented in receivers of finite bandwidth. Later we investigate the performance of our proposed skywave delay estimation techniques which can be used to overcome these limitations.

1.1 Loran Transmission

Loran employs pulsed transmissions, as shown in Figure 1. The pulses are bursts of 100 kHz signal of a precisely-defined carrier phase and envelope shape. Each of the stations that constitute a Loran chain radiates a group of such pulses in a precise time sequence. The interval between the groups of pulses from any single station is the Group Repetition Interval (GRI), a time between 40 μs and 100 μs which characterises and identifies the chain.

A traditional Loran-C receiver measures the time differences between the arrivals of corresponding pulses from pairs of stations and uses these time-difference measurements to compute its own position. A modern "all-in-view" eLoran receiver measures the times of arrival of pulses from all stations it can receive. Traditionally the timing point on each pulse is the "standard zero-crossing" (see Figure 1), the third positive-going zero-crossing of the 100 kHz, 30 µs after the start of the pulse. The receiver distinguishes this particular zero-crossing by identifying the corresponding point on the pulse envelope, which has the appropriate gradient.

1.2 Skywave Rejection Principle

In Loran signal propagation, there are two principal routes by which the signals can travel between the transmitters and the receiver: directly along the earth's surface (the groundwave) and by *unwanted* reflections from the ionosphere (skywaves). Thus, the receiver is presented with the sum of groundwave and skywave signal components, as shown in Figure 2. The computation of an accurate position by a Loran receiver relies on measurements made on the groundwaves only. The skywave signals can contaminate the wanted groundwave signal and may cause large errors in the computed position.

The skywave signals will always arrive at the receiver later than the groundwaves because they have had further to travel. The time difference between the arrival of the groundwave and the first skywave component from a station is known as the skywave delay. Within the coverage area of a chain of Loran transmitters the skywave delay will be at least 35 µs. Thus, the strategy which has been used traditionally to minimise skywave errors has been to choose a zero-crossing that precedes the arrival of the earliest skywave component. The zero-crossing usually employed is the one that occurs 30 µs after the start of the pulse (Figure 2); that is, the receiver essentially makes use of the groundwave of the Loran transmission only and by so doing it reject the skywave components.

1.3 Other Techniques for Skywave Identification

Some traditional Loran-C receivers do check for an incorrect skywave lock by means of a guard sample, for example, at 37.5 µs before the zero-crossing sample (USCG, 1985). If the receiver is correctly tracking the 30 µs point, no signal

Figure 1. Pulse shape of Loran transmission. The time reference point at 30 µs is marked "standard zero-crossing".

Figure 2. Loran groundwave pulse followed, 37.5 µs later, by skywave pulse 12 dB stronger. Received composite pulse is sum of skywave and groundwave signals.

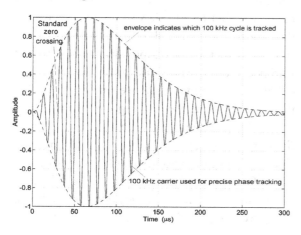

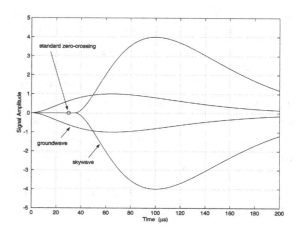

Figure 3. Amplitude spectrum of the 5ᵗʰ order Butterworth bandpass filter typical of front-end filters used in Loran receivers

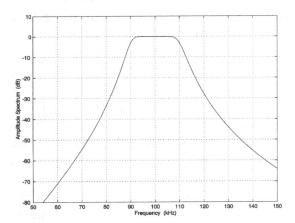

should be detected by the guard sample. If a signal is detected, the receiver may be tracking the skywave, or synchronous carrier-wave interference. Although this technique may detect the cycle-slips caused by skywave, it cannot measure the skywave parameters when tracking the correct zero crossing, even when experiencing intolerable phase-tracking errors. Neither can the technique distinguish skywave from carrier-wave interference.

The problem of skywave interference, including measuring skywave parameters, has also been studied (Peterson & Dewalt, 1992; Lievin & Hamaide, 1981) where the groundwave was first modelled, assuming that it was an ideal Loran pulse distorted during propagation and by the front-end filters of the receiver. A least-squares best fit to the leading edge of the received groundwave revealed its parameters. Then the groundwave was subtracted from the composite signal to disclose the skywave. The same least-squares procedure was then applied to the skywave to determine its parameters. It is believed that this method works well when the skywave delay is long, but it may be difficult to apply to short-delay skywaves when relatively little of the groundwave is available on which to attempt the least-squares fit.

Reference (Lievin & Hamaide, 1981) and others have measured skywave parameters in Europe, again by modelling the groundwave and subtracting it from the received signal to reveal the skywave. These sources point out that accurate modelling of skywave parameters depends on an accurate knowledge of the groundwave; they measured the groundwave, free of skywave contamination, close to the transmitter. However, this still begs the question of how to model the groundwave accurately, or measure skywave parameters, in a receiver.

1.4 Limitation of Skywave Rejection Capability

Loran receivers work in noisy environments and in the presence of strong interfering signals. Therefore, in addition to phase-decoding and subsequent integration, a major operation of the receiver signal processing is front-end filtering. Front-end filters provide substantial improvements in signal-to-noise ratio (SNR) and signal-to-interference ratio (SIR) at the timing measurement point in the receiver. However, they also cause distortion of the Loran pulses, which reduces the ability of the receiver to reject skywave interference (Last & Farnworth et al., 1992).

The amplitude-frequency transfer function of a filter is a precise measure of the attenuation of interference it provides. Figure 4 shows the amplitude spectrum of the 5ᵗʰ order Butterworth bandpass filter of 20 kHz width which is used in the simulations and analysis in this chapter. This filter shape is the one adopted by the North West European Loran Technical Working Group as typical of filters in widespread use, and employed by them in predicting the coverage of Loran chains.

Because of their finite bandwidth and non-linear phase transfer function, receiver bandpass filters distort Loran pulses (Last & Farnworth et al., 1992). The distortion may be examined in either the time or the frequency domain, as shown in Figure 4. Figure 4a shows the amplitude spec-

Figure 4. The effect of front-end filtering on Loran in (a) Frequency domain and (b) Time domain. Circle at 30 µs is standard zero-crossing point of the unfiltered pulse.

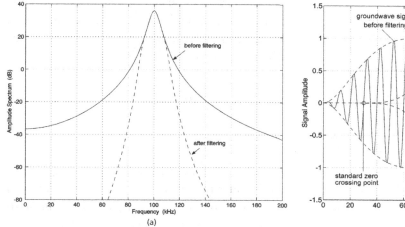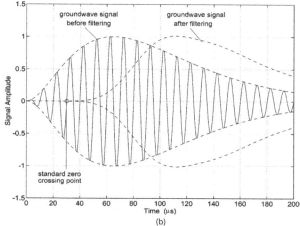

trum of a standard Loran pulse before and after filtering: the filter greatly reduces the signal energy outside the 90-110 kHz Loran band. This filtering operation, however, causes the precisely-defined, and relatively steep, leading edge of the pulse to be stretched out in time, as shown in Figure 4b. Although the peak magnitude of the pulse remains almost unchanged, its amplitude 30 µs after the start has been reduced so substantially that a much later zero-crossing must be selected for timing measurements. In practice, therefore, receiver design is a compromise between filter bandwidth and skywave tolerance.

When a later zero-crossing is chosen, the signal there may be contaminated by skywave interference. The solid line in Figure 5 represents a groundwave pulse and the dashed line is a delayed skywave pulse at the output of the filter. The strength (+12 dB) and time delay (37.5 µs) of the skywave relative to the groundwave are limiting values cited in the Minimum Performance Standards (MPS) for Loran receivers (IEC, 1989; RTC, 1977). Even by 80 µs, the filtered groundwave pulse has only regained an amplitude of half its peak value (equivalent to the envelope amplitude of an unfiltered pulse at 30 µs); however, the skywave signal is four times higher than the

groundwave at this epoch! Such a skywave signal would cause a quite unacceptable disturbance of up to 0.40 µs in the time-of-arrival measurements. Thus an important, and undesirable, result of pulse distortion caused by filtering is the reduction of signal amplitude at the sampling point.

1.5 Receivers with Adaptive Sampling Point Capability

An attractive solution to the skywave problem is a receiver that adjusts its sampling point adaptively to the optimal value in a constantly-changing skywave interference environment. It would sample well up the leading edges of the pulses during the substantial proportion of time in which skywaves are delayed by more than the 37.5 µs minimum. However, for such a receiver to work, it would need to be able to estimate skywave delays in real-time. If that were possible, we could significantly improve the accuracy and reliability of positioning under adverse skywave conditions. This chapter presents a novel application of signal processing techniques that employ both classical Fourier analysis and modern high-resolution algorithms for identifying skywaves and measuring their

Figure 5. Groundwave (solid line), and skywave (dashed line), signals after passing through typical bandpass filter

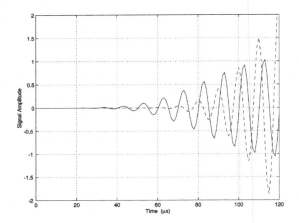

delays in Loran receivers (Mohammed & Last, 1994; Mohammed & Last, 1995; Yi & Last, 1992).

2 MATHEMATICAL FORMULATION OF THE SKYWAVE DETECTION PROBLEM

This section sets up a signal model that will be used when we attempt to identify skywave parameters.

2.1 Loran Signal Model

Loran receivers must process signals that contain a groundwave and skywaves, plus noise and interference. These components are added together at the receiver antenna. In the time domain, the composite received signal $x_c(t)$ is (Mohammed & Last, 1994; Mohammed & Last, 1995):

$$x_c(t) = x_g(t) + x_s(t) + i_T(t) \tag{1}$$

where $x_g(t)$ and $x_s(t)$ represent the ground-wave and skywaves, respectively, and $i_T(t)$ is the total noise and interference.

The skywave signal may contain many components that have arrived via different paths. Let us assume that these components have the same shape as the groundwave signal; they may then be viewed as delayed and linearly-scaled versions of the ground-wave. That is,

$$x_s(t) = \sum_{n=1}^{N} k_n x_g(t - \tau_n) \tag{2}$$

where N is the number of skywave components, and k_n and τ_n represent the amplitude and delay of the nth skywave component relative to the groundwave. Hence, Equation (1) can be rewritten as:

$$x_c(t) = x_g(t) + \sum_{n=1}^{N} k_n x_g(t - \tau_n) + i_T(t). \tag{3}$$

In the frequency domain, the composite signal can be represented as

$$X_c(f) = X_g(f)\left[1 + \sum_{n=1}^{N} k_n \exp(j2\pi f \tau_n)\right] + I_T(f) \tag{4}$$

where $X_c(f)$, $X_g(f)$ and $I_T(f)$ are the Fourier transforms of $x_c(t)$, $x_g(t)$ and $i_T(t)$, respectively.

Equation (3) and (4) constitute the signal model, in the time and frequency domains, which will be used throughout this chapter for the purpose of estimating Loran skywave parameters.

2.2 The Spectral-Division Concept

A key concept in all the methods we propose, except cepstral analysis, is a process we call a "spectral-division" operation. In this, the spectrum of the composite received signal is divided by the spectrum of a standard Loran pulse; ideally one that has been passed through the same receiving channel. This process can be represented mathematically as (Mohammed & Last, 1994; Mohammed & Last, 1995):

$$\frac{X_c(f)}{X_0(f)} = k_g \left[1 + \sum_{n=1}^{N} k_n \exp(j2\pi f \tau_n) \right] + \frac{I_T(f)}{X_0(f)} \tag{5}$$

where $X_0(f)$ is the spectrum of a normalised standard Loran pulse $x_0(t)$ and k_g is a constant representing the amplitude of the groundwave. Equation (5) establishes the principle used in determining skywave parameters. In the rest of this chapter, the analysis concentrates on the delays of skywave components, since estimating their amplitudes is relatively straightforward once the delays are known.

3 CLASSICAL SKYWAVE DETECTION TECHNIQUES

The classical techniques employ the fast Fourier transform algorithm and are the most computationally-efficient spectral estimation methods. Two techniques will be presented briefly: the power-cepstral and inverse fast Fourier transform (IFFT) methods (Mohammed & Last, 1994).

3.1 The IFFT Method

The principle of IFFT method for skywave estimation is simple (Mohammed & Last, 2004): we only need to take the inverse Fourier transform (F^{-1}) of the spectral-division operation of Equation (5). That is,

$$F^{-1}\left[\frac{X_c(f)}{X_0(f)} \right] = k_g \left[\delta(t) + \sum_{n=1}^{N} k_n \delta(t - \tau_n) \right] + F^{-1}\left[\frac{I_T(f)}{X_0(f)} \right] \tag{6}$$

The time-domain expression of Equation (6) contains impulses at the arrival times of the groundwave and skywave components which

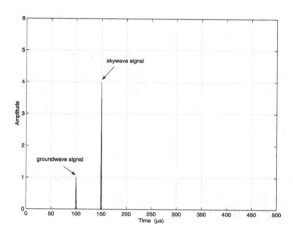

Figure 6. Groundwave and skywave components separated by the IFFT method. SGR=12 dB and skywave delay=50 µs. The impulse at 100 µs is the unit amplitude groundwave.

allow them to be separated and their relative magnitudes estimated. Figure 6 shows an example of the result of applying Equation (6) to a noise-free composite signal in a simulation. The skywave delay, which has been arbitrarily set to 50µs, is seen to be estimated successfully. *The* skywave-to-groundwave ratio (SGR) is found to be 4 (or 12 dB), which is correct.

3.2 Power-Cepstral Method

Cepstral analysis is a nonlinear processing technique for analysing data containing an arbitrary unknown signal and its echoes. It is known to be superior to conventional time-domain analysis (Bogart & Healy et al., 1963; Childers & Skinner et al., 1977). The principle of cepstral analysis is that the logarithm of the power spectrum of a signal that contains an echo has an additive periodic component due to that echo. Thus, an inverse Fourier transform of this log-power spectrum exhibits a peak at the delay of the echo (which is what interests us). This log-power spectrum is termed the cepstrum (Bogart & Healy et al., 1963), a play on the word spectrum. Mathematically, the power cepstrum of a signal $g(t)$ is defined as

$$g_p(\tau) = F^{-1}\left\{ \ln |G(f)| \right\} \qquad (7)$$

where $G(f)$ is Fourier transform of $g(t)$, ln represents the natural logarithm operation.

The power-cepstral analysis turned out to be very sensitive to noise and, consequently, we did not pursue this method further in our studies but instead concentrate on the more promising IFFT technique.

4 MODERN HIGH-RESOLUTION SKYWAVE ESTIMATION TECHNIQUES

The better of the Fourier techniques, the IFFT method, not only works successfully but also is computationally efficient and robust. However, both Fourier methods have poor time-domain resolution (Brigham, 1988). This results directly from the need to filter the signal in the frequency domain using a window of limited width, in order to achieve good performance at low SNR. It makes for a poor ability to distinguish between the groundwave signal and the shortest-delayed skywaves which, of course, are those of greatest practical importance. Although our simulation allowed us to trade off the conflicting requirements of SNR, window width and resolution, even the best solution was found to give inadequate resolution in adverse propagation conditions. This prompted a search for higher-resolution estimation methods (e.g., Kaveh, 1979; Candy, 1988; Kesler, 1986; Proakis, 1988; Rao & Aron, 1992; Johnson, 1992).

Two modern, high-resolution, estimation techniques were considered: eigendecomposition, employing the MUltiple SIgnal Classification (MUSIC) algorithm (Rao & Aron, 1992; Johson, 1992), and parametric AutoRegressive Moving Average (ARMA) modelling (Candy, 1988; Pro-

akis, 1988; Kesler, 1986). Considerable research has been devoted to these techniques in recent years because of their superior ability to resolve multiple, superimposed, sinusoids of closely spaced frequencies in noisy environments. When using them for Loran skywave analysis we first carry out the spectral-division operation (as with the IFFT method above) and then employ the MUSIC or ARMA algorithm to estimate the arrival times of the Loran components in the time domain.

4.1 Parametric Models

Many discrete-time random processes encountered in practice are well approximated by the following linear difference equation (Candy, 1988; Proakis, 1988; Kesler, 1986; Mohammed & Last, 1995):

$$x(n) = -\sum_{k=1}^{p} a_k x(n-k) + \sum_{k=0}^{q} b_k v(n-k) \qquad (8)$$

where $v(n)$ is the input sequence (a white noise process with zero mean and variance σ^2) and $x(n)$ is an output sequence that models the data. This most general model is called an *ARMA(p, q)* model (that is, an AutoRegressive Moving Average of orders p and q), in which the a_k's and b_k's denote, respectively, the AR (autoregressive) and the MA (moving average) coefficients.

All parametric (or model-based) methods employ the assumption that the data to be analysed have been produced by a process that can be represented by such a model; the analysis then identifies the parameters of this model. The ARMA spectrum estimation is obtained by performing of the following steps (Candy, 1988; Kesler, 1986, Kaveh, 1979; Mohammed & Last, 1995):

Use the Akaike information criterion (AIC) to determine the ARMA model order:

$$AIC(p, q) = -\ln \sigma^2 + 2\left(\frac{p+q}{N}\right) \qquad (9)$$

where N is the number of samples.

Use ARMA Yule-Walker normal equations to determine the model coefficients:

$$r_z(m) = \begin{cases} -\sum_{k=1}^{p} a_k r_z(m-k) + \sigma^2 \sum_{k=m}^{q} b_k h(k-m) & 0 \le m \le q \\ -\sum_{k=1}^{p} a_k r_z(m-k) & m \ge q+1 \end{cases}$$
$$(10)$$

where $r_x(k)$ is the autocorrelation estimate, defined as:

$$r_x(k) = \frac{1}{N} \sum_{n=0}^{N-1-k} x(n) x^*(n+k). \qquad (11)$$

Compute the spectrum of the ARMA model from the following equation:

$$P_{ARMA}(\omega) = \sigma^2 \left| 1 + \sum_{k=1}^{q} b_k \exp(-j\omega k) \right|^2 \Bigg/ \left| 1 + \sum_{k=1}^{p} a_k \exp(-j\omega k) \right|^2$$
$$(12)$$

ARMA models the data as a pole-zero system. When we investigated the suitability of the simpler variants of ARMA for analysing our skywaves, MA (which uses only zeroes) was found to require too many coefficients when analysing the relatively narrow Loran spectrum, and so was discarded. The AR model (which uses only poles) was found to be simpler than ARMA but nevertheless to retain good resolution and its parameters could be estimated in a variety of computation-ally-efficient ways. We chose to implement and test both ARMA and AR in the initial simulations. In the case of AR, we obtained the model coefficients and spectrum by setting all the b_k's to zero in Equations (10) and (12), respectively.

An important issue in the design of an ARMA or AR model is choosing its order, that is, the numbers of poles and zeros. This is especially critical when receiving noisy data. Because the optimum order is not generally known *a priori*, in practice it is necessary to try several values. If the value chosen is too low, the resolution of the result will be poor. Too high an order may lead to spurious peaks in the time-domain result.

Selecting the order for an ARMA model is not a simple procedure and the little information on the subject in the literature covers only simple cases. With an AR model, however, the problem is much simpler because only the number of poles is required. Several alternative criteria for selecting the optimum order for Loran sky-wave estimation were investigated and tested. Of these, the AIC in Equation (9) which is used for ARMA processes is one of the best-known and most widely-used. The order of a pure AR model can be obtained by simply letting $q=0$ in Equation (9); that is

$$AIC(p, q) = -\ln \sigma^2 + 2\left(\frac{p}{N}\right). \qquad (13)$$

In the cases of both ARMA and AR, the order that minimises the AIC is the optimal one. The value of this order, though, was found to depend on signal strength, skywave delays and the numbers of samples (Mohammed & Last, 1994; Mohammed & Last, 1995).

4.2 The MUSIC Algorithm

Consider a time domain signal $x(t)$ that contains K sinusoids of angular frequencies $\omega_1, \omega_2, .., \omega_K$ and amplitudes $A_1, A_2, ..., A_K$. Let the signal be sampled, with $x(n)$ denoting the *n-th* sample:

$$x(n) = \sum_{k=1}^{K} A_k \cos(\omega_k + \theta_k) + v(n) \quad n=1,...,N \tag{14}$$

where θ_k is the phase of the *k-th* sinusoidal component and $v(n)$ is the noise, which has zero mean and a variance σ^2.

Define an $(M \times M)$ correlation matrix of the signal $x(t)$ as:

$$R_x = \begin{bmatrix} r_x(0) & r_x(1) & \cdots & r_x(M-1) \\ r_x(1) & r_x(0) & \cdots & r_x(M-2) \\ \vdots & \vdots & \ddots & \vdots \\ r_x(M-1) & r_x(M-2) & \cdots & r_x(0) \end{bmatrix} \tag{15}$$

where $r_x(m)$ is the autocorrelation estimate at lag m, $m=0,1,2,...,M-1$.

The angular frequencies of the sinusoidal components of $x(t)$ may then be estimated using the MUSIC algorithm as follows:

Arrange the N signal samples $x(1), x(2),...,x(N)$ into a signal data matrix:

$$X = \begin{bmatrix} x(M) & x(M-1) & \cdots & x(1) \\ x(M+1) & x(M) & \cdots & x(2) \\ \vdots & \vdots & \ddots & \vdots \\ x(N) & r_x(N-1) & \cdots & x(N-M+1) \end{bmatrix} \tag{16}$$

2. Perform a singular value decomposition (SVD) operation on this data matrix. The eigenvalues of the correlation matrix R_x are the squares of the singular values of X, and the eigenvectors are the right singular vectors of X.

Let $v_1, v_2,...,v_{2K+1},...,v_M$ denote the right singular vectors corresponding to the eigenvalues $\lambda_1 \geq \lambda_2 \geq ... \geq \lambda_M$. Identify the right singular vectors that are associated with the $(M-2K)$ smallest eigenvalues. Define the matrix

V_N and the frequency-search vector e_ω as follows:

$$V_N = \begin{bmatrix} v_{2K+1}, ..., v_M \end{bmatrix} \tag{17}$$

and

$$e_\omega = \begin{bmatrix} 1, e^{j\omega}, ..., e^{j(M-1)\omega} \end{bmatrix}^T \quad -\pi \leq \omega \leq \pi \tag{18}$$

Compute the eigenspectrum, $P_{MUSIC}(\omega)$, with respect to ω in the interval $-\pi \leq \omega \leq \pi$.

$$P_{MUSIC}(\omega) = \frac{1}{e^H(\omega)V_N V_N^T e(\omega)} \tag{19}$$

The peaks of the resulting spectrum, $P_{MUSIC}(\omega)$, show the angular frequencies of the sinusoidal components of the composite signal.

For estimating the time-delays of Loran skywave components we apply this MUSIC method to the result of the spectral-division process which was, of course, carried out in the frequency, rather than the time, domain.

5 NOISE AMPLIFICATION DUE TO SEPECTRAL-DIVISION OPERATION

The spectral-division operation leads to a noise amplification problem that seriously detracts from its ability to determine the delays of skywave components. Thus, Equation (6) demands a very high SNR if the skywave delay is to be identified correctly. A spectral window of finite width may be imposed to overcome this problem. This problem is illustrated in Figure 7, where the skywave delay is *50* µs. SGR 12 dB and the SNR 40 dB. Despite this very high SNR, the skywave component is

Figure 7. Illustration of noise amplification due to the spectral-division operation. SNR=40 dB, skywave delay=50 μs and SGR=12 dB. (a) No window: skywave impulse is lost in the noise, and (b) 50 kHz rectangular spectral window is used: skywave impulse is detected successfully.

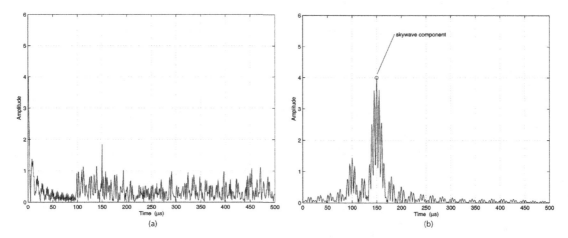

lost in the noise (Figure 7a) when Equation (6) is applied directly (i.e., no windowing). In contrast, when a rectangular window of 50 kHz bandwidth is employed (Figure 7b) the skywave component is easily identified.

6 SIMULATION SET-UP AND RESULTS

This section evaluates the performance of the skywave detection algorithms under noisy conditions. The evaluation is conducted by means of computer simulation employing a Monte Carlo method and by testing off-air signals. In addition, in section 8, we also test the AR algorithm on real-time signals in a prototype receiver. The simulation set-up is discussed briefly first, followed by the simulation results.

6.1 Simulation Setup

The functional block diagram of the simulation programs for skywave identification is shown in Figure 8. The Program Control sets up the initial parameters in all the functional blocks and

controls their operations. Simulated Atmospheric Noise (SAN), generated in accordance with the standard defined in the Loran Minimum Performance Standards (MPS) (RTC, 1977), is added to the separately generated Loran groundwave and skywaves. The composite signal is then fed into a program block which simulates the filtering effect of the front-end of the receiver, then input to the

Figure 8. Functional block diagram of simulation programs to test the operation of the proposed skywave detection techniques under noisy conditions

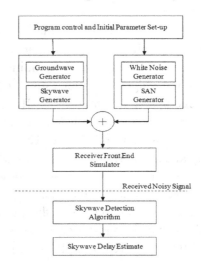

Figure 9. Results of estimation by: (a) IFFT, (b) MUSIC, (c) ARMA and (d) AR. The peak at 100 μs is the groundwave pulse. The skywave is 12 dB stronger and arrives 50 μs later. SNR=-13 dB at antenna.

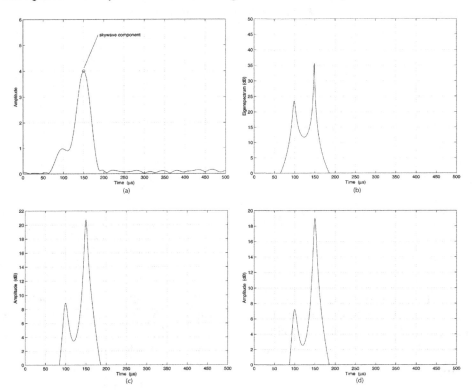

Skywave Detection Algorithm under test which analyses it to determine the times of arrival of the skywave components.

6.2 Simulation Results

A skywave delay estimation simulator has been developed to evaluate the performance of the various algorithms under noisy conditions and the results of the high-resolution estimation techniques are compared with the classical IFFT-based method. The value of the ARMA model order (p, q) used in the simulation was (10, 10). The size of the signal data matrix analysed by the MUSIC algorithm was (10×10). A Hanning window of 50 kHz was used in these simulations.

Figure 9 shows the results of the simulation with a signal consisting of a groundwave pulse at time 100 μs followed by a skywave pulse 50

μs later and 12 dB stronger (a typical night-time skywave condition). The SNR at the antenna input was -13dB, 3 dB below the minimum SNR specified for Loran signals by the United States Coast Guard (USCG, 1985). At this SNR, all methods work and successfully determine skywave delays. However, the MUSIC, ARMA and AR algorithms clearly provide superior estimation performance of the groundwave and skywave components (and accordingly a more accurate skywave delay estimate) than does the IFFT method as indicated by the position and the sharpness of the peaks; that is, they achieve the higher resolution we are seeking.

When the sensitivity of high-resolution algorithms to increasing noise is measured, the MUSIC algorithm fails at −19dB (SNR at the antenna). In addition, it is found to be very sensitive to small variations in signal parameters, producing incon-

Figure 10. (a) Off-air Loran pulse. (b) Estimated arrival times of its groundwave and skywave components isolated by AR algorithm.

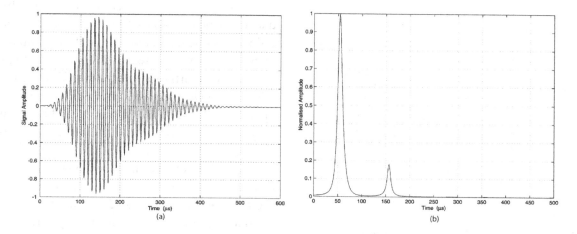

sistent results at low SNRs. The ARMA and AR methods continue to work down to −23dB, more than 13 dB below the lower limit required of receivers by the USCG. The IFFT method is computationally efficient and robust. However, it has a poor time-domain resolution particularly in adverse propagation conditions. This results in a poor capability to differentiate between the groundwave signal and the shortest-delayed skywaves which, of course, are those of greatest practical importance. This is a major reason that prompted the search for higher-resolution estimation methods as discussed in section 5. The AR model (which uses only poles) was found to be simpler than ARMA but nevertheless to retain good resolution, as evident from comparing Figures 9c and 9d, and its parameters could be estimated in a variety of computationally-efficient ways. In addition, selection of the optimal AR model order is much simpler than ARMA since only the number of poles is needed (see section 5.1).

Considering the above facts, and after extensive simulation studies of the various methods, we concluded that the AR technique is the most suitable technique for practical implementation since it gives the best combination of resolution, noise

immunity and computational complexity. Accordingly, we decided to test the AR algorithm using stored off-air data first, and then to implement it in a prototype Loran receiver and demonstrate its effectiveness in detecting skywaves and estimating their delays, in real-time.

7 A PROTOTYPE LORAN SKYWAVE DETECTION SYSTEM

The receiver to which the prototype Loran skywave estimation system was added was a high-quality, digital, Loran receiver designed for operation in a dense noise environment. This receiver comes with interface software that runs on a PC under Windows. It is equipped with programmable fixed or automatic notch filters driven by a data-base of all European Loran interferers known at the time of the measurement. Both the executable program and the database can be manipulated, and off-air data logged, using the PC.

Figure 11 shows the prototype Loran skywave detection system formed by the combination of a conventional Loran receiver and software implemented skywave detection channel (employing the AR algorithm) running on a PC. Additional

Figure 11. A Prototype Loran skywave detection receiver

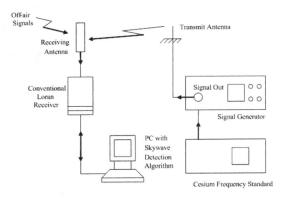

code was written to allow the PC to communicate with the receiver. The receiver fed raw data to the PC for use by the skywave analysis software. Then, having determined the minimum skywave delay, the PC was able to adjust the receiver's sampling points accordingly, simulating as closely as possible the goal of having the code run in the receiver's own processor. In addition, we were able to record large volumes of off-air data for testing the algorithms off-line under widely varying conditions of skywave delay and strength.

The spectral-division operation ideally requires a standard Loran pulse that has been passed through the same receiving channel as the off-air signals to be analysed. This was obtained as follows. A signal generator, controlled by a cesium frequency standard, was set-up to simulate an additional secondary station within the chain being received. It was sampled, and the data fed to the PC together with the samples of the off-air signals. All were then processed using the same FFT operation, the simulator's transform being employed as the divisor for the off-air signals' transforms.

Figure 12 shows an on-line analysis performed by an AR algorithm on a real-time skywave-corrupted signal received at 02:30 am from station in Vaerlandet, Norway, using the prototype receiver described above. The groundwave component is seen to start at 347 μs, a first skywave at 422 μs and a later (and much smaller) skywave at 547 μs. The delay between the groundwave and the onset of the first skywave – what this research is all about – is measured as 75 μs, sufficient to allow the sampling point to be safely moved back to the peak of the pulse and a 6 dB SNR benefit obtained. The analysis also shows this first skywave component to be about 10 dB stronger than the groundwave. The IFFT technique, in contrast, has failed to resolve the arrival times of these

Figure 12. Off-air signal received from Vaerlandet, Norway: (a) Pulse waveform, (b) Arrival times of groundwave and skywave components estimated by AR algorithm in real-time.

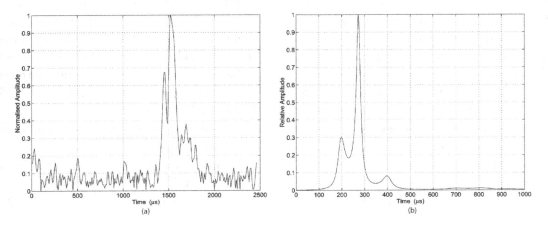

off-air Loran data, demonstrating the superiority of the high-resolution algorithms.

The adaptive skywave estimation techniques presented above establishes a basis on which to design a Loran receiver capable of adjusting its sampling point adaptively to the optimal value in a constantly-changing skywave environment. Such techniques would enable the receiver to minimize the errors due to skywave interference while maximizing the signal-to-noise and signal-to-interference ratios, and significantly improve the accuracy and reliability of positioning under adverse operational conditions.

8 CONCLUSION AND FUTURE RESERACH DIRECTIONS

Skywave interference commonly affects Loran receivers. This chapter has discussed the merits and limitations of a number of skywave estimation techniques, assessing their effectiveness by theoretical analysis, computer simulations under a range of realistic propagation conditions and by the use of real-time signals in a prototype receiver. Incorporating these skywave estimation techniques in receivers should significantly improve the accuracy and repeatability of the positions they measure. The ability to measure skywave parameters is not only valuable for improving the receiver performance, it also provides a means of making ionospheric measurements at this frequency which is so important for navigation and positioning systems.

The question of skywave contamination, and the receivers' ability to deal with it, are currently of particular interest to those studying the question of the "exceptionally early skywaves". This phenomenon appears to affect signals received at higher latitudes, such as at sites in Alaska, and has only come to attention recently. Therefore, a future research direction will be to investigate the propagation scenarios behind these early skywaves and their effects on performance, and to tune-up the estimation algorithms for reliable detection of these exceptionally early skywaves. In addition, investigating other higher-resolution estimation techniques which might be more suitable and reliable in these exceptionally adverse propagation conditions is also a topic of great interest. For example, in the past decade the ESPRIT (Estimation of Signal Parameters via Rotational Invariance Techniques) algorithm gained considerable interest and was applied to many diverse problems because of its superior ability to resolve multiple closely-spaced, super-imposed, signals in noisy environments. Initial investigations of the high-resolution ESPRIT algorithm have shown promising preliminary results in Loran skywave delay estimation problem. However, further research is necessary in order to evaluate the performance limits, and the practical aspects of using ESPRIT in long range navigation systems.

REFERENCES

Bogert, B., Healy, M., & Tukey, J. (1963). The Quefrency analysis of time series for echoes: Cepstrum, pseudo-autocovariance, cross cepstrum, and saphe cracking. In M. Rosenblatt (Ed.), *Proceedings of the Symposium on Time Series Analysis*. (pp. 209-243). New York: John Wiley and Sons Inc.

Brigham, E. (1988). *The fast fourier transform and its applications*. NJ: Prentice-Hall Inc.

Candy, J. (1988). *Signal processing: The modern approach*. NJ: McGraw-Hill Inc.

Childers, D., Skinner, D., & Kemerait, R. (1977). The cepstrum: A guide to processing. *Proceedings of the IEEE, 65*(10), 1428–1443. doi:10.1109/PROC.1977.10747

IEC. (1989). *Draft standard-Loran receivers for ships*, International Electrotechnical Commission, IEC Technical Committee No. 80.

ILA. (2007). *Enhanced Loran (eLoran) definition document V1.0.* International Loran Association.

Johnson, R. (1992). An experimental investigation of three Eigen DF techniques. *IEEE Transactions on Aerospace and Electronic Systems, 28*(3), 852–860. doi:10.1109/7.256305

Kaveh, M. (1979). High resolution spectral estimation for noisy signals. *IEEE Transactions on Acoustics, Speech, and Signal Processing, 27,* 286–287. doi:10.1109/TASSP.1979.1163243

Kesler, S. (1986). *Modern spectral analysis - II.* New York: IEEE Press.

Last, D., Farnworth, R., & Searle, M. (1992). Effects of Skywave interference on the coverage of Loran. *IEE Proceeding-F, 139*(4), 306–314.

Lievin, J., Hamaide, J., Scholiers, W. & Lechien, J. (1981). *Measurements of ECD, time of arrival, amplitude and phase of both ground and reflected Loran pulses.* AGARD-CP-305.

Lo, S., Morris, P., & Enge, P. (2005). *Early Skywave detection network: Preliminary design and analysis.* International Loran Association, 34th Annual Convention and Technical Symposium, Paper 5-1.

Mohammed, A., & Last, D. (1995). *Loran Skywave delay detection using rational modelling techniques* (pp. 100–104). Bath, UK: Radio Receivers and Associated Systems.

Mohammed, A., Yi, B., & Last, D. (1994). *Rational modelling techniques for the identification of Loran Skywaves.* Wild Goose Association, 23rd Annual Convention and Technical Symposium, (pp. 184-191). Rhode Island, USA.

Peterson, B. & Dewalt, D. (1992). *Loran and the effects of terrestrial propagation.* USCG Academy, Technical Report.

Proakis, J., & Manolakis, D. (1988). *Introduction to digital signal processing.* New York: MacMillan Publishing Company.

Roa, B., & Arun, K. (1992). Model based processing of signals: A state space approach. *Proceedings of the IEEE, 80*(2), 283–309. doi:10.1109/5.123298

RTC. (1977). *Minimum Performance Standards (MPS)-Marine Loran receiving equipment.* Radio Technical Commission for Marine Services, U.S. Federal Communication Commission. Report of Special Committee No. 70.

USCG. (1985). *US Coast Guard Academy: Loran Engineering Course.* USCG.

van Willigen, D. (1989). *Eurofix: Differential hybridized integrated navigation.* Wild Goose Association, 18th Annual Convention and Technical Symposium, Hyannis, MA, November 1989.

Volpe. (2001). *Vulnerability assessment of the transportation infrastructure relying on the global positioning system.* Volpe National Transportation Systems Center.

Yi, B., & Last, D. (1992). Novel techniques for the identification of Loran Skywaves. *Proceeding of the 21st Annual Technical Symposium, Wild Goose Association,* Birmingham, UK, (pp. 239-246).

Chapter 3
Application of Space– Time Signal Processing and Active Control Algorithms for the Suppression of Electromagnetic Fields

Tommy Hult
Lund University, Sweden

Abbas Mohammed
Blekinge Institute of Technology, Sweden

ABSTRACT

Several studies have been conducted on the effects of radiation on the human body. This has been especially important in the case of radiation from hand held mobile phones. The amount of radiation emitted from most mobile phones is very small, but given the close proximity of the phone to the head it might be possible for the radiation to cause harm. The suggested approach involves the use of adaptive active control algorithms and a full space-time processing system setup (i.e. multiple antennas at both the transmitter and receiver side or MIMO), with the objective of reducing the possibly harmful electromagnetic radiation emitted by hand held mobile phones. Simulation results show the possibility of using the adaptive control algorithms and MIMO antenna system to attenuate the electromagnetic field power density.

DOI: 10.4018/978-1-60960-477-6.ch003

BACKGROUND AND INTRODUCTION

There have been several studies done, with conflicting results, on the effects of cell-phone radiation on the human body (Christensen, Schüz, Kosteljanetz, Poulsen, Thomsen & Johansen, 2004; Lai, 1998; Sienkiewicz & Kowalczuk, 2005). The amount of radiation emitted from most cell phones is very minute. However, given the close proximity of the phone to the head, it is entirely possible for the radiation to cause harm. If you want to be on the safe side, the easiest way to minimize the radiation you are exposed to is to position the antenna as far from your head as possible. Utilizing a hands-free kit, a car-kit antenna or a cell phone whose antenna is even a couple of inches farther from the head can do this most effectively. This chapter makes a contribution to that discussion by proposing a new approach employing adaptive active control algorithms combined with a Multiple-Input Multiple-Output (MIMO) antenna system to suppress the electromagnetic field at a certain volume in space. In addition, this system can be applied to other applications such as heavy electric machinery (electric engines, generators) and power lines or when performing maintenance and testing on high power radio transmitters (e.g., broadcasting or radar).

Active methods for attenuating acoustic pressure fields have been successfully used in many applications. In this chapter we investigate if these methods can be applied to an electromagnetic field in an attempt to lower the power density at a specified volume in space.

The cancelling out of a signal can be achieved by employing the principle of superposition. For example, if two signals are superimposed, they will add either constructively or destructively. The objective of our study is to investigate the possibility of applying adaptive active control algorithms with the goal of reducing the electromagnetic field power density at a specific volume using the superposition principle and MIMO antenna system. Initially, the application we evaluate is a model of a mobile phone equipped with one ordinary transmitting antenna and a number of actuator-antennas which purpose is to cancel out the electromagnetic field at a specific volume in space (e.g. at the human head) (Hult & Mohammed, 2004, 2005) using power level information obtained by an sensor antenna array. Later, we investigate the effects of the size and number of MIMO antenna elements on the performance of the system (Hult & Mohammed, 2004).

It is worth stressing at this point that the purpose of this MIMO system is not to improve the capacity or quality of transmission between the mobile unit and base station, but to predict the channel response or sense the radiated field which can then be controlled by using the active control algorithms. For this purpose, a class of algorithms called Filtered-X (Widrow, 1985; Johansson, 2000; Kuo, 1996), which are well known from the area of acoustic noise cancellation are employed and evaluated to assess their behaviour and performance in this electromagnetic type of environment. By constraining these adaptive algorithms we also try to make the total output power level transmitted by the antenna elements, locked to a predefined value. This power constraint is achieved through the use of a quadratic constraint on the active control algorithms (Hult & Mohammed, 2004).

The modelling of the antenna elements and the electromagnetic field calculations are performed using the simulation package FEMLAB (currently COMSOL Multiphysics) (COMSOL, 2006a, 2006b). This software is also used in combination with MATLAB to implement and test the adaptive algorithms used to control the electromagnetic field.

THE MODEL

The FEM Model

The application used in this chapter is a three-dimensional (3D) model of a physical system consisting of eight vertical antenna elements and of

Figure 1. 2D model representing the tested physical system of the interaction between the antennas and the human head

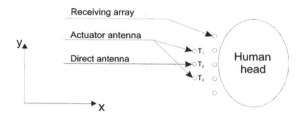

a human head a two-dimensional representation of the 3D model is shown in Figure 1. The simulation of the radio waves is performed numerically by using the finite element method (FEM) in COMSOL Multiphysics software package for solving the electromagnetic field Equations. The operating carrier frequency used in the investigation is 900 MHz (a wave length λ of approximately 0.33 m).

In a simple medium where we have no external sources except inside the transmitting antenna elements, we can write Maxwell's Equations in time-harmonic form, where the vector arguments and the term $exp\{-j\omega t\}$ are omitted for simplicity.

To model materials that contain both conductive and dielectric properties, a complex valued permittivity ε_c is defined as,

$$\varepsilon_c \triangleq \varepsilon_r - j\frac{\sigma}{\omega\varepsilon_0}, \tag{1}$$

where σ is the conductivity and ε_r is the relative permittivity of the material when there is an incident time-harmonic wave with an angular frequency ω.

Several authors have suggested permittivity and conductivity values of the human brain tissues as a function of frequency (examples in (Gabriel, 1996; Kritikos & Schwan, 1976; Weil, 1975)). The data have been assessed through measurements or by deriving values from the intensity levels of

magnetic resonance images (MRI) and thermographies, or by theoretical analysis. This variety of methods produce a wide range of geometrical and dielectric properties of the brain tissue. The most common data was published 1957 by Schwan (Schwan, 1957). Schwan and other authors (Foster, Schepps, Stoy, & Schwan, 1979) validated these values up to microwave frequencies (Kritikos & Schwan, 1976; Weil, 1975), and proposed analytical expressions derived from Debye's model of molecular dipole moment in dispersive materials. The standardization seems to converge on data published by Gabriel (Gabriel, 1996).

For simplicity an average of the electric properties of the brain and skull is used here; for example at a frequency of 900 MHz the following parameters are used, $\varepsilon_r = 45.8$ and $\sigma = 0.766$ [S/m]. These values are based on the 4-Cole-Cole Equation as described in by (Gabriel, 1996). The antenna elements are assumed to be made out of copper and will have the following electric properties, $\varepsilon_r = 1$ and $\sigma = 5.99 \cdot 10^7$ [S/m]. If we assume that there are no ferromagnetic materials in this FEM model, it will be sufficient to set the permeability equal to the free space permeability $\mu = \mu_0$.

In order to simulate an electromagnetic wave travelling out towards infinity, it is necessary to define the outer boundary of the modeled area so that it does not reflect any signal back towards the antennas (i.e. total absorption at the outer boundary). When the time-harmonic solution of the electric field components $\vec{E}(x, y, z)$ is calculated, FEM simulator solves the other fields automatically using *Maxwell's* Equations to get the magnetic $\vec{H}$, the magnetic flux density $\vec{B}$ and the electric displacement $\vec{D}$ fields, respectively. The $\vec{E}(x, y, z)$, is then used to calculate the *Poynting* vector $\vec{S}(x, y, z)$ and the time averaged power density. In this case with a stationary wave, the time average of this vector can be defined as (Cheng, 1989),

Figure 2. A surface plot showing the calculated relative magnitude of the power density based on the electric field solution

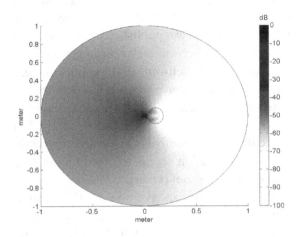

$$\left\langle \vec{S} \right\rangle = \mathrm{Re}\left\{ \frac{1}{2}\, \vec{E} \times \vec{H}^* \right\}, \qquad (2)$$

where $\langle \cdot \rangle$ denotes a time average.

The best way to visualize the power density is by showing a surface plot of a two-dimensional slice from the three-dimensional model to show the magnitude of $\left\langle \vec{S} \right\rangle$ in decibels as illustrated in Figure 2.

Further, from the numerical FEM solution $\vec{E}(x,y,z)$ we can also calculate the maximum specific absorption rate (SAR) value according to,

$$SAR = \sigma \cdot \frac{\left| \vec{E} \right|^2}{\rho}, \qquad (3)$$

where σ is the conductivity and ρ is the density of the material. The SAR value calculated in the simulations is the maximum SAR value in each point inside the human head. This value differs from the value used in the standards for mobile telephones, which is the mean value of Equation (3) taken over 10 gram of tissue.

The SAR limit recommended by the International Commission of Non-Ionizing Radiation Protection (ICNIRP) is 2 W/kg averaged over 10 gram of tissue. Thus to fulfill this limit with certainty we need to have a maximum SAR below 2 W/kg.

The MIMO Model

To reduce the electromagnetic field within a certain volume in the FEM-modeled space, a MIMO radio channel is modeled in order to compensate for the spatial displacement. In this chapter, the FEM simulation program is used to simulate the physical MIMO antenna system, which (initially in this Section) consists of 3 transmitting antennas and 5 receiving antennas as shown in Figure 1. The spacing between the antenna elements used in this application is $\lambda = 0.02$m; thus this arrangement cannot be seen as an ordinary beamformer as the antenna elements are working in the radiated near-field. The input signals to this system are three separate currents in a complex-valued phasor notation, one for each transmitting antenna. The simulated output current from the 5 receiving antennas form a complex-valued data vector denoted as the error signal vector of the system.

By changing the amplitudes and phases of the currents assigned to the three transmitting antennas it is possible to control the transmitted power from the separate antenna elements. The calculated time-harmonic electromagnetic wave in the model will then generate a current density inside the receiving antenna elements. According to Ampere's law for a time-harmonic wave in a simple conductive media we have the following Equation,

$$\nabla \times \vec{B} = j\omega\mu\varepsilon_c \vec{E} \qquad (4)$$

The total output current I_{out} from each receiving antenna element can be calculated by integrating both sides over the cross section area S of the antenna element,

$$I_{out} = j\omega\varepsilon_0\left(\varepsilon_r - j\frac{\sigma}{\varepsilon_0\omega}\right)\iint_S \vec{E}\, dS \qquad (5)$$

The result from each antenna element is then stored in a complex-valued data vector **e**.

If we have a system of three transmitter antennas and five receiver antennas the transmitter antenna in the middle (T_2) (see Figure 1) is the one we want to cancel, then the two flanking transmitter antennas are denoted as the actuator-antennas (T_1, T_3). Since the simulated model is experimentally confirmed to be linear, and this is a weak-stationary problem with a narrow bandwidth time-harmonic signal, it is in this case sufficient to describe the parameters as a 5×3 complex-valued matrix **H**,

$$\mathbf{H} = \begin{bmatrix} H_{11}(\omega) & H_{12}(\omega) & H_{13}(\omega) \\ H_{21}(\omega) & H_{22}(\omega) & H_{23}(\omega) \\ H_{31}(\omega) & H_{32}(\omega) & H_{33}(\omega) \\ H_{41}(\omega) & H_{42}(\omega) & H_{43}(\omega) \\ H_{51}(\omega) & H_{52}(\omega) & H_{53}(\omega) \end{bmatrix} \qquad (6)$$

These complex-valued numbers H_{ij} describe the amplitude and phase due to the distance between the different combinations of transmitting and receiving antennas. Each column in **H** represents the time-harmonic frequency response functions between one of the transmitting antennas and each of the receiving antennas. The superimposed signals received by the antenna array, constitutes a vector **e** with five complex-valued elements, $\mathbf{e} = [e_1 e_2 e_3 e_4 e_5]^T$. If we now divide the matrix **H** into two separate complex-valued matrices (**F** and **g**), we get,

$$\mathbf{F} = \begin{bmatrix} \mathbf{H}_1 & \mathbf{H}_3 \end{bmatrix} = \begin{bmatrix} F_{11} & F_{12} \\ F_{21} & F_{22} \\ F_{31} & F_{32} \\ F_{41} & F_{42} \\ F_{51} & F_{52} \end{bmatrix}, \qquad (7)$$

and,

$$\mathbf{g} = \mathbf{H}_2 = \begin{bmatrix} g_1 \\ g_2 \\ g_3 \\ g_4 \\ g_5 \end{bmatrix}. \qquad (8)$$

The two columns in **F** represent the frequency response functions of the *actuator*-antennas and are denoted as the *forward* channels. The vector **g** is denoted as the *direct* channel and represents the frequency response function of the antenna with the signal we want to cancel out. The total noise of the model is described by vector **v**, and is modeled as a complex-valued additive white Gaussian noise vector.

If the carrier signal transmitted through the *direct* channel **g** is to be suppressed at the receiving antenna array, a phase-shifted and amplified copy of the same carrier signal could be transmitted through the *forward* channels to superimpose the signal in the *direct* channel. This could be achieved by incorporating a filter **w** which would allow control over the signals going through the *forward* channels.

To achieve the best possible attenuation in energy sense, the total energy output ξ of the signal **e** at the receiving antennas must minimized. The minimum energy with respect to the filter **w** is,

$$\min_{\mathbf{w}} \xi = \min_{\mathbf{w}} \mathrm{E}\left\{|\mathbf{e}|^2\right\} = \min_{\mathbf{w}} \mathrm{E}\left\{\mathbf{e}^H\mathbf{e}\right\}, \qquad (9)$$

where ^{H} denotes a conjugate transpose. With the noise vector **v** included in the system, the residual error signal **e** in Equation (9) is given by the super-position of the signals from the three transmitting antenna elements,

$$\mathbf{e} = s\mathbf{g} + s\mathbf{F}\mathbf{w} + \mathbf{v}. \tag{10}$$

If the input signal s and the noise **v** are assumed to be uncorrelated, then the mean energy can be written as,

$$\xi = r_d + \mathbf{w}^H\mathbf{p} + \mathbf{p}^H\mathbf{w} + \mathbf{w}^H\mathbf{R}_F\mathbf{w} + r_v, \tag{11}$$

where $\mathbf{R}_F = \mathbf{F}^H\mathbf{F}$, is the covariance matrix of the *forward* channels, $\mathbf{p} = \mathbf{g}^H\mathbf{F}$, is the cross-correlation between the *direct* channel and the *forward* channels, r_d and r_v are the signal power of the *direct* channel and the noise, respectively.

The minimum energy ξ_{min} is found by differentiating ξ with respect to the complex conjugate of filter coefficients **w*** and then setting the derivative equal to zero,

$$\nabla_{\mathbf{w}^*}\xi = 0 \quad \Rightarrow \quad \mathbf{p} + \mathbf{R}_F\mathbf{w} = 0. \tag{12}$$

This is the *Least Mean Square* (LMS) solution to the problem and is the optimal solution in mean energy sense. Figure 3 show the surface plot of the power density solution when the filter coefficients controlling the signals going through the forward channels are the optimal least mean square coefficients $\mathbf{w}_{opt}$ obtained from Equation (12).

THE ADAPTIVE ALGORITHMS

The least mean square solution in Equation (12) describes a quadratic form in the complex valued **w**-domain, which has only one optimum point. The gradient of the quadratic performance surface

Figure 3. The power density of the model in Figure 1 when employing the optimal filter coefficients $\mathbf{w}_{opt}$

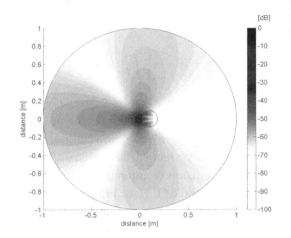

will be evaluated with respect to the conjugate filter coefficients, $-\nabla_{\mathbf{w}^*}\xi$. This will give the local steepest descent direction towards the minimum point of the performance surface. If the filter coefficients $\mathbf{w} = [w_0, w_1]$ has the energy ξ, then a new point in direction of the negative gradient vector must be closer to the minimum point of the surface. So this will give an iterative update Equation of the filter-coefficients as,

$$\mathbf{w}_{n+1} = \mathbf{w}_n + (-\nabla_{\mathbf{w}^*}\xi(n)). \tag{13}$$

The mean-energy, ξ, of the error function can according to Equations (9) and (10) be expressed as,

$$\xi = \mathrm{E}\left\{|\mathbf{e}|^2\right\} = \mathrm{E}\left\{\mathbf{e}^H\mathbf{e}\right\} = \mathrm{E}\left\{(s\mathbf{g} + s\mathbf{F}\mathbf{w} + \mathbf{v})^H\mathbf{e}\right\} \tag{14}$$

If we differentiate Equation (14) with respect to the conjugate of the filter coefficients **w*** and define $\mathbf{X}^H \triangleq \left(s\mathbf{F}\right)^H$, we get the gradient of the mean energy,

$$-\nabla_{\mathbf{w}^*}\xi = \mathrm{E}\left\{-\mathbf{X}^H\mathbf{e}\right\}. \tag{15}$$

The expected value is generally unknown, so this can be estimated by a sample mean instead; that is,

$$-\hat{\nabla}_{\mathbf{w}^*}\xi = -\mathbf{X}^H\mathbf{e}, \tag{16}$$

where $\hat{\nabla}$ denotes the estimated gradient. If this is substituted into the weight-updating Equation (13) we get,

$$\mathbf{w}_{n+1} = \mathbf{w}_n + \mu(-\nabla_{\mathbf{w}^*}\xi(n)) = \mathbf{w}_n - \mu\mathbf{X}^H\mathbf{e}. \tag{17}$$

This is the so-called Filtered-X LMS (Widrow & Stearns, 1985; Kuo & Morgan, 1996) (FX-LMS), since $\mathbf{X}$ is the input signal filtered through the *forward* channels $\mathbf{F}$.

The step-length μ in FX-LMS is a constant value and therefore the stability range and convergence rate will change with the change of input power as,

$$\left\{0 < \mu < \frac{2}{\mathrm{Tr}\left\{\mathbf{R}_F\right\}}\right\}. \tag{18}$$

To get around the change of convergence rate, consider that,

$$\mathbf{R}_F = s^*\mathbf{F}^H\mathbf{F}s = |s|^2 \cdot \begin{bmatrix} \sum_{m=1}^{5}|F_{m1}|^2 & \sum_{m=1}^{5}F_{m1}F_{m2}^* \\ \sum_{m=1}^{5}F_{m2}F_{m1}^* & \sum_{m=1}^{5}|F_{m2}|^2 \end{bmatrix} \tag{19}$$

Then the trace of the matrix $\mathbf{R}_F$ will be,

$$\mathrm{Tr}\left(\mathbf{R}_F\right) = |s|^2 \cdot \sum_{m=1}^{5}|F_{m1}|^2|F_{m2}|^2, \tag{20}$$

where m designates the five different receiving antennas. The above range for convergence (Equation (18)) can then be written as,

$$\left\{0 < \mu < \frac{2}{|s|^2 \cdot \sum_{m=1}^{5}|F_{m1}|^2|F_{m2}|^2}\right\}. \tag{21}$$

If we introduce a new step-length parameter β $(0 < \beta < 2)$ and normalize by the trace of the matrix $\mathbf{R}_F$, we get,

$$\mu = \frac{2}{|s|^2 \cdot \sum_{m=1}^{5}|F_{m1}|^2|F_{m2}|^2}. \tag{22}$$

Then the range of the step-length will be fixed within the range $0 < \beta < 2$. If this substitution is made in the FX-LMS weight-updating algorithm (Equation (17)), we get the Normalized FX-LMS algorithm,

$$\mathbf{w}_{n+1} = \mathbf{w}_n + \beta\left(\frac{\mathbf{X}_n^H}{\alpha + \mathrm{Tr}\left(\mathbf{R}_F\right)}\right)\mathbf{e}, \tag{23}$$

where α is a noise regulating parameter (Widrow & Stearns, 1985; Kuo & Morgan, 1996).

Another approach to an adaptive algorithm is by using the optimal least mean square solution from Equation (12) in combination with the gradient vector of the quadratic performance surface (Widrow & Stearns, 1985; Kuo & Morgan, 1996),

$$\nabla\xi = \mathbf{R}_F\mathbf{w} + \mathbf{p}, \tag{24}$$

If we multiply both sides of the gradient by $\mathbf{R}_F^{-1}$, we get,

$$\mathbf{R}_F^{-1} \nabla \xi = \mathbf{w} - \mathbf{w}_{opt}. \tag{25}$$

Rearrange Equation (25) into an iterative Equation where $\mathbf{w}_n = \mathbf{w}$ is the present position (or iteration) and $\mathbf{w}_{n+1} = \mathbf{w}_{opt}$ is the next position, we get,

$$\mathbf{w}_{n+1} = \mathbf{w}_n - \mathbf{R}_F^{-1} \nabla_n \xi. \tag{26}$$

If the expression of the gradient vector is inserted into Equation (26) we obtain,

$$\mathbf{w}_{n+1} = \mathbf{w}_n - \mathbf{R}_F^{-1} \left(\mathbf{R}_F \mathbf{w}_n + \mathbf{p} \right) = \mathbf{w}_{opt}. \tag{27}$$

This is the FX-Newton algorithm and its iterative Equation moves from any arbitrary point $\mathbf{w}_n$ on the performance surface to the minimum point in one single step. If the noise level is high (i.e., low SNR), this can give a very erratic search of the minimum point with a large misadjustment (i.e. noise). One approach to smooth the misadjustment noise is by using a step-length variable μ as a smoothing regulator,

$$\mathbf{w}_{n+1} = \mathbf{w}_n - \mu \mathbf{R}_F^{-1} \left(\mathbf{R}_F \mathbf{w}_n + \mathbf{p} \right), \tag{28}$$

where $0 < \mu < 1$. This solution is still going to give an erratic search with a large misadjustment, unless a very small step-length is used which will also slow down the rate of convergence. However, if the gradient vector $(\mathbf{R}_F \mathbf{w}_n + \mathbf{p})$ is estimated by the sample mean as was done in the FX-LMS algorithm (Widrow & Stearns, 1985), we get,

$$-\hat{\nabla}_{\mathbf{w}^*} \xi = \hat{E} \left\{ \mathbf{X}^H \mathbf{e} \right\} = \mathbf{X}^H \mathbf{e}, \tag{29}$$

where $\wedge$ denotes an estimation. Using Equation (29), the new weight update Equation is given by,

$$\mathbf{w}_{n+1} = \mathbf{w}_n - \mu \mathbf{R}_F^{-1} \left(\hat{\nabla}_{\mathbf{w}^*} \xi(n) \right) = \mathbf{w}_n - \mu \mathbf{R}_F^{-1} \mathbf{X}_n^H \mathbf{e}. \tag{30}$$

This is the so-called FX-Newton/LMS algorithm, which is a compromise between the two adaptive approaches. This will result in a greatly enhanced smoothing of the gradient-noise. The main problem with both the FX-Newton and the FX-Newton/LMS algorithms is the need to calculate the inverse of the covariance matrix, which is computationally inefficient. However, if the diagonal elements of the covariance matrix $\mathbf{R}_F$ are large compared to the off-diagonal values, then the covariance matrix can be estimated from,

$$\hat{\mathbf{R}}_F \approx \text{diag} \left\{ \mathbf{R}_F \right\}. \tag{31}$$

By inserting this estimate into the weight updating Equation and using a separate step-length for each matrix element, we get,

$$\mathbf{w}_{n+1} = \mathbf{w}_n - \mathbf{M} \mathbf{X}_n^H \mathbf{e}, \tag{32}$$

where

$$\mathbf{M} = \begin{bmatrix} \dfrac{\mu_1}{|s|^2 \cdot \sum_{m=1}^{5} |F_{m1}|^2} & 0 \\ 0 & \dfrac{\mu_2}{|s|^2 \cdot \sum_{m=1}^{5} |F_{m2}|^2} \end{bmatrix}$$

Equation (32) is known as the Actuator Individual Normalized FX-LMS algorithm (Johansson, 2000).

If the eigenvalues of $\mathbf{R}_F$ are disparate, then the Actuator Individual Normalized FX-LMS will outperform the Normalized FX-LMS since each filter weight will be controlled and normalized separately. On the other hand, if the eigenvalues of the covariance matrix roughly have the same value, then the Normalized FX-LMS and the Actuator Individual Normalized FX-LMS behave in a similar way.

SIMULATION RESULTS OF THE UNCONSTRAINED SOLUTION

In the previous section we presented the different adaptive algorithms to suppress the power density of the electromagnetic field and thereby decreasing the maximum SAR value inside the human head. These algorithms are unconstrained; that is there is no control over the total output power from the mobile phone. The constrained solution will be presented in the next section. In this section we evaluate and compare the different unconstrained adaptive algorithms.

The ordinary FX-LMS algorithm is the simplest to implement of the evaluated algorithms, but this algorithm has some disadvantages when the input signal is non-stationary. The Normalized FX-LMS algorithm normalizes the input signal with its signal power, resulting in a more robust algorithm at the expense of higher computational complexity. Another approach to the adaptive search is Newton's method where it is possible to solve the problem in one single step under ideal conditions. This single-step algorithm, however, is very sensitive to noise and is therefore impractical. To improve the noise insensitivity of the Newton algorithm, a gradient vector estimate is used to smooth the algorithm. This algorithm is called the FX-Newton/LMS. Both the FX-Newton and FX-Newton/LMS algorithms require a matrix inversion of the covariance matrix, resulting in high computational complexity. The Actuator Individual Normalized FX-LMS algorithm only uses the diagonal of the covariance matrix to simplify the problem of calculating the inverse of the covariance matrix.

From the above discussion and by testing the algorithms by simulations, it was concluded that the

Figure 4. The maximum SAR level inside the human head. Plots (from top to bottom): (a) One transmitting antenna only. (b) 5 passive sensor elements as a passive reflector. (c) FX-LMS. (d) Actuator Individual FX-NLMS. (e) Least Mean Square solution. (f) Actuator Individual FX-NLMS with constraint.

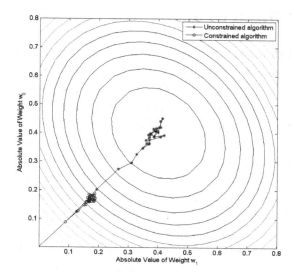

Normalized FX-LMS and the Actuator Individual Normalized FX-LMS are the preferred algorithms since they are both robust and noise-insensitive. Figure 4 show the calculated maximum SAR level inside the human head relative to the maximum SAR level of a single transmitting antenna for both adaptive algorithms.

The figure also show the corresponding maximum SAR level attained by employing a passive five element reflector and the least mean square solution which is used as a benchmark for comparisons. The amount of SAR attenuation achieved by the least mean square solution is approximately 12 dB relative to the maximum SAR level produced by a single antenna system (i.e., by using the direct transmitting antenna only, as shown in Figure 1). It is clear from Figure 4 that the adaptive algorithms after convergence give about 10 dB more attenuation compared to using the five receiving antenna elements as a passive reflector. It can also be seen that the Actuator Individual FX-NLMS converges about 40% faster than the FX-NLMS towards the least mean square solution, since each diagonal element of the covariance matrix is normalized separately.

POWER CONSTRAINTS

In the previous sections we presented the different unconstrained adaptive algorithms to suppress the power density of the electromagnetic field and their respective simulation results. There is however a major drawback with these adaptive algorithms: that is although the SAR is attenuated by approximately 5 dB (as shown in plots c-e in Figure 4) inside the human head, there is no control over the total output power from the mobile phone. This means that the total output power changes when the filter adapts, which is unfortunate since the magnitude of the total output power from the mobile phone depends on the distance from the base station. For example, if we take the case of three transmitting antennas and five receiving antennas, this would result in an increase of the total output power by approximately 20% (although this still gives a suppression of 5 dB inside the human head). However, with some other antenna spacing the mobile phone might lose the connection when the adaptive suppression filter converges towards the optimum value.

To alleviate this problem, some form of power constraint (Hult & Mohammed, 2005) could be used on the minimization process; that is,

$$\min_{\mathbf{w}^H} \left(r_d + \mathbf{w}^H \mathbf{p} + \mathbf{p}^H \mathbf{w} + \mathbf{w}^H \mathbf{R}_F \mathbf{w} + r_v \right)$$
$$\text{subject to} : \left| s\mathbf{w} \right|^2 + \left| s \right|^2 = C, \qquad C \in \Re$$
$$(33)$$

where the symbol $\Re$ denotes a set of positive real numbers. This optimization problem can then be solved by forming a Lagrange Equation (Tian, Bell & Van Trees, 1998) defined as,

$$L\left(\mathbf{w}, \lambda \right) = \mathbf{w}^H \mathbf{R}_F \mathbf{w} + \mathbf{w}^H \mathbf{p} + \mathbf{p}^H \mathbf{w} - \lambda \left(C - s^* \mathbf{w}^H \mathbf{w} s - s^* s \right).$$
$$(34)$$

By differentiating this Lagrange Equation and setting it to zero, we get a suboptimal solution of $\mathbf{w}$ which is dependent on the variable λ,

$$\nabla_{\mathbf{w}^H} L(\mathbf{w}, \lambda) = 0 \quad \Leftrightarrow \quad \mathbf{R}_F \mathbf{w}_{co} + \mathbf{p} + \lambda \left| s \right|^2 \mathbf{w}_{co} = 0$$
$$(35)$$

$$\left(\mathbf{R}_F + \lambda \left| s \right|^2 \mathbf{I} \right) \mathbf{w}_{co} = -\mathbf{p}, \qquad (36)$$

where $\mathbf{w}_{co}$ denote the *constrained* values of the filter coefficients. If we multiply Equation (36) by $\mathbf{R}_F^{-1}$ we get,

$$\left(\mathbf{I} + \lambda \left| s \right|^2 \mathbf{R}_F^{-1} \right) \mathbf{w}_{co} = -\mathbf{R}_F^{-1} \mathbf{p}. \qquad (37)$$

The right hand side of Equation (37) is the unconstrained optimal solution $\mathbf{w}_{opt}$ which was derived earlier in this paper (see Equation (12)). Using this information and rearranging Equation (37), we get,

$$\mathbf{w}_{co} = -\left(\mathbf{I} + \lambda \left| s \right|^2 \mathbf{R}_F^{-1} \right)^{-1} \mathbf{w}_{opt}. \qquad (38)$$

It can be clearly seen from Equation (38) that it is now possible to adjust the unconstrained solution by using a diagonal loading of the covariance matrix. The parameter λ can be chosen so that Equation (38) satisfies the constraint. Unfortunately there are no closed form solutions for the optimal value of the loading variable λ. However, Equation (38) can be simplified by employing a *Maclaurin* expansion of the first

Figure 5. An example of using the power constraint in combination with the FX-LMS algorithm. The trace with rings shows the convergence of the unconstrained filter coefficients. In the trace with stars we have a constraint that allows for half the power needed to reach the optimal point.

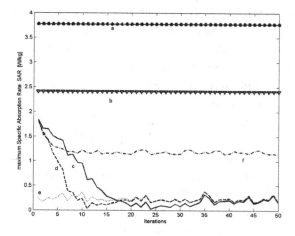

is possible to derive an approximate expression where we only need to perform a single matrix inversion operation,

$$\left(\mathbf{I} + \lambda \left| s \right|^2 \mathbf{R}_F^{-1} \right)^{-1} \approx \mathbf{I} - \lambda \left| s \right|^2 \mathbf{R}_F^{-1}. \qquad (39)$$

When this approximation is substituted into the solution of the constrained minimization, we get the constrained values of the filter coefficients as,

$$\mathbf{w}_{co} = -\mathbf{w}_{opt} - \lambda \left| s \right|^2 \mathbf{R}_F^{-1} \mathbf{w}_{opt}. \qquad (40)$$

To find out which value of λ we need, the constraint (Equation (41)) should be solved for the value of the constrained filter coefficients $\mathbf{w}_{co}$ and the required power constraint level C,

$$\left| s \mathbf{w}_{co} \right|^2 + \left| s \right|^2 = C. \qquad (41)$$

term on the right hand side, for values of λ that are close to zero. If we use the first two terms of the *Maclaurin* expansion (see Equation (39)), it

Figure 6. Power density surface plots inside the human head. The figure shows the attenuation of electromagnetic energy inside the human head and the radiated power from the antenna system without constraint (bottom left figure) and with a constraint (bottom right figure). All three figures have the distance in x and y directions measured in meters.

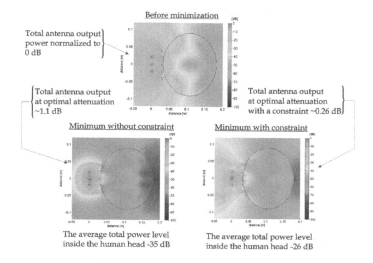

This will yield a quadratic Equation which has the following solution:

$$\lambda = \frac{-2 \cdot \Re\left\{\mathbf{w}_{opt}^H \mathbf{q}\right\} \pm \sqrt{\left(2 \cdot \Re\left\{\mathbf{w}_{opt}^H \mathbf{q}\right\}\right)^2 - 4 \cdot \mathbf{q}^H \mathbf{q}\left(\mathbf{w}_{opt}^H \mathbf{w}_{opt} + 1 - \frac{C}{|s|^2}\right)}}{2 \cdot |s|^2 \mathbf{q}^H \mathbf{q}} \tag{42}$$

where $\mathbf{q} = \mathbf{R}_F^{-1} \mathbf{w}_{opt}$.

So, by setting the constraining power level C and using the unconstrained optimal values of the filter coefficients $\mathbf{w}_{opt}$, we can now use Equation (42) to calculate what the value of λ should be. This value is then inserted into Equation (40) in order to calculate the constrained filter coefficients $\mathbf{w}_{co}$, which (for convenience) is re-stated again here,

$$\mathbf{w}_{co} = -\mathbf{w}_{opt} - \lambda |s|^2 \mathbf{R}_F^{-1} \mathbf{w}_{opt}. \tag{43}$$

As an example, if the constrained filter coefficients of Equation (43) are used in the iterative FX-LMS adaptive algorithms, it can be seen in Figure 5 that it will converge at the non-optimal solution that satisfy the constraint and has the shortest distance to the unconstrained optimal point.

The convergence of this non-optimal constrained least square solution for the Actuator Individual FX-NLMS can also be observed in Figure 4 and in Figure 6 is shown the maximum specific absorption rate (SAR) inside the human head when the Actuator Individual FX-NLMS algorithm have reached the constrained least square solution.

In Figure 6 and Figure 7 we show the effect of using an adaptive algorithm with a power constraint imposed on the solution. The purpose of this constraint is to allow for the minimum SAR value inside the human head while keeping the radiated power from the antenna system at a level consistent with the specified radiated output power from a mobile phone. The adaptive algorithm used in Figures 6, 7 and 4 is the Actuator Individual Normalized FX-LMS algorithm with an imposed constraint according to Equation (51).

Figure 7. The same results as in Figure 6, but the three plots are zoomed out to show more of the far-field. The figure shows the attenuation of electromagnetic energy inside the human head and the radiated power from the antenna system without constraint (bottom left figure) and with a constraint (bottom right figure). All three figures have the distance in x and y directions measured in meters.

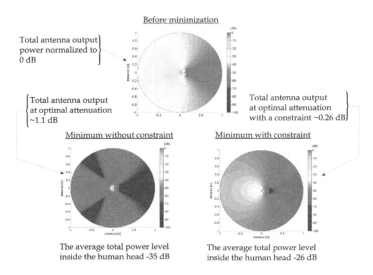

THE EFFECTS OF SPATIAL MIMO-ANTENNA PARAMETERS

The Least Mean Square solution obtained is the optimal solution in energy sense for this problem. This particular solution (Figure 6) is only valid assuming the position of each element does not change. However, there might be positions of the antenna elements that are more favorable with respect to the power density and SAR level inside the head. By changing the spacing of the antenna elements during calculations of the attenuated SAR level inside the human head we will investigate if there exist some optimal spacing between the different antenna elements.

In this FEM model setup, see Figure 1, we assume three degrees of freedom (DOF), as shown in Figure 8. The spacing between the sensor elements, denoted Δy, the distance d between the sensor element array (the receiving elements) and the actuator element array (the transmitting elements). The third DOF is the spacing between the actuator elements, denoted Δa.

In the first analysis we look at how the spacing of the sensor elements and the distance between the transmitter and receiver antennas affect the SAR level inside the head, see Figure 9.

It is clear from Figure 9 that the farther apart the transmitter and receiver antennas are located the lower the SAR level is inside the head. This is due to the increase of the distance d (see Figure 8) between the transmitter and receiver antennas. An increase of this distance will also increase the distance between the transmitter antennas and the head which will decrease the SAR level inside the head. By analyzing Figure 9 we can see that the two-dimensional cost function $J(\Delta y, d)$ is flattening out at a distance d of approximately 25 cm. The spacing between the sensor elements (receiving elements) Δy at the distance d=25 cm should be approximately 5 cm. With this spacing the sensor element array will cover a larger portion of the head. By using these values as a good approximation of the optimal displacement of the actuator elements and the distance between the sensor elements this would result in an SAR attenuation of approximately 7 dB. Using these values as a starting point, Figure 10 is showing how the separation of the actuator antenna elements Δa affects the SAR level of the cost function $J(\Delta y, d)$ inside the head.

Figure 9. The maximum SAR level inside the head plotted as a 2-dimensional cost function $J(\Delta y, d)$ with respect to the spacing between sensor elements and distance between the transmitting and receiving antennas. These SAR levels refer to a system with N=5 sensor elements and M=2 actuator elements.

Figure 8. The MIMO antenna system showing the three variables tested in the simulation: distance d, actuator separation Δa and sensor separation Δy

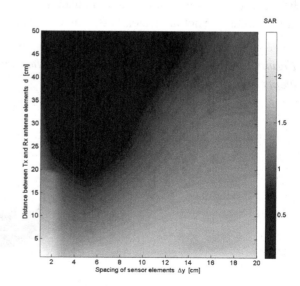

Figure 10. The effect of increasing the spacing Δa between the actuator antenna elements, see Figure 8. This result was calculated with a sensor array spacing of Δy = 5 cm and a separation between sensor array and actuator array of d = 25 cm.

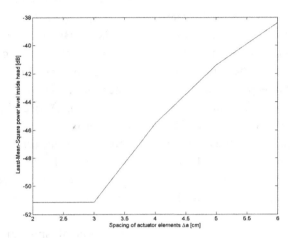

From Figure 10 we can see that the attenuation inside the head will increase as the spacing between the actuator elements decrease. This is a consequence of the electromagnetic waves being transmitted from almost the same point in space. The theoretical extreme of this is to place all actuator antennas in the exact same position, which will give a complete cancellation of the waves and would give a zero power and SAR level inside the head. According to this analysis we need a MIMO antenna system that has a spacing of 5 cm between the sensor elements and a spacing of 3 cm between the actuator elements. The distance between the sensor elements and actuator elements should be about 25 cm or more. This would result in a MIMO antenna system with a size of approximately 25 by 20 cm which is not practical to place on top of a mobile phone. However, we foresee other applications (heavy electric machinery (electric engines, generators, etc.) and power lines or when performing maintenance and testing on high power radio transmitters, e.g., for broadcasting or radar), where this size would be more practical. Studying Figure 9 and Figure 10 we observe that if the original positions of the antenna elements, with an actuator antenna spacing

Figure 11. The maximum SAR level inside the human head plotted as a function of the number of elements in the actuator and sensor array

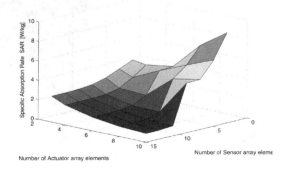

of 2 cm is used and we increase the spacing of the sensor elements from 2 cm to 3 cm we would get an extra 2 dB SAR attenuation inside the head compared to the original positioning of the antenna elements in Figure 1. We have also investigated the impact on the systems performance as a result of changing the number of antenna elements in the actuator and sensor arrays, see Figure 11. In these simulations we have calculated the least mean square solution as a function of the number of antenna elements in the actuator array M and the sensor array N.

Figure 11 shows, as expected, a decrease in the SAR level inside the head as long as every new added sensor element cover more of the head. Although, when the sensor array extends outside the length of the human head we attain no further improvement in the attenuation. Another interesting observation from Figure 11 is that if the number of actuator elements is larger than the number of sensor elements the system becomes unstable.

CONCLUSION AND FUTURE RESEARCH DIRECTIONS

In this chapter we have presented a FEM model which, solves the partial differential Equation of an electromagnetic field, and simulate the physical MIMO antenna system which is controlled by various adaptive signal processing algorithms in order to suppress the field at a specified volume in space. We have also presented the solution for constraining the total output power of the system to a predefined level. In addition, we have investigated the effects of the size and number of MIMO antenna elements on the performance of the system. The SAR attenuation levels achieved from these simulations suggest the possibility of using an active antenna system for the reduction of electromagnetic field density. However, our result also show some limitations associated with implementing these antenna arrays in mobile

phones, for which further research is needed to find practical solutions. For example, this array size and arrangement would be suitable for practical implementation in other applications such as heavy electric machinery (e.g., electric engines, generators) and power lines or when performing maintenance and testing on high power radio transmitters (e.g., for broadcasting or radar).

REFERENCES

Cheng, D. K. (1989). *Field and wave electromagnetics*. Addison-Wesley Publishing.

Christensen, H. C., Schüz, J., Kosteljanetz, M., Poulsen, H. S., Thomsen, J., & Johansen, C. (2004). Cellular telephone and risk of acoustic neuroma. *American Journal of Epidemiology, 159*(3), 277–283. doi:10.1093/aje/kwh032

COMSOL AB. (2006a). *FEMLAB electromagnetics module (Version 3.1)*.

COMSOL AB. (2006b). *FEMLAB reference manual (Version 3.1)*.

Fletcher, R. (2000). *Practical methods of optimization*. John Wiley & Sons Ltd.

Foster, K. R., Schepps, J. L., Stoy, R. D., & Schwan, H. P. (1979). Dielectric properties of brain tissue between 0.01 and 10 GHz. *Physics in Medicine and Biology, 24*(6), 1187–1197. doi:10.1088/0031-9155/24/6/008

Gabriel, C. (1996). *Compilation of the dielectric properties of body tissues at RF and microwave frequencies*. (Brooks Air Force Technical Report AL/OE-TR-1996-0037).

Hayes, M. H. (1996). *Statistical digital signal processing and modeling*. John Wiley & Sons Inc.

Hult, T., & Mohammed, A. (2004). Suppression of EM fields using active control algorithms and MIMO antenna system. *Radioengineering Journal, 13*(3), 22–25.

Hult, T., & Mohammed, A. (2005). Power constrained active suppression of electromagnetic fields using MIMO antenna system. *Journal of Communications Software and Systems, 1*(1).

Johansson, S. (2000). *Active control of propeller-induced noise in aircraft*. Unpublished doctoral dissertation, Blekinge Institute of Technology.

Kritikos, H. & Schwan, H. (1976). Formation of hot spots in multilayered spheres. *IEEE Transactions on BME,* 168–172.

Kuo, S. M., & Morgan, D. R. (1996). *Active noise control systems*. John Wiley & Sons Inc.

Lai, H. (1998). Neurological effects of radiofrequency electromagnetic radiation. *Proceedings of the Workshop on Possible Biological and Health Effects of RF Electromagnetic Fields, Mobile Phone and Health Symposium, October 25-28.*

Schwan, H. P. (1957). Electrical properties of tissue and cell suspension. *Advances in Biological and Medical Physics,* 5.

Sienkiewicz, Z.J. & Kowalczuk, C.I. (2005). *A summary of recent reports on mobile phones and Health* (2000-2004). (National Radiological Protection Board Report, January).

Tian, Z., Bell, K. L., & Van Trees, H. L. (1998). A recursive least squares implementation for adaptive beamforming under quadratic constraint. *Proceedings of 9th IEEE Signal Processing Workshop on Statistical Signal and Array Processing, Portland OR USA,* (pp. 9-12).

Weil, C. (1975). Absorption characteristic of multilayered sphere models exposed to UHF / Microwave radiation. *IEEE Transactions on BME, 22*(6), 468–476. doi:10.1109/TBME.1975.324467

Widrow, B., & Stearns, S. D. (1985). *Adaptive signal processing*. Prentice-Hall.

Chapter 4
Data Broadcast Management in Wireless Communication:
An Emerging Research Area

Seema Verma
Bansathali University, India

Rakhee Kulshrestha
Birla Institute of Technology and Science, India

Savita Kumari
University of Seventh April, Libya

ABSTRACT

Recently, the wireless data broadcast has been receiving a lot of attention, both from industries and academia. The recent advances in the field of Wireless Communication Technology (WCT) have increased the functionality of mobile information services. Various technologies like CDMA, 3G, and smart personal technologies (SPOT) and the advent of new gadgets of communication have made many mobile communication applications a reality. The prime feature of Wireless Communication Technology is that users can retrieve information from wireless channels anytime, anywhere. Attempting to disseminate data effectively to a large number of users with minimum consumption of time and physical resources of client in WCT environment is a challenge to system. There are two modes to disseminate data in WCT: (i) broadcast mode, where the client can retrieve data by simply listening to channel and (ii) on–demand mode, where the client can send the request to the operator to get data, and the operator in turn can serve data to the former in response. Both modes have significance in their domain, and sometimes, overlap of both modes may bring out better performance. The WCT is hindered by factors like low battery power, frequent disconnection, asymmetric and heterogeneous broadcast, scalability, et cetera. To overcome these difficulties, various data management strategies, along with broadcast mode of dissemination, are implemented. These data management strategies involve scheduling, hashing, indexing, and replication of data to be broadcasted. Building an index of broadcast data can help the user to find out when and where the desired data item will be available. If the arrival time of demanded data items is known a

DOI: 10.4018/978-1-60960-477-6.ch004

priori, clients can go to doze (power saving) mode to save energy, and if data are indexed, clients may have direct access to desired data to save time.

The data broadcast policies have been developed for single channel and multi channel with various scheduling and indexing techniques. For the data management policies which consider the different broadcast cycles for different broadcast operators, it can be said that traditional types of data management policies are known previously, and the policies of Central Server (CS) and Unified Index Hub (UIH), which consider single broadcast cycle for all operators, are recent. This chapter presents both strategies very simply for better understanding, discusses the work done in the past and present on data broadcast management, along with suggestions for the future possibilities to explore the field.

1. INTRODUCTION

In recent years, the use of wireless technology devices has been growing at an exponential rate. Most people are now able to access information systems located in wired networks anywhere and anytime using portable size wireless computing devices like notebooks, tablet PCs, personal digital assistants (PDAs) and GPRS-enabled cellular phones, laptops, palmtops which are powered by small batteries. These portable computing devices communicate with a central stationary server via a wireless channel and become the integral part of the existing distributed computing environment. These mobile clients can have access to database information systems located at the static network while they are traveling and this type of computing is known as wireless computing or mobile computing. Figure 1, shows the architecture of wireless computing network.

Wireless computing provides database applications with useful aspects of wireless technology and a subset of mobile computing that focuses on querying central database servers is referred as wireless databases. Mobile service providers have established a number of information services including weather information services, news, stock indices information, foreign exchange rates, election results, tourist services, airlines, railways schedules etc. Apart from this there is enormous number of operators transmitting data on consecutive band widths. The major

shortcoming with broadcast data items in a wireless environment is that data are accessed sequentially; the increasing number of broadcast items causes mobile clients to wait for a large time before receiving desired data item. Consequently, dependence of mobile devices on rechargeable batteries, which has limited capacities, is a drawback for mobile data retrieval. To study these drawbacks and develop remedies for them, it is necessary to visualize the following two parameters of mobile client:

- *Access time:* It is the time elapse from the moment a request is initiated until all data items of interest are received. Access time represents the fastness or delay in retrieving the desired data from broadcast channel.
- *Tuning time:* It is the time spent by the client to listen for the desired broadcast data item. The tuning time represents the power consumption in filtering required data from broadcast channel.

Tuning time comprises time taken in two modes viz: ***active and doze mode.***

For successful retrieval of information from wireless network the mobile clients have to tune to broadcast channel. The proliferation of mobile computing environment demands that the queries of client must be satisfied, by whatever way it may be. Queries in a mobile environment can be

Figure 1. Architecture of wireless computing network

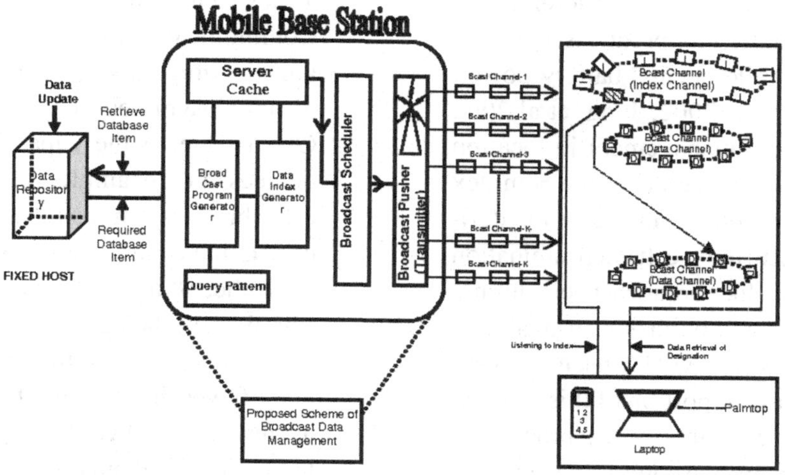

classified into two categories *traditional queries and location-dependent queries.* The queries that invoke in traditional wired environment are traditional queries, while when these are transmitted over a wireless communication network, are called *location-dependent queries*. For example queries evoked in point to point contact through the fixed telephone is traditional queries, while the queries like train information for a person while in traveling or information of availability of tiger for a tourist wandering in Tiger Reserve is *location dependent queries*. However, queries on wireless database system have much more variety on the basis of different envisaging parameters, which do not exist in traditional wired databases. It is realized that location-dependent queries are common and of great interest because of their essence and comfortness in information retrieval in situations where wired network is either not possible or too costly. Henceforth, providing efficient and effective mobile information services that cover both traditional queries and location-dependent queries are highly desirable. Due to sluggish demand of traditional queries here, in this chapter, they will only be discussed in short and data broadcast

management for location dependent queries will be discussed in detail.

1.1 Data Broadcast Management for Location-Dependent Queries

When the mobile client's location is relevant to the information requested or the information requested is based on a particular location, the utilization of broadcast mechanism is very challenging. Data organization for mobile client queries will be much more complex, as the calculation may become obsolete as soon as the client moves. The response to client query also changes naturally, depending on the location of the clients. To render satisfactory service mobile client should be informed when the relevant data will be available on the channel while considering the current location of the client as the parameter. By its hierarchy mobile users may spread into different regions. The central database server must be intelligent enough to organize its data so that mobile clients in different regions can still efficiently retrieve the relevant data; otherwise, mobile clients will waste a lot of power unnecessarily during the tuning and listening processes. The problem is more complex when the mobile users frequently move from one region to another, while requiring. Therefore, an efficient data organization for broadcast cycles is critical in saving mobile clients' battery power. Xu, et al 2003; Xu, et al 2002; Jung, et al 2002 have explored the data management in location dependant queries which is much more complex than that in traditional environment and is definitely of great interest to pursue. It is a combination of various techniques like broadcast scheduling, indexing, replication and assorting of available bandwidth and data, which will be discussed with great detail in subsequent portion of this chapter.

Despite the complexity involved in processing the mobile queries, a mobile computing environment also possesses several novel characteristics, which make it more challenging and desirable than a traditional distributed system. These characteristics include:

- **Resource constrained mobile devices:** To provide better portability and improve attractiveness, mobile devices are becoming smaller and lighter. To make the device cheaper one its size has to be reduced. In such designs usually some trade-offs between battery life and size to be made.

- **Low network bandwidth:** Mobile clients can connect to the stationary server via various wireless communication networks like wireless radio, wireless LAN, wireless cellular and satellite network etc. Each of the wireless networks provides a different bandwidth capacity. However, this wireless bandwidth is too small compared with a fixed network such as ATM. Also, due to increase in broadcast operators in the fray the available band width per operator available on air is also shrinking. The low band width worsens the data access time and energy consumption.

- **Asymmetric communication environment:** The difference between bandwidth capacities of downstream communication and upstream communication has created a new environment called Asymmetric Communication Environment. In fact, there are two situations that can lead to communication asymmetry (Acharya, et al, 1995). One is due to the capability of physical devices. For example, servers have powerful broadcast transmitters, while mobile clients have little transmission capability. The other is due to the patterns of information flow in the applications. For instance, in a situation where the number of servers is far fewer than the number of clients, it is asymmetric as there is not enough capacity to handle simultaneous requests from multiple clients.

- **Heterogeneity of mobile devices:** Mobile telecommunication industries have developed a large variety of mobile computing devices such as Laptops, Tablet PCs, Handheld PCs, Pocket PCs, and internet enabled Mobile Phones. However, these mobile devices have also various features and capabilities such as operating system, computational power, display and network capability. This heterogeneity raises some challenges in content management and content delivery to the mobile service providers.

- **Mobility:** Wireless technology enables mobile users to move freely and independently from one place to another. A service handoff occurs when a user moves from one cell to another cell. It is essential to ensure service handoffs transparently to the users.

- **Frequent disconnections:** Mobile users are frequently disconnected from the network. This may be due to several reasons including signal failures, empty network coverage, and power saving. The later reason is advantageous since active mode requires thousand times more power than doze or power saving mode (Imielinski,

Viswanathan and Badrinath, 1994). Wireless radio signals may also be weakened due to the client's further distance from the base station or with the motion of client.

1.2 Data Management and Query Processing in Mobile Environments

The data for mobile environment can be managed from two sides, either from client side or from server side. The client side data management involves cache management, device receiving or transmitting power management and battery capacity management where as server side data management involves scheduling, indexing and partitioning the data, arranging available bandwidth and partitioning the channel. The document classification and bandwidth issues are interleaved, simply because a given bandwidth division determines the performance of a document classification choice and a given document classification in turn determines a bandwidth split that optimizes performance. Any way, both document classification and bandwidth division depend on the popularity of data items because download latency is smaller when hot items are assigned to multicast push, cold items to unicast pull, and the bandwidth is divided appropriately between various channels.

Query optimization techniques have to consider the effects of mobility. It has severe effect on the data retrieval properties. Due to fast changing location, queries may be answered in an approximate way. Another major issue is querying the broadcast data on the air is to find the best execution plan for a query response. It involves data broadcast on selecting single or multichannel mode and organizing the broadcast data so that the consumed energy and retrieval time can be kept as per demand.

In the light of the complexity of query processing, as well as the above retained characteristics of a mobile environment, it is essential to have an effective data delivery mechanism that may be able to manage all of the above issues. The processing of queries retrieval in a wireless environment can be generally classified into: (i) on-demand queries or pull based queries, and (ii) broadcast-based queries or push based queries.

On-demand queries are those where the client initiates the query and sends it to the server. The server processes the query and responds back to the client. Such applications include information retrieval in weather and traffic information systems. For instance a broadcast server at airport records data for all flights in the near future. The client sends queries and the server responds to them according to its available recourses. It is one to one mode of information delivery.

Sometimes, demand for some information is higher than for others. In such situation the server has to deliver the same record many times and some clients are likely to receive information lately. Here data broadcaster stands to an efficient solution in which broadcast of single item is likely to satisfy a possibly large number of users. Such delivery mechanism is discussed under subsequent sub part.

In *broadcast-based queries* the server broadcasts the data items periodically over one or more broadcast channels. Mobile clients tune to it and select data items of interest and capture them. With broadcast-based queries, a mobile client is able to retrieve information without wasting power to transmit a request to the server. Also it supports a large number of queries at a time and the number of users in a cell or request rate for a query does not affect query performance and it is effective despite a large number of users' request simultaneously. In broadcast mode the behavior of information system is unidirectional which means the server disseminates a set of data periodically to a multiple number of users without getting any request from them. In this mechanism, the requests from the clients are not known a priori.

Apart from above delivery mechanism client's procedure of retrieving data also contribute to its efficiency. While retrieving data from broadcast channel mobile client can be in two modes – active and doze. In *active mode* client listens to the channel for the desired data item. In this mode power consumption is more because client remains in tuned to broadcast channels for whole time and filters all the data till desired data arrive. While in *doze mode* the clients simply turned into a power saving mode i.e. sleep for the time between getting link to data and down of data starts. In doze mode power consumption is much less than the active mode. The data on channel may be repeated according to the demand. Due to various constraints of business, economy and technical reasons mobile service providers generally use combination of low bandwidth to obtain combined high bandwidth channel instead of single high bandwidth channel.

1.3 Caching

Cache management plays an important role in mobile computing because of its capacity to alleviate the performance and availability limitations during weak network connections. It can reduce contention on limited bandwidth networks. This improves query response time and supports disconnected or weakly connected operations. If a mobile user has cached a portion of the shared data, different levels of cache consistency may be requested. In a strongly connected mode, the user may get the current values of the database items belonging to its cache. During weak connections, the user may require weak consistency when the cached copy is a quasi-copy of the database items. Each type of connection may have different degree of cache consistency associated with it, namely weak connection corresponds to weaker level of consistency. Cache consistency is severely hampered by both disconnections and mobility, since a server may be unaware of the current locations and connection status of clients.

1.4 Replication

The ability to replicate data objects is essential in mobile computing in order to increase availability and performance. Shared data items have different synchronization constraints depending on their semantics and particular use. These constraints should be enforced on an individual basis. Replicated systems need to provide support for disconnected mode, data divergence, application defined reconciliation procedures, and optimistic concurrency control.

1.5 Mobile Databases

Mobility and portability pose new challenges to mobile database management and distributed computing. In conventional database systems, there is one common characteristic: All components, especially the processing units, are stationary. The first research efforts for mobile databases concentrated on relational databases. The presence of personal and terminal mobility incurs several problems related to the maintenance of the ACID (Atomicity, Consistency, Isolation and Durability) properties. Naturally, the ACID properties of a transaction must be maintained in all data management activities. The major dominating characteristics of mobile data bases are: -

- Computation and communication have to be supported by stationary hosts.
- The transactions are prolonged due to the mobility of both data and users, and due to frequent disconnections.
- The models should support and handle concurrency, recovery, disconnection and mutual consistency of the replicated data objects.
- As mobile hosts move from one cell to another, the states of transaction and accessed data objects, and the location information also change.

- Computations might have to be split into sets of operations executed on mobile and stationary hosts.

1.6 Channel Allocation for Data Dissemination

In a mobile computing system, the communication load between mobile computers and the server varies over time and space because of the mobility of mobile computers. With the limited and fixed bandwidth available in a cell, mobile computing system should adjust the allocation of channels for broadcast and on-demand purposes in order to improve the overall communication performance in a cell. In the following, four classes of channel allocation methods are defined:

On-demand dissemination: In on-demand broadcast, clients make explicit requests for data. If multiple clients request the same data at approximately the same time, the server may match these requests and deliver the data only once. Here data requests and results are delivered through point-to-point connections. This method is desirable when the number of queries is small compared to the number of channels available and energy efficiency is not an issue for the mobile computer to transmit uplink requests. Being a user-oriented methodology, it provides interactive capability to users for accessing the information through query. Users do not have to search in the wireless information space by tuning several channels. However, this approach has many disadvantages. First of all, it is resource intensive. Users require a separate channel to send requests to the server (upstream). The server, after receiving the request, composes the result and sends it to the user on a back channel (downstream) known to the user. Thus, every pull needs two channels for completing the process.

Moreover, incoming requests are usually not identical; the server cannot always efficiently group requests in order to exploit the advantages of broadcast. Obviously, this depends on the volume and the context of the incoming workload. When the number of incoming requests becomes too high, the server fails to keep up and non scalability of client-server architectures notoriously worsens this problem.

Push-based dissemination: Contrary to the exclusive on demand method, all of the channels are in broadcast mode. Data items are broadcast periodically over broadcast channels. This method is useful when a small number of data items are of interest to a large group of users. In push-based systems, the server employs point-to-multipoint communication and sends data items without the explicit client requests. In order to achieve that, the server maintains a broadcast schedule, which determines the order and the frequency in which data items are broadcast. For example, the scheduler handles three data items (A, B and C), out of which B and C are broadcast with the same frequency and A twice more frequently, resulting in the transmission schedule: (A, A, B, C, A, A, B, C...). The major feature of such systems is scalability. By this way server can scale up to millions of users simultaneously. Client population does not influence the dissemination process because clients do not issue requests. The addition of new clients does not influence the server's incoming load or the client perceived access time. In addition, clients need few resources. Mechanisms such as wireless indexing enable clients to efficiently locate data in the broadcast channel. Moreover, data can be kept actually, since the server can simply broadcast any update.

The major problem of push-based systems is lack of self organization and adaptiveness. Since the server does not receive explicit client requests, it remains unaware of possible changes in client population or querying characteristics. This encounters several problems. Bandwidth for instance, can be unnecessarily utilized for a relatively low number of end clients. Apart from that, the push service requires more powerful hardware.

Hybrid data dissemination: It is a combination of on-demand and push-based approaches. Data items are classified into popular (hot) and unpopular (cold) data. Popular data items are delivered via push-based channels, while unpopular data items are disseminated via on-demand channels. This data dissemination technique is a way to balance push and pull to effectively utilize the available band width. The main idea is to make broadcast and on-demand channels complement each other. The typical design issues in hybrid system involves channel allocation for push and on-demand dissemination, data classification in to cold or hot data items and item scheduling as per requirement.

Dynamic dissemination: This method dynamically allocates broadcast and on-demand channels to achieve optimal data access performance. In contrast to the hybrid method, the dynamic method allocates channels based on different workloads at the server. When the load is heavy, the broadcast channels may significantly relieve the load on on-demand channels by taking care of frequent accesses to hot data items. When the load is light, on-demand channels can take over to provide instantaneous access to data. Data access methods, require no prior knowledge on how the requested data item is provided by the server, a mobile computer has to send data access request to the server. In response, the server will either directly deliver the data item, if it is only available through on-demand service, or reply with the broadcast channel access information, such as the channel frequencies, data identifier, the decryption key, expiration time, and estimated access time. When the mobile computer receives the broadcast channel access information, it may decide if it wants to get the data through the broadcast channel or right away through on-demand channels. In the former case, the mobile computer will terminate the connection and monitor the broadcast channels, whereas in the latter it returns a confirmation to the server, which will then deliver the data and terminate the connection. If the mobile computer already has the broadcast channel access information regarding the data item of interest, it does not need to query the server unless the information is expired. The mobile computer may use the information to monitor broadcast channels to receive the data item. Since broadcast channels allow simultaneous access by an arbitrary number of users, the charge for accessing data items from the broadcast channels may be based on a different pricing scheme from that of the on-demand channels. For example, the price may be based on the valid period of the broadcast channel access.

2. RELATED WORK ON SINGLE CHANNEL AND MULTICHANNEL

There is a lot of works done on data management over single broadcast channel. The multichannel is technically the bandwidth partition to speed up the data dissemination process.

For single channel: Acharya et.al. (1995) proposed the broadcast disk in which *hot data* items are allocated more frequently than *cold data* items from which average access time decreases. Acharya et.al. (1997) discussed the problem of *push* and *pull* based broadcast. Hu and Lee et. al. (1999) suggested scheduling and clustering of data with indexing. The distributed indexing method was proposed to efficiently replicate and distribute the index tree in broadcast by Imielinski, Viswanathan (1997). Lo and Chen (2000) examined the issues of allocating data at broadcast channel to minimize access time et. al. (2003) and Shivkumar and Venkata Subramanian (1996) considered unbalanced tree structure to optimize energy consumption for a non-uniform data access. Xu, Lee and Tang (2004) proposed exponential indexing of data on air for single channel. Wiesmann and Schiper (2005) compared database replication techniques for total order broadcast.

*For multi channel:*Prabhakara et. al. (2000) considered index and data allocation on separate broadcast channels. Navethe et. al. (2003) proposed indexing structure for multiple broadcast channels. Huang et. al. (2004) addressed the issues of data scheduling over multiple channel. Amarmend et. al. (2006) proposed index allocation over multichannel of reducing access latency and tuning time. Chi, Chun and Lee (2006) examined the data allocation for sufficient and insufficient channels and developed algorithms for allocating data over insufficient number of channels without indexing for single page access at a time and a few others use it to multichannel. Some have developed indexing pattern to minimize error. This author group (2008) has developed Unified Index Hub (UIH) for amalgamating index information of various broadcast operators. This is further extended to deliver better result in worse situation.

As the number of channels on air is also limited, the data scheduling for insufficient channels is a way, the client can receive data with minimum power consumption and time is critical for wireless computing environment and this has yet to be explored properly. Since the wireless broadcast is asymmetric and heterogeneous, the problem of discussing exponential indexing over insufficient multichannel, in non-clustered flat broadcast can yield a better result in terms of both power consumption and access latency. With the development of various miniaturized gadgets like laptop, palmtops, mobile phones etc. the issues related to their power and time in which client can receive information are a major concern for mobile communication community. In nutshell, all the concerns like extending existing indexing techniques to multichannel, optimal partitioning of index and data space on air, utilizing maximum of channel space, managing data in worse conditions and unbiasely comparing various indexing strategies and dynamic scheduling for single and multichannel are open areas in this field.

3. BROADCAST PROGRAM GENERATION ON MULTIPLE CHANNELS

The broadcast cycle is a fixed length and fixed interval paradigm. According to generic paradigm of broadcast, the server partitions the database into group and each group is broadcasted cyclically in round robin manner in a channel.

Let database D has n equal sized data items i.e. $D = \{d_1, d_2, d_3, \underline{\hspace{2cm}}, d_n\}$ and a set of broadcast channel C having k channels is $C = \{c_1, c_2, c_3, \underline{\hspace{1cm}}, c_k\}$ with access probability $p_1, p_2, p_3, \underline{\hspace{2cm}} p_n$, then $P = \{p_1, p_2, p_3, \underline{\hspace{2cm}} p_n\}$.

Without loss of generality, the channels with low index item will store more data items. Generating a broadcast program for k channel can be viewed as a partition problem. Here the straight meaning of generation of broadcast program is to develop a program, which can place n data item over k channels, i.e. to establish $D_n \rightarrow C_k$ relation. Let:

$$\sum_{i=1}^{n} p_i = \int Total\ Access\ \Pr obability$$

Let s_i is interval between two consecutive broadcasts of same items, then broadcast frequency is,

$$b_i = \frac{1}{s_i}$$

When there are k broadcast channels to broadcast data then,

$$k = \frac{1}{\sum_{i=1}^{n} b_i}$$

Let average access time for $d_i = a_i$ then minimum average access time for the data broadcast program for k channels is given by the formula

$$a_{\min imum} = \frac{1}{2k}\left(\sum_{i=1}^{n} \sqrt{p_i}\right)^2 \qquad because \qquad b_i = \sqrt{p_i}$$

The optimal scheduling to place n data items over k channel has time complexities in application. Where the access frequencies change quite frequently or new items are to be inserted in broadcast this program is not found applicable. So various new methods to generate fast partitioning have to be developed which may optimize partitioning of space between index and data.

4. BROADCAST SCHEDULING

Access time and power consumptions are two major issues, which affect the efficiency of data retrieval in mobile computing environment. To increase the efficiency, low access time and minimum power consumption are required, which can be achieved by scheduling of data on *either server side or client side*. The server side treatment of data is generally more economical because all clients will be benefited by it. This is being used widely.

Data scheduling can be done in two ways. When time is the major concern, a pure schedule of data items without any index information is performed on broadcast channel. This is called pure data scheduling. Its efficiency is further increased by periodic data scheduling where data is periodically transmitted over broadcast channel. In periodic data scheduling, hot data are placed with high frequency while cold data are placed with low frequency on channels. The broadcast in which a single data item is placed once in a single broadcast channel is called flat broadcast and a broadcast in which a single data item appears more

than once is called skewed broadcast. Periodic data scheduling demands skewed broadcast.

Scheduling is done on broadcast server by broadcast scheduler. The broadcast scheduler periodically looks up the access profit and selects those data items that have been accessed frequently during recent past. The broadcast scheduler constructs index tree based on key values for hot data items. Then it generates a broadcast program, and broadcast pusher (transmitter) sends it over channel.

These transmitted items are accessed by mobile client called broadcast data access. This can be done by identifying key values. The access time for a data request on broadcast channel can be divided into two parts: *probe wait* (the average duration for getting to next index information after making initial probe), and *broadcast wait* (the average duration from the point the index information relevant to the required data is encountered to the point when required record is downloaded). The server, which has only one receiver, can tune only to single data or index at a time. It makes first probe and then follows the index root till desired data item is searched.

5. INDEXING

When energy consumption is a major concern in broadcast data access, the index is provided with data on broadcast channel. An index of broadcast data helps users to decide when to allow the mobile device to turn into power saving (doze) mode while waiting for the desired data. The minimized average tuning time for a data request is expected to require a skewed index tree that has more popular data item located at higher levels.

Further, replication can provide fault tolerance capability and can even speed up data searching while increasing the average access time due to large number of redundant data being broadcasted .The average access time can be lessened by increasing the number of broadcast channels. If

Figure 2. Architecture of UIH over AHME

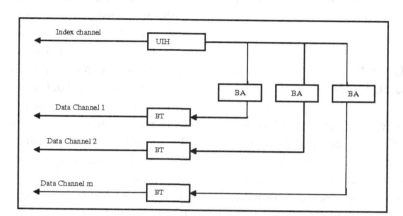

same data is broadcast over k channels rather than single channel the average access time reduces by factor k. But, on air, number of channels is limited. So, in practice, the tradeoff between access time and tuning time is adjusted and to achieve optimal efficiency, both parameters need to be treated.

6. UNIFIED INDEX HUB

It is an emerging technique of data broadcast management, which is at infancy of its development. It may be defined as combined index for all broadcast operators. Here, we define the broadcast agent (BA) as any individual operator, which has data to be broadcast on the broadcast channel. Each BA is connected to the UIH and Broadcast Transmitter (BT), which put data of BA on its subscribed channels, through a wired network. The key point in data retrieval on Asymmetric Heterogeneous data retrieval in Multichannel Environment (AHME) is the Unified Indexing Hub (UIH), which manages and broadcasts index information about the data being broadcast on all of the broadcasting channels. In AHME each BA is connected to the UIH and Broadcast Transmitter (BT), which put data of BA on its subscribed channels, through a wired network. The UIH is responsible for broadcasting index information

on a dedicated wireless channel for the whole broadcast, while the BT is only responsible for broadcasting a data message (DM) on its own wireless channels for BAs who subscribe to it. Figure 2, shows the architecture of AHME.

When a BA has to broadcast the data then at first it will send a "data-to-broadcast" notification (DTB) to the UIH through the wired network. The contents of the DTB are in standard format as prescribed by the UIH. The major contents of DTB are BA's ID, the channel ID that the BA has subscribed to, the message ID, a list of key attributes describing the data and data type ID. When the UIH receives the DTB, it extracts the information from the DTB and converts it into the index message (IM) format, which contains information on BA's ID, channel ID, message ID, the IM size, the information of attribute list keys, and pointer to the starting of data. Next, the UIH puts the IM into the broadcast queue for index broadcast. Once the UIH receives the DTB, it will reply to the BA so that it can broadcast its data. This replied message is called the "earliest-send" notification (ES). The ES can have maximum value equal to end time of IM of BA, which guides BA that in how much time it has to place its data on its subscribed channel so that MC becomes able to receive data. When BA upgrades information on its channel it informs the UIH through message so that it can update its index.

7. CONCLUSION AND FUTURE PROJECTION

With the advent of modern gadgets of cellular communication and mobile computers the paradigms of communication shift towards wireless communication. Example applications include intelligent navigational systems, wearable battlefield computers, and computerized interactive TV cable boxes etc. The significant asymmetry between downstream and upstream communication capacities and the significant disparity between server and client storage capacities have prompted researchers to explore the field to uncover its vast. The execution of critical tasks in such asymmetric client server environments requires that data retrievals be successfully completed within little time and with less power consumption. Previous work on broadcast did not deal explicitly with the fault tolerance and timeliness constraints imposed by such critical limitations. In this chapter, outlined almost all possible contents needed for broadcast organizations which are helpful in strategy building to alleviate such constraints have been outlined. In particular, we have presented novel real-time, fault-tolerant, secure broadcast technique based on the state-of-the-art traditional policies and UIH. The UIH is still a new proposal in comparison to traditional data broadcast policies by keeping different broadcast agent as separate entities for index development. It can be a better alternate of traditional policies to solve the problem of market cost, increasing number of operators and shrinking bandwidth.

Here two possible organizations for broadcast programs are considered: flat and skewed, out of which the skewed organization strikes the right balance between efficient bandwidth utilization and ease of programming. One problem that is not considered in this chapter but which is the subject of immense research is the optimal specification of the broadcast periods for the various broadcast disks (and the allocation of data items thereon) in a skewed organization. The problem of dynamic updates to data stored on broadcast disks is also an interesting topic that receives severe attention from every group of research community. Current work ignores this problem under the assumption that modified data will eventually be rebroadcast, thus invalidating client caches, and possibly restarting transactions or computations carried on the old stale data. Delays from such restarts could be fatal in a real-time environment. Techniques that allow for very frequent broadcasting of invalidation messages coupled with speculative client processing policies could be useful in solving such problems. Other related issues include real-time database concurrency control and indexing. Broadcast disks are likely to be used by clients in retrieving information from a large body of data. Despite the abundance of downstream bandwidth, it is likely that this bandwidth is not going to be enough to broadcast all the information that clients may ever need. One approach to deal with this problem is for clients to dispatch agents by sending appropriate control messages through the limited upstream bandwidth. In a real-time environment, there are a number of issues to be considered. For example, broadcast agents must be designed to meet timing constraints established by clients. This could be done through the use of imprecise computation techniques. Also, admission control and scheduling strategies must be incorporated in broadcast protocols to ensure that agents are able to transmit results to clients in due time.

Other related issues include client-initiated caching and pre fetching strategies. The selection of data to be placed on broadcast channel and discussion of broadcast frequency are interesting problems, reminiscent of the speculative data dissemination strategies discussed in literature needed to enhance further. The use of broadcast disks, however, poses new challenges for the implementation of these strategies. In particular, when broadcast disks are used, servers cannot keep track of the access patterns necessary for data dissemination. Therefore, new protocols must be devised to allow servers to reconstruct these access patterns. The problems such as security of broadcasted data are also a concern in broadcast environment, which is critical to be solved before further enhancement of existing schemes. If data is broadcast to a client, then it is available to all. More importantly, if clients are to rely on agents they dispatch, then mechanisms must be devised to authenticate messages received from such agents. The security/privacy of broadcast disks through the use of secret dispersal keys is one of the possible solutions of this which is not discussed in this chapter. More work needs to be done to embed authentication of broadcasted data. In order to evaluate real-time fault-tolerant broadcast disks protocols, it is necessary to develop a test bed and a set of benchmarks.

We believe that the current limitations of mobile computing should not be considered only as such. Instead, it should be considered how these are expected to evolve in the near future because of expansion rate of mobile / wireless services. In this context, we have presented a new approach like UIH has been presented because it is believed that such architectures are capable of coping with these limitations. The limitations are not expected to vanish in near future, but the data management community needs to resolve these issues as serious research problem.

REFERENCES

Acharya, S., Alonso, R., Franklin, M., & Zdonik, S. (1995). Broadcast disks: Data management for asymmetric communication environments. In *Proceedings of ACM Sigmod*, (pp. 199-210).

Acharya, S., Alonso, R., Franklin, M., & Zdonik, S. (1996). Prefetching from a broadcast disk. In *Proceedings of the International Conference on Data Engineering* (ICDE), (pp. 276-285).

Acharya, S., Franklin, M., & Zdonik, S. (1997). Balancing push and pull for data broadcast. In *Proceedings of ACM Sigmod Conference*, (pp. 183-194).

Aksoy, D., & Franklin, M. "Scheduling for Large-Scale On-Demand Data Broadcasting", In Proceedings of IEEE Info COM Conference, pp.651-659, 1998.

Amermend, D., & Aristugi, M. (2006). An index allocation method for data access over multiple wireless broadcast channel. *IPSJ Digital Courier*, *2*, 852–862. doi:10.2197/ipsjdc.2.852

Barbara, D. (1999). Mobile computing and databases-a survey. *IEEE Transactions on Knowledge and Data Engineering*, *11*(1), 108–117. doi:10.1109/69.755619

Cao, G. (2003). A scalable low-latency cache invalidation strategy for mobile environments. *IEEE Transactions on Knowledge and Data*, *15*(5), 1251. doi:10.1109/TKDE.2003.1232276

Chen, H., Xiao, Y., & Shen, X. (2006). Update–based cache access and replacement in wireless data access. *IEEE Transactions on Mobile Computing*, *5*(12), 1734–1748. doi:10.1109/TMC.2006.188

Chung, Y., Chen, C. C., & Lee, C. (2006). Design and performance evaluation of broadcast algorithms for time constrained data retrieval. *IEEE Transactions on Knowledge and Data Engineering*, *18*(11), 1526–1543. doi:10.1109/TKDE.2006.171

Hameed, S., & Vaidya, N. H. (1997). Log-time algorithms for scheduling single and multiple channel data broadcast. In *Proceedings of the 3rd ACM MOBICOM*, (pp. 90-99).

Hameed, S. & Vaidya, N.H. (1999). Efficient algorithms for scheduling data broadcast. *ACM/Baltzer Journal of Wireless Network, 5*(3), 183-193.

Hu, Q., Lee, W. C., & Lee, D. L. (1999). Indexing techniques for wireless data broadcast under data clustering and scheduling. In *Proceedings of the 8th ACM International Conference on Information and Knowledge Management*, (pp. 351-358).

Hu, Q., Lee, W.-C., & Lee, D. L. (2001). A hybrid index technique for power efficient data broadcast. *Distributed and Parallel Databases, 9*, 151–177. doi:10.1023/A:1018944523033

Huang, J.-L., & Chen, M.-S. (2004). Dependent data broadcasting for unordered queries in a multiple channel mobile environment. *IEEE Transactions on Knowledge and Data Engineering, 16*(9), 1143–1156. doi:10.1109/TKDE.2004.39

Huang, Y., Sistla, P., & Wolfson, O. (1994). Data replication for mobile computers. In *Proceedings of the ACM SIGMOD*, (pp. 13-24).

Imielinski, T., Viswanathan, S., & Badrinath, B. R. (1994). Energy efficient indexing on air. *Proceedings of the ACM Sigmod Conference*, (pp. 25-36).

Imielinski, T., Viswanathan, S., & Badrinath, B. R. (1997). Data on air: Organization and access. *IEEE Transactions on Knowledge and Data Engineering, 9*(3), 353–371. doi:10.1109/69.599926

Lee, C. K., Leong, H. V., & Si, A. (2002). Semantic data access in an asymmetric mobile environment. In *Proceedings of the 3rd Mobile Data Management*, (pp. 94-101).

Lee, D. K., Xu, J., Zheng, B., & Lee, W.-C. (2002). Data management in location-dependent information services. *IEEE Pervasive Computing / IEEE Computer Society [and] IEEE Communications Society, 2*(3), 65–72.

Lee, G., Chen & Lo, S-C. (2003). Broadcast data allocation for efficient access on multiple data items in mobile environments. *Mobile Networks and Applications, 8*, 365–375. doi:10.1023/A:1024579512792

Lee, W. C., Hu, Q., & Lee, D. L. (1997). Channel allocation methods for data dissemination in mobile computing environments. In *Proceedings of the 6th IEEE High Performance Distributed Computing*, (pp. 274-281).

Lin, K.-F., & Liu, C.-M. (2006). *Broadcasting dependent data with minimized access latency in a multi-channel environment*. IWCMC'06, July 3–6, (pp. 809-814).

Lo, S.-C., & Chen, A. L. P. (2000). An adaptive access method for broadcast data under an error-prone mobile environment. *IEEE Transactions on Knowledge and Data Engineering, 12*(4), 609. doi:10.1109/69.868910

Navathe, S. B., Yee, W. G., Omiecinski, E., & Jermaine, C. (2002). Efficient data allocation over multiple channels at broadcast servers. *IEEE Transactions on Computers, 51*(10), 1231–1236. doi:10.1109/TC.2002.1039849

Prabhakara, K., Hua, K. A., & Oh, J. H. (2000). A new broadcasting technique for an adaptive hybrid data delivery in wireless environment. In *Proceedings of 19th IEEE International Performance, Computing and Communications Conference*, (pp. 361-367).

Rakhee, V. S., & Savita, K. (2008). Two level signature model for multiple broadcast channel using unified index hub (UIH). *Proceeding of International Conference on Soft Computing (ICSC – 2008)*, November 8-10, 2008 at IET, Alwar (Rajasthan) – India, (pp. 422-431).

Savita, K. (2008). Dynamic broadcast scheduling at Unified Index Hub (UIH). *International Journal of Intelligent Information Processing, 2*(2), 243–250.

Shivkumar, N., & Venkatasubramanian, S. (1996). Energy efficient indexing for information dissemination in wireless systems. *MONET, 1*(4), 433–446.

Tran, D. A., Hua, K. A., & Jiang, N. (2001). A generalized design for broadcasting on multiple physical channel air-cache. In *Proceedings of the ACM SIGAPP Symposium on Applied Computing* (SAC'01), (pp. 387-392).

Verma, S., et al. (2008). Two level signature model for multiple broadcast channel using Unified Index Hub (UIH). *Proceeding of 2nd International Conference on Soft Computing ICSC - 2008, IET, Alwar* (Rajasthan) – India, (pp. 422-431).

Verma, S., Rakhee, S., & Sheoran, K. (2008). Signature model for heterogeneous multiple broadcast channel using Unified Index Hub (UIH). *Proceeding of NCRAET, Govt. Engg. College, Ajmer* (Rajasthan) – India, (pp. 45-46).

Weismann, M., & Schiper, A. (2005). Comparison of database replication techniques on total order broadcast. *IEEE Transactions on Knowledge and Data Engineering, 17*(4), 551–566. doi:10.1109/TKDE.2005.54

Wong, J. W. (1988). Broadcast delivery. *Proceedings of the IEEE, 76*(12), 1566–1577. doi:10.1109/5.16350

Xu, J., Lee, W. C., & Tang, X. (2004). Exponential index: A parameterized distributed indexing scheme for data on air. [Boston.]. *MobiSYS, 04*(June), 6–9.

Xu, J., Lee, W.-C., Tang, X., Gao, Q., & Li, S. (2006). An error- resilient and tunable distributed indexing scheme for wireless data broadcast. *IEEE Transactions on Knowledge and Data Engineering, 18*(3), 392–404. doi:10.1109/TKDE.2006.37

Chapter 5

Blind Equalization for Broadband Access using the Constant Modulus Algorithm

Mark S. Leeson
University of Warwick, UK

Eugene Iwu
DHL Supply Chain, UK & Ireland Consumer Division, Solstice House, 251

ABSTRACT

The cost of laying optical fiber to the home means that digital transmission using copper twisted pairs is still widely used to provide broadband Internet access via Digital Subscriber Line (DSL) techniques. However, copper transmission systems were optimally designed for voice transmission and cause distortion of high bandwidth digital information signals. Thus equalization is needed to ameliorate the effects of the distortion. To avoid wasting precious bandwidth, it is desirable that the equalization is blind, operating without training sequences. This chapter concerns the use of a popular blind adaptive equalization algorithm, namely the Constant Modulus Algorithm (CMA) that penalizes deviations from a fixed value in the modulus of the equalizer output signal. The CMA is set in the context of blind equalization, with particular focus on systems that sample at fractions of the symbol time. Illustrative examples show the performance of the CMA on an ideal noiseless channel and in the presence of Gaussian noise. Realistic data simulations for microwave and DSL channels confirm that the CMA is capable of dealing with the non-ideal circumstances that will be encountered in practical transmission scenarios.

INTRODUCTION

It is widely recognized that broadband access to the Internet has implications beyond the communication technology. In addition to the growing number of government and other services accessible on line, it is believed that advanced communication capabilities play a role in increased economic growth (Paltridge, 2001). The average

DOI: 10.4018/978-1-60960-477-6.ch005

penetration of fixed broadband services in the EU had reached almost 22% in July 2008 but this hid a range from over 37% in Denmark to less than 10% in Bulgaria (European Commission, 2008). There was also a gap between urban and rural areas with the latter running at typically half the value found in the former (Falch & Henten, 2009). Continued growth in demand and desires to increase social inclusion mean that growth in broadband access is likely to continue for the foreseeable future. Given the economic realities of installing optical fiber cables to reach millions of homes (Koonen, 2006) it is likely that fast digital transmission will be via copper twisted pairs using Digital Subscriber Line (DSL) technologies. The family of xDSL-techniques, where "x" is commonly H for High Rate, A for Asymmetric or V for Very High Rate, allows fast implementation of many new digital services without changing the transmission medium. The growing demand for high speed digital services ensures that xDSL technologies will continue to be the favorable bridge to high speed backbone optical networks for many years to come (Walkoe & Starr, 1991).

A major drawback for xDSL technologies is that these copper transmission systems were optimally designed for voice transmission. When such systems are used to transmit high bandwidth information signals distortion, interference and attenuation occur. In digital communications, a critical manifestation of distortion is inter-symbol interference (ISI), whereby symbols transmitted before and after a given symbol corrupts the detection of that symbol.

Equalization

Since ISI is a problem that is common to many communication channels, its removal has formed and continues to form a subject of major interest and research. To restore a sequence of received symbols distorted by an unknown system to those that were transmitted is the purpose of equalization (Qureshi, 1985). In the sense understood in recent times, equalization began with the use of linear prediction by Wiener (Makhoul, 1975) and has since found wide application in many fields. Linear channel equalization is often utilized to ameliorate the effects of linear channel distortion and can be considered to be the application of a linear filter (the equalizer) to the received signal (Proakis & Salehi, 2007). The equalizer attempts to estimate the transmitted symbol sequence by counteracting the effects of ISI, thus improving the probability of correct symbol detection. Since channel characteristics change over time the equalizer has to be adaptive in structure, usually involving the use of training signals known in advance by the receiver. The receiver then adapts the equalizer, for example via a least mean square (LMS) approach (Haykin, 2001), so that its output closely matches the known reference (training) signal.

Fractionally Spaced Equalizers

By the early 1970s equalizers with a tap spacing of less than the symbol rate had been implemented, and these have become known as fractionally spaced equalizers (FSEs) (Gitlin & Weinstein, 1981). Standard communication theory indicates that the best receiver for a linear modulation channel is a filter matched the channel followed by a T-spaced equalizer, where T denotes the symbol duration. (Forney, 1972). The use of an FSE offers the chance to synthesize the characteristics of an adaptive matched filter and a T-spaced equalizer in a way that is not possible using symbol rate sampling with the constraints of filter length and delay (Qureshi, 1985). Sampling at the symbol rate causes aliasing with constructive or destructive interference between the overlapping components. As a result, changes in the sampler phase produce variations in amplitude and phase characteristics in the overlapping regions. This means that the minimum mean squared error (MSE) achieved by a T-spaced equalizer depends on the sampler phase. Using fractional spacing on the other hand does not produce spectral overlap and greatly reduces the sensitivity of the minimum MSE to the sampler phase compared to symbol rate equalization. Of particular relevance to the xDSL case are the results of (Qureshi & Forney, 1977), who presented numerical results for Quadrature Amplitude Modulation (QAM) systems operating over representative voice-grade telephone circuits. The results showed that a $T/2$-spaced equalizer that utilizes the same number of coefficients as a T-spaced equalizer delivers approximately the same performance but in half the time span without the need for a pre-equalizer receiver shaping filter. Furthermore, a $T/2$-spaced equalizer outperforms a T-spaced equalizer in channels with severe band-edge delay distortion for all choices of sampler phase. The advantages outlined above mean that adaptive FSEs applied to a distortive DSL channel form the focus of this chapter.

Blind Equalization

The availability of *a priori* channel knowledge clearly makes the equalization process more straightforward but this is generally not the case for a particular xDSL implementation. Furthermore, the use of some of the available bandwidth for training signals to enable receiver adaptation is not optimum for channels of limited bandwidth. It is thus of considerable interest to consider adaptation algorithms without training (i.e. blind equalization) that utilize only the data and thus offer bandwidth savings (Tugnait et. al., 2000). Although signal recovery lacking knowledge of both the transmitted channel appears counterintuitive, the use of the statistical and constellation properties provides a successful route for the realization of blind equalization. The approaches developed for blind equalization of single input single output (SISO) channels such as the xDSL case can generally be categorized based on the type of statistics that they utilize (Zaouche et. al., 2008). These include maximum likelihood (ML) methods (Alberge et. al., 2002), techniques using second-order statistics (SOS) (Delmas et.

al., 2000) and approaches employing higher-order statistics (HOS), (Zhu et. al., 1999). One may also divide the methods up into indirect types that estimate the channel and then compensate for it, and those that directly implement equalizer based on the received signal (Tugnait et. al., 2000). By directly estimating the coefficients required, the latter avoid the computational burden associated with channel estimation.

The use of SOS produces an approach that has no information concerning the phase of the unknown transmission system (Benveniste & Goursat, 1984). This means that in their basic form SOS-based schemes are not sufficient to equalize systems operating with a phase greater than the minimum possible (Shalvi & Weinstein, 1990). This inadequacy in the case of T-spaced systems led to the employment of HOS. Nevertheless, since the work of (Tong et. al., 1994) there have been continual developments in FSEs employing SOS-based approaches. In particular, the cyclical variation of a signal's statistical properties, or *cyclostationarity*, in an oversampled channel output offers new information for blind channel identification and equalization (Tong et. al., 1994), (Gardner, 1991).

SOS-based channel estimation methods appeared for use in subsequent maximum-likelihood (ML) estimates of the transmitted sequence using the Viterbi algorithm (Hua, 1996). Such techniques provided a blind estimate of the channel for estimation of the minimum probability of error; the computational expense of the ML approach led to the development of approximation methods (Liu and Xu, 1994), (Tong, 1995). Direct filter estimation methods also appeared using SOS-based approaches leading to reduced calculation burdens and adaptive methods for tracking time-varying channels (Slock & Papadias, 1995), (Giannakis & Halford, 1997).

Although HOS-based methods were first used because SOS are insufficient in symbol rate systems, their delivery of phase information in addition to magnitude information led to their widespread use from the middle of the 1980s. HOS-based schemes include the super exponential algorithm (SEA) (Kawamoto et. al., 2005), inverse filter criteria-(IFC-) based algorithms (Chen et. al., 1996), polyspectra-based algorithms (Bessios & Nikias, 1995) and Bussgang algorithms (Mathis & Douglas, 2003). These rely on optimizing multimodal nonlinear and non-convex cost functions with many local minima. This can make the global optimization task time consuming and so in recent years, the fact that many digital communication schemes utilize constant modulus (CM) signals is exploited by iterative gradient-based blind equalization algorithms (Maricic et. al., 2003). These seek to minimize a cost defined by the Constant Modulus (CM) criterion, which penalizes deviations in the modulus of the equalizer output signal away from a fixed value.

The Constant Modulus Algorithm (CMA), developed independently by (Godard, 1980) and (Treichler & Agee, 1983), is the most popular and most studied of the CM-based blind adaptive equalization algorithms (Chi et. al., 2003). It exhibits quite different features to LMS despite achieving similar convergent performance (Treichler, et. al., 1998) and remains a subject of active research (Abrar & Nandi, 2010). Although several algorithms were proposed in the literature (Sato, 1975), (Godard, 1980), (Porat & Friedlander, 1991), (Treichler & Larimore, 1985) based on data sampled at the baud rate, it is the implementation of blind FSE with the advantages discussed above that has made the approach particularly successful.

In recent years, work on SOS methods has included the use of Neural Networks (Burse et. al., 2010) and advanced linear prediction (Abed-Meraim et al, 1997) but the CMA retains a simplicity amongst HOS-based methods that helps to explain its continued appeal. Moreover, the requisite technical conditions for successful application of SOS-based FSEs are not always present in practice (Tugnait, 1995). As hardware advances permit increased information rates,

physical channel delay-spread remains unchanged. This means that the relative length of the channel impulse response grows proportionally and there is a corresponding need for increased equalizer lengths to combat ISI. This is a primary justification for the continued development and study of a truly adaptive equalizer, which is robust to modest channel under-modeling with the CMA providing an effective route to achieve this.

The Constant Modulus (CM) Cost Function

To implement a blind equalization scheme, it is necessary to (a) produce an estimate of the received signal and (b) to determine and appropriate cost function to ascertain the quality of the estimate. Turning to the first of these, the estimation of a distorted signal in noise is a classic problem that may be modeled using an observed vector $\mathbf{r}$, which is the result of a signal vector $\mathbf{s}$ distorted by a function $\mathbf{H}(\cdot)$ and corrupted by additive noise, $\mathbf{w}$:

$$\mathbf{r} = \mathbf{H}\!\left(\mathbf{s}\right) + \mathbf{w} \qquad (1)$$

For the CM-equalization scenario we assume the following:

- $\mathbf{H}(\cdot)$ is a linear channel but otherwise unknown.
- The signal vector $\mathbf{s}$ is composed of statistically independent and identically distributed (i.i.d.) random variables with sub-Gaussian distributions.
- The noise vector $\mathbf{w}$ is composed of i.i.d. Gaussian random variables statistically independent of $\mathbf{s}$.

This assumption set accurately describes the estimation scenario encountered in electrical access network environments by using the Central Limit Theorem to justify the Gaussian noise assumption for $\mathbf{w}$ (Schniter & Johnson, 2000b). Moreover, in digital communications the source symbols are usually finite with sub-Gaussian distributions (Johnson et. al., 1998). For the second element of estimation quality we consider a coefficient vector **f**, which is adjustable to generate a linear estimate $y_n = \mathbf{f}^T\mathbf{r}$. The CM-minimizing estimates are defined by the set of estimators **f** that locally minimize the expected CM cost

$$J_{CM}\left\{y_n\right\} = \mathrm{E}\left[\left(\left|y_n\right|^2 - \gamma\right)^2\right] = \mathrm{E}\left[\left(\left|\mathbf{f}^T\mathbf{r}\right|^2 - \gamma\right)^2\right] \tag{2}$$

where E[ ] denotes expectation and γ is known as the dispersion constant, set to $\mathrm{E}\left\{\left|s_n\right|^4\right\}/\sigma_s^2$ for optimum algorithm operation (Goddard, 1980). Thus, the CM criterion penalizes the dispersion of estimates $\{y_n\}$ from γ. The popularity of the CM criterion (2) is usually attributed to (Schniter & Johnson, 2000b): (a) the existence of the Constant Modulus Algorithm (CMA), a simple adaptive algorithm for estimation and tracking of the CM-minimizing estimator $\mathbf{f}_{CM}(z)$; (b) the excellent mean squared error (MSE) performance of CM-minimizing estimators.

The Constant Modulus Algorithm (CMA)

The implementation of schemes to utilize the cost function J_{CM} above is clearly necessary to deliver a CM blind equalization scheme. However, although the cost function in (2) appears simple, there are no closed form expressions for its local minimizers given general **s**, **w** and **H**(·) (Schniter & Johnson, 2000b). This leads to the use of gradient descent (GD) methods to locate the estimators and the CMA is the most widely employed CM gradient descent method because of its particularly simple implementation. The stochastic update equation

for the CMA using a small step size μ is (Johnson & Anderson, 1995):

$$\mathbf{f}_{new} = \mathbf{f}_{old} - \frac{\mu}{4}\frac{\partial}{\partial\mathbf{f}}\left\{\left(\left|y_n\right|^2 - \gamma\right)^2\right\} = \mathbf{f}_{old} - \mu\mathbf{r}y_n\left(\left|y_n\right|^2 - \gamma\right) \tag{3}$$

The CMA is termed "stochastic", since we take the gradient of the instantaneous CM function (J_{CM}) rather than the 'true' cost function (the expectation) that we wish to minimize. This approximation to the dynamics of the smooth gradient descent in (2) is a close fit when the step size is suitably small (Johnson & Anderson, 1995). As will be seen in subsequent sections, the mean squared error (MSE) performance of CM-minimizing estimates has been shown to closely approximate that of the minimum mean squared error criterion (MMSE) in (4) given a signal instance s_n.

$$J_M\left(y_n\right) := \mathrm{E}\left\{\left|y_n - s_n\right|^2\right\} \tag{4}$$

Although MMSE equalization is in general, not optimal in the sense of minimizing symbol error rate (SER), it serves as a performance measure for CM estimates, since it is perhaps the most widely used method in modem designs for ISI limited channels (Johnson & Anderson, 1995). Theoretically the combination of coding and linear MMSE equalization offers a practical way to achieve channel capacity (even when the SER is not minimized) (Cioffi et. al., 1995).

FRACTIONALLY SPACED CM-EQUALIZATION

As discussed above, a fractionally spaced (FS) system with a sampling rate of T/N has the ability to "perfectly equalize" a moving average channel with a finite order equalizer (FIR) under certain conditions that will be discussed below.

Fractionally Spaced System Model

A $T/2$-spaced FSE is considered although all results are extendible to the more general T/N-spaced case. Considering the single-channel model illustrated in Figure 1(a), a (possibly complex-valued) T-spaced symbol sequence $\{s_n\}$ is transmitted through a pulse shaping filter, modulated onto a propagation channel and demodulated. All processing between the transmitter and receiver is assumed linear and time invariant (LTI) and can thus be described by the continuous-time impulse response $c(t)$. The received signal $r(t)$ is also corrupted by additive channel noise, whose baseband equivalent is denoted by $w(t)$. The received signal is then sampled at $T/2$-spaced intervals and filtered by a $T/2$-spaced finite impulse response filter (FIR) equalizer of length $2N$. Finally, the FSE output $\{x_k\}$ is decimated by a factor of 2 to create the T-spaced output sequence $\{y_n\}$. Decimation is accomplished by disregarding alternate samples, thus producing the baud-spaced "soft decisions" y_n. Equalizer taps are then updated based on the output of the CM error function J_{CM} in (2). It has been shown in (Johnson et. al., 1998) that the fractionally spaced system may be alternatively described by the sub-channel model in Figure 1(b), in which all sub-blocks are considered as baud-spaced discrete-time systems. This formulation eases introduction of the necessary conditions for "perfect equalization" (under no channel noise). When the time-span of the equalizer is greater than time-span of the channel and the sub-channels (as polynomials in z) have no common roots, then an equalizer parameterization exists that is capable of setting the combined channel-equalizer system to a pure delay with no intersymbol interference (Tong et. al., 1994).

This may be illustrated with a simple example where the over sampled channel $\mathbf{h}(z)$, and equalizer $\mathbf{f}(z)$ (with z^{-1} as the $T/2$ delay operator) are

$$\mathbf{h}\left(z\right) = h_0 + h_1 z^{-1} + h_2 z^{-2} + h_3 z^{-3}$$
$$\mathbf{f}\left(z\right) = f_0 + f_1 z^{-1} + f_2 z^{-2} + f_3 z^{-3} \tag{5}$$

then the half-baud $(T/2)$ combined equalizer response (denoted $\mathbf{q}_{T/2}$) is represented by the product of the channel convolution matrix and the equalizer tap column vector:

$$\mathbf{q}_{T/2} = \begin{bmatrix} h_0 & 0 & 0 & 0 \\ h_1 & h_0 & 0 & 0 \\ h_2 & h_1 & h_0 & 0 \\ h_3 & h_2 & h_1 & h_0 \\ 0 & h_3 & h_2 & h_1 \\ 0 & 0 & h_3 & h_2 \\ 0 & 0 & 0 & h_3 \end{bmatrix} \begin{bmatrix} f_0 \\ f_1 \\ f_2 \\ f_3 \end{bmatrix} \tag{6}$$

Decimation of the equalizer output yields a baud-spaced channel-equalizer combination (denoted $\mathbf{q}$ and resulting from striking out every other row of $\mathbf{q}_{T/2}$)

$$\mathbf{q} = \begin{bmatrix} h_0 & 0 & 0 & 0 \\ h_2 & h_1 & h_0 & 0 \\ 0 & h_3 & h_2 & h_1 \\ 0 & 0 & 0 & h_3 \end{bmatrix} \begin{bmatrix} f_0 \\ f_1 \\ f_2 \\ f_3 \end{bmatrix} \tag{7}$$

This noise-free multichannel baud-spaced system response is written compactly as:

$$\mathbf{q} = \mathbf{h}\mathbf{f} \tag{8}$$

It leads to what are commonly referred to as the "length and zero" conditions for perfect fractionally spaced equalization, when the equalizer estimate is $y_n = s_{n-\delta}$ for some fixed delay δ and any source sequence $\{s_n\}$. In addition to the absence of noise, perfect equalization requires the "zero-forcing" system impulse response with position δ containing a nonzero coefficient thus:

Figure 1. Basic channel models (a) fractionally spaced equalization; (b) subchannel

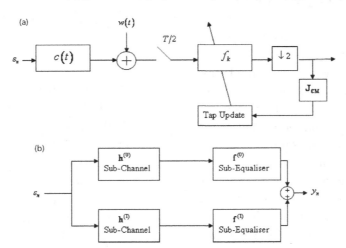

$$\mathbf{q}_\delta = \begin{bmatrix} 0 & ... & 0 & 1 & 0 & ... & 0 \end{bmatrix}^T \qquad (9)$$

This response characterizes a system that merely delays the transmitted symbols by δ baud intervals. To achieve this particular response, the system of linear equations described by $\mathbf{q}_\delta = \mathbf{hf}$ must have a solution. The matrix $\mathbf{h}$ must be full rank to achieve perfect equalization for arbitrary δ (Tong et. al., 1995), a condition sometimes referred to as *strong perfect equalization* (SPE) (Johnson et. al., 1998) implying two conditions:

C1: Matrix $\mathbf{h}$ must have at least as many columns as rows. Considering a fractionally spaced $T/2$ channel $\mathbf{h}$ with a finite length $2M$ (or at least having a response magnitude that decays below some sufficiently small threshold for all times $t \geq MT$, this results in the following equalizer *length* requirement (Johnson et. al., 1998):

$$2N \geq M + (N - 1) \Rightarrow N \geq M - 1 \qquad (10)$$

C2: The sub channels $(\mathbf{h}^{(0)}(z) = h_0 z + h_2, \mathbf{h}^{(1)}(z) = h_1 z + h_3)$ must have no common roots (Bitmead et. al., 1978), (Johnson et. al., 1998). With these two conditions satisfied, a parameterization realizing a strict delay is possible.

In the case of CMA equalizers two further conditions apply:

C3: Sources are zero-mean, independent (and circularly-symmetric if complex valued) and sub-Gaussian with finite alphabets.

C4: No channel noise is present.

When conditions **C1-C4** are satisfied, the fractional equalizer adapted under the CMA algorithm is globally convergent to a perfect equalization setting in the absence of ISI (Li & Ding, 1996). As one might imagine, these conditions are rarely satisfied in practice so there has been work to ascertain the impact of their violations. For example (LeBlanc et. al., 1996) break condition (**C3**), (Fijalkow et. al., 1997) deal with condition (**C2**), (Endres et. al., 1999) break condition (**C1**) and (Zeng et. al., 1998) attack the problem in condition (**C4**).

CM COST SURFACE EXAMPLES

To illustrate the behavior of the CMA, we now present a two tap example using a zero mean real source in the presence of real valued white, zero mean Gaussian noise. Under these conditions, we denote the signal and noise variances by

Figure 2. CMA-FSE cost function for channel $\mathbf{h}_1$*; (a) in the absence of noise and (b) with SNR = 7.0dB. Optimum delay MMSE solutions are denoted by "*".*

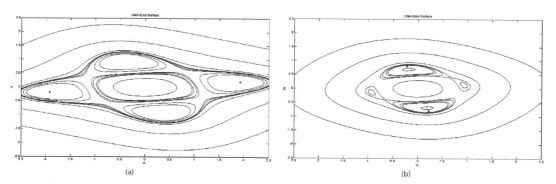

$E\left\{s^2\right\} = \sigma_s^2$ and $E\left\{\mathbf{w}^2\right\} = \sigma_w^2$. The normalized signal kurtosis is defined by $\kappa_s = E\left\{s^4\right\}\big/\left(\sigma_s^2\right)^2$ and is permitted to take an arbitrary value. The cost function with the constraints stated may be derived based on the literature, where $\|\ \|_2$ denotes the Euclidean Norm, as (Johnson et. al., 1998):

$$J_{CM} = \sigma_s^4\left(\kappa_s - 3\right)\sum_i q_i^4 + 3\sigma_s^4 \left\|\mathbf{q}\right\|_2^4 + 3\sigma_w^4 \left\|\mathbf{f}\right\|_2^4 + 6\sigma_s^2\sigma_w^2 \left\|\mathbf{q}\right\|_2^2 \left\|\mathbf{f}\right\|_2^2$$
$$- 2\kappa_s\sigma_s^2\left(\sigma_s^2 \left\|\mathbf{q}\right\|_2^2 + \sigma_w^2 \left\|\mathbf{f}\right\|_2^4\right) + \sigma_s^4\kappa_s^2$$

$$(11)$$

Using (11) we now present contour plots of the CMA surface for a two-tap *T*/2-spaced FSE under various operating conditions, following studies presented in (Johnson et. al., 1998). The length condition is satisfied for channels with impulse response $[h_0 h_1 h_2 h_3]$ and shorter, and the results were generated using MATLAB.

Ideal Zero Cost Equalization

Without noise: In the first example, a white signal with unit kurtosis travels through a channel with impulse response $\mathbf{h}_1=[0.2,0.5,1.0,-0.1]$ in the absence of channel noise. The CMA-FSE cost function is shown in Figure 2(a). In these rather ideal conditions, the CMA minima are global and can perfectly equalize the channel (Li & Ding, 1996).

The function J_{CM} has 9 stationary points, viz. a maximum at the origin, four minima for overall channel-FSE combinations $\pm(1,0)$ and $\pm(0,1)$ and saddlepoints between each of the minima.

With Additive White Gaussian Noise (AWGN): Adding substantial white noise (giving 7 dB SNR to disrupt the behavior of the idealized channel) to the channel moves the minima towards the origin and compresses the CMA ``bowl'', shrinking the convex regions about the minima and increasing their costs as shown in Figure 2(b). Since noise gain is proportional to the norm of the equalizer, the minima nearest to the origin sustain the least increase in cost, remaining global in nature. Consequently, the other pair of minima becomes local minima and equalizer initializations at $\pm(1,0)$ will converge to these false minima.

Near Common Subchannel Roots

Without noise: The effect of near common subchannel roots is illustrated by retaining the same source and choosing a channel with impulse response $\mathbf{h}_2=[0.01,1,0.01,-1]$ possessing the roots $\{-99.98, -1.0102, 0.9901\}$ and an ill-conditioned convolution matrix requiring an equalizer with *large coefficients*. Since the subchannel roots are very nearly but not quite repeated, the channel is still equalizable, and the CMA cost surface shown

Figure 3. CMA-FSE cost function for channelh_2; (a) in the absence of noise and (b) with SNR = 7.0dB. Optimum delay MMSE solutions are denoted by "*".

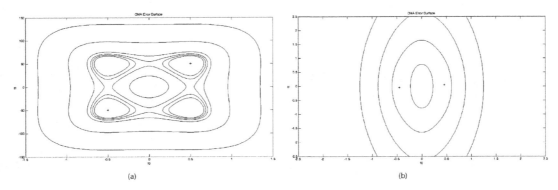

(a) (b)

in Figure 3(a) resembles a "stretched" version of that in Figure 2(a).

With AWGN: The addition of noise leads to shrinking of the CMA "bowl" as previously observed. Adding the same amount of noise as for h_1 to give an SNR of 7 dB gives the distorted CMA cost function shown in Figure 3(b). The four distinct minima that were apparent are reduced to a single pair; the equalizer performance becomes compromised and tends more towards noise reduction than zero forcing equalization.

Under-Modeling of Channel Length

Without noise: Given hardware constraints on equalizer length, residual ISI is unavoidable in practice. Mild contributions from uncompensated portions of channel response typically result in mild surface deformation. The questions arising from this scenario relate to whether local minima appear, the distance of MMSE solutions from CMA solutions and initialization so that convergence is to the global minimum. Using channel $h_3 = [0.1, 0.3, 1.0, -0.1, -0.5, 0.2]$, which cannot be perfectly equalized by a two-tap equalizer with the source as before, we obtain Figure 4(a). The CMA minima are no longer perfect: there exist a pair of global minima and a pair of false minima to which the equalizer initializations $f = \pm(1,0)$ will converge. Simulation suggests here that for

the under-modeled FSE, the global minima are located near the MMSE minima, corresponding to the optimum delays.

With AWGN: Once again, as shown in Figure 4(b), the addition of noise such that the SNR becomes 7 dB shrinks the size of the CMA bowl. As in the near common channel subroots case, two minima are lost and only the true ones remain. So when there is substantial channel noise, a modest amount of under-modeling will be accommodated by the CMA which will converge to minima close to the MMSE minima for optimum delay. What is less clear is whether spurious local minima that are not clearly associated with any MMSE solution appear when the FSE is under-modeled. This would be in contrast to channels h_1 and h_2, whose false minima are close to MMSE solutions with sub-optimum delays.

BOUNDING THE STEADY STATE PERFORMANCE OF CM-EQUALIZERS

The previous section considered the effect of imperfections on CM-minimizing equalizers (CM-equalizers) and this part explores how the likely equalizer performance under non-ideal conditions can be quantified. The complicated relationship between bit error rate (BER) and the

Figure 4. CMA-FSE cost function for channelh_3; (a) in the absence of noise and (b) with SNR = 7.0dB. Optimum delay MMSE solutions are denoted by "".*

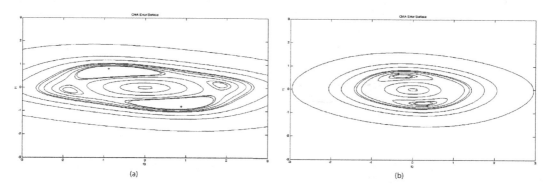

CM-equalizer coefficients means that the input-output mean squared error (MSE) is commonly utilized as a performance measure since in the limit of a small amount of Gaussian noise, the MSE and BER are monotonically related (Fijalkow et. al., 1997). A considerable amount of work has been performed in the past giving evidence of the "robustness" (using MSE performance as a robustness proxy) of CM-equalizers in non-ideal conditions. However, this has focused on particular non-ideal system model features such as the presence of Additive White Gaussian Noise (AWGN) (Fijalkow et. al., 1997), (Zeng et. al., 1998), common or nearly common sub-channel roots (Fijalkow et. al., 1997), (Li & Ding, 1996) and under-modeling (Endres et. al., 1999), (Regalia & Mboup 1999). Two notable exceptions are the works of (Zeng et. al., 1999) and (Schniter & Johnson, 2000b), of which the latter is the most general and derives tight, closed-form, bounds for the MSE of CM-equalizers supporting arbitrary additive interference, complex-valued channels, and both IIR and FIR equalizers. CM-equalizers seek to minimize the cost surface of the CM criterion (2), which is a multimodal surface with multiple stationary points. Hence, false minima can exist that are locally optimum but do not provide globally optimal performance. This false convergence first appeared over 30 years ago (Messerschmitt, 1974) and was later confirmed

by (Ding et. al., 1991). Fortunately, however, the presence of false minima does not eliminate the existence of global minima even in the presence of severe channel noise and common or nearly common sub-channel roots (Johnson et, al., 1998), (Ding et. al., 1991). This section considers a bound on the MSE performance of the CM-minimizing equalizers in the context of a global minimum. The following section will addresses the question of how we ensure that the CMA converges to the desired global minimum.

General System Model

In this section we now describe a more general, time-invariant multi-channel model shown in Figure 5 (Schniter & Johnson, 2000b). The desired symbol sequence $\left\{s_n^{(0)}\right\}$ and K sources of interference $\left\{s_n^{(1)}, \ldots, s_n^{(k)}\right\}$ each pass through separate linear "channels" before being observed at the receiver. In addition, the receiver uses a sequence of P-dimensional vector observations $\{\mathbf{r}_n\}$ to estimate (a possibly delayed version of) the desired source sequence, where the case $P>1$ corresponds to a receiver that employs multiple sensors and/or samples at an integer multiple of the symbol rate. The observations $\mathbf{r}_n$ can be written as

Figure 5. Linear system model with K sources of interference (Schniter & Johnson, 2000b)

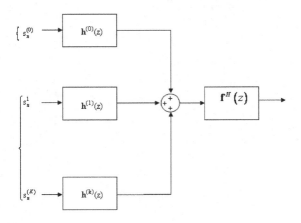

$$\mathbf{r}_n = \sum_{k=0}^{K} \sum_{i=0}^{\infty} \mathbf{h}_i^{(k)} s_{n-i}^{(k)} \qquad (12)$$

The sequence $\left\{ \mathbf{h}_i^{(k)} \right\}$ denotes the impulse response coefficients of the linear time-invariant (LTI) channel, which is such that $\mathbf{h}^{(k)}(z)$ is causal and bounded-input bounded-output (BIBO) stable.

From the vector-valued observation sequence $\{\mathbf{r}_n\}$, the receiver generates a sequence of linear estimates $\{y_n\}$ of $\left\{ s_{n-\nu}^{(k)} \right\}$, where ν is a fixed integer. Using $\{\mathbf{f}_n\}$ to denote the impulse response of the linear estimator $\mathbf{f}(z)$, the estimates are formed as:

$$y_n = \sum_{i=-\infty}^{\infty} \mathbf{f}_i^H \mathbf{r}_{n-i} \qquad (13)$$

where $\mathbf{f}_i^H$ denotes the Hermitian transpose. We will assume that the linear system $\mathbf{f}(z)$ is BIBO stable with constrained autoregressive moving average (ARMA) structure (Schniter & Johnson, 2000b). The impulse response coefficients may be collected into a (possibly infinite dimensional) vector

$$\mathbf{f} \equiv \left(..., \mathbf{f}_{-2}^T, \mathbf{f}_{-1}^T, \mathbf{f}_0^T, \mathbf{f}_1^T \mathbf{f}_2^T, ... \right)^T \qquad (14)$$

and the corresponding observations into a vector

$$\mathbf{r}(n) \equiv \left(..., \mathbf{r}_{n+2}^T, \mathbf{r}_{n+1}^T, \mathbf{r}_n^T, \mathbf{r}_{n-1}^T \mathbf{r}_{n-2}^T, ... \right)^T \qquad (15)$$

so that

$$y_n = \mathbf{f}^H \mathbf{r}(n)$$

The global channel-plus-estimator is denoted by $q^{(k)}(z) \equiv \mathbf{f}^H(z) \mathbf{h}^{(k)}(z)$, where the impulse response coefficients can be written as

$$q_n^{(k)} = \sum_{i=-\infty}^{\infty} \mathbf{f}_i^H \mathbf{h}_{n-i}^{(k)} \qquad (16)$$

which allows the estimates to be expressed as

$$y_n = \sum_{k=0}^{K} \sum_{i=-\infty}^{\infty} q_i^{(k)} s_{n-i}^{(k)} \qquad (17)$$

For conciseness, the following vector notation is adopted:

$$\mathbf{q}^{(k)} = \left(..., q_{-1}^{(k)}, q_0^{(k)} q_1^{(k)}, ... \right)^T$$

$$\mathbf{q} \equiv \left(..., q_{-1}^{(0)}, q_{-1}^{(1)}, ..., q_{-1}^{(k)}, q_0^{(0)}, q_0^{(1)}, ..., q_0^{(k)}, q_1^{(0)}, q_1^{(1)}, ..., q_1^{(k)} ... \right)^T$$

$$\mathbf{s}^{(k)} = \left(..., s_{n+1}^{(k)}, s_n^{(k)} s_{n-1}^{(k)}, ... \right)^T$$

$$\mathbf{s}(n) \equiv \left(..., s_{n+1}^{(0)}, s_{n+1}^{(1)}, ..., s_{n+1}^{(k)}, s_n^{(0)}, s_n^{(1)}, ..., s_n^{(k)}, s_{n-1}^{(0)}, s_{n-1}^{(1)}, ..., s_{n-1}^{(k)} ... \right)^T$$

Thus, for example, the estimates can be re-written as

$$y_n = \mathbf{q}^T \mathbf{s}(n) \qquad (18)$$

The relevant and important properties of $\mathbf{q}$ are: i) a particular channel and set of estimator constraints will restrict the set of attainable global responses, denoted by Q_a; ii) the BIBO stability of $\mathbf{f}(z)$ and $\mathbf{h}^{(k)}(z)$ implies that the p-norms of $\mathbf{q}$ exist for all $\mathrm{p} \geq 1$.

The following assumptions are made about the $K+1$ source processes:

A1. $\left\{ s_n^{(k)} \right\}$, $\forall k$ is zero-mean i.i.d. (two points should be noted: i) although each source process must be identically distributed, different sources may have different distributions; ii) to introduce non-white interference in the observed interference it is possible to utilize appropriate construction of $\mathbf{h}^{(k)}(z)$ for $k \geq 1$).

A2. Processes $\left\{ s_n^{(0)} \right\}, ..., \left\{ s_n^{(K)} \right\}$ are jointly statistically independent.

A3. For all k, $\mathrm{E} \left\{ \left| s_n^{(k)} \right|^2 \right\} = \sigma_s^2 \neq 0$ (this does not lead to loss of generality since interference power may be absorbed into the channels $\mathbf{h}^{(k)}(z)$).

A4. The fourth-order (auto-) cumulant, or kurtosis, $K(s_n^{(0)})$, of the source should be less than zero. For zero-mean random processes (Schniter & Johnson, 2000b):

$$\mathrm{K}_s^{(k)} \equiv \mathrm{K} \left(s_n^{(k)} \right) = \mathrm{E} \left\{ \left| s_n^{(k)} \right|^4 \right\} - 2\,\mathrm{E}^2 \left\{ \left| s_n^{(k)} \right|^2 \right\} - \left| \mathrm{E} \left\{ \left(s_n^{(k)} \right)^2 \right\} \right|^2 \qquad (19)$$

This condition imposes a sub-Gaussian requirement on the signal but *not* on the interferers ($k \neq 0$).

A5. If for any k, $q^{(k)}(z)$ or $\left\{ s_n^{(k)} \right\}$ is not real-valued, then $\mathrm{E} \left\{ \left(s_n^{(k)} \right)^2 \right\} = 0$. This is a requirement that all sources be "circularly-symmetric" in the complex plane when any of the global responses or sources is complex-valued.

The normalized kurtosis of zero-mean quantities is defined as previously in the cost surface examples:

$$\kappa_s^{(k)} \equiv \frac{\mathrm{E} \left[\left| s_n^{(k)} \right|^4 \right]}{\mathrm{E}^2 \left[\left| s_n^{(k)} \right|^2 \right]} \qquad (20)$$

Noting the conditions A1) and A5), the normalized kurtosis of a Gaussian source, κ_g, is equal to 2 for complex signals and 3 for real signals.

The Conditionally Unbiased Mean-Squared Error (UMSE)

To quantify the performance of the CMA, it is necessary to obtain an unbiased estimate of the MSE. This begins with the (MSE) criterion

$$\mathbf{J}_{m,\frac{1}{2}}\left(y_n\right) \equiv \mathrm{E}\left\{\left|y_n - s_{n-\frac{1}{2}}^{(0)}\right|^2\right\} \qquad (21)$$

which constitutes a well-known and useful measure of estimate performance. As a means of quantifying the performance of the CM estimates, it is good to compare their MSE to the MMSE (Wiener estimates) given identical sources, channels, and equalizer constraints. Since both symbol power and channel gain are unknown in the CM scenario, CM estimates are bound to suffer gain ambiguity that may be dealt with by normalization by the receiver gain $q_{\frac{1}{2}}^{(0)}$. Decomposing the CM estimate y_n into signal and interference terms,

$$y_n = q_\nu^{(0)} s_{n-\nu}^{(0)} + \overline{\mathbf{q}}^T \overline{\mathbf{s}}\left(n\right) \qquad (22)$$

using $\overline{\mathbf{q}}$ to denote $\mathbf{q}$ without the $q_\nu^{(0)}$ term and $\overline{\mathbf{s}}\left(n\right)$ to denote $\mathbf{s}(n)$ without the $s_{n-\frac{1}{2}}^{(0)}$ term, makes it easy to see that the normalized estimate $y_n/q_{\frac{1}{2}}^{(0)}$ is conditionally unbiased since $\mathrm{E}\left\{y_n/q_{\frac{1}{2}}^{(0)}\middle|s_{n-\frac{1}{2}}^{(0)}\right\} = s_{n-\frac{1}{2}}^{(0)}$. The conditionally unbiased MSE (UMSE) associated with y_n is then defined as

$$\mathbf{J}_{u,\frac{1}{2}}\left(y_n\right) \equiv \mathrm{E}\left\{\left|y_n/q_{\frac{1}{2}}^{(0)} - s_{n-\frac{1}{2}}^{(0)}\right|^2\right\} \qquad (23)$$

Substituting (22) into (23) and using the assumptions A1)-A3) we define the UMSE in terms of $\mathbf{q}$ by

$$\mathbf{J}_{u,\frac{1}{2}}\left(\mathbf{q}\right) \equiv \frac{\mathrm{E}\left\{\left|\overline{\mathbf{q}}^T \overline{\mathbf{s}}\left(n\right)\right|^2\right\}}{\left|q_\nu^{(0)}\right|^2} = \frac{\left\|\overline{\mathbf{q}}\right\|_2^2}{\left|q_\nu^{(0)}\right|^2} \sigma_s^2 \qquad (24)$$

The UMSE provides a measure of the performance of CM algorithms and it has been upper bounded by (Schniter & Johnson, 2000b), where a full proof is given. Here we reproduce their Theorem 1 which states that when a Wiener estimator associated with the desired user at delay ν generating estimates with kurtosis κ_{y_m} exists obeying

$$\frac{\rho_{\min}+1}{4} < \frac{\kappa_g - \kappa_{y_m}}{\kappa_g - \kappa_s^{(0)}} \le 1 \qquad (25)$$

where $\kappa_s^{\min} \equiv \min_{0 \le k \le K} \kappa_s^{(k)}$ and $\rho_{\min} \equiv \left(\kappa_g - \kappa_s^{\min}\right)/\left(\kappa_g - \kappa_s^{(0)}\right)$, the UMSE of a CM-minimizing estimator associated with the same user/delay can be upper bounded by

$$\mathbf{J}_{u,\frac{1}{2}}\Big|_{c,\frac{1}{2}}^{\max,\kappa_{y_m}} \equiv \frac{1 - \sqrt{\left(\rho_{\min}+1\right)\dfrac{\kappa_g - \kappa_{y_m}}{\kappa_g - \kappa_s^{(0)}} - \rho_{\min}}}{\rho_{\min} + \sqrt{\left(\rho_{\min}+1\right)\dfrac{\kappa_g - \kappa_{y_m}}{\kappa_g - \kappa_s^{(0)}} - \rho_{\min}}} \sigma_s^2 \qquad (26)$$

The existence of a CM-minimizing estimator associated with this user/delay is guaranteed by the condition (26) when $\mathbf{q}$ is FIR, and the theorem delivers a closed-form CM-UMSE bounding expression in terms of the kurtosis of the MMSE estimates.

THE TRANSIENT PERFORMANCE OF CM-EQUALIZERS

One of the problems that can occur is that the global minima reached by the CMA are those for the desired sources rather than any interferers. To avoid this, it is important to correctly initialize the GD algorithm which determines the stationary point to which the trajectory will eventually converge (Gu & Tong, 1999). This is made more difficult because in general, closed-form expressions for the gradient of the CM cost function do not exist. Thus we must employ gradient descent (GD) to solve for the global minimum. However, exact GD requires statistical knowledge of the received process, which is not usually available in practical situations. Therefore stochastic GD algorithms such as the CMA are used to estimate and track the cost function gradient since small step-size GD algorithms exhibit mean transient behavior very close to those of exact GD under typical operating conditions (Ljung, 1999).

Various approaches to initialization have been proposed based on the 'multi-user' model to jointly estimate all sub-Gaussian sources present. For example, it is possible to add a non-negative term to the CM criterion penalizing correlation between pairs of parallel equalizer outputs (Papadias & Paulraj, 1997) or to use the CM criterion in a successive interference cancellation scheme Treichler & Larimore, 1985). Such techniques implicitly incorporate the relationship between initialization and correct convergence but this was explicitly addressed by (Schniter & Johnson, 2000a) who provided three sufficient conditions under which the CMA will generate an estimator for the desired source which will now be discussed.

Sufficient Conditions for Local Convergence of CMA

The convergence conditions derived by (Schniter & Johnson, 2000a) are in terms of CM cost, kurtosis and signal to interference-plus-noise ratio (SINR):

$$\text{SINR}_v = \frac{\text{E}\left[\left|q_\nu^{(0)} s_{n-\nu}^{(0)}\right|^2\right]}{\text{E}\left[\left|\overline{\mathbf{q}}^{-\text{T}}\overline{\mathbf{s}}(n)\right|^2\right]} = \frac{\left|q_\nu^{(0)}\right|^2}{\left\|\overline{\mathbf{q}}\right\|_2^2} \qquad (27)$$

The set of global responses associated with the desired source ($k = 0$) at estimation delay v is denoted $Q_\nu^{(0)}$ and defined as follows:

$$Q_\nu^{(0)} \equiv \left\{\mathbf{q} \text{ s.t. } \left|q_\nu^{(0)}\right| > \max_{(k,\delta)\neq(0,\nu)}\left|q_\delta^{(k)}\right|\right\} \qquad (28)$$

Under assumptions A1)-A3) above, the definition of $Q_\nu^{(0)}$ ensures that an estimator is associated with a particular {source, delay} combination if and only if it contributes more energy to the estimate than any other choice. The initial estimates of the desired source at delay v are $y_n = \mathbf{q}_{\text{init}}^T \mathbf{s}(n)$ for $\mathbf{q}_{\text{init}} \in Q_\nu^{(0)} \cap Q_a$, and the conditions are (Schniter & Johnson, 2000a):

CM cost:

$$\mathbf{J}_c\left(y_n\right) < \gamma^2\left(1 - \frac{4}{\kappa_s^{(0)} + \kappa_s^{\min} + 2\kappa_g}\right) \qquad (29)$$

Variance:

$$\sigma_y^2 = \sigma_y^2\Big|_{\text{crit}} \equiv \gamma\left(\frac{4}{\kappa_s^{(0)} + \kappa_s^{\min} + 2\kappa_g}\right) \qquad (30a)$$

and normalized kurtosis:

$$\kappa_y < \kappa_y^{\text{crit}} \equiv \frac{1}{4}\left(\kappa_s^{(0)} + \kappa_s^{\min} + 2\kappa_g\right) \qquad (30b)$$

The estimates $\{y_n\}$ have variance $\sigma_y^2\Big|_{\text{crit}}$ and

$$\kappa_s^{(0)} \leq \left(\kappa_s^{\min} + 2\kappa_g\right)\Big/3, \text{ then provided}$$

$\max_{0\leq k\leq K}\kappa_s^{(k)} \leq \kappa_g$ and $\text{SINR}_\nu\left(y_n\right) > \text{SINR}_{\min,\nu}$:

$$SINR_{\min,\nu} = \frac{\sqrt{1 + \rho_{\min}}}{2 - \sqrt{1 + \rho_{\min}}} \qquad (31)$$

In the operating conditions commonly encountered in data communications, $SINR_{\min,\nu}$ has a simple form. This occurs since the sources of interference are generally not super-Gaussian (ensuring that $\kappa_s^{\max} \leq \kappa_g$) and none have kurtosis less than the desired source. As a consequence of the latter point, $\kappa_s^{\min} = \kappa_s^{(0)}$ so $\rho_{\min} = 1$ and thus $SINR_{\min,\nu} = 1 + \sqrt{2}$ or 3.8dB (Schniter & Johnson, 2000a). These conditions also ensure that $\kappa_s^{(0)}$ meets the conditions required for condition (iii).

The gain constraint $\sigma_y^2 = \sigma_y^2\big|_{\mathrm{crit}}$ that occurs in the second two conditions above is present because it is possible to construct scenarios where satisfaction of all except the gain condition results in incorrect convergence (Schniter & Johnson, 2000a). Hence, the kurtosis cannot be a sole indicator of CM-GD convergence, as suggested in (Li & Ding, 1995). Such scenarios are nevertheless rare unless σ_y^2 is far from $\sigma_y^2\big|_{\mathrm{crit}}$ or the SINR and/or kurtosis conditions are themselves near violation (Schniter & Johnson, 2000a) so in practice, convergence is robust to small violations of $\sigma_y^2 = \sigma_y^2\big|_{\mathrm{crit}}$. Thus, there exist statistical properties of initial estimates, which guarantee that subsequent CM gradient descent will produce an estimator of the same source at the same delay. The theorems suggest that when a particular source or delay is desired, initialisation schemes should be designed to either (i) maximise SINR or (ii) minimise CM cost or kurtosis when the initial estimates are known to correspond to a desired source/delay combination (Schniter & Johnson, 2000a). This informs our choice of initialization scheme for the blind equalizers described in the next section.

SIMULATION OF CM-EQUALIZERS

In this section, the simulation of a CMA-adapted fractionally spaced equalizer is described. From the theory presented so far, one can deduce that equalizer design can be reduced to the selection of three key parameters: initialization, equalizer length and step size. Ideally, to efficiently choose equalizer design parameters an adequate knowledge of the particular channel characteristics is required. For example an idea of the channel delay spread would be required in order to estimate equalizer length accurately. The Signal Processing Information Base (SPIB) (Rice University, 2000) provides a good source of empirical impulse responses for various channels. In this section, samples of the received signal for a V.29 modem and a digital microwave radio channel are used to assess the CMA performance over typical broadband access channels. An impulse noise model based on the measurements of (Henkel & Kessler, 1999) at customer premises within an xDSL test environment is employed. The received signal power is assumed to be normalized to unity by automatic gain control prior to equalization.

Simulation Program

A MATLAB® simulation program was written to process the received signal in which filter coefficients are first initialized and updated using the CMA update equation

$$\mathbf{f}_{\mathrm{new}} = \mathbf{f}_{\mathrm{old}} + \mu \mathbf{r} y \left(\gamma - |y|^2 \right) \qquad (32)$$

equivalent to (3) but slightly rearranged for software implementation. The small step size μ is chosen depending on the simulation, the design parameter γ is chosen as the normalized kurtosis and $T/2$ sampling implemented at this stage. After initial convergence, reinitialization is performed based on the strategy described below and ultimate convergence to the final CM-equalizer filter coefficients ensues.

INITIALIZATION

For T/2-spaced FSEs, double spike initialization is employed (Johnson et. al., 1998). This entails all but two of the taps being set to zero, with the two exceptions set to $1/\sqrt{2}$ ensuring that a unity equalizer norm results. The unity norm is a compromise to ensure that initial convergence is not too slow (Johnson et. al., 1998). The received signal power is assumed to be normalized by automatic gain control prior to equalization. After, initial convergence using the double taps placed in a central location, the equalizer is reinitialized as described below.

Reinitialization

The strategy for reinitialization employed is an extension of the idea introduced in (Tong & Zeng, 1997). This provides an adaptive method to reinitialize the CMA after convergence to improve its performance by exploiting the channel structure and the link between the CMA and MMSE (Wiener) equalizers. As discussed previously, if a Wiener equalizer for a particular delay has reasonably good MSE performance, there exists a CMA equalizer in its immediate neighborhood. Using the initially converged CMA equalizer as an estimate for the MMSE equalizer, we estimate a column of the channel matrix via well-known Wiener results. Since the channel matrix is Toeplitz, an estimate of a different column of the channel can be determined by shifting the column estimate up or, down a number of times and inserting zeros at one end. With this technique we get estimates of different columns of the channel matrix. The link between CMA and Wiener equalizers is further exploited by calculating the Wiener MSE cost for the different channel estimates. The channel with

the least MSE cost is then deduced. The equalizer coefficients corresponding to this estimate are then used to reinitialize the CMA. This intuitive "surfing" through the columns of the channel matrix is formalized below.

The well-known Wiener MSE cost function for delay ν may be written down from (4)

$$\mathbf{J}_v\left(y_n\right) := \mathrm{E}\left(\left|y_n - s_{n-v}\right|^2\right) \qquad (33)$$

For each v, the MMSE equalizer $\mathbf{f}_v$ that minimizes $J_v(y_n)$ can be deduced by equating the differential (with respect to $\mathbf{f}_v$) of (33) to zero thus:

$$\mathbf{J}_v\left(y_n\right) = \mathrm{E}\left\{y_n y_n^*\right\} - \mathrm{E}\left\{y_n s_{n-v}^*\right\} - \mathrm{E}\left\{s_{n-v} y_n^*\right\} + 1 \qquad (34)$$

$$\frac{d}{d\mathbf{f}}\left(\mathrm{E}\left\{y_n y_n^*\right\} - \mathrm{E}\left\{y_n s_{n-v}^*\right\} - \mathrm{E}\left\{s_{n-v} y_n^*\right\}\right) = 0 \qquad (35)$$

Substituting $y_n = \mathbf{f}_v^H \mathbf{r}$ and $\mathbf{r} = \mathbf{h}_v s_{n-v} + \mathbf{w}$ and using the properties of $\mathbf{s}$ and $\mathbf{w}$ gives:

$$2\mathbf{f}_v \, \mathrm{E}\left\{\mathbf{r}\mathbf{r}^*\right\} = \mathrm{E}\left\{\left(\mathbf{h}_v s_{n-v} + \mathbf{w}\right)s_{n-v}^*\right\} + \mathrm{E}\left\{\left(\mathbf{h}_v^* s_{n-v}^* + \mathbf{w}^*\right)s_{n-v}\right\} = 2\,\mathrm{Re}\left(\mathbf{h}_v\right) \qquad (36)$$

$$\mathbf{f}_v = \mathbf{R}^{-1}\,\mathrm{Re}\left(\mathbf{h}_v\right) \qquad (37)$$

where $\mathbf{R} = \mathrm{E}(\mathbf{r}\mathbf{r}^*)$ is the covariance matrix of the observation. This expression reduces to the oft quoted one ($\mathbf{f}_v = \mathbf{R}^{-1}\mathbf{h}_v$) when $\mathbf{h}_v$ is real (Tong, & Zeng, 1997).

Thus, if we say that the converged CMA equalizer $\mathbf{f}_{CM}$ is an estimate of $\mathbf{f}_v$ then a simplified estimate of the channel column is given by $\mathbf{h}_v \approx \mathbf{R}\mathbf{f}_{CM}$. Other columns can be estimated by shifting the nonzero components of $\mathbf{R}\mathbf{f}_{CM}$ up or down k entries to approximate $\mathbf{h}_{v+k}$. The Wiener cost function is then applied to the $\mathbf{h}_{v+k}$ estimates

Box 1.

$$\mathbf{J}_v\left(y_n\right) = \mathrm{E}\left\{y_n y_n^*\right\} - \mathrm{E}\left\{y_n s_{n-v}^*\right\} - \mathrm{E}\left\{s_{n-v} y_n^*\right\} + 1$$

$$= \mathrm{E}\left\{\mathbf{f}_\nu^H \mathbf{r}\left(\mathbf{f}_\nu^H \mathbf{r}\right)^*\right\} - \mathrm{E}\left\{\mathbf{f}_\nu^H \mathbf{r} s_{n-v}^*\right\} - \mathrm{E}\left\{s_{n-v}\left(\mathbf{f}_\nu^H \mathbf{r}\right)^*\right\} + 1$$

$$= \mathrm{E}\left\{\mathbf{f}_\nu^H\left(\mathbf{h}_\nu s_{n-\nu} + \mathbf{w}\right)\left(\mathbf{f}_\nu^H\left(\mathbf{h}_\nu s_{n-\nu} + \mathbf{w}\right)\right)^*\right\} - \mathrm{E}\left\{\mathbf{f}_\nu^H\left(\mathbf{h}_\nu s_{n-\nu} + \mathbf{w}\right)s_{n-v}^*\right\} - \mathrm{E}\left\{s_{n-\nu}\left(\mathbf{f}_\nu^H\left(\mathbf{h}_\nu s_{n-\nu} + \mathbf{w}\right)\right)^*\right\} + 1$$

$$\mathbf{J}_v\left(y_n\right) = \left(\mathbf{f}_\nu^H \mathbf{h}_\nu\right)^2 - \mathbf{f}_\nu^H \mathbf{h}_\nu - \left(\mathbf{f}_\nu^H \mathbf{h}_\nu\right)^* + 1 + \mathrm{E}\left\{\mathbf{w}^2\right\}$$

(38)

and the one with least MSE is used to deduce the equalizer coefficients, which are used to reinitialize the CMA. The technique works well enough even when the channel has complex coefficients. To calculate the MSE of the different channel estimates and expanded Wiener cost function is used thus see Equation 38 in Box 1.

Since $\mathrm{E}\left\{\mathbf{w}^2\right\} = \sigma_w^2 \ll 1$ and $\mathbf{f}_v = \mathbf{R}^{-1}\mathrm{Re}(\mathbf{h}_\nu)$, combining the $\mathbf{f}_\nu^H \mathbf{h}_\nu$ with its conjugate:

$$\mathbf{J}_v\left(y_n\right) \approx \left(\left[\mathbf{R}^{-1}\mathrm{Re}\left(\mathbf{h}_\nu\right)\right]^H \mathbf{h}_\nu\right)^2 - 2\mathrm{Re}\left(\left[\mathbf{R}^{-1}\mathrm{Re}\left(\mathbf{h}_\nu\right)\right]^H \mathbf{h}_\nu\right) + 1$$

(39)

Equalizer Length and Step Size

Length is one of the most important equalizer design parameters and is directly related to both performance and implementation cost. The MSE performance of CMA-adapted equalizers does not improve with FSE length (Treichler et. al., 1998). Although a longer FSE is better at mitigating ISI and noise effects of it incurs the penalty of additional MSE due to increased adaptation noise. This excess MSE (EMSE) also occurs when CMA is used with a non-CM source (such as QAM), which makes it impossible to achieve zero CM cost causing oscillation of the CMA-adapted FSE coefficients around the optimal CM solution (Jablon, 1989). Thus, the FSE length should be long enough to mitigate channel induced noise and ISI but short enough to prevent the EMSE from dominating. An FSE length of twice the channel delay spread (the duration of the channel response region containing 99.9% of the energy) has been found empirically to give satisfactory results (Treichler et. al., 1996).

Another critical design parameter is the step size, with an increased value offering convergence but at the price of increased EMSE. Nevertheless, an increased step size is necessary for time varying channels to facilitate rapid adaptation but care must be exercised in the tradeoff between convergence rate and EMSE.

Distorted Signal in Impulse Noise

A considerable amount of the analysis on the CMA is based on a Gaussian noise model. This is not directly applicable to many broadband access techniques. Considering DSL transmission, a statistical impulse noise model is needed. A suitable approach was developed by (Henkel & Kessler, 1999) whose statistical impulse-noise generator is based on measurements carried out at subscriber premises within an xDSL test environment. The model produces a distribution of impulse voltage amplitudes that follows a generalized exponential distribution:

Figure 6. (a) received V.29 sequence; (b) equalizer output

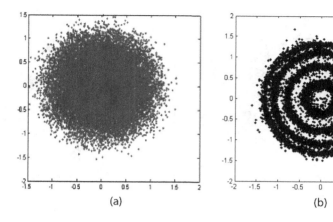

(a)   (b)

$$f(u) = \frac{1}{240u_0} e^{-[u/u_o]^{0.2}}, \quad u_0 = 3.1\,\text{nV} \qquad (40)$$

and of pulse lengths that is lognormal:

$$f(t) = \frac{1}{\sqrt{2\pi}s_1 t} e^{-\frac{1}{2s_1^2}\ln^2(t/t_1)}, \quad s_1 = 1.15, \ t_1 = 18.1\,\mu\text{s}$$

$$(41)$$

For the simulations two 10,000 sample random sequences were generated from the above distributions using the rejection method (Gentle, 2004), one for the voltages and the other for the impulse-lengths. The random sequence representing the voltages was then windowed by a Hamming window with a variable length provided by the sequence of impulse-lengths. This represents a worse case where impulses follow immediately without any inter-arrival gap. A random sequence of 5,000 16-QAM signals (normalized to unit power) was then generated and up sampled by zero insertion to produce 10,000 signal points. The received signal was generated by convolving the sequence of fractionally sampled 16-QAM signals with the channel characteristics and the impulse noise was added with an SNR of 50 dB. Although determination of the precise SINR is

non-trivial for such systems (Chen and Beaulieu, 2005), this level of noise ensures that it is above the 3.8 dB previously determined and thus the CMA will function well.

Voiceband Modem Results

The first simulation makes use of samples of received signal for a V.29, 9600bps voice band modem captured at subscriber premises from the SPIB modem database. The carrier frequency (~1700Hz) has been removed but some offset remains and the power level has been normalized to unity. The resulting complex baseband samples are plotted in Figure 6(a) showing complete eye closure. An equalizer length of 16 T/2-spaced taps and a step size of 0.0005 were selected and the resultant signal is shown in Figure 6(b). Although the eye has been opened, the algorithm does not distinguish distinct constellation points of the V.29 signal. To resolve concentric rings of different moduli into the distinct constellation points, phase recovery is needed because the algorithm depends only on the error modulus. Addition of a phase locked loop after equalization to acquire the carrier phase would permit phase recovery and, effected by decoupling the mitigation of channel effects from carrier phase tracking (Treichler et. al., 1998).

Figure 7. (a) Received signal after ISI microwave channel; (b) equalizer output

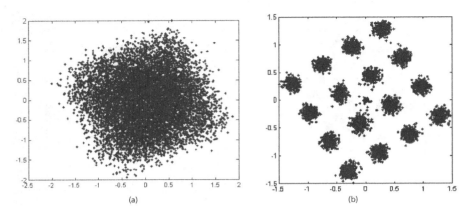

(a)
(b)

Microwave Channel

To compare xDSL with an alternative broadband access possibility, a microwave channel from the SPIB database (an FIR model of a digital microwave radio channel impulse response) (Rice University, 2000) was chosen. The complex-valued baseband channel response has been fractionally sampled at twice the baud rate (which is 30 Mbaud) and extracted from the processing of field measurements of received signals. The database channel 7 shows a significant secondary pulse in the impulse response indicating ISI and thus forms a suitable comparator. For direct comparison with the modem results, a random sequence 16-QAM signals was convolved with the microwave channels and impulse noise (as discussed previously) added to generate the equalizer input signal. Empirical testing produced a best length of 80 T/2-spaced taps for the channel and step size of 0.0005 was suitable. Figure 7(a) shows the unequalized received signal, where no constellation points are visible. As may be observed in Figure 7(b), equalizer was successful in its task but with evidence of jitter around the signal points, resulting from the EMSE present in the equalizer output. This arises from violations of the perfect equalization conditions, use of a non-CM source and a non-vanishing step size. This level

of jitter, where the constellation points are clearly separated ma be overcome by use of a decision device. The CMA also systematically produces rotated solutions, since it cannot compensate for phase offset introduced by the channel.

FUTURE RESEARCH DIRECTIONS

Although optical fiber runs close to many premises, the presence of installed copper means that there is still a large incentive to utilize it for broadband access. With DSL speeds in the Gbps range on the horizon (Lee et. al., 2007), such systems continue to be of ongoing interest and look likely to be so for many years to come. Moreover, there have also been moves to adopt fiber access architectures as part of the progress in DSL access solutions (Cioffi et. al., 2007). These advanced systems still face the same issues regarding equalization meaning that CMA will remain an option into the future. As a result, activities continue in the development of CMA algorithms themselves. For example, (Kreutz-Delgado & Isukapalli., 2008) have employed Newton's method to tackle problems arising from the presence of complex arguments in the real CMA cost function making it non-analytic. In addition a CMA based on fractional lower-order statistics has been explored by (Lin & Qui, 2009) in the context of the impulsive noise environment. In recent years, methods from the Computational Intelligence area have also been applied to the CMA. For example (Wang & Wang, 2009) have employed a neural network to enhance convergence performance. The CMA has also made its way into modern modulation methods with applications in Multiple Input Multiple Output (MIMO) systems (Ikhlef & Le Guennec, 2007) and OFDM (Cheng, 2009) transmission. Blind equalization and the CMA have also expanded beyond the traditional electrical or radio areas of application. For example, a modified CMA has been successfully investigated for the multipath underwater acoustic channel (Guo et. al., 2006). Furthermore, the increased usage of advanced modulation formats in optical communications has led to the consideration of the CMA for adaptive electronic equalizers in optical transmission systems (Vgenis et. al., 2010). Thus, in addition to its ongoing use in DSL and radio systems, the blind equalization has a significant future in a diverse range of transmission media.

CONCLUSION

This chapter has investigated the modeling and application of the CMA to the blind equalization of DSL channels. It has discussed the concept of equalization and the use of blind methods to avoid wasting bandwidth on training sequences followed by the advantages of FSEs. The diverse range of blind equalization techniques has been outlined to provide a context for the CMA. In particular the use of SOS and HOS has been introduced with a range of methods and developments highlighted. Given the popularity and advantages of FSEs adapted by the CMA, this has formed the major focus of the work. The consideration of synthetic channel examples has enables the demonstration of "perfect equalization" by the CMA given an ideal channel with an unknown but fixed delay

and unavoidable magnitude scaling ambiguities. In practice, the ideal conditions will always be violated and the CMA is robust in mitigating the effects of the violations, as was shown through a study of its cost surface characteristics. The most severe test for the CMA is the under-modeled case which introduces false minima. Given hardware constraints on equalizer length this scenario is likely to unavoidable in practice. Establishing that all CMA minima are close to a particular MMSE minimum makes it possible to devise a way to initialize the CMA so as to converge to the global minimum (the minimum supposedly associated with the MMSE solution optimized over all possible system delays). An outline of the bounds for the MSE performance of the CMA compared to MMSE counterparts and sufficient conditions for convergence of CMA to its global minimum have been discussed and have provided the motivation for the initialization scheme employed; the bounds presented for UMSE were a function of the kurtoses of the signal, the interference and the MMSE equalizer. There are many situations in which the steady-state performance of CM equalizers is comparable to that of the MMSE equalizers but the CMA does not require the statistical channel knowledge needed by MMSE equalizers. Despite the good steady-state performance of CM equalizers, it is possible that the CMA converges to an equalizer for a source of interference rather than for the desired signal. If this happens, the recovered signal is of no practical use so conditions on the CMA initialization sufficient for convergence to the desired user were presented. These are a function of the signal to interference-plus-noise ratio (SINR), which has a critical value of approximately 3.8dB for typical digital communication scenarios. To complete the study, simulations using realistic data were performed for DSL and a microwave channel. These showed that CMA was capable of equalizing the channels in a rigorous environment using a statistical impulse noise model rather than a Gaussian model.

REFERENCES

Abed-Meraim, K., Moulines, E., & Loubaton, P. (1997). Prediction error method for second-order blind identification. *IEEE Transactions on Signal Processing, 45*(3), 694–705. doi:10.1109/78.558487

Abrar, S., & Nandi, A. K. (2010). An adaptive constant modulus blind equalization algorithm and its stochastic stability analysis. *IEEE Signal Processing Letters, 17*(1), 55–58. doi:10.1109/LSP.2009.2031765

Alberge, F., Duhamel, P., & Nikolova, M. (2002). Adaptive solution for blind identification/equalization using deterministic maximum likelihood. *IEEE Transactions on Signal Processing, 50*(4), 923–936. doi:10.1109/78.992140

Benveniste, A., & Goursat, M. (1984). Blind equalizer. *IEEE Transactions on Communications, 32*(8), 871–883. doi:10.1109/TCOM.1984.1096163

Bessios, A. G., & Nikias, C. L. (1995). POTEA: The power cepstrum and tricoherence equalization algorithm. *IEEE Transactions on Communications, 43*(11), 2667–2671. doi:10.1109/26.481216

Bitmead, R. R., Kung, S. Y., Anderson, B. D. O., & Kailath, T. (1978). Greatest common divisors via generalized Sylvester and Bezout matrices. *IEEE Transactions on Automatic Control, 23*(6), 1043–1047. doi:10.1109/TAC.1978.1101890

Burse, K., Yadav, R. N., & Shrivastava, S. C. (2010). Channel equalization using neural networks: A review. *IEEE Transactions on Systems, Man and Cybernetics. Part C, Applications and Reviews, 40*(3), 352–357. doi:10.1109/TSMCC.2009.2038279

Chen, C.-H., Chi, C.-Y., & Chen, W. T. (1996). New cumulant-based inverse filler criteria for deconvolution of nonminimum phase systems. *IEEE Transactions on Signal Processing, 44*(5), 1292–1297. doi:10.1109/78.502346

Chen, Y., & Beaulieu, N. C. (2005). NDA estimation of SINR for QAM signals. *IEEE Communications Letters, 9*(8), 688–690. doi:10.1109/LCOMM.2005.1496583

Cheng, Q. (2009). A constant-modulus algorithm for carrier frequency offset estimation in OFDM systems. *Proceedings of the 2009 IEEE Region 10 Conference (TENCON 2009)*, Singapore.

Chi, C.-Y., Chen, C.-Y., Chen, C.-H., & Feng, C.-C. (2003). Batch processing algorithms for blind equalization using higher order statistics. *IEEE Signal Processing Magazine, 20*(1), 25–49. doi:10.1109/MSP.2003.1166627

Cioffi, J. M., Dudevoir, G. P., Eyuboglu, M. V., & Forney, G. D. Jr. (1995). MMSE decision feedback equalisation and coding- Part I and II. *IEEE Transactions on Communications, 43*(10), 2582–2604. doi:10.1109/26.469441

Cioffi, J. M., Jagannathan, S., Mohseni, M., & Ginis, G. (2007). CuPON: The copper alternative to PON 100 Gb/s DSL networks. *IEEE Communications Magazine, 45*(6), 132–139. doi:10.1109/MCOM.2007.374437

Delmas, J.-P., Gazzah, H., Liavas, A. P., & Regalia, P. A. (2000). Statistical analysis of some second-order methods for blind channel identification/equalization with respect to channel undermodeling. *IEEE Transactions on Signal Processing, 48*(7), 1984–1998. doi:10.1109/78.847785

Ding, Z., Kennedy, R. A., Anderson, B. D. O., & Johnson, C. R. (1991). IllConvergence of Godard blind equalizers in data communications systems. *IEEE Transactions on Communications, 39*(9). doi:10.1109/26.99137

Endres, T. J., Anderson, B. D. O., Johnson, C. R., & Green, M. (1999). Robustness to fractionally-spaced equalizer length using the constant modulus criterion. *IEEE Transactions on Signal Processing, 47*(2), 544549. doi:10.1109/78.740141

European Commission. (2008). *Broadband access in the EU: Situation at 1 July 2008*. Document COCOM08-41 FINAL.

Falch, M., & Henten, A. (2009). Achieving universal access to broadband. *Informatica Economică, 13*(2), 166–174.

Fijalkow, I., Touzni, A., & Treichler, J. R. (1997). Fractionally spaced equalization using CMA: Robustness to channel noise and lack of disparity. *IEEE Transactions on Signal Processing, 45*(1), 5666. doi:10.1109/78.552205

Forney, C. D. Jr. (1972). Maximum-likelihood sequence estimation of digital sequences in the presence of intersymbol interference. *IEEE Transactions on Information Theory, 18*(3), 363–378. doi:10.1109/TIT.1972.1054829

Gardner, W. A. (1991). Exploitation of spectral redundancy in cyclostationary signals. *IEEE Signal Processing Magazine, 8*(2), 14–36. doi:10.1109/79.81007

Gentle, J. E. (2004). *Random number generation and Monte Carlo methods* (2nd ed.). New York: Springer.

Giannakis, G. B., & Halford, S. D. (1997). Blind fractionally spaced equalization of noisy FIR channels: Direct and adaptive solutions. *IEEE Transactions on Signal Processing, 45*(9), 2277–2292. doi:10.1109/78.622950

Gitlin, R. D., & Weinstein, S. B. (1981). Fractionally spaced equalisation: An improved digital transversal equalizer. *The Bell System Technical Journal, 60*(2), 275–296.

Godard, D. N. (1980). Selfrecovering equalization and carrier tracking in twodimensional data communication systems. *IEEE Transactions on Communications, 28*(11), 1876–1875. doi:10.1109/TCOM.1980.1094608

Gu, M., & Tong, L. (1999). Geometrical characterizations of constant modulus receivers. *IEEE Transactions on Signal Processing, 47*(10), 27452756.

Guo, Y., & Han, Y. Zhou, Q. & Duo, Z. (2006). Decision circle based dual-mode constant modulus blind equalization algorithm. *Proceedings of 8ᵗʰ International Conference on Signal Processing*, Beijing.

Haykin, S. (2001). *Adaptive filter theory* (4th ed.). Englewood Cliffs, NJ: Prentice-Hall.

Henkel, W., & Kessler, T. (1999). *A simplified impulse-noise model for the xDSL test environment*. Paper presented at the Broadcast Access Conference (BAC'99), Cracow, Poland.

Hua, Y. (1996). Fast maximum likelihood for blind identification of multiple FIR channels. *IEEE Transactions on Signal Processing, 44*(3), 661–672. doi:10.1109/78.489039

Ikhlef, A. & Le Guennec, D. (2007). A simplified constant modulus algorithm for blind recovery of MIMO QAM and PSK signals: A criterion with convergence analysis. *Eurasip Journal on Wireless Communications and Networking*.

Jablon, N. K. (1989). Carrier recovery for blind equalization. *Proceedings of the International Conference on Acoustics, Speech, and Signal Processing (ICASSP-89), 2*, (pp. 1211-1214).

Johnson, C. R. Jr, & Anderson, B. D. O. (1995). Godard blind equalizer error surface characteristics: White, zeromean, binary case. *International Journal of Adaptive Control and Signal Processing, 9*(4), 301324.

Johnson, C. R. Jr, Schniter, P., Endres, T. J., Behm, J. D., Brown, D. R., & Casas, R. A. (1998). Blind equalization using the constant modulus criterion: A review. *Proceedings of the IEEE, 86*(10), 19271950. doi:10.1109/5.720246

Kawamoto, M., Ohata, M., Kohno, K., Inouye, Y., & Nandi, A. K. (2005). Robust super-exponential methods for blind equalization in the presence of Gaussian noise. *IEEE Transactions on Circuits and Wystems. II, Express Briefs, 52*(10), 651–655. doi:10.1109/TCSII.2005.852174

Koonen, T. (2006). Fiber to the home/fiber to the premises: What, where, and when? *Proceedings of the IEEE, 94*(5), 911–934. doi:10.1109/JPROC.2006.873435

Kreutz-Delgado, K., & Isukapalli, Y. (2008). Use of the Newton method for blind adaptive equalization based on the constant modulus algorithm. *IEEE Transactions on Signal Processing, 56*(8), 3983–3995.

LeBlanc, J. P., Fijalkow, I., & Johnson, C. R., Jr. (1996). Fractionally-spaced constant modulus algorithm blind equalizer error surface characterization: Effects of source distributions. *Proceedings of the IEEE International Conference on Acoustics, Speech, and Signal Processing*, Atlanta, GA, (pp. 2944-2947).

Lee, B., Cioffi, J. M., Jagannathan, S., & Mohseni, M. (2007). Gigabit DSL. *IEEE Transactions on Communications, 55*(9), 1689–1692. doi:10.1109/TCOMM.2007.904374

Li, S., & Qiu, T. Sh. (2009). Tracking performance analysis of fractional lower order constant modulus algorithm. *Electronics Letters, 45*(11), 545–546. doi:10.1049/el.2009.0561

Li, Y., & Ding, Z. (1995). Convergence analysis of finite length blind adaptive equalizers. *IEEE Transactions on Signal Processing, 43*(9), 21209.

Li, Y., & Ding, Z. (1996). Global convergence of fractionally spaced Godard (CMA) adaptive equalizers. *IEEE Transactions on Signal Processing, 44*(4), 818–826. doi:10.1109/78.492535

Liu, H., & Xu, G. (1994). A deterministic approach to blind symbol estimation. *IEEE Signal Processing Letters, 1*(12), 205–207.

Ljung, L. (1999). *System identification: Theory for the user* (2nd ed.). Upper Saddle River, NJ: Prentice Hall.

Makhoul, J. (1975). Linear prediction: A tutorial review. *Proceedings of the IEEE, 63*(4), 561–580.

Maricic, B., Luo, Z.-Q., & Davidson, T. N. (2003). Blind constant modulus equalization via convex optimization. *IEEE Transactions on Signal Processing, 51*(3), 805–818. doi:10.1109/TSP.2002.808112

Mathis, H., & Douglas, S. C. (2003). Bussgang blind deconvolution for impulsive signals. *IEEE Transactions on Signal Processing, 51*(7), 1905–1915. doi:10.1109/TSP.2003.812836

Messerschmitt, D. (1974). *Design of a finite impulse response for the Viterbi algorithm and decision-feedback equalizer.* Paper presented at IEEE International Conference on Communications, Minneapolis, MN.

Paltridge, S. (2001). *The development of broadband access in OECD countries.* OECD Working Party on Telecommunication and Information Services Policies. (Document DSTI/ICCP/TISP(2001)2/FINAL).

Papadias, C. B., & Paulraj, A. J. (1997). A constant modulus algorithm for multi-user signal separation in presence of delay spread using antenna arrays. *IEEE Signal Processing Letters, 4*(6), 17881. doi:10.1109/97.586042

Porat, B., & Friedlander, B. (1991). Blind equalisation of digital communication channels using high-order moments. *IEEE Transactions on Signal Processing, 39*(2), 522–526. doi:10.1109/78.80846

Proakis, J. G., & Salehi, M. (2007). *Digital communications* (5th ed.). New York: Mc-Graw-Hill.

Qureshi, S. U. H., & Forney, G. D., Jr. (1977). Performance and properties of a T/2 equalizer. *Proceedings of the National Telecommunications Conference*, Los Angeles, CA, (pp. 1-14).

Qureshi, S. U. H. (1985). Adaptive equalizers. *Proceedings of the IEEE*, *73*(9), 1349–1387. doi:10.1109/PROC.1985.13298

Regalia, P., & Mboup, M. (1999). Undermodeled equalization: A characterization of stationary points for a family of blind criteria. *IEEE Transactions on Signal Processing*, *47*(3), 760770.

Rice University. (2000). Signal processing information base. Retrieved on December 8, 2009, from http://spib.rice.edu/

Sato, Y. (1975). A method of self-recovering equalisation for multilevel amplitude modulation. *IEEE Transactions on Communications*, *23*(6), 679–682. doi:10.1109/TCOM.1975.1092854

Shalvi, O., & Weinstein, E. (1990). New criteria for blind deconvolution of nonminimum phase systems (Channels). *IEEE Transactions on Information Theory*, *36*(2), 312–321. doi:10.1109/18.52478

Schniter, P., & Johnson, C. R. Jr. (2000a). Bounds for the MSE performance of constant modulus estimators. *IEEE Transactions on Signal Processing*, *48*(10), 2785–2796.

Schniter, P., & Johnson, C. R. Jr. (2000b). Bounds for the MSE performance of constant modulus estimators. *IEEE Transactions on Information Theory*, *46*(7), 2544–2560. doi:10.1109/18.887862

Slock, D. T. M., & Papadias, C. (1995). Further results on blind identification and equalization of multiple FIR channels. *Proceedings of the IEEE International Conference on Acoustics, Speech, and Signal Processing*, Detroit, MI, (pp. 1964–1967).

Tong, L., Xu, G., & Kailath, T. (1994). Blind identification and equalization based on secondorder statistics: A timedomain Approach. *IEEE Transactions on Information Theory*, *40*(2), 340–349. doi:10.1109/18.312157

Tong, L. (1995). Blind sequence estimation. *IEEE Transactions on Communications*, *43*(12), 2986–2994. doi:10.1109/26.477501

Tong, L., Xu, G., & Kailath, T. (1995). Blind identification and equalization based on secondorder statistics: A frequencydomain approach. *IEEE Transactions on Information Theory*, *41*(1), 329–334.

Tong, L., & Zeng, H. H. (1997). Channel surfing reinitialization for the constant modulus algorithm. *IEEE Signal Processing Letters*, *4*(3), 85–87.

Treichler, J. R., & Agee, M. G. (1983). A new approach to multipath correction of constant modulus signals. *IEEE Transactions on Acoustics, Speech, and Signal Processing*, *31*(2), 459–472. doi:10.1109/TASSP.1983.1164062

Treichler, J. R., & Larimore, M. G. (1985). New processing techniques based on the constant modulus adaptive algorithm. *IEEE Transactions on Acoustics, Speech, and Signal Processing*, *33*(2), 420–431.

Treichler, J. R., Larimore, M., & Harp, J. C. (1998). Practical blind demodulators for high-order QAM signals. *Proceedings of the IEEE*, *86*(1), 1907–1926. doi:10.1109/5.720245

Tugnait, J. K. (1995). On blind identifiability of multipath channels using fractional sampling and second-order cyclostationary statistics. *IEEE Transactions on Information Theory*, *41*(1), 308–311.

Tugnait, J. K., Tong, L., & Ding, Z. (2000). Single-user channel estimation and equalization. *IEEE Signal Processing Magazine*, *17*(3), 17–28. doi:10.1109/MSP.2000.841720

Vgenis, A., Petrou, C. S., Papadias, C. B., Roudas, I., & Raptis, L. (2010). Nonsingular constant modulus equalizer for PDM-QPSK coherent optical receivers. *IEEE Photonics Technology Letters*, *22*(1), 45–47. doi:10.1109/LPT.2009.2035820

Walkoe, W., & Starr, T. J. J. (1991). High bit rate digital subscriber line: A copper bridge to the network of the future. *IEEE Journal on Selected Areas in Communications*, *9*(6), 765–768. doi:10.1109/49.93087

Wang, D., & Wang, D. (2009). Generalized derivation of neural network constant modulus algorithm for blind equalization. *Proceedings of the 5th International Conference on Wireless Communications, Networking and Mobile Computing, (WiCOM 2009)*, Beijing.

Zaouche, A., Dayoub, I. Rouvaen, J.M. & Tatkeu1, C. (2008). Blind channel equalization using constrained generalized pattern search optimization and reinitialization strategy. *EURASIP Journal on Advances in Signal Processing*.

Zeng, H. H., Tong, L., & Johnson, C. R. Jr. (1998). Relationships between the constant modulus and Wiener receivers. *IEEE Transactions on Information Theory*, *44*(4), 152338. doi:10.1109/18.681326

Zeng, H. H., Tong, L., & Johnson, C. R. Jr. (1999). An analysis of constant modulus receivers. *IEEE Transactions on Signal Processing*, *47*(11), 29902999. doi:10.1109/78.796434

Zhu, J., Cao, X.-R., & Liu, R.-W. (1999). A blind fractionally spaced equalizer using higher order statistics. *IEEE Transactions on Circuits and Systems II*, *46*(6), 755–764. doi:10.1109/82.769783

ADDITIONAL READING

Bellanger, M. (2007). On the Performance of Two Constant Modulus Algorithms in Equalization with non-CM Signals, *Proceedings of the IEEE International Symposium on Circuits and Systems (ISCAS 2007)*, 3475-3478.

Bergmans, J. W. M., & Janssen, A. J. E. M. (1987). Robust data equalisation, fractional tap spacing and the Zak transform. *Philips Journal of Research*, *42*(4), 351–398.

Diniz, P. S. R. (2008). *Adaptive Filtering: Algorithms and Practical Implementation* (3rd ed.). New York, NY: Springer.

Duryea, T., Saria, I., & Serpedin, E. (2006). Blind carrier recovery for circular QAM using nonlinear least-squares estimation. *Digital Signal Processing*, *16*(4), 358–368. doi:10.1016/j.dsp.2006.03.001

Fatadin, I., Ives, D., & Savory, S. J. (2009). Blind equalization and carrier phase recovery in a 16-QAM optical coherent system. *Journal of Lightwave Technology*, *27*(15), 3042–3049. doi:10.1109/JLT.2009.2021961

Goupil, A., & Palicot, J. (2007). New algorithms for blind equalization: The constant norm algorithm family. *IEEE Transactions on Signal Processing*, *55*(4), 1436–1444. doi:10.1109/TSP.2006.889398

Johnson, C. R., Jr., Schniter, P., Fijalkow, I., Tong, L., Behm, J. D., Larimore, M. G., et al. (2000). The core of FSECMA behaviour theory. In. S. Haykin (Ed.), *Unsupervised Adaptive Filtering, Volume 2: Blind Deconvolution* (pp. 13-112). New York, NY: Wiley.

Lee, E. A., & Messerschmitt, D. G. (2003). *Digital Communications* (3rd ed.). Boston, MA: Kluwer.

Lei, Z., Franzen, O., Brakensiek, J., & Schroder, H. (2001). Constant modulus algorithm for blind equalization of multipath transmitted video signals in cable networks. *Signal Processing Image Communication, 16*(5), 413–420. doi:10.1016/S0923-5965(00)00006-0

Li, X.-L., & Zhang, X.-D. (2006). A family of generalized constant modulus algorithms for blind equalization. *IEEE Transactions on Communications, 54*(11), 1913–1917.

Miranda, M. D. Silva, M.T.M. & Nascimento, V.H. (2008). Avoiding divergence in the constant modulus algorithm, *Proceedings of the IEEE International Conference on Acoustics, Speech and Signal Processing (ICASSP 2008)*, 3565-3568.

Özen, A., Güner, A., Çakır, O., Tuğcu, E., Soysal, B., & Kaya, I. (2008). A Novel Approach for Blind Channel Equalization, Lecture *Notes. Artificial Intelligence, 5227*, 347–357.

Pandey, R. (2005). Complex-valued neural networks for blind equalization of time varying channels. *International Journal of Signal Processing, 1*(1), 1–85.

Sharma, V., & Raj, V. N. (2005). Convergence and Performance Analysis of Godard Family and Multimodulus Algorithms for Blind Equalization. *IEEE Transactions on Signal Processing, 53*(4), 1520–1533. doi:10.1109/TSP.2005.843725

Widrow, B., & Stearns, S. D. (1995). *Adaptive Signal Processing Algorithms: Stability and Performance*. Englewood Cliffs, NJ: Prentice-Hall.

Yao, Y., & Giannakis, G. B. (2005). On regularity and identifiability of blind source separation under constant-modulus constraints. *IEEE Transactions on Signal Processing, 53*(4), 1272–1281.

Zaouche, A., Dayoub, I., & Rouvaen, J. M. (2008). Baud-spaced constant modulus blind equalization via hybrid genetic algorithm and generalized pattern search optimization. *AEÜ. International Journal of Electronics and Communications, 62*(2), 122–131.

Zarzoso, V., & Comon, P. (2008). Optimal step-size constant modulus algorithm. *IEEE Transactions on Communications, 56*(1), 10–13. doi:10.1109/TCOMM.2008.050484

Zhu, H., Chen, X., Zhou, W., Li, Z., Zhou, X., & Zhang, Z. (2009). A modified CMA for blind equalization and phase recovery in optical coherent receivers, *Asian Communications and Photonics Conference and Exhibition (ACP 2009), Supplement*, 1-6.

Chapter 6
Field Asymmetric Ion Mobility Spectrometry Based Plant Disease Detection:
Intelligent Systems Approach

Fu Zhang
School of Engineering, University of Warwick, UK

Evor Hines
School of Engineering, University of Warwick, UK

Reza Ghaffari
School of Engineering, University of Warwick, UK

Mark Leeson
School of Engineering, University of Warwick, UK

Daciana Iliescu
School of Engineering, University of Warwick, UK

Richard Napier
Warwick HRI, University of Warwick, UK

ABSTRACT

This chapter presents the initial studies on the detection of two common diseases and pests, the powdery mildew and spider mites, on greenhouse tomato plants by measuring the chemical volatiles emitted from the tomato plants as the disease develops using a Field Asymmetric Ion Mobility Spectrometry (FAIMS) device. The processing on the collected FAIMS measurements using PCA shows that clear increment patterns can be observed on all the experimental plants representing the gradual development of the diseases. Optimisation on the number of dispersion voltages to be used in the FAIMS device shows that reducing the number of dispersion voltages by a factor up to 10, preserves the key development patterns perfectly, though the amplitudes of the new patterns are reduced significantly.

INTRODUCTION

Ion Mobility Spectrometry (IMS) has become one of the most successful technologies widely used for detection and analysis of chemical substances (Eiceman & Karpas, 2005). This chapter introduces the potential application of a new generation of IMS, Field Asymmetric Ion Mobility Spectrometry (FAIMS), on greenhouse tomato

DOI: 10.4018/978-1-60960-477-6.ch006

plant health monitoring and the processing of the collected data sets. A continuous experiment was performed using three identical glass boxes simulating greenhouses and each glass box contains a tomato plant. One of the tomato plants was used as the health control and the others were infected with powdery mildew (*Oidium neolycopersici*) and two-spotted spider mites manually at the early stage of the experiment respectively. During the experiment, daily measurements were collected using a commercial FAIMS instrument. The post-processing on the collected data sets indicates that e-nose can be used as a tool to monitoring tomato plant disease. The data processing techniques include, Principal Component Analysis (PCA), Grey Incidence (GI), a key technique in Grey System Theory (GST), and Hierarchical Clustering (HC).

BACKGROUND

IMS Instrument

IMS generally refers to the principles, techniques and equipments designed to analysing gaseous chemical substances based on the transport of ions in electric fields. The foundational studies of IMS started in late 1940s followed by the development of practical IMS instruments in the 1970s (Eiceman & Karpas, 2005). IMS has been used as an inexpensive and powerful technique for the detection of many chemical compounds. For example, commercial IMS units are used at airports worldwide to detect explosives in carryon luggage for aviation security (Eiceman, 2002; Eiceman et al., 2004); tens of thousands of IMS units have been used by military on battlefield to determine chemical warfare agents (Eiceman, 2002); IMS can also be used to characterise drugs (Lawrence, 1989). Over the last decades, the interests of scientific researchers and engineers have been shifted from the conventional IMS to FAIMS, also known as Differential Ion Mobility Spectrometry (DIMS) or Nonlinear Ion Mobility Spectrometry (NIMS) (Eiceman & Karpas, 2005; Shvartsburg, 2009).

Conventional IMS

Conventional IMS, also called linear IMS as linear voltage gradients were used in the IMS units, was based on the determination of the velocities of the ions (Creaser et al., 2004). The basic components of a conventional IMS unit include ionization unit, drifting tube and detection plate. General ionization sources used in the ionization unit include corona discharge, photoionisaiton, electrospray ionisation, and radioactive source (Creaser et al., 2004). In the drifting tube, purified drift gas flows from the detection plate to the drafting tube entrance and a homogeneous electric field, which is generated by a series of charged rings of different voltages, is applied along the drifting tube to attract ions towards the detection plate. The electric field gradient can be alternated to allow the detection of both positive and negative ions generated in the ionization unit (Borsdorf &

Figure 1. Structure of a simplified drifting tube (Borsdorf & Eiceman, 2006)

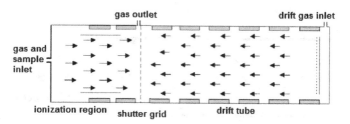

Figure 2. Asymmetric voltage waveform in FAIMS

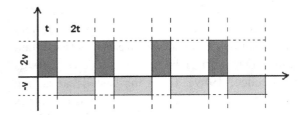

Eiceman, 2006). Figure 1 illustrates the systematic structure of a simple drifting tube.

The sample molecules are taken into the functional unit by carrier gas and the molecules are ionised to carry positive or negative charges when passing through the ionization unit. The ionised sample molecules and the non-ionised molecules are separated at the entrance of the drifting tube by the drifting gas flow and the electric field applied in the tube. The ions move towards the detection plate along the electric field gradient at random paths and hit the detection plate at different times depending on their shapes, sizes and the strength of electric field gradient. When the ions hit the detection plate, weak current is generated; the strength of the current and the time of occurrence are recorded as the signature of the test sample. The current strength is proportional to the number of ions hit the detector.

Field Asymmetric Ion Mobility Spectrometry (FAIMS)

A FAIMS unit consists of three major components, which are ionisation unit, filter and detector. Instead of using the electric field gradient generated by charged rings to attract ions towards the detection plate, DMS applies carrier gas to blow ions toward detection plate. A high-voltage asymmetric waveform termed Dispersion Voltage (DV) or Separation Voltage (SV) is applied to a set of parallel electrodes perpendicular to the detection plate. The voltages operate at high frequency asymmetric waveform and the time-voltage integrals above and below the time axis are equal. Figure 2 illustrates the asymmetric voltage waveform (Borsdorf & Eiceman, 2006; Guevremont, 2003; Krylov et al., 2007).

Ions are separable due to the different mobilities in different electric fields. As the voltage on the electrodes alternates, the ions oscillate along the electric field gradient depending on the applied voltage. The different mobilities in different electric field lead to displacement towards one of the electrodes. Ions with positive mobility dependence are displaced in a direction opposite to the ions with negative mobility dependence. Ions that hit the electrodes are neutralised and filtered out; ions reach the detection plate generate weak signals and the signals are recorded for processing. Figure 3 illustrates the possible movements of ions in the filter region. (Borsdorf & Eiceman, 2006)

Apart from the DV, a Compensation Voltage (CV) can be superimposed with the asymmetric voltage waveform to restore ions that are displaced from the centre of the gap between two electrodes. A certain CV is only efficient for ions of specific characteristics, such as charges, masses and shapes. A scan of various CVs and DVs can provide a complete measure of different ions.

Figure 3. Routes of ions in filter region

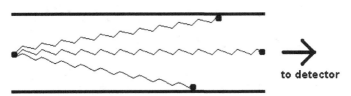

to detector

The OwlStone® LoneStar FAIMS device used in the experiment applies a comb-like ion filter, which carries the asymmetric voltages, to increase the number of filter regions and hence increases the sensitivity of the device. One FAIMS collection circle normally takes about 5 minutes and it consists of a full scan of positive ions and a full scan of negative ions. The dimensionalities of the positive scan and negative scan are identical, the compensation voltages were applied at 512 steps in the range [-6V, 6V] and the dispersion voltages were applied at various scales of a constant voltage in the range [1%, 100%]. Figure 4 presents a typical reading collected by the LoneStar FAIMS device.

Tomato Plant Diseases

Tomato plants suffer from many kinds of pests and diseases, which may cause severe damages to the tomato fruits and reduce the tomato yields. This work concentrates the initial studies on the development of a common tomato disease, powdery mildew, and a major pest, spider mite.

Powdery Mildew

Powdery mildew, caused by the fungus *Oidium neolycopersici*, is one of the most common diseased of tomato (both in greenhouse and fields) around the world. Powdery mildew can cause large

damage on tomato production if not controlled properly, especially in the greenhouse tomatoes. The symptoms of powdery mildew are the white patches on the upper surface of leaves in early stages (few days after infection). The disease causes defoliation as the powdery spots develop into brown lesions. Severe infections can lead to premature senescence and significant reduction in fruit size, quality, and yield. Effective control of powdery mildew can be achieved by applying fungicides at early stages of the disease development. (Jacob et al., 2008; Jones et al., 2001)

Spider Mite

There are over 1500 species of spider mites around the world causing damages to hundreds of species of plants. They feed on plant cells and the symptoms are the typical small, yellowish, speckled feeding marks on leaves. Spider mites are tiny, usually less than 1mm. The most well known species of spider mite is *Tetranychus urticae*, also called glasshouse red spider mite or two-spotted spider mite. Spider mites are serious pests in tomato crops due to their great reproduction rate. They can cause severe damages to tomato plants and reduce the yields. If left uncontrolled, they can destroy a plant within a short period of time.

Figure 4. A typical FAIMS reading

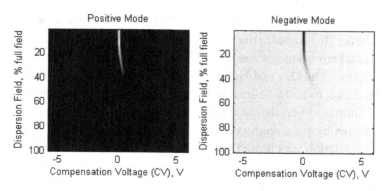

Figure 5. Systematic structure of the experiment

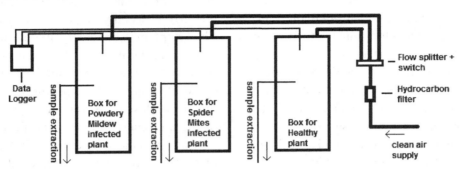

EXPERIMENT AT WARWICK

Initial studies on the early detection of tomato diseases and pests were carried out at the Engineering School of the University of Warwick in 2008. The objective of experiment was to study the development of powdery mildew and spider mites on tomato plants by measuring the chemical volatiles emitted by the infect plants using a FAIMS device. Three Espresso tomato plants were used in the experiment and the plants were provided by the Warwick Horticultural Research Institute. The plants were growing in a controlled greenhouse for 6 weeks since seedling and then moved into three glass boxes (150cm H, 50cm W, 50cm D), namely Box 1, Box 2 and Box 3, in a controlled chamber. The tomato plants in the glass boxes were under different treatments. Box 1 contains the plant which will be infected by powdery mildew; Box 2 contains the plant which will be infected by spider mites and Box 3 the plant which will be used as the health control. The glass boxes are specifically made for this experiment using glass panels and alloy frames to minimize the gas exchange with the external environment and the possibility of contamination. The Day and Night times in the chamber were set to be 16 hours and 8 hours respectively. Continuous ventilation was provided to the glass boxes by pressured gas at equal flow rates. The ventilation was turned off one hour before taking measurements and switched back on after the measurements. The plants were watered manually when necessary. Each glass box has been fitted with a humidity and temperature logging device which saves the current humidity and temperature inside the box every 5 minutes. During the experiment, repetitive experimental readings were taken daily during the Day hours in the chamber. Figure 5 illustrates the systematic structure of the experiment.

PROCESSING TECHNIQUES

As mentioned in the previous section, the data processing techniques presented in this work include Grey Incidence (GI), Hierarchical Clustering (HC) and Principal Component Analysis (PCA). PCA is used to extract the key information from a batch of collected FAIMS readings. GI and HC are used together to optimise the FAIMS data collection process by reducing the amount of data to be collected.

Principal Component Analysis (PCA)

PCA is a common mathematical technique which is used to reduce the dimensionality of a data set consisting of a large number of variables, while retaining the variation of the original data set as much as possible. PCA transforms a dataset into a new system of coordinates linearly with the possibility of reducing the number of dimensions. The first dimension of the new coordinates system is called the first principal component, which retains the highest variations of the original data set; the second dimension retains the second highest variation, and so on (Jolliffe, 2002). The principal components are determined by following four steps:

Step 1: subtract mean value

$$X' = X - \overline{X} \qquad (1)$$

where $\overline{X}$ is the mean of a data set X, and X' is the new data set.

Step 2: calculate covariance matrix

$$\text{cov}(X) = \frac{XX^T}{N-1} \qquad (2)$$

where X is the matrix consisting of data sets, X^T is the transpose of matrix X and N is the size of each data set.

Step 3: calculate eigenvectors and eigenvalues

$$Ax = \lambda Ix \qquad (3)$$

$$(A - \lambda I)x = 0 \qquad (4)$$

$$\det(A - \lambda I) = 0 \qquad (5)$$

where A is the square covariance matrix derived in step 2, I is the identity matrix, λ is an eigenvalue and x is the eigenvector.

Step 4: reorder the eigenvectors based on their associated eigenvalues, highest to lowest, which represent the explained variance of the eigenvectors.

The matrix of the ordered eigenvectors represents the original data transformed into the new coordinate system. The eigenvector of the largest eigenvalue is the first dimension (the first principal component) in the new coordinate system and it accounts for the most information (variance) of the dataset; the eigenvector with the second largest eigenvalue is the second dimension (the second principal component) of the new coordinate system; and so on. As the first few principal components usually account for the most information (variances) of the original dataset, they could be used to represent the original dataset. For visualisation purposes, two or three sets of principal components with the most significance are normally selected to express the original dataset graphically.

In this work, PCA is used to reduce the complexity of the FAIMS readings by extracting the first principal component from each measurement and visualise the development trend of the complete dataset collected during the experiment.

Grey Incidence (GI)

GI is one of the key techniques under the scope of Grey System Theory (GST). GST is a series of techniques that initially appeared in the 1980s including Grey Equations, Grey Incidence, Grey Systems Modelling, Grey Prediction, Grey Decisions and Grey Control. The theory states that the information available is always uncertain and limited due to noise. After over 2 decades of development, GST had been applied successfully in various scientific areas, including industry, agriculture, economics, etc. (Deng, 1989; Liu & Lin, 2005).

As mentioned previously, the LoneStar FAIMS was used to collect data under 100 different dispersion voltages and it has the ability to operate under customised dispersion voltages. In this work, GI analysis is used to find the closeness between two signals collected at different dispersion voltages, represented by a single GI value. Higher similarity (closeness) levels would generate higher GI values, and vice versa. An upper triangular GI matrix can be generated to represent the similarities of all possible pairs of signals. In the case of this work, the size of the GI matrix is 100 by 100 holding the effective GI values of 4950 pairs of signals.

The GI values can be calculated by the following steps. Assume that there exist two signals, each of n elements.

$$Y = (y(1), y(2), ..., y(n))$$

$$X = (x(1), x(2), ..., x(n))$$

where Y and X are the signals collected at different dispersion voltages.

Step 1

The first step is to find the initial image of each sequence using

$$Y' = Y/y(1) = (y'(1), y'(2), ..., y'(n)) = (y(1)/y(1), y(2)/y(1), ..., y(n)/y(1))$$

$$X' = X/x(1) = (x'(1), x'(2), ..., x'(n)) = (x(1)/x(1), x(2)/x(1), ..., x(n)/x(1))$$

where Y' and X' are the image sequences or the initial sequences of the original signals, Y and X.

Step 2

Find the difference sequences between the target sequence and the influencing factors using

$$\Delta = X' - Y' = (\Delta(1), \Delta(2), ..., \Delta(n)) = (x'(1)-y'(1), x'(2) - y'(2), ..., x'(n) - y'(n))$$

Step 3

Compute the global maximum and minimum differences using

$$M = max(max(\Delta(j)))$$

$$m = min(min(\Delta(j)))$$

Step 4

Calculate the incidence coefficients using Equation (6)

$$\lambda(k) = \frac{m + \zeta M}{\Delta(k) + \zeta M} \tag{6}$$

where $\lambda(k)$ represents the incidence coefficients of the kth element and ζ is a user defined factor in the range [0, 1] and generally taken to be 0.5.

Step 5

Calculate the GI using equation (7)

$$r = \frac{1}{n} \sum_{k=1}^{n} \lambda(k) \tag{7}$$

Thus the GI between the signals X and Y can be derived as

$$\gamma = (\gamma(1) + \gamma(2) + ... + \gamma(n))/n$$

Hierarchical Clustering Analysis (HCA)

HCA is a powerful exploratory methods widely used in many disciplines. It consists of a set of statistical techniques which is capable of finding the underlying structure of objects and separating the objects into constituent groups. The grouping of objects is based on a multilevel hierarchical tree or dendrogram, which is constructed from the similarities or distances between the objects

(Almeida et al., 2007; Leung et al., 1999). In this work, HCA is performed using the built-in commands provided by the Statistics Toolbox in Matlab®. The HCA in Matlab® takes the following three steps:

Step 1: Find Dissimilarities (Distances) between Data Pairs

The distance of each data pair is represented by a numerical value and a distance matrix is used to hold the distances of all the data pairs. Similar data would generate lower distance. There exist many algorithms calculating distance under various metrics. Common distance metrics include Euclidean distance, city block distance (Manhattan distance) and so on. Different algorithms generate different distance values. For example, the distance between points (1, 1) and the origin is $\sqrt{2}$ under Euclidean metric and 2 under city block metric (MathWorks, 2009). Application of various metrics may lead to distinct HCA results. In this work, the distance between the data pair is calculated using the GI instead of the conventional metrics. We may call it the Grey metric. As the GI is a measure of the data closeness in the range [0, 1], the complement of GI, 1-GI, is a better representation of the distance and is used in the following procedures.

Step 2: Group Data into a Binary Hierarchical Cluster Tree

In this step, the distance matrix generated in Step 1 is used by the linkage function to construct the hierarchical cluster tree. The linkage is an iterative program, it links single objects to each other in groups gradually. The first connection is the data pair of the highest similarity. Once the first group is constructed, the similarities between the new group and the remaining data need to be recalculated. The resulting hierarchical tree is usually illustrated using a dendrogram where each linkage step is represented by a U-shaped connection line and the height of the connection line represents the distance between the two objects connected. (MathWorks, 2009)

Step 3: Determine Clusters

The hierarchical cluster tree generated in Step 2 contains the completed cluster division information. As the connection lines represent the distances between the connected nodes or sub-trees, by selecting the appropriate cut-off distance, the data can be divided into the desired number of groups. (MathWorks, 2009)

In this work, the HCA is used to group the signals collected under various dispersion voltages based on their similarities. The signals in the same group are believed to have similar patterns and the entire group can be represented by one of the signals within the group.

RESULTS AND DISCUSSIONS

The tomato plants were moved in the dedicated glass boxes three days before the infection. Since then, repetitive measurements were taken every day. On the day of infection, the tomato plant in Box 1were infected manually by shaking the powdery mildew source (a tomato branch severely infected) over the plant for several seconds; thirty spider mites were handpicked using a small brush from the spider mite source (a pea plant used to feed the spider mites) and placed on the tomato plant. During the experiment, powdery mildew spots became visible on several leaves about 5 days after infection. However, the 30 spider mites cannot be found after the day of infection and the signs of their survival and reproduction were not obvious either. The health control plant was healthy throughout the experiment.

Figure 6. Development of the powdery mildew infected plant

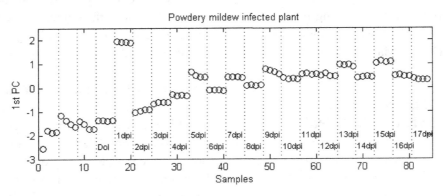

Figure 7. Development of the spider mites infected plant

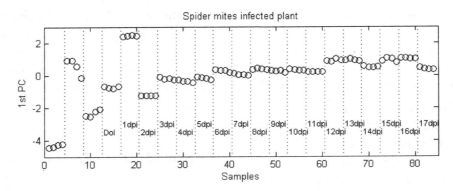

Disease Development Patterns

To discover the development trends of the powdery mildew and spider mites on the tomato plants, 4 readings were selected from each set of the daily measurements and the PCA was used to extract the first principal components from the readings. The PCA results show that, on average, the first principal components carry about 51.43% variability; the second principal components carry only

Figure 8. Development of the health control plant

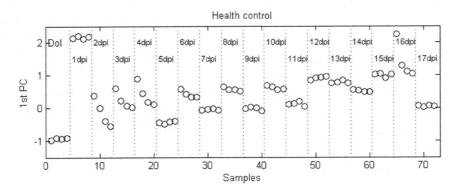

Figure 9. Grey incidence matrix

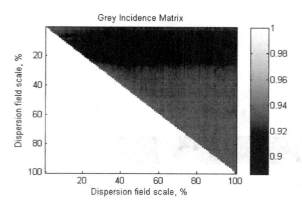

9.85% variability; the third principal component carry only 1.16% variability and so on. By placing all the first principal components together in the same graph, the development trend of the infection can be discovered. Figures 6, 7 and 8 illustrate such trends of the three tomato plants.

In Figure 6, the powdery mildew infected plant, a clear increment pattern can be observed. The pattern can be divided into three periods, which are the healthy period, the development period and the stable period. On and before the day of infection (DoI), the healthy period, the values of the 1st PCs are about the same. After the infection, the development period, the 1st PCs fluctuate and follow a gradual increasing pattern, expect the data collected on the 1st day post infection (DPI), which is invalid due to instrumental errors. The

increasing pattern may due to the development of the disease; the condition of the plant was getting worse day by day. Since the 10th DPI, the stable period, the 1st PCs fluctuate around a constant value. In these days the condition of the plant is severe and the relevant volatiles in the glass box reached a certain density.

The data of the spider mites infection plant illustrated in Figure 7 can be divided into three periods as well. During the healthy period, the 1st PCs are not stable and show distinct differences. After the infection, a weak increasing pattern is noticeable until the 12th DPI. From the 12th DPI onwards, the 1st PCs show a flat pattern.

In Figure 8, the health control plant, the data points form a gradual increasing pattern. This may be due to the fact that as the tomato plant grows the leaves at the bottom of the stem turn yellow and fall off eventually. The leaves turned yellow may emit volatiles different from those emitted by the healthy green leaves.

Dispersion Voltages Optimisation

As mentioned in the previous section, the dispersion voltages applied by the FAIMS can be set manually and a full scale scan (100 different dispersion voltages) takes about 5 minutes. By reducing the number of dispersion voltages, the scanning speed of the FAIMS can be increased dramatically. The GI is used to calculate the

Figure 10. HCA dendrogram calculated using the GI distance matrix

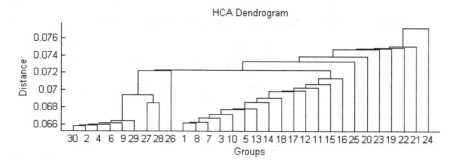

Figure 11. Clustering of dispersion fields

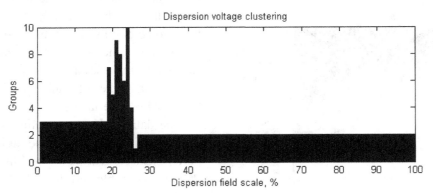

similarities between the signals collected under various dispersion voltages. Figure 9 illustrates the Gray Incidence matrix (similarity matrix) derived using all the data.

As the linkage procedure of HCA acquires the distance matrix rather than the similarity matrix, the complement of the GI matrix is used in the linkage procedure. Figure 10 illustrates the dendrogram generated based on the linkage.

The original data was collected under 100 different dispersion voltages. If we aim to reduce the number of dispersion voltages by a factor of 10, we need to divided the 100 different dispersion voltages into 10 (100/10) groups base on the dendrogram and find a representative dispersion voltage for each group. Figure 11 illustrates the resulting clusters.

The first element of each group can now be used as the representative of that group. By extracting the signals collected under the selected

dispersion voltages and perform the PCA again, patterns similar to the ones obtained from the original can be produced. Figure 12 illustrates the disease development pattern of the powdery mildew infected plant generated using 10 dispersion voltages, instead of 100. The new development retains about 96% of the original pattern.

Table 1 summarises the performance of the HCA with the application of various numbers of dispersion voltages. The reduced number of dispersion voltages work well as it always retains over 90% of the similarities of the original development patterns.

CONCLUSION

This chapter presents the initial studies on the responses of greenhouse tomato plants to two kinds of diseases and pests, the powdery mildew

Figure 12. Disease development pattern generated using reduced number of dispersion voltages

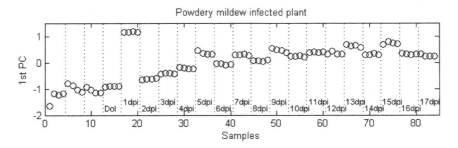

Table 1. Performance of the selection of various numbers of dispersion voltages

	Various numbers of remaining DVs				
	10 DVs	**20 DVs**	**30 DVs**	**40 DVs**	**50 DVs**
Powdery mildew infected plant	96.0%	92.7%	95.6%	96.9%	97.4%
Spider mites infected plant	95.7%	96.9%	96.4%	96.4%	96.9%
Health control	96.2%	93.5%	95.3%	95.7%	96.5%
Average performance	96.0%	94.4%	95.8%	96.3%	96.9%

and spider mites, by measuring the chemical volatiles emitted from the tomato plants using a FAIMS device. Post processing of the collected data shows that clear increment patterns can be observed using PCA. This may be due to the development of the diseases.

Further analysis on the collected data using GI and HCA shows good results. By grouping the signals collected under the full FAIMS scan (100 various dispersion voltages) into various numbers of groups and selecting one representative dispersion voltage from each group, the amount of measurement can be reduce effectively by various factors. Statistical results show that reducing the amount of dispersion voltages by a factor between 2 and 10 can always retain over 90% of the variation of the original development patterns.

REFREENCES

Almeida, J.A.S., Barbosa, L.M.S., Pais, A.A.C.C. & Formosinho, S.J. (2007). Improving hierarchical cluster analysis: A new method with outlier detection and automatic clustering. *Chemomotrics and Intelligent Laboratory Systems, 87*.

Borsdorf, H., & Eiceman, G. A. (2006). Ion mobility spectrometry: Principles and applications. *Applied Spectroscopy Reviews, 41*, 323–375. doi:10.1080/05704920600663469

Creaser, C. S., Griffiths, J. R., Bramwell, C. J., Noreen, S., Hill, C. A., & Thomas, C. L. (2004). Ion mobility spectrometry: A review. Part 1: Structural analysis by mobility measurement. *Analyst (London), 129*, 984–994. doi:10.1039/b404531a

Deng, J. (1989). Introduction to Grey system theory. *Journal of Grey System, 1*, 1–24.

Eiceman, G. A. (2002). Ion-mobility spectrometry as a fast monitor of chemical composition. *Trends in Analytical Chemistry, 21*, 259–275. doi:10.1016/S0165-9936(02)00406-5

Eiceman, G. A., & Karpas, Z. (2005). *Ion mobility spectrometry* (2nd ed.). CRC Press. doi:10.1201/9781420038972

Eiceman, G. A., Krylov, E. V., Krylova, N. S., Nazarov, E. G., & Miller, R. A. (2004). Separation of ions from explosives in differential mobility spectrometry by vapor-modified drift gas. *Analytical Chemistry, 76*, 4937–4944. doi:10.1021/ac035502k

Guevremont, R. (2003). High-field asymmetric waveform ion mobility spectrometry (FAIMS). *Canadian Journal of Analytical Cciences and Spectroscopy, 49*, 105–113.

Jacob, D., David, D. R., Sztjenberg, A., & Elad, Y. (2008). Conditions for development of powdery mildew of tomato caused by Oidium neolycopersici. *Ecology and Epidemiology, 98*, 270–281.

Jolliffe, I. T. (2002). *Principal component analysis* (2nd ed.). Springer.

Jones, H., Whipps, J. M., & Gurr, S. J. (2001). The tomato powdery mildew fungus Oidium neolycopersici. *Molecular Plant Pathology, 2,* 303–309. doi:10.1046/j.1464-6722.2001.00084.x

Krylov, E. V., Nazarov, E. G., & Miller, R. A. (2007). Differential mobility spectrometer: Model of operation. *International Journal of Mass Spectrometry, 266,* 76–95. doi:10.1016/j.ijms.2007.07.003

Lawrence, A. H. (1989). Characterization of benzodiazepine drugs by ion mobility spectrometry. *Analytical Chemistry, 61,* 343–349. doi:10.1021/ac00179a012

Leung, S. C., Fung, W. K., & Wong, K. H. (1999). The identification of credit card encoders by hierarchical cluster analysis of the jitters of magnetic stripes. *Science & Justice, 34,* 231–238. doi:10.1016/S1355-0306(99)72054-X

Liu, S., & Lin, Y. (2005). *Grey information: Theory and practical applications.* Springer.

MathWorks T. (2009). *Statistics toolbox 7: User's guide.* The MathWorks.

Shvartsburg, A. A. (2009). *Differential ion mobility spectrometry: Nonlinear ion transport and fundamentals of FAIMS.* CRC Press.

KEY TERMS AND DEFINITIONS

Field Asymmetric Ion Mobility Spectrometry (FAIMS): Is a new technology for quick and accurate detection of a broad range of chemicals at low quantities (below part per billion) with high confidence. The FAIMS operates by applying high frequency asymmetric electric fields on the samples carried by appropriate carrier gas passing through the filtering channel. Ions are separable due to the different motilities under different electric fields. Ions that hit the electrodes in the filtering channel are neutralised and filtered out; those reach the detection plane generate weak signals, which are recorded as the fingerprint of the sample.

Grey Incidence (GI): Is one of the key techniques under the scope of Grey System Theory (GST). It is an alternative way to measure the distances or similarities between data sequences.

Hierarchical Clustering Analysis (HCA): Is a powerful exploratory methods widely used in cluster analysis. It consists of a set of statistical techniques capable of finding the underlying structure of objects and group together objects that are 'close' to one another.

Principal Component Analysis (PCA): A common mathematical technique developed in 1901. It can reduce the dimensionality of a data set by transforming it to a new coordinate system and retaining the dimensions of high variances. The first dimension of the new coordinate system holds the highest variation of the original data set and the second dimension retains the second highest variation, and so on.

Powdery Mildew: (*Oidium neolycopersici*): A type of common fungal disease around the world that affects a wide range of plants, including grapes, onions, tomatoes, and so on. It appears as dusty gray or white coating on leaf surfaces and other parts. It can cause large damage on fruit production if not controlled properly.

Spider Mites: Are tiny pests, usually less than 1mm. There are over 1500 species of spider mites causing damages to hundreds of species of plants. The most well known species is *Tetranychus urticae,* also known as glasshouse red spider mite or two-spotted spider mite.

Chapter 7

The Analysis of Plant's Organic Volatiles Compounds with Electronic Nose and Pattern Recognition Techniques

Reza Ghaffari
School of Engineering, University of Warwick, UK

Fu Zhang
School of Engineering, University of Warwick, UK

Daciana Iliescu
School of Engineering, University of Warwick, UK

Evor Hines
School of Engineering, University of Warwick, UK

Mark Leeson
School of Engineering, University of Warwick, UK

Richard Napier
Warwick HRI, University of Warwick, UK

ABSTRACT

In this chapter, the authors introduce the principles of some of the most widely used supervised and unsupervised Pattern Recognition (PR) techniques and assess behaviour and performances. A dataset acquired from a set of experiments conducted at University of Warwick is employed to construct a case study in which the techniques will be applied. The chapter will also evaluate the integration of PR methods with an Electronic Nose (EN) device to develop and implement a plant diagnosis tool based on discriminating the Organic Volatile Compounds (VOC) released by plants when attacked by pest. The chapter concludes with a performance comparison and a brief discussion of how an appropriate PR technique can be coupled with an EN to produce a greenhouse plant pest and disease diagnosis system for day-to-day utilisation. Some consideration of further work is also presented.

DOI: 10.4018/978-1-60960-477-6.ch007

INTRODUCTION

Engineering systems often comprise a sensory subsystem of one or multiple sensors which collect or detect sensory data and produces measurement signals. These signals contain a raw data therefore in most cases the direct usage of these sensory signals is impossible, inefficient or even pointless. According to Heijden (2004), this can have several causes:

a. The information in the signals is represented in an inexplicit or ambiguous way making it harder to be recognizable without further processing.
b. The information is often hidden and only available in an encoded form prior to processing.
c. Sensors always produce measurement signals which come with a substantial noise and other complex disturbances therefore needs noise reduction techniques.

This indicates that a sensory signal which has been processed is more precise and more complete than information brought forth by empirical knowledge alone (Heijden et al., 2004). For the system to be able to make sensible and accurate decisions, the measurement signals should be used in combination with previous knowledge or pattern. Several techniques and methods have been used to process the measurement signals in order to suppress the noise and disclose the advantageous information required for the task at hand. *Pattern Recognition* (PR) is one of the most widely used techniques which have been implemented within various engineering systems.

In principal, *PR* is the scientific discipline whose goal is the classification of *objects* into categories or *classes*. These objects can be anything from a simple image, signal or any other type of sensory data, depending on the application (Theodoridis & Koutroumbas, 2006).

PR techniques have gained their popularity by being the brain behind the recent Handwritten Character Recognition and Speech Recognition tools built in various machines and software such as Navigation and Call Center Systems. They are designed to mainly do complex feature selection, classification or data clustering.

There have been several sub-categories for PR techniques based on their characteristics, learning method and mathematical algorithms. However, PR techniques are normally based on three basic and well-known approaches: (a) Statistical (b) Structural and (c) Neural. Moreover, PR learning methods are often grouped into two more general categories although a combination of both can be used: (1) Supervised learning and (2) Unsupervised learning. Supervised learning is one of the most commonly undertaken analyses of the PR problems in which the learning phase will be adjusted according to a *target* dataset. In unsupervised learning, however, the classifier will not have any information regarding the subsets (classes or categories) of the sample data.

In PR approach, supervised learning is often associated to classification whereas unsupervised learning is mostly used for data clustering purposes. Several mathematical algorithms and optimisation techniques were used previously to enhance the performance of the classic PR methods and customise it for a specific application. The performance of the PR techniques is often rated by their ability in correct classification/clustering of provided training data samples. High processing power and memory capabilities of the recent computers allow PR algorithms to analyse the samples in fraction of a second and provide reliable solution depending on the application.

In this chapter few statistical and Artificial Neural Network (ANN) based PR techniques will be discussed and applied on the Electronic Nose (EN) generated dataset. In the next section, we will explain the case study, data collection and experimental setup.

CASE STUDY

As mentioned earlier, we will use PR techniques to investigate a dataset acquired by an array of sensors (i.e. Electronic Nose). We will then analyse the sensor responses collected by EN to discriminate the Organic Volatile Compounds (VOCs) released from the healthy and artificially infected tomato plants.

The main purpose of this study is to evaluate and recognize the ability of EN (Bloodhound model ST214, Scensive Technologies Ltd., Normanton, UK) in diagnosing the diseases and subsequently discriminate between healthy plants from the infected one in a timely manner.

Plant's VOC

Plants naturally release VOCs under normal conditions (Baldwin, 2002), but they also emit a diverse range of VOCs in response to either physical and biotic stress or infection (Holopainen, 2004). These compounds combat the infection directly, attract natural biological control agents and function as signals to induce indirect defence responses (Kessler, 2001). The VOC profile emitted from plants usually changes in response to environmental and ecological factors, and by examining the change of such profiles a non-destructive means of plant health evaluation could be offered (Kant et al., 2009). Investigating the visual appearance of the plant part (i.e. leaf surface) by image processing techniques (Cameron & Smith, 2009; Parsons et al., 2009) and examining the Volatile Organic Compounds (VOCs) of plant and pathogen (Moalemiyan et al., 2007; Schütz, 1996) are two options which might produce an attractive means of rapid and non-destructive plant diagnosis testing. These need to feed information to knowledge-led decision support software to advise growers of options for intervention and control.

Pests and Diseases

Plant diseases can have several symptoms and may occur in a number of ways but manifest mainly via infections by fungi, virus and bacterial infection (Agrios, 2004). Two very common diseases in tomato plants, powdery mildew and spider mites will be investigated in this study.

Recent reports of powdery mildews invading Europe from other continents emphasises the need for accurate identification of the disease as well as a method to control it in the shortest period of time (Braun, 2009).

Spider mites on the other hand, are the most common mites attacking commercial plants and are considered to be one of the most economically important diseases which threaten the tomato plants. This mite has been reported to be infesting over 200 species of plants around the world. A number of vegetable plants such as tomatoes, squash, eggplant and cucumber are subject to spider mite infestations and damage (Helbert, Hodges & Sapp, 2007).

Electronic Nose (EN)

EN is developed to mimic the human olfactory system and was evolved dramatically as different types of sensor arrays were built in to it to make it suitable for specific odors and applications. EN was introduced in 1982 by Dodd and Persaud from the Warwick Olfaction Research Group, UK and several applications of EN were investigated (Gardner, Hines, & Pang, 1996). The EN technology has been used in a variety of applications, including food and fruit quality measurements (Peris & Escuder-Gilabert, 2009), animal disease diagnosis (Dutta, Morgan, Baker, Gardner, & Hines, 2005), automotive industry (Kalman, Löfvendahl, Winquist, & Lundström, 1999) and plant health monitoring (Baratto, et al., 2005).

EN is internationally well known for being able to solve a wide variety of problems with a high precision at a potentially low cost. Nowadays, most EN devices have built in PR system or a mean of data processing algorithm (i.e. PCA) which will increase the portability as well as the performance of such devices for diverse range of applications. The EN's portability, cost of running and simplicity to operate make it an attractive solution compare to other alternatives such as Gas chromatography-mass spectrometry (GC-MS).

Later in this chapter, we will investigate the response of EN sensors and consequently its capacity in classifying plants VOCs.

Experimental Setup

In order to replicate the greenhouse environment, we used disinfected clear glass boxes to house one plant each. Three clean glass boxes (150cm * 50cm * 50cm) simulated the greenhouse environment. A control healthy plant was kept healthy throughout the experiment. Humidity and temperature were logged at all times with the interval set at 10 minutes between each reading. Clean air was filtered and pumped into each box to create positive pressure inside the boxes which decreased the possibility of cross contamination between the boxes as well as maintaining environmental parameters constant throughout the experiment. The light system was precisely controlled by a timer to make sure that the plants had 16 hours of artificial daylight. Each plant was watered daily. Prior to sampling, air inflow was switched off for 3 hours to allow volatile concentration around the plants to build up. Individual box tubes were connected to the EN to take readings of volatile concentrations. Sampling tubes kept separated to reduce the possibility of cross contamination. A solution of butan-2-ol (2% in distilled water) was used as a reference sample and also acted as a sensor wash to regenerate sensor surfaces.

A various range of algorithms have been proposed by researchers and applied on EN generated dataset such as Principle Component Analysis (PCA), Linear Discriminant Analysis (LDA).

Artificial Neural Network (ANN) techniques such as Multi-Layer Perceptron (MLP), Learning Vector Quantization (LVQ), Radial Basis Function (RBF) and Probabilistic Neural Network (PNN) are also among the favorite analysis algorithms. For clustering purposes, Self-Organizing Map (SOM), K-Means Clustering, Fuzzy C-Mean clustering and Support Vector Machine (SVM) are often employed.

TECHNIQUES AND METHODOLOGIES

Data Pre-Processing

During the experiment, the EN continuously recorded the responses from its array of 13 sensors and the data was saved as a data matrix. The data comprised of the profile of the VOC determined by the EN at the following time intervals: 7 s absorption, 0 s pause, 20 s desorption and 5 s flush. The key dataset parameters were determined and extracted from the measurements: (a) Divergence (b) Absorption (c) Desorption and (d) Area; thus forming a 52 component (13×4) matrix.

These values later formed the final dataset. Four days within the dataset was selected: 4 days post infection (DPI), 6 DPI, 8 DPI and 9 DPI. All the data processing and analysis were performed in MATLAB® environment and *PRTools* toolbox was employed for some of the analysis.

These datasets were normalized by subtracting the mean value using Equation 1.

$$X' = X - \overline{X} \tag{1}$$

Each dataset is a matrix of 60 x 52. Next, the four datasets are divided into two sets one for training (60% of data), and the other for validation and testing purposes (40% of data).

Some initial analyses were applied to estimate the location of the clusters inside the dataset which

Figure 1. 6 DPI – (Left) Probability densities of the signal responses (Right) the single-link dendrogram of hieratical clustering

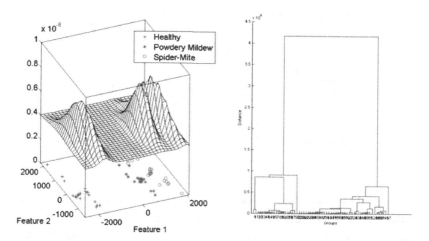

will later enable us to have a better understanding of the dataset characteristics. Figure 1 (left) is a graphical representation of the probability densities of the sensory data generated by EN on 6 DPI after normalization. The probability densities exposed are estimates obtained from the pre-processed sensory samples.

In Figure 1 (right) a simple, single-link dendrogram was constructed using hieratical clustering to visualize the possible clusters within the dataset. The dendrogram demonstrates a large gap in distances which indicates that the two feature clusters are far apart from each other. Some minor sub-clusters are also visible within the dataset. Nevertheless, in this case, it is rather complicated to estimate the actual number of clusters from the figures. We can conclude that there are differences within the dataset though. A more intelligent clustering techniques in necessary and will be applied later to reveal the actual number of clusters inside the dataset.

Linear and Quadratic Discriminate Analysis

As a statistical technique, LDA is one of the most widely used classification procedures (Hai &

Wang, 2006). The method maximizes the variance between categories and minimizes the variance within categories by means of a data projection from a high dimensional space to a low dimensional space (Zhang, 2008; Yongwei, 2009). In other words, LDA simply looks for a sensible rule to discriminate between sample points by forming linear functions of the data maximizing the ratio of the between-group sum of squares to the within-group sum of squares (Zhang, 2007).

Figure 2 shows two scatter diagram and the linear decision boundaries separating the clusters. The linear discriminant function was able to classify the dataset with 95% success rate. However, the performance enhances when the function employed the lease Square (LS) error classifier. The LS classifier's performance reached 97% (Figure 2- Right). It is clear that LDA coupled with LS function attempted to adjust the decision boundaries so more sample points can be accommodated into the correct classes. However, the classification is still linear and can only show good degree of a classification if the dataset is spreadable.

To enhance the classification, the same dataset was analysed with Quadratic Discriminant Analysis (QDA). QDA which uses quadratic decision function was added to the classifier to

Figure 2. 6 DPI - Left: 95% correct classification with linear discriminant Analysis - Right: 97% Classification with LDA and least squared error classifier

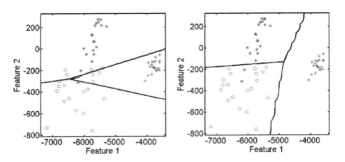

decrease the error rate. QDA was able to correctly classify the data with an excellent 98% success rate. Figure 3 illustrates the Quadratic decision boundaries which is categorizing the classes. Each class represents a status of the plant (i.e. Healthy, Powdery-Mildew infected and Spider Mite infested)

Nearest Neighbour Classification

K-nearest neighbor (KNN) is a non-parametric classical classification technique widely used in pattern recognition problems. KNN classifier method is used for performing general, non-parametric classifications (Wu, 2002). KNN as a fast, reliable and flexible method is a popular approach to perform the classification task in EN enabled applications.

Briefly, in the learning phase of KNN algorithms the presented training dataset will be stored until a new instance of k is encountered, then a set of similar training instances is retrieved from memory and used to make a local approximation of the target function (García-Laencina, 2009). Subsequently, to classify a new pattern, the Euclidean distance between the new pattern and each pattern in the training set is computed. The Euclidean distance metric is given by the following equation:

$$d_i = \sqrt{\sum_{j=1}^{n} (x1_j - y1_j)^2} \qquad (2)$$

Figure 3. 6 DPI -Quadratic discriminant analysis with 98% correct classifications

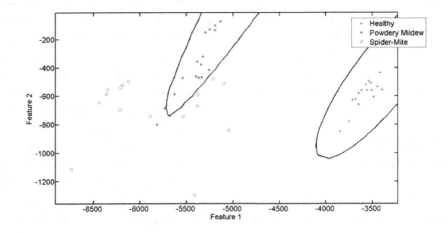

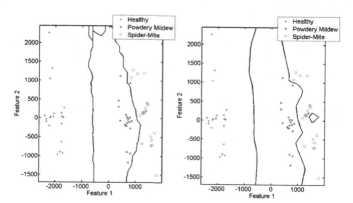

Figure 4. K-Nearest Neighbour – 8 DPI - With k = 1 (left) and k= 9 (right) Optimum Value

where d is the Euclidean distance between the calculated attributes x and the data points y_i, with n variables (Zhang, 2008). Usually in KNN algorithms, the Euclidean distance is used for finding the Nearest Neighbor, but for strongly correlated variables, it is also recommended to use correlation-based measures (Berrueta, 2007).

In this case, the initial value of K was set to 1. However, the best results were obtained when K =9. Normally, increasing the value of K improves the classification performance, because of various spreads of data belonging to various classes. However, k should not exceed the optimum value (Ciosek, 2006; Dragovic, 2007).

Figure 4 illustrates the K-Nearest result when k is equal to 1 (left) and when k is equal to the optimum value which is 9 (right). It is evident that K-Nearest performed better with the optimal k value with the error rate of 0.02%.

Artificial Neural Network (ANN)

Among statistical classification methods, ANNs are generally considered as the most promising PR routine to process the sensory signals from a chemical sensor array of ENs (Fu, 2007). ANNs are able to build a non-linear multivariate model of a presented training dataset (Nils Paulsson, 2000). They can recognize patterns within vast datasets and then generalize those patterns into advantageous information (Yu & Wang, 2007). ANN algorithms are multipurpose and with appropriate training and customization, a single technique could solve several problems and become adopted for different environments (Escuder-Gilabert, 2009).

Initially, an ANN network needs to be created with suitable input and output layers. The neurons and weights are then initialized and adjusted ready for the learning phase. The numbers of inputs and outputs layers of the system are often determined by the system features. The final stage of learning is the network testing and validation. Testing procedure will test the accuracy of the trained NN model in discriminating and hence classifying the samples.

The EN data from sensor responses were used to train the ANN for the purpose of comparing and identifying the correct class that each sample belongs to. Once the weights have been adjusted using the samples in the training and target matrix, the network can be used to predict the class membership of unknown samples.

RBF and Parzen classifier based network were trained and tested for this case study. RBF networks are relatively suitable to be used with EN generated data because they have advantages at convergence rate, probability of reaching global

Figure 5. Left: Parzen Classifier with 94% Correct Classification - Right: RBF with 88% Correct Classification

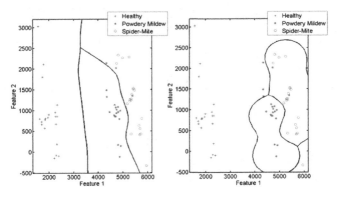

points and local sensitivity compared to other techniques such as MLP (Daqi, Shuyan & Yan, 2004). In this study, a RBF network was formed with the following profile: the network consisted of two layers: a hidden radial basis layer and an output linear layer and contained 2 bias vectors. The SPREAD constant was set as an initial value of 1.0. The spread value did not change throughout the simulation and training.

Finally, the testing is performed using the testing data set which contained 40% of the original dataset. In later stages, in order to get better training results, the testing dataset size was increased. The decision boundaries are illustrated in Figure 5. Parzen classifier was able to achieve 94% correct classification while RBF was only successful when classifying 88% of the dataset.

K-Means Clustering

The k-means algorithm has been publicized to be effective in producing respectable clustering results for many practical applications including EN. As an unsupervised clustering method, K-Means follow the following four steps before it reaches the optimum decision: (1) Randomly Assign each sample to one of the clusters $k = 1,... K$. (2) Calculate the means of each of the pre-defined clusters with the following formula:

$$\mu_k = \frac{1}{N_k} \sum_{z_i \in C_k} z_i \tag{3}$$

(3) Considering the N data points, it reassign each object z_i to the cluster with the closest mean m_k. and (4) Repeat step 2 until the means of the clusters do not change anymore (Heijden et al., 2004).

The four datasets were subjected to k-means technique so the differences between clusters could be investigated based on each DPI. This algorithm managed to clarify three categories especially in the 6 DPI dataset allocated to the healthy, powdery mildew and spider mite infected plants. The centroids are visibly separated favorably (Figure 6).

Support Vector Machine (SVM)

In the last decade, a new classification and regression technique called SVM which is based on Statistical Learning Theory (SLT) and has been proposed in the wide machine learning field and have been successfully applied to a number of problems ranging from face identification and text categorization to bioinformatics and data mining (Pardo & Sberveglieri, 2005).

The SVM, as a supervised classification technique, has been broadly used to process and

Figure 6. K-means clustering: 4, 6, 8 and 9 days post infection

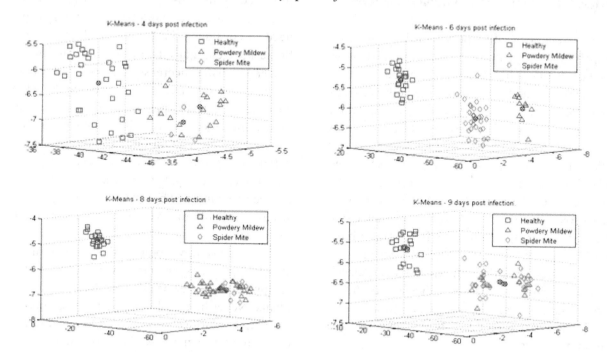

analyze the EN datasets (Pardo et al., 2005; Du & Sun, 2005). The objective of the classifier is to find optimal hyperplane for separating clusters in the non-linearly separable context (Distante, Ancona, & Siciliano, 2003).

In this study, the SVM with two kernels was applied on a 6 DPI dataset. As well as polynomial kernels, the Gaussian kernel has been employed.

Our SVM was able to classify the dataset with 89% performance rate when a polynomial kernel was used. With the Gaussian kernel however, a 98% classification was achieved. SVM has shown a great ability in discriminating between healthy and diseased tomato plants. Figure 7 illustrated the perfect decision boundaries separating healthy and diseased plants.

Figure 7. 6 DPI - Two support vector classifiers. Left: Polynomial kernel with 89% correct Classification - Right: Gaussian kernel with 98% correct Classification

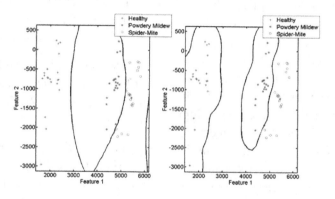

DISCUSSIONS AND CONCLUSION

In this chapter, we attempted to evaluate the EN's capability in discriminating plants VOC samples and analyzing it by several PR methods.

Statistical and ANN based methods were applied on 4 datasets gathered by the EN and their performances were reviewed. LDA and QDA managed to separate the samples with 97% and 98% success rate respectively. The ANN based classifiers were capable of discriminating the dataset with the relatively lower performance. Two ANN based network were constructed and trained using a Parzen and RBF algorithms. RBF managed to classify the 6 DPI dataset with 88% correct classification while Parzen classifier categorised the dataset with a slightly better performance (94%). The modest performance of ANN classifier is partially due to the small size of training dataset.

It was evident that the methods have all exposed a better performance when classifying 6 DPI and 8 DPI datasets. From these results, we can conclude that Powdery Mildew and Spider-Mites infected tomato plants can be discriminated from the healthy plant by 6 DPI. It is also clear that the powdery mildew disease had a major effect on the VOCs emitted from the plant. After 6 days of infection there were no gross visual changes on the leaves either from powdery mildew or spider mite plants. However, EN was able to discriminate between them.

SVM and QDA managed to discriminate between the healthy and infected plant with the highest performance rate (98%) and can be a perfect choice to be integrated with EN.

This chapter attempted to explain the capabilities of few supervised and unsupervised PR techniques in classifying a sensory dataset. Each technique offers a unique ability in categorizing the data but clearly they all have their own drawbacks. When designing an automated system, it is crucial to evaluate the overall performance of PR algorithm before being integrated into the proposed system. Reliability, classification rate and speed are few parameters that engineers need to consider prior to system development. Some PR techniques demand a high processing power which may decrease their popularity when the hardware resources are limited. On the other hand, few PR techniques require extensive preprocessing procedure as well as rapid calibration which make them an unsuitable candidate for an automated system. In overall, a fully computerized PR enabled system such as EN diagnosis tool can be improved by optimisation techniques widely available for every method discussed in this chapter.

In conclusion, we believe that PR classifiers and EN provides an attractive means of discrimination between healthy and diseased tomato plants. Nevertheless, due to the modest number of samples, further sampling and enhancement of methods can increase the classification rate as well as the overall performance of the system.

ACKNOWLEDGMENT

This project is co-funded by the European Commission, Directorate General for Research, within the 7th Framework Programme of RTD, Theme 2 – Biotechnology, Agriculture & Food.

REFERENCES

Agrios, G. (2004). Plant pathogens and disease: General introduction. In *Encyclopedia of Microbiology* (3rd ed.). (pp. 613-646).

André Kessler, I. T. (2001). Defensive function of herbivore-induced plant volatile emissions in nature. *Science, 291*, 2141–2144. doi:10.1126/science.291.5511.2141

Baratto, C., Faglia, G., Pardo, M., Vezzoli, M., Boarino, L., & Maffei, M. (2005). Monitoring plants health in greenhouse for space missions. *Sensors and Actuators. B, Chemical, 108*(1-2).

Berrueta, L. A., Alonso-Salces, R. M., & Heberger, K. (2007). Supervised pattern recognition in food analysis. *Journal of Chromatography. A*, 196–214. doi:10.1016/j.chroma.2007.05.024

Braun, R. C. (2009). Conidial germination patterns in powdery mildews. *Mycological Research, 113*(5), 616–636. doi:10.1016/j.mycres.2009.01.010

Camargo, A., & Smith, J. S. (2009). An image-processing based algorithm to automatically identify plant disease visual symptoms. *Biosystems Engineering, 102*(1), 9–21. doi:10.1016/j.biosystemseng.2008.09.030

Ciosek, P., & Wroblewski, W. (2006). The analysis of sensor array data with various pattern recognition techniques. *Sensors and Actuators. B, Chemical, 114*(1). doi:10.1016/j.snb.2005.04.008

Daqi, G., Shuyan, W., & Yan, J. (2004). An electronic nose and modular radial basis function network classifiers for recognizing multiple fragrant materials. *Sensors and Actuators. B, Chemical, 97*, 391–401. doi:10.1016/j.snb.2003.09.018

Distante, C., Ancona, N., & Siciliano, P. (2003). Support vector machines for olfactory signals recognition. *Sensors and Actuators. B, Chemical*, 30–39. doi:10.1016/S0925-4005(02)00306-4

Dragovic, S., & Onjia, A. (2007). Classification of soil samples according to geographic origin using gamma-ray spectrometry and pattern recognition methods. *Applied Radiation and Isotopes, 65*(2). doi:10.1016/j.apradiso.2006.07.005

Du, C.-J., & Sun, D.-W. (2005). Pizza sauce spread classification using colour vision and support vector machines. *Journal of Food Engineering, 66*(2). doi:10.1016/j.jfoodeng.2004.03.011

Dutta, R., Morgan, D., Baker, N., Gardner, J. & Hines, E. (2005). *Identification of Staphylococcus aureus infections (MRSA, MSSA and C-NS) in hospital environments: Electronic nose based approach.*

Escuder-Gilabert, M.P. (2009). *A 21st century technique for food control: Electronic noses.*

Garcia-Laencina, P. J., Sancho-Gomez, J. L., Figueiras-Vidal, A. R., & Verleysen, M. (2009). K nearest neighbours with mutual information for simultaneous classification and missing data imputation. *Neurocomputing, 72*(7-9). doi:10.1016/j.neucom.2008.11.026

Gardner, J.W., Hines, E. & Pang, C. (1996). *Detection of vapours and odours from a multisensor array using pattern recognition: self-organising adaptive resonance techniques.*

Hai, Z., & Wang, J. (2006). Electronic nose and data analysis for detection of maize oil adulteration in sesame oil. *Sensors and Actuators. B, Chemical, 119*(2). doi:10.1016/j.snb.2006.01.001

Heijden, F. v., Duin, R., Ridder, D. d., & Tax, D. (2004). *Classification, parameter estimation and state estimation.* The Netherlands: John Wiley & Sons Ltd. doi:10.1002/0470090154

Holopainen, J. K. (2004). Multiple functions of inducible plant volatiles. *Trends in Plant Science, 9*(11), 529–533. doi:10.1016/j.tplants.2004.09.006

Ian, T., & Baldwin, A. K. (2002). Volatile signaling in plant–plant–herbivore interactions: What is real? *Current Opinion in Plant Biology, 5*(4), 351–354. doi:10.1016/S1369-5266(02)00263-7

Jun Fu, G. L. (2007). A pattern recognition method for electronic noses based on an olfactory neural network. *Sensors and Actuators. B, Chemical*, 489–497.

Kalman, E.-L., Löfvendahl, A., Winquist, F., & Lundström, I. (1999). Classification of complex gas mixtures from automotive leather using an electronic nose. *Analytica Chimica Acta, 403*(1-2).

Merijn, R., & Kant, P. M. (2009). Plant volatiles in defence. *Advances in Botanical Research, 51*, 613–666. doi:10.1016/S0065-2296(09)51014-2

Moalemiyan, M., Vikram, A., & Kushalappa, A. C. (2007). *Detection and discrimination of two fungal diseases of mango (cv. Keitt) fruits based on volatile metabolite profiles using GC/MS.* (pp. 117–125).

Pardo, M., & Sberveglieri, G. (2005). Classification of electronic nose data with support vector machines. *Sensors and Actuators. B, Chemical*, 730–737. doi:10.1016/j.snb.2004.12.005

Parsons, N. R., Edmondson, R. N., & Song, Y. (2009). Image analysis and statistical modeling for measurement and quality assessment of ornamental horticulture crops in glasshouses. *Biosystems Engineering, 104*(2), 161–168. doi:10.1016/j.biosystemseng.2009.06.015

Paulsson, N., Larsson, E. & Winquist, F. (2000). *Extraction and selection of parameters for evaluation of breath alcohol measurement with an electronic nose.*

Peris, M., & Escuder-Gilabert, L. (2009). A 21st century technique for food control: Electronic noses. *Analytica Chimica Acta, 638*(1). doi:10.1016/j.aca.2009.02.009

Schütz, S., Weißbecker, B., & Hummel, H. E. (1996). Biosensor for volatiles released by damaged plants. *Biosensors & Bioelectronics, 11*(4), 427–433. doi:10.1016/0956-5663(96)82738-2

Theodoridis, S., & Koutroumbas, K. (2006). *Pattern recognition.* Academic Press.

Wang, Y., Wang, J., Zhou, B., & Lu, Q. (2009). Monitoring storage time and quality attribute of egg based on electronic nose. *Analytica Chimica Acta, 650*(2).

Yingquan, W. (2002). Improved k-nearest neighbor classification. *Pattern Recognition, 35*(10).

Zhang, H., Chang, M., Wang, J. & Ye, S. (2008). *Evaluation of peach quality indices using an electronic nose by MLR, QPST and BP network.*

Zhang, H., & Wang, J. (2007). Discrimination of LongJing green-tea grade by electronic nose. *Sensors and Actuators. B, Chemical, 122*(1).

Zhang, H., & Wang, J. (2007). Detection of age and insect damage incurred by wheat, with an electronic nose. *Journal of Stored Products Research, 43*(4), 489–495. doi:10.1016/j.jspr.2007.01.004

Zhang, Q. (2008). Sensory analyses of Chinese vinegars using an electronic nose. *Sensors and Actuators. B, Chemical, 128*(2). doi:10.1016/j.snb.2007.07.058

Chapter 8
Advanced Signal Processing Techniques in Non-Destructive Testing

Ali Al-Ataby
University of Liverpool, UK

Waleed Al-Nuaimy
University of Liverpool, UK

ABSTRACT

Non-destructive testing (NDT) is commonly used to monitor the quantitative safety critical aspects of manufactured components and forms one part of quality assurance (QA) procedures. The strategic trend in the development of NDT has changed towards the issue of safety in the broadest sense, to the protection of the population and the environment against man-made and natural disasters. Industrial NDT of manufactured items is usually undertaken when a product is likely to be placed under extreme or long periods of stress or wear, or if any component failure is liable to result in a major incident. While the emphasis in NDT has long been on the hardware technology, there has been an increased realisation of the potential benefits of applying advanced signal processing techniques to the signals resulting from an NDT examination.

This chapter describes some recent advances in signal processing as applied to NDT problems. This is an area that has made progress for over twenty years and its importance is gaining attention gradually, especially since the new advanced techniques in signal processing and pattern recognition.

DOI: 10.4018/978-1-60960-477-6.ch008

INTRODUCTION

NDT is a technique widely used in industry to detect, size, classify and evaluate different types of defects in materials, and it plays an important role whenever the integrity and safe operation of engineered components and structures are critical. Efficient and reliable NDT techniques are essential to ensure the safe operation of complex parts and construction in an industrial environment for evaluating service life, acceptability, verification and validation and risk. Automating the evaluation and inspection process can potentially lead to a reduction or elimination of the impact of human error, thus making the inspection process more reliable, reproducible, and faster. The most widely used conventional NDT techniques are ultrasonics, radiography, infrared thermography and computed tomography (CT) techniques.

NDT is not a direct measurement method, thus the nature and size of defects must be obtained through analysis of the signals obtained from inspection. Signal processing has provided powerful techniques to elicit information on defect detection, sizing, positioning and characterisation. Inspection signals were in general 1-D, utilising the very basic signal processing techniques. In case of 2-D inspection signals (images), the main processing methods include operations like image restoration and enhancement, morphological operators, Wavelet transforms, image segmentation, as well as object and pattern recognition. These methods facilitate the extraction of special information from the original images, which would not, otherwise, be obtainable. Moreover, 3-D image processing can provide advance information if an image sequence is available. Currently, NDT techniques have developed greatly due to recent advances in microelectronic systems and signal and image processing and analysis. Many image processing and analysis techniques can now be readily applied at standard video rates, in particular, to methods that generate image sequence (TV-type), such as real-time radiography, ultrasonic-phased array,

laser ultrasonics, pulse-video thermography and shearography.

Signal processing for NDT has many different approaches that may not be well correlated. The role of computational intelligence (CI) methodologies for NDT applications is vital and is expected to receive more attention in the future. Advanced signal processing techniques, such as Wavelet transform and independent component analysis (ICA), broadly listed as CI methods, can be used in solving NDT problems such as feature extraction, de-noising and the identification of defects. Perhaps the most influential signal processing development for NDT is split spectrum processing (SSP), which is one of the important and powerful treatments to automatically detect multiple flaws embedded in non-stationary grain noise. The nonlinearities of the SSP algorithm effectively change the flaw and grain echo distribution to enhance the separation of their amplitudes beyond that of simple envelope detection technique. Ultrasonic non-destructive characterisation of thin layered composite materials or structures can be difficult because the reflected signals are highly overlapped. The classical signal processing approach would have difficulty separating the layers and determining the thickness of each layer, necessitating the need for advanced approaches (Cacciola, Morabito & Versaci, 2007).

Advanced signal processing techniques allow extracting of information not easily available from the NDT measurements and thus essentially extend the resolution of the measurement beyond what is offered by the physical system. These techniques are expected to address the main problems in NDT when performing the inspection or examination. Mainly, these problems are: high levels of noise, high reflectivity of the material under test, defects orientation, high attenuations and/or low amplitudes of the received inspection signals, cladding thickness effect, grain structure of the material under test and the low accuracy of sizing, positioning and characterisation of flaws, among others.

This chapter discusses some advanced signal processing approaches able to emphasise features in NDT signals. The chapter also illustrates a suggestion for an automatic NDT interpretation system.

Background

For over sixty years, the NDT of materials has been an area of continued growth. The need for NDT has increased dramatically in recent years for various reasons such as product safety, in-line diagnostics, quality control, health monitoring, and security testing. Besides the practical demands, the progress in NDT has a lot to do with its inter-disciplinary nature. NDT is an area closely linked to the engineering disciplines of aerospace, civil, electrical, material science, mechanical, nuclear and petroleum, among others. There are at least two dozen of NDT methods in use. In fact, any sensor that can examine the inside of material non-destructively is useful for NDT.

Examples of NDT testing techniques are: magnetic particle, ultrasonic, Eddy current and radiographic. No single NDT method works for all flaw detection or measurement applications. Each of the methods has its own advantages and disadvantages when compared to other methods (Cartz, 1995).

In magnetic particle testing for example, a magnetic field is established in a component made from ferromagnetic material. The magnetic lines of force travel through the material and exit and re-enter the material at the poles. Defects such as crack or voids cannot support as much flux, and force some of the flux outside of the part. Magnetic particles distributed over the component are attracted to areas of flux leakage and produce a visible indication.

For the case of ultrasonic testing, high frequency sound waves are sent into a material by the use of a transducer. The sound waves travel through the material and are received by the same transducer or a second transducer. The amount of energy transmitted or received and the time the energy is received are analysed to determine the presence of flaws. Changes in material thickness and properties can also be measured (Popovics, Bilgutay, Karaoguz & Akgul, 2000).

In Eddy current testing, alternating electrical current is passed through a coil producing a magnetic field. When the coil is placed near a conductive material, the changing magnetic field induces current flow in the material. These currents travel in closed loops and are called Eddy currents. Eddy currents produce their own magnetic field that can be measured and used to find flaws and characterise conductivity, permeability, and dimensional features.

Radiographic testing is another well-known technique in the NDT field. X-rays (or Gamma-rays) are used to produce images of objects using film or other detector that is sensitive to radiation. The test object is placed between the radiation source and detector. The thickness and the density of the material that X-rays must penetrate affect the amount of radiation reaching the detector. This variation in radiation produces an image on the detector that often shows internal features of the test object.

Ultrasonic NDT methods are the most popular because of its capability in detecting and sizing cracks in a wide variety of locations and orientations in many materials used in engineering and even for considerable thickness of material (e.g., greater than 300 mm in steel) (Charlesworth & Temple, 2001). Other features of ultrasonic testing are flexibility, safety, and relative cost effectiveness. Ultrasonic testing can be quantitative and non-invasive or minimally invasive. Threshold-specific performance measures such as sensitivity and specificity, and threshold-independent performance measures based on the area under the receiver operating characteristics (ROC) curve can be used for computer-based detection. Recently, there has been increased interest in ROC analysis for NDT problems in general (Krautkramer, & Krautkramer, 1990).

Sonic infrared imaging is an emerging inspection technology that employs short pulses of ultrasound, typically less than a second, via a 20–40 kHz ultrasonic transducer. The increase in temperature in the vicinity of cracks in the components being tested can then be detected with an infrared camera (Han, 2007).

Guided waves play a strong role in NDT particularly for identifying cracks or defects in industrial applications for judging the integrity of structures. Recently, interest has arisen in using NDT to inspect the structural integrity of bars embedded in a surrounding material, or of curved pipelines. Elastic wave theory has been used extensively to tackle the underlying difficulties in the application of guided waves by using numerical (finite elements), experimental and semi-analytical methods (ray-theory) to describe the effects of curvature upon guided elastic waves (Zhu, Rose, Barshinger and Agarwala, 1998).

Many other advanced methods of NDT have also been well developed in recent years, such as the use of terahertz waveforms, and the thermography method. Terahertz waveforms represent a new method for NDT. The pulsed terahertz technology is an excellent example of emerging techniques that provide NDT of difficult-to-inspect material such as the sprayed-on foam insulation of the space shuttle external tank. Terahertz NDT is useful also for metallic surface roughness evaluation and characterisation of corrosion under paint and corrosion under tiles (Anastasi, Madaras, Seebo & Winfree, 2007).

Signal processing can help human to get a better visualization of the damage. Filtering techniques aimed at improving the damage visualization capabilities have been studied comprehensively. Frequency domain methods are very common in this regard, where the presence of reflections associated with damage can be easily observed and separated from the main signal generated for inspection. Signal enhancement and de-noising were studied and applied in frequency domain as well, using Fourier transform (FT), where signals and filter impulse response are decomposed into a sum of sinusoids with varying frequencies, amplitudes and phases. FT duality property makes the temporal and frequency signal descriptions needful and insufficient because the information are displayed in a complementary way and the display is rather far from the exploitable physical reality. These limitations have been overcome by the use of the Wavelet transform (WT), which has the characteristics of a pass band filter bank. Hence, because the energies of an NDT signals are concentrated in a frequency band, all other frequencies are represented by very low amplitudes in the transform domain, and can be scattered without any loss of information (Abbate, Koay, Frankel, Schroeder, & Das, 1997). The enhancement of the flaw detection performance and noise suppression has been verified experimentally for several types of artificial and natural flaws (Al-Ataby, Al-Nuaimy, Brett, & Zahran, 2009).

ADVANCED SIGNAL PROCESSING APPROACHES ABLE TO EMPHASISE

Features in NDT Signals

Signal processing techniques are nowadays used to emphasise peculiar features of one- and two-dimensional signals. The referred techniques involve, for instance, extraction of statistical features, but they are not sufficient to describe the signal trend in many cases, above all when the signal is produced by a non-linear phenomenon. Thus, advanced signal processing must be introduced to extract maximum information content from the reading, or else to reduce complexity of the extracted data. The following are some techniques useful to achieve these aims.

Multi-Resolution Analysis Using Wavelet Transform

There are several tools for signal processing, Fourier Transform is one of the most important. The Fourier Transform consists of breaking down a signal into constituent sinusoids of different frequencies, and this is enough for the characterisation of a number of signals. However, Fourier analysis has a serious drawback. When transformed to the frequency domain, time information is lost. If a signal does not change much over time, that is, if it is what is called a stationary signal, then this drawback is not very important. However, most NDT signals contain numerous non-stationary or transitory characteristics. These characteristics are often the most important part of the signal, and Fourier analysis is not suited to detecting them (Rallabandi, 2008).

The Wavelet Transform was developed especially to overcome these deficiencies. It is a windowing technique with variable-sized regions, which allows the use of long time intervals where more precise low frequency information is required, and shorter regions where high frequency information is required (Rallabandi, 2008; Graps, 1995).

In mathematics, Wavelets, Wavelet analysis, and the Wavelet transform refer to the representation of a signal in terms of a finite length or fast decaying oscillating waveform (known as the mother Wavelet). This waveform is scaled and translated to match the input signal (Hubbard, 1998). In this way, it is possible to split local and global dynamics of a signal (i.e. low- and high-frequency contribute in a frequency domain) by a multi-resolution (or multi-scale) analysis (MRA) in the Wavelet domain, less sensitive to noise than Fourier transform (Chui, 1992). MRA through Wavelet transform enables decomposition of a signal into different frequency subbands, similar to the way the human visual system operates. This property makes it especially suitable for the segmentation and classification of signals (Sun & Huang, 2004; Huang & Aviyente, 2008). MRA can be applied to analyse both one- and two-dimensional signals, in order to retrieve features useful to characterise relative trends into the time-, spatial- or frequency-domain. The global dynamics of a NDT-related signal $f(x)$ can be condensed in the Wavelet approximation coefficient (WACs) at the higher multi-resolution level J; if the inspected signal is measured in time- or spatial domain then it is possible to affirm that WACs are related to the low frequencies. On the other hand, local oscillations of $f(x)$ are depicted in a set of so called Wavelet details coefficients (WDCs) at the different scales $j = 1, 2, ..., J$; they are related to the high frequencies of a time- or spatial-domain signal. Moreover, it should be remarked that noising effects such as lift-off are restricted to low-frequencies; therefore, a suitable MRA could be useful in NDT. In formal terms, the above representation is a Wavelet series, which is the coordinate representation of a square integrable function with respect to a complete, orthonormal set of basic functions for the Hilbert space of square integrable functions. By selecting the optimal basis function needed to project the NDT-related signal, Wavelet transform can guarantee the possibility of specifying the key signal features that would be used in further processing steps (Cacciola et al., 2007).

Wavelet transform, and hence MRA, can be utilised (in conjunction with other techniques) in the area of de-noising and enhancement of inspection signal automatic detection and classification of weld flaws (Al-Ataby, et al., 2009; Kaya, Bilgutay & Murthy, 1994). Other examples are the application of Wavelet analysis in crack identification of frame structures and of beams. (Vafadar & Saroukhani, 2009; Saroukhani & Vafadari, 2009).

Independent Component Analysis

In NDT, it is possible to find such problems involving mixed signals and blind source separation, with special reference to noise reduction in both Eddy current and ultrasonic techniques. For the related non-linear phenomena, and for non-Gaussian signals (i.e., the lift-off effect in Eddy current technique), it is very difficult to separate the informative signal from the noise. In this case, blind source separation techniques and independent component analysis (ICA) could be an interesting aid to NDT field. The general framework for ICA was introduced in 1986. In 1995, a fast and efficient ICA algorithm based on infomax was introduced (Hyvarinen, Karhunen & Oja, 2001). In 1997, infomax ICA was improved by using the natural gradient. However, the original infomax ICA algorithm with sigmoidal nonlinearities was only suitable for super-Gaussian sources (Morabito, 2000). Let us consider a signal $s(.)$ in time or spatial domain, which is registered by j receivers and is the result of mixing of j sources, such that $s(.) = \{s_1(.), s_2(.), ..., s_j(.)\}$, where $s_k(.) = \sum_{h=1}^{j} a_{kh} X_k(.)$, $X_h(.)$ being the h-th source. Under the assumption of mutual statistical independence of the non-Gaussian source signals, it is possible to recover the set of j sources by calculating a suitable mixing matrix A, i.e. the matrix with elements a_{kh}. Each mixture s_k as well as each independent component (IC) x_h is a random variable. Mixtures must have zero mean and unitary variation (whitening process). Once A is calculated, it is possible to obtain its inverse A^{-1} and retrieve ICs having non-Gaussian distributions. There are many algorithms available in literatures which implement ICA (Simone & Morabito, 2000). An often used one (especially in industrial applications) is the FastICA algorithm, based on the fixed-point algorithm. Since it is possible to demonstrate that ICs must be non-Gaussian, the FastICA exploits kurtosis as cost function measuring the non-Gaussianity. But, since kurtosis is not the best parameter to evaluate the non-Gaussianity of the ICs, FastICA algorithm uses also the concept of differential en-

tropy H, also known in this framework as negentropy $\hat{H}$. It is so defined: $\hat{H}(y) = H(y_{Gauss}) - H(y)$, where y_{Gauss} is a Gaussian random variable of the same covariance matrix as y (Cacciola, et al., 2007). ICA can be also considered as an extension of the well-known principal components analysis (PCA) (Morabito, 2000). ICA/PCA is very relevant in the area of feature extraction/reduction that is applied to flaw classification problems for different NDT techniques.

Computational Intelligence Techniques to Solve Inverse Problems

Computational intelligence (CI) techniques, like artificial neural networks (ANNs), fuzzy logic (FL) and evolutionary computation, are nowadays widespread in engineering. Starting from large database, they can give the key to solve ill-posed inverse problems, by optimising the mapping between physical quantities, or by minimising suitable functional as in the case of genetic algorithms (GAs) or particle swarm optimisation (PSO). As far as NDT is concerned, CI offers excellent capabilities in the field of flaw detection and characterisation.

The following paragraphs give a brief theoretical description of some computational intelligence techniques relevant to NDT purposes.

Fuzzy Inference System and Adaptive Neuro-Fuzzy Inference System

The fuzzy inference systems (FISs) and the adaptive neuro-fuzzy inference systems (ANFISs) are popular computing frameworks based on the concepts of fuzzy sets theory, fuzzy if-then rules and fuzzy reasoning. FISs could be useful in order to implement algorithms able to solve inverse problems based on experts' suggestions. The basic structure of a fuzzy system consists of three conceptual components: a rule base, which contains a selection of fuzzy rules; a database or dictionary, which defines the membership functions used in the fuzzy rules and also called fuzzy membership functions (FMFs); a reasoning mechanism, which performs the inference procedure to derive a reasonable output or conclusion (Zadeh, 1996). A FIS implements a nonlinear mapping from its input space to output space. The inputs of the procedure are considered as fuzzy variables, and characterised by FMFs. Each FMF expresses a membership measure for each of the linguistic properties. FMFs are usually scaled between zero and unity, and they overlap.

The method usually used to build a FIS, i.e. the fuzzy modelling, defines the optimal number of clusters and their initial values for initialising an iterative optimisation algorithm that minimises a suitable cost function in order to reach a high quality representation of the original data. Once the optimal cluster configuration has been selected, the set of input/output data can be used for model identification, by considering each cluster centre as fuzzy rule. The ANFIS is a way of applying to FISs learning techniques, as it can be found in NNs literature (Jang & Sun, 1993).

FL can also be interpreted as a pattern classification problem-solver (a pattern defect to belong to different classes at the same time, according to the concept of partial membership $0 \leq u_{jk} \leq 1$). According to Kosko (1997), the concept of fuzzy entropy (FE), H, can be introduced as follows: $H = \sum_{k=1}^{N} u_{jk} \ln(u_{jk})$, where $\ln(u_{jk}) = 0$ when $u_{jk} = 0$. A typical fuzzy operator, called subsethood operator (SO), measures how a particular pattern belongs to a given fuzzy set; therefore, $0 \leq SO \leq 1$. This operator can be helpful for the discrimination of middle-depth defect in metallic plates (Cacciola, et al., 2007). FIS can be used in automatic flaw detection for flaws that are embedded inside material structure or noise, by providing excellent tools of image segmentation (Al-Ataby, et al., 2009). Other potential application is the classification of embedded flaws (Shekhar, Shitole, Zahran & Al-Nuaimy, 2006).

Figure 1. SVM classification concept (a) Evaluation of an optimal hyperplane; (b) Feature mapping (non-linear hyperplane becomes linear)

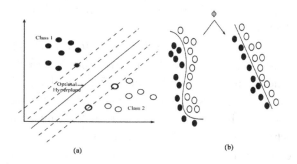

Support Vector Machines

Support vector machines (SVMs) are another useful soft computing technique applicable to the solution of ill-posed classification or regression problems. SVMs were introduced by Vapnik (1995) within the framework of the statistical learning theory (SLT), which describes the principle of structural risk minimisation (SRM). It has been demonstrated to be more effective than the traditional empirical risk minimisation (ERM) principle employed by artificial neural networks (Vapnik, 1995). A SVM-based classifier attempts to locate a hyperplane that maximises the distance from the members of each class to the same hyperplane (see Figure 1 (a)). Due to SRM, the error probability is upper bounded by a quantity depending by both the error rate achieved on the training set and a measure of the "richness" of the set of decision functions it can implement (named "capacity", or Vapnik-Chervonenkis dimension). SVMs can be also applied to regression problems (Vapnik, 1995).

The concepts of SVM are not limited to a linear classification case. These are generalisable to a nonlinear case where a mapping function Φ is used to map the input space into a higher dimensional feature space such that the nonlinear hyperplane becomes linear (Figure 1 (b)). To avoid the increased computational complexity and curse of dimensionality, a kernel-trick or kernel function is employed, which, in essence, computes an equivalent kernel value in the input space such that no explicit mapping is required (Vapnik, 1995).

SVM classifier gives high classification rates with problems related to classification of defects when compared to ANN classification. The use of the SVM classifier is robust and promising specifically when there is lack of training data which appeared to be not a severe problem (as compared with other classifiers) (Al-Ataby, et al., 2009; Cacciola, et al., 2007).

Particle Swarm Optimisation

Particle swarm optimisation (PSO) is a population based stochastic optimisation technique, inspired by social behaviour of bird flocking and used with nonlinear continuous functions. PSO-based systems learn from the scenario how to solve the optimisation problems (Eberhart & Kennedy, 1995). In PSO, each single solution is defined as a particle in the multidimensional search space. The PSO algorithm evolves a population of particles called swarm. All of the particles have fitness values which are evaluated by the fitness function to be optimised, and have velocities which

direct the convergence of the particles. During iterations, each particle adjusts its position making use of the best position encountered by itself (the so called *pbest*) and its neighbours (*gbest*). Each particle updates its current velocity and position in the search space using historical information regarding its own previous best position as well as the best position discovered by all other particles or neighbouring particles (Cacciola et al., 2007). PSO can be used in flaw classification problems through training ANNs.

Signal Processing and Flaw Reconstruction Techniques in NDT

In NDT of structural material defects, the size, shape and orientation are important flaw parameters in structural integrity assessment. To illustrate flaw reconstruction, a multi-viewing transducer system is needed along with signal processing techniques. A single probe moved sequentially to achieve different perspectives would work equally as well.

Figure 2 is an example of multi-probe system consisting of a sparse array of seven unfocused immersion transducers. This system can be used to "focus" onto a target flaw in a solid by refraction at the surface. The six perimeter transducers are

Figure 2. Flaw reconstruction scheme

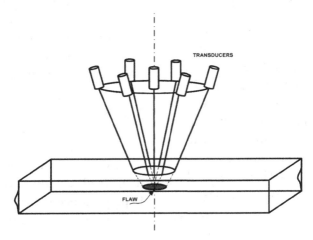

equally spaced, surrounding a centre transducer. Each of the six perimeter transducers may be independently moved along its axis to allow an equalisation of the propagation time for any pitch-catch or pulse-echo combinations. The axis of the aperture cone of the transducer assembly normally remains vertical and perpendicular to the part surface. The example used here employs the concept of ultrasonic transducers (Cartz, 1995).

The data acquisition and signal processing system in this case has four basic steps:

1. The first step involves the experimental setup, the location and focusing on a target flaw, and acquisition (in a predetermined pattern) of pitch-catch and pulse-echo back-scatter waveforms.

2. The second step employs a measurement model to correct the backscatter wave-forms for effects of attenuation, diffraction, interface losses, and transducer character-istics, thus resulting in absolute scattering amplitudes.

3. The third step employs a one-dimensional inverse approximation to extract a tangent plane to centroid radius estimate for each of the scattering amplitudes.

4. The fourth step estimates the radius, and their corresponding look angles are used in a regression analysis technique to determine the six ellipsoidal parameters, three semi-axes, and three Euler angles, defining an ellipsoid which best fits the data.

The inverse approximation sizes the flaw by computing the characteristic function of the flaw (defined as unity inside the flaw and zero outside the flaw) as a Fourier transform of the ultrasonic scattering amplitude. The one-dimensional inverse algorithm treats scattering data in each interrogation direction independently and has been shown to yield the size of ellipsoidal flaws (both voids and inclusions) in terms of the distance from the centre of the flaw to the wavefront that is tangent to the front surface of the flaw. Using the multi-probe ultrasonic system, the 1-D inverse technique is used to reconstruct voids and inclusions that can be reasonably approximated by an equivalent el-lipsoid. The angular scan method described here is capable of locating the bisecting symmetry planes of a flaw. The utility of the multi-probe system is, therefore, expanded since two-dimensional elliptic reconstruction may now be made for the central slice. Additionally, the multi-probe system is well suited for the 3-D flaw reconstruction technique using 2-D slices.

The model-based reconstruction method has been previously applied to voids and incursion flaws in solids. Since the least-squares regression analysis leading to the "best fit" ellipsoid is based on the tangent plane to centroid distances for the interrogation directions confined within a finite aperture. The success of reconstruction depends on the extent of the flaw surface "illuminated" by the various viewing directions. The extent of coverage of the flaw surface by the tangent plane is a function of the aperture size, flaw shape, and the flaw orientation. For example, a prolate spheroidal flaw with a large aspect ratio oriented along the axis of the aperture cone has only have one tip illuminated (i.e., covered by the tangent planes) and afford a low reconstruction reliability. For the same reason, orientation of the flaw also has a strong effect on the reconstruction accuracy (Cartz, 1995).

Figure 3 shows the difference in surface cover-age of a tilted flaw and an untilted flaw subjected to the same insonification aperture. From a flaw reconstruction standpoint, an oblate spheroid with its axis of rotational symmetry perpendicular to the part surface represents a high leverage situation. Likewise, a prolate spheroid with its symmetry axis parallel to the part surface also affords an easier reconstruction than a tilted prolate spheroid.

The orientation of a flaw affects reconstruction results in the following ways (Cartz, 1995):

Figure 3. Surface coverage of a (a) tilted flaw; and (b) an untilted flaw

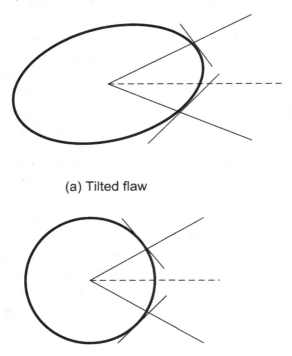

(a) Tilted flaw

(b) Untilted flaw

1. For a given finite aperture, a change in flaw orientation changes the insonified surface area and hence changes the "leverage" for reconstruction.
2. The scattering signal amplitude and the SNR for any given interrogation direction depends on the flaw orientation.
3. Interference effects, such as those due to tip diffraction phenomena or flash points may be present at certain orientations.

Signal Processing for Automatic NDT Detection, Sizing and Classification of Flaws

This following section shows the concept of an automatic system that performs detecting, sizing and classification of flaws. The discussion is mainly applied to ultrasonic NDT technique (more specifically, the time-of-flight diffraction (TOFD) method), but can be generalised to other NDT techniques after taking into consideration the merits of each technique and the nature and dimension of the inspection signal. The suggested system uses some of the previously discussed signal processing techniques in conjunction with other advanced methods.

Although the data acquisition configuration lends itself conveniently to automation, and methods such as robotic scanning and computer-conditioned data acquisition are routinely used, the critical stages of data processing and interpretation are still performed off-line manually depending heavily on the skill, experience, alertness and consistency of a trained operator. The NDT data acquisition and display configurations themselves introduce a host of errors that cannot be accounted for by manual interpretation. Increased industrial pressure to complete the interpretation of inspection data in near real-time has motivated researchers to develop computational tools capable of aiding the operator by automating the processing and interpretation (Zahran & Al-Nuaimy, 2005).

The automation of processing NDT data is an essential stage of a comprehensive inspection and interpretation aid. Signal and image processing tools have been specifically developed for use with data and adapted to function autonomously without the need for continuous intervention through automatic configuration of the critical interpretation parameters according to the nature of the data and the data acquisition settings. This section suggests several automation methods based on MRA through Wavelet transform and texture analysis for de-noising and enhancing the quality of the collected data to help in automatic and accurate detection, sizing, positioning and classification of weld defects. The automatic classification is implemented via the support vector machines (SVM) method which is considered faster and more accurate than artificial neural networks (ANN).

Pre-Processing and Data Quality Enhancement

In NDT, detection, sizing, positioning and classification of flaws are often made difficult by the superimposed noise due to the structure of the material and external noise sources. This noise can sometimes mask indications due to a small but potentially dangerous defect. On the other hand, noise can increase the false alarm rate by changing the envelope of the received signal at certain places, hence introducing false flaw signatures in the data. It is quite obvious that the quality of data has a major impact on successful NDT interpretation system, since all the subsequent processes utilise this data and the accuracy of the analysis is affected by the nature of the collected data. This, in fact, necessitates the need to intelligent and effective techniques to enhance data quality as possible without affecting or changing the original data. MRA is considered as a one powerful tool

to enhance SNR of ultrasonic NDT signal. For de-noising the NDT signal, MRA method based on the Wavelet transform is used. The Wavelet transform (as explained previously) is a MRA technique that can be used to obtain time-frequency representation of ultrasonic signal. Filtering procedure is based on decomposition of signal using DWT in N levels using band pass filtering and decimation to obtain the approximation and detail coefficients.

Next step is thresholding of detail coefficients and reconstruction of signal from detail and approximation coefficients using inverse transform (IDWT). The method is a generalisation of wavelet decomposition that offers a larger range of possibilities for signal analysis. In wavelet analysis, a signal is split into an approximation and detail coefficients. The approximation is then itself split into a second-level approximation and detail, and the process is repeated. In this analysis, the detail coefficients as well as the approximation coefficients can be split. Hard thresholding is used for thresholding detail coefficients. Hard thresholding can be described as the process of setting to zero the elements whose absolute values are lower than the threshold. For thresholding of detail coefficients, the local threshold value based on standard deviation is used (Al-Ataby, et al., 2009).

De-noising through adaptive Wiener filter technique is another suggested technique in this context. It applies a Wiener filter (a type of linear filter) to scan images adaptively, tailoring itself to the local image variance. Where the variance is large, the filter performs little smoothing, and when the variance is small, the filter performs more smoothing.

Other pre-processing procedures can be applied based on the nature of the used NDT method. These may include adding gain to the received inspection signal, drift correction, scan alignment and enhancing reference signals (e.g. lateral wave and backwall echo in TOFD) that are used for accurate sizing and position purposes, among others.

Automatic Flaw Detection

Discrete MRA along with texture analysis is used to select signal(s) (image(s) for ultrasonic TOFD method) to be used and processed in detection. The concept of image segmentation is used to segment and highlight the potential parts (defects in this case) in the scan image. Image segmentation can be done by statistical methods (e.g. variance thresholding) or through computational intelligence methods (e.g. fuzzy logic). The method mentioned below has shown a probability of detection that reaches up to 99.40% for the fully embedded defects in steel plates.

Discrete MRA-Wavelet Packet Transform

MRA-WPT is a signal analysis tool that has the frequency resolution power of the Fourier transform and the time resolution power of the Wavelet transform. It can be applied to time varying signals, where the Fourier transform does not produce useful results, and the Wavelet transform does not produce sufficient results. WPT can be considered as an extension to the DWT, which performs better reconstruction process than the discrete one (Robini, Magnin, Benoit & Baskurt, 1997). The detection process in the current work is based on MRA-WPT. First, the scan image is decomposed by the WPT. For each of the obtained images (4 per level), the statistical parameters of texture will be calculated. The image with the weakest textural information is discarded and the one with the highest statistical contents is selected. Next, another decomposition level runs to generate new set of images. The process of calculating statistical contents commences again, and so on. The process of decomposition proceeds until level 6 as the required coarse resolution is obtained and going further does not add any significance to the analysis. The final image is to be reconstructed from three images carrying the highest statistical

textural information. This image will be the crucial input for the next stages (segmentation).

Segmentation through Fuzzy Logic

The suggested segmentation technique is based on the use of fuzzy c-mean clustering classifier or the fuzzy c-mean iterative (FCMI) algorithm. It is critical to have some automatic method to highlight the defects in situations where it so complicated and difficult to detect and distinguish between adjacent defects by normal means (via trained operator or statistical methods). This algorithm is used to distinguish two classes or pattern space components: Flaw or background. The input to this algorithm is the final image after MRA-WPT. The algorithm will act as a binarisation operation, with black (pixel values set to zeros) and white (pixel values set to ones) output, where the black areas represent background and white areas represent flaws. This image (or mask) is used to segment the scan image after some further processing (namely, defect blob analysis).

Automatic Flaw Sizing and Positioning

The sizing and positioning techniques in the current work are mainly based on the standards and codes for the quantification of flaw sizes in ultrasonic inspection (e.g. the British standard BS or the European standard EN), which are increasingly used to aid the interpretation of NDT data. Sizing and positioning techniques are very dependent on the used method of NDT, and there is no general approach, hence each method is considered on individual basis. Some methods (which are considered less accurate) use the amplitude of the received inspection signal to estimate the size and position of the flaw, while others may use the time of flight of the diffracted signal from the tips of flaw to do that task. Signal processing techniques like successive sub-sampling or cross-correlation, along with other mathematical operations, can be applied on the model of sizing and positioning to obtain accurate measurements. An accuracy of ±1mm has been achieved in sizing and positioning of embedded flaws in steel plates.

Automatic Flaw Classification

Defects can be classified into broad categories according to their location, size, geometry and visual signature. Generally speaking, classes of defects can be categorised into either apparent or internal defects (Zahran, 2006). Apparent defects may be classified by using geometrical and phase information alone (can be done after sizing and positioning). Internal defects, which are clearly embedded and not open to or approaching either surfaces of the workpiece, require advanced visual processing for interpretation. The classification therefore can be done in two-stage process, whereby defects are first classified as belonging to one of the two categories.

Classification of internal defects requires computational intelligence techniques for classification. This classification is based on extracting a set of discriminating features between these defect classes which are applied to a data set representing the defect classes in question and then both the data set and the selected features form input to a SVM classifier to discriminate between internal defects (Al-Ataby, et al., 2009).

Classification of defects is a typical pattern classification problem in which computational intelligence turns out very useful. Heuristic machines do not have to suffer of the curse of dimensionality in order to work in real time. In this sense, it is possible to jointly use MRA-WPT and textural analysis (with PCA) as feature extractors, and SVM as classifier. Firstly, suitable MRA-WPT levels are carried out on each defect blob to generate the WAC's and WDC's. The considered WAC's or WDC's are subsequently reduced to less elements (these are actually form inputs of SVM) by means of PCA, cutting out all the principal components whose contributions to total variation of the whole set of these components are less than a threshold (2% was found to be satisfactory). WAC's of the defect blobs are selected as input candidates to the SVM classifier. Other texture analysis features can be added to WAC's to form the complete set of input features to the SVM. Candidates are variance, skewness, kurtosis, fifth

Figure 4. Output report of the suggested automatic system (for ultrasonic TOFD)

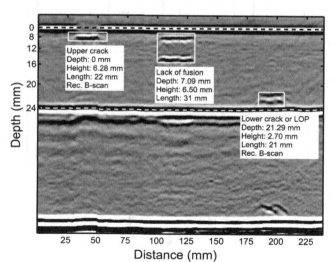

moment, maximum probability co-occurrence, homogeneity and contrast, among others.

Different defect samples must be used as the training patterns that should cover all the relevant flaws under consideration. Part of the available samples is used for training and the other part composes the test set. The output of the SVM is a number which codifies each kind of defect (for example, 1 for internal crack, 2 for lack of fusion, 3 for lack of penetration ...etc.). SVM can be trained by using a 2-degree polynomial kernel.

At the end of the training phase, SVM returns a particular matrix, so called confusion matrix (CM), which evaluates the goodness of a trained classifier. In fact, generally speaking, the element CM_{ij} of a CM is the probability that a single pattern belonging to the i-th class could be classified as belonging to the j-th class (sum of elements of each rows is therefore equal to 1). Thus, the more the CM is similar to the identity matrix, the better classification performances are. A classification rate of 98.90% was obtained after the application of the built SVM classifier on the fully embedded defects in steel plates.

Post-Processing

Post-processing applies general steps to integrate a comprehensive output for the entire stages by generating a user-friendly visual report which contains all the relevant information. The report contains suggestions and recommendations to be considered by the inspectors when they do scans in the future. Figure 4 shows an example of a report from the suggested automatic system, where the potential defects are highlighted with their corresponding names (classes), sizing and positioning details and recommendations to the inspectors to

Figure 5. Automatic interpretation system for ultrasonic TOFD NDT technique

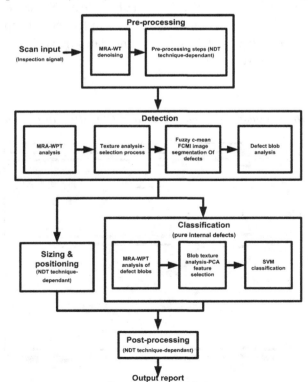

do different type of scans (for example, parallel B-scan) for more accurate results.

Post-processing can apply acceptance criteria to check whether the obtained results match the standards or not. Other steps may be included depending on the NDT method under consideration.

Figure 5 summarises the previous steps toward an automatic interpretation system for ultrasonic TOFD NDT technique.

FUTURE RESEARCH DIRECTIONS

Future trend will be focusing mainly on the utilisation of the computational intelligence methods in the area of NDT to expose the unseen details with less number of scans and less complicated practical approaches. This will make the developed NDT systems suitable for implementation in situations requiring near real-time processing of large volumes of data.

Also, as more than one sensor is likely to be used in NDT in the future, the basic question is how to best integrate the information from different sensors. Image fusion framework for interpreting NDT images is a powerful candidate in this regard.

The recent advancement of wireless sensor networks for NDT allows for distributed sensing and improved decision making by the utilisation the fibre-optic based sensor networks, which requires higher level of signal processing and automated processes.

Finally, the next generation of thick wall stainless steel plates and high-pressure, high-temperature pipes (that will be used in power generation field for example) will lead to difficult and complicated inspection procedures, hence the need for more advanced signal processing techniques to detect, size and characterise flaws.

CONCLUSION

The chapter has presented different signal processing techniques relevant to NDT field. Also, the chapter has introduced an automatic interpretation system used to detect, size and classify flaws for ultrasonic TOFD NDT technique.

The work in this area is still ongoing, and the coming years will carry further advanced work in NDT signal processing, as NDT is about to enter a new era (thick walls, high-pressure and high-temperature plates and pipes), which may lead to the fact that available techniques would be inappropriate to NDT. Of course, the solution in this case would be between the hands of advanced signal processing.

REFERENCES

Abbate, A., Koay, J., Frankel, J., Schroeder, S., & Das, P. (1997). Signal detection and noise suppression using a Wavelet transform signal processor: Application to ultrasonic flaw detection. *IEEE Transactions on Ultrasonics, Ferroelectrics, and Frequency Control, 44*(1), 14–26. doi:10.1109/58.585186

Al-Ataby, A., Al-Nuaimy, W., Brett, C. R., & Zahran, O. (2009). *Automatic detection and classification of weld flaws in TOFD data using Wavelet transform and support vector machines.* Non-Destructive Testing 2009 conference. Blackpool, UK: BINDT.

Anastasi, R., Madaras, E., Seebo, J., & Winfree, W. (2007). Terahertz NDE for aerospace applications. In Chen, C. H. (Ed.), *Ultrasonic and advanced methods for nondestructive testing and material characterization* (pp. 279–302). Dartmouth, RI: World Scientific. doi:10.1142/9789812770943_0012

Cacciola, M., Morabito, F. C., & Versaci, M. (2007). Computational intelligence methodologies for non-destructive testing/evaluation applications. In Chen, C. H. (Ed.), *Ultrasonic and advanced methods for nondestructive testing and material characterization* (pp. 493–516). Dartmouth, RI: World Scientific. doi:10.1142/9789812770943_0021

Cartz, L. (1995). *Nondestructive testing: Radiography, ultrasonics, liquid penetrant, magnetic particle, eddy current.* Illinois: ASM International.

Charlesworth, J. P., & Temple, J. (2001). *Engineering applications of ultrasonic time-of-flight diffraction* (2nd ed.). Baldock, UK: Research Studies Press.

Chui, C. K. (1992). *An introduction to wavelets.* San Diego: Academic Press.

Eberhart, R. C., & Kennedy, J. (1995). A new optimizer using particle swarm theory. *Proceedings of the Sixth International Symposium on Micromachine and Human Science*, (pp. 39-43). Japan: Nagoya.

Graps, A. L. (1995). An introduction to wavelets. *IEEE Computational Science & Engineering, 2*(2), 50–61. doi:10.1109/99.388960

Han, X. (2007). Sonic infrared imaging: A novel NDE technology for detection of cracks/delaminations/disbands in materials and structures. In Chen, C. H. (Ed.), *Ultrasonic and advanced methods for nondestructive testing and material characterization* (pp. 369–384). Dartmouth, RI: World Scientific. doi:10.1142/9789812770943_0016

Huang, D., & Aviyente, S. (2008). Wavelet feature selection for image classification. *IEEE Transactions on Image Processing, 17*(9), 1709–1720. doi:10.1109/TIP.2008.2001050

Hubbard, B. (1998). *The world according to wavelets: The story of a mathematical technique in the making.* Wellesley, MA: AK Peters.

Hyvarinen, A., Karhunen, J., & Oja, E. (2001). *Independent component analysis.* New York: John Wiley & Sons. doi:10.1002/0471221317

Jang, R., & Sun, C. (1993). Neuro-fuzzy modeling and control. *IEEE Transactions on Systems, Man, and Cybernetics, 23*(3), 378–406.

Kaya, K., Bilgutay, N., & Murthy, R. (1994). Flaw detection in stainless steel samples using wavelet decomposition. *Proceeding of the 1994 Ultrasonics Symposium, 2*(4), 1271-1274.

Kosko, B. (1997). *Fuzzy engineering.* New Jersey: Prentice Hall.

Krautkramer, J., & Krautkramer, H. (1990). *Ultrasonic testing of materials* (4th ed.). New York: Springer.

Morabito, F. C. (2000). Independent component analysis and feature extraction techniques for NDT data. *Materials Evaluation, 58*(1), 85–92.

Popovics, S., Bilgutay, N., Karaoguz, M., & Akgul, T. (2000). High-frequency ultrasound technique for testing concrete. *ACI Materials, 97*(1), 58–65.

Rallabandi, V. P. (2008). Enhancement of ultrasound images using stochastic resonance-based wavelet transform. *Computerized Medical Imaging and Graphics Journal, 32*(1), 316–320. doi:10.1016/j.compmedimag.2008.02.001

Robini, M., Magnin, I., Benoit, H., & Baskurt, A. (1997). Two-dimensional ultrasonic flaw detection based on the wavelet packet transform. *IEEE Transactions on Ultrasonics, Ferroelectrics, and Frequency Control, 44*(6), 1382–1394. doi:10.1109/58.656642

Saroukhani, S., & Vafadari, R. (2009). *Application of wavelet analysis in crack identification of beams.* Non-Destructive Testing 2009 conference. Blackpool, UK: BINDT.

Shekhar, C., Shitole, N., Zahran, O., & Al-Nuaimy, W. (2006). *Combining fuzzy logic and neural networks in classification of weld defects using ultrasonic time-of-flight diffraction*. Non-Destructive Testing 2006 conference. Derby, UK: BINDT.

Simone, G., & Morabito, F. C. (2000). ICA-NN based data fusion approach in ECT signal restoration. *Proceedings of the International Joint Conference on Neural Networks (IJCNN2000), 5*(1), 59-64.

Sun, B., & Huang, D. (2004). Texture classification based on support vector machine and Wavelet transform. *Proceedings of the 5th World Congress on Intelligent Control and Automation.* Hangzhou, China, (pp. 1862-1864).

Vafadari, R., & Saroukhani, S. (2009). *Application of wavelet analysis in crack identification of frame structures*. Non-Destructive Testing 2009 conference. Blackpool, UK: BINDT.

Vapnik, V. N. (1995). *The nature of statistical learning theory*. New York: Springer.

Zadeh, L. A. (1996). *Fuzzy sets, fuzzy logic, and fuzzy systems*. New Jersey: World Scientific Publishing.

Zahran, O. (2006). Automatic detection, sizing and characterisation of weld defects using ultrasonic time-of-flight diffraction. Unpublished doctoral dissertation, University of Liverpool, Liverpool.

Zahran, O. & Al-Nuaimy, W. (2005). Automatic data processing and defect detection in time-of-flight diffraction images using statistical techniques. *Insight, the Journal of the British Institute of Non-Destructive Testing, 47*(9), 538-542.

Zhu, W., Rose, J., Barshinger, J., & Agarwala, V. (1998). Ultrasonic guided wave NDT for hidden corrosion detection. *Research in Nondestructive Evaluation, 10*(4), 205–225.

ADDITIONAL READING

Al-Ataby, A. Al-Nuaimy and Zahran, O. (2010). Towards Automatic Flaw Sizing Using Ultrasonic Time-Of-Flight Diffraction, *The Seventh International Conference on Condition Monitoring and Machinery Failure Prevention Technologies CM2010/MFPT2010,* Stratford-upon-Avon:BINDT and MFPT.

Al-Nuaimy, W. and Zahran, O. (2005). Time-of-flight diffraction – from semi-automatic inspection to semi-automatic interpretation. *Insight - Non-Destructive Testing and Condition Monitoring (The Journal of the British Institute of Non-Destructive Testing), 47(*10), 639-644.

Shekhar, C. Shitole, N., Zahran, O. and Al-Nuaimy, W. (2007). Combining fuzzy logic and neural networks in classification of weld defects using ultrasonic time-of-flight diffraction. *Insight - Non-Destructive Testing and Condition Monitoring (The Journal of the British Institute of Non-Destructive Testing), 49*(2), 538-542.

Zahran, O. Shihab, S. and Al-Nuaimy, W. (2002). Recent developments in ultrasonic techniques for rail-track inspection. *Non-Destructive Testing 2002 conference.* United Kingdom.

Zahran, O. Al-Nuaimy, W. (2004). Automatic defect classification in time-of-flight-diffraction data using fuzzy logic, *Non-Destructive Testing 2004 conference*, United Kingdom.

Zahran, O. and Al-Nuaimy, W. (2004). Automatic segmentation of time-of-flight diffraction images using time-frequency techniques – application to rail-track defect detection. *Insight - Non-Destructive Testing and Condition Monitoring (The Journal of The British Institute of Non-Destructive Testing), 46*(6), 338-343.

Zahran, O., & Al-Nuaimy, W. (2004). Automatic classification of defects in time-of-flight-diffraction data. *The 16th World conference of NDT, WCNDT 2004.* Montreal, Canada.

Zahran, O., & Al-Nuaimy, W. (2004). *Rail-track inspection using time-of-flight diffraction*. United Kingdom: Railway Engineering.

Zahran, O. and Al-Nuaimy, W. (2005). Automatic data processing and defect detection in time-of-flight diffraction images using statistical techniques. *Insight - Non-Destructive Testing and Condition Monitoring (The Journal of the British Institute of Non-Destructive Testing), 47*(9), 538-542.

Zahran, O., & Al-Nuaimy, W. (2005). Automatic defect classification in time- of-flight-diffraction data using neural networks. *Non-Destructive Testing 2005 conference*, United Kingdom.

Zahran, O., & Al-Nuaimy, W. (2005). Utilising phase relationships for automatic weld flaw categorisation in time-of-flight diffraction images. *The 11th International conference on fracture ICF.* Turin, Italy.

Zahran, O., & Al-Nuaimy, W. (2006). Image Processing for Accurate Sizing of Weld Defects Using Ultrasonic Time-Of-Flight Diffraction. *European Federation for Non-Destructive Testing conference proceeding ECNDT2006.* Berlin, Germany.

Zahran, O., & Shihab, S. Al-Nuaimy, W. (2003). Automatic segmentation of time-of-flight-diffraction images using time-frequency techniques – application to rail track defect detection. *Non-Destructive Testing 2003 conference.* United Kingdom.

Zahran, O., Shihab, S. and Al-Nuaimy, W. (2002). Comparison between surface impulse ground penetrating radar signals and ultrasonic time-of-flight diffraction signals. *The 7th IEEE High Frequency Postgraduate Student Colloquium.* IEEE catalogue number: 02TH8642.

Zahran, O., Shihab, S., & Al-Nuaimy, W. (2003). Time-frequency techniques applied to TOFD for the automation of rail-track inspection. *The 2003 Railway Engineering Conference Proceeding,* United Kingdom.

Zahran, O., Shihab, S., & Al-Nuaimy, W. (2004). Discussion of the ability of defect classification in weld inspection using ultrasonic time-of-flight diffraction technique. *PREP 2004,* United Kingdom.

KEY TERMS AND DEFINITIONS

A-Scan: A method of data presentation on a cathode-ray tube (CRT) using a horizontal baseline that indicates distance, or time, and a vertical deflection from the baseline, which indicates amplitude.

Guided Waves Testing: An emerging NDT testing technology that can quickly survey a large area of a structure for defects and can provide comprehensive condition information by using relatively low-frequency (typically in the range of kHz). This technology performs 100% volumetric examination of large pipe areas detecting and locating both internal and external defects. In aboveground pipelines, tens of meters of pipeline can be inspected in both directions from a single sensor location.

Independent/Principal Component Analysis (ICA/PCA): ICA is a computational method for separating a multivariate signal into additive subcomponents supposing the mutual statistical independence of the non-Gaussian source signals. It is a special case of blind source separation. PCA involves a mathematical procedure that transforms a number of possibly correlated variables into a smaller number of uncorrelated variables called principal components (PC). The first principal component accounts for as much of the variability in the data as possible, and each succeeding component accounts for as much of the remaining variability as possible. In signal processing, ICA/PCA can be used for feature extraction/reduction purposes.

Multi-Resolution Analysis (MRA): A signal processing technique to split local and global dynamics for a signal into low- and high-frequency contribute in a frequency domain (in the wavelet domain). MRA is less sensitive to noise than Fourier Transform, and can be applied to the analysis of both one- and two- dimensional signals in order to retrieve features useful to characterise relative trend into the time-, spatial- or frequency-domain.

Non-Destructive Testing/Evaluation (NDT/ NDE): Is a wide group of analysis techniques used in science and industry to evaluate the properties of a material, component or system without causing damage. Because NDT does not permanently alter the article being inspected, it is a highly-valuable technique that can save both money and time in product evaluation, troubleshooting, and research. NDE is an interdisciplinary field of study which is concerned with the development of analysis techniques and measurement technologies for the quantitative characterisation of materials, tissues and structures by noninvasive means

Support Vector Machines (SVM): A set of related supervised learning methods used for classification and regression. Given a set of training samples, each marked as belonging to one of two categories, an SVM training algorithm builds a model that predicts whether a new sample falls into one category or the other. Intuitively, an SVM model is a representation of the samples as points in space, mapped so that the samples of the separate categories are divided by a clear gap that is as wide as possible. New samples are then mapped into that same space and predicted to belong to a category based on which side of the gap they fall on. SVM can be extended to multi-class problems.

Split Spectrum Processing (SSP): A frequency diversity approach, which involves subband filtering the inspection signals (A-scans) into ensembles of frequency diverse signals and applying novel processing schemes over the spectral bands. It is used when inspecting materials that composed of large-grained structures, which pose significant challenges for ultrasonic pulse-echo techniques because scattering from the grain microstructure often obscures material flaw echoes.

Time-of-Flight Diffraction (TOFD) Testing: NDT testing method that was developed in the 70's by the Atomic Energy Authority (AEA), UK. This method differs from traditional pulse echo technique in that it monitors diffracted signals at the edges of defects which are directly related to the true position and size of the defect, as opposed to the reflection on defects according to a reference reflector. The TOFD technique uses two probes in a transmitter-receiver arrangement. When sound is introduced into the material via the transmitter the defect will oscillate. Each defect edge works as a source point of ultrasound signals. These very weak signals are called diffracted waves and their appearance does not relate to the orientation of flat or spherical defects. These diffracted signals are received via the receiver probe.

Terahertz Waveform NDT Method: Is an emerging technology with significant potential for inspecting nonconductive aerospace materials for hidden flaws and damage. This new technology is potentially useful for inspections that have previously been identified as either extremely difficult or impossible with current inspection technologies. Characteristic of a difficult-to-inspect material is the Sprayed-On Foam Insulation of the Space Shuttle External Tank. Terahertz NDT/NDE system is able to detect voids and unbonds in materials. Other viable applications of terahertz technology include examination of metallic surface roughness, measurement of paint thickness, and corrosion detection under Space Shuttle tiles.

Ultrasonic: Pertaining to mechanical vibrations having frequency greater than approximately 20,000 Hz.

Ultrasonic Testing (UT): NDT testing method that uses high frequency sound energy to conduct examinations and make measurements. Ultrasonic inspection can be used for flaw detection/ evaluation, dimensional measurements, material characterization, and more. A typical UT inspection system consists of several functional units, such as the pulser/receiver, transducer, and display devices.

Wavelet Transform (WT): A signal processing transform that has the additional advantages (over Fourier transform) of preserving the temporal information after the transformation and to use wavelets as the basis function instead of sinusoids. Wavelets are periodic waves of short duration that allow better reproduction of a transient signal and to use different scales or resolutions. Extensions to t he Wavelet transform is the discrete Wavelet transform (DWT) and that Wavelet packet transform (WPT).

Chapter 9
Low Frequency Array (LOFAR) Potential and Challenges

Mark J. Bentum
ASTRON, The Netherlands & University of Twente, The Netherlands

André W. Gunst
ASTRON, The Netherlands

Albert Jan Boonstra
ASTRON, The Netherlands

ABSTRACT

The Low Frequency Array (LOFAR) is a large radio telescope based on phased array principles, distributed over several European countries with its central core in the Northern part of the Netherlands. LOFAR is optimized for detecting astronomical signals in the 30-80 MHz and 120-240 MHz frequency window. LOFAR detects the incoming radio signals by using an array of simple omni-directional antennas. The antennas are grouped in so called stations mainly to reduce the amount of data generated. More than forty stations will be built, mainly within a circle of 150 kilometres in diameter. But LOFAR stations will also be built in other European countries. The signals of all the stations are transported to the central processor facility, where all the station signals are correlated with each other, prior to imaging. In this chapter, the signal processing aspects on system level will be presented. Methods to image the sky will be given and the mapping of these concepts to the LOFAR phase array radio telescope will be presented. Challenges will be addressed, and potentials for further research will be presented.

INTRODUCTION

Astronomy is one of the oldest sciences in the world. In the early days, astronomers performed methodical observations by looking at the sky. It was the Dutch invention of the telescope in 1609,

used by Galileo Galilei, which brought astronomy into modern science. Observational astronomy was purely optical until the serendipitous discovery of radio emission from the center of our Galaxy by Karl Jansky in 1932 (Jansky, 1932). He built an antenna, designed to receive terrestrial radio waves at a frequency of 20.5 MHz. After recording signals from all directions, Jansky categorized

DOI: 10.4018/978-1-60960-477-6.ch009

them into three types of signals: static from nearby thunderstorms, static from distant thunderstorms, and a faint steady signal of unknown origin. This was the discovery of extra-terrestrial radio signals and in fact the start of radio astronomy science. It took some time before these results were taken seriously and before radio astronomy started to build new instruments.

After World-War-2 new radio-astronomical instruments were built all over the world. To improve important properties such as sensitivity and spatial resolution, telescopes became larger. Since resolution is linear with respect to wavelength, new telescopes also operated at much higher frequencies. Another reason for observing at higher frequencies is the occurrence of spectral lines, such as the Hydrogen emissions line at 1420.4 MHz predicted by van de Hulst in 1944, and detected shortly after. After the development of earth-rotating synthesis techniques in the sixties large telescope arrays were deployed, such as the Westerbork Synthesis Radio Telescope in Westerbork, The Netherlands and the Very Large Array in Socorro, New Mexico, Unites States of America.

Since the nineties, extension of observing frequencies has been an important aspect of radio astronomy. At high frequencies new observatories have been developed and built, such as the eSMA in Hawaii (Bentum, 2006). Currently the Atacama Large Millimeter Array (ALMA) is being constructed consisting of fifty parabolic reflectors for the frequency range of 31 to 950 GHz (Wootten 2005). Research at low frequencies is one of the major topics at this moment in radio astronomy and several Earth-based radio telescopes are constructed at this moment. It is considered as one of the last unexplored frequency ranges (Weiler, 2000). A number of major new facilities for low frequency operation are being developed or under construction, such as the Giant Metre wave Radio Telescope (GMRT, Swarup et al., 1991), Long-Wavelength Array (LWA, Kassim et al., 2005), Murchison Widefield Array (MWA, formerly

known as the Mileura Widefield Array, Morales, 2006), the 21 Centimeter Array (21CMA5 ; formerly called Primeval Structure Telescope, PAST 6, Peterson et al., 2005), Precision Array to Probe the Epoch of Reionization (PAPER, Bradley et al., 2005), and the Square-Kilometre Array (SKA, Schilizzi, 2004). For frequencies below ~50 MHz, space-based instruments must be considered due to the ionosphere (Bentum et al, 2009 and Jester et al, 2009).

Our approach is LOFAR, the Low Frequency Array. LOFAR will be a wide-area sensor network for astronomy, geophysics and precision agriculture (Gunst et al 2007, 2008, Bregman 2000). The LOFAR infrastructure will consist of a collection of over thirty-six sensor fields, also referred to as "stations". At least eighteen sensor fields will be concentrated in a central area, further referred to as core stations. The rest of the stations (remote stations) will be distributed over a larger area of about 150 kilometers in diameter. The stations will cover the spectrum from 30 to 240 MHz. A dedicated supercomputer, the Central Processor, will combine and process the sensor data. Data will be transported over optical fiber connections from the sensor fields to the Central Processor. The total digitized data rate from the sensors is about 0.5 Tb/s at each sensor field. Station level processing reduces this rate to roughly 2 Gb/s by combining data from multiple sensors into phased array beams. LOFAR is fully based on phased array principles. This gives LOFAR the ability to operate in multiple directions simultaneously. LOFAR is one of the first radio telescopes in which RFI mitigation techniques are an integral part of the system design, which will be presented in this chapter as well as some practical issues.

A signal processing view of the instrument should starts with a description of the signal. In radio astronomy we are interested in signals from cosmic sources very far away. A fundamental property of the radio waves emitted by these cosmic sources is that they are stochastic in nature. The signal strength of these signals is very small

Figure 1. Signal reception in a radio astronomical interferometer

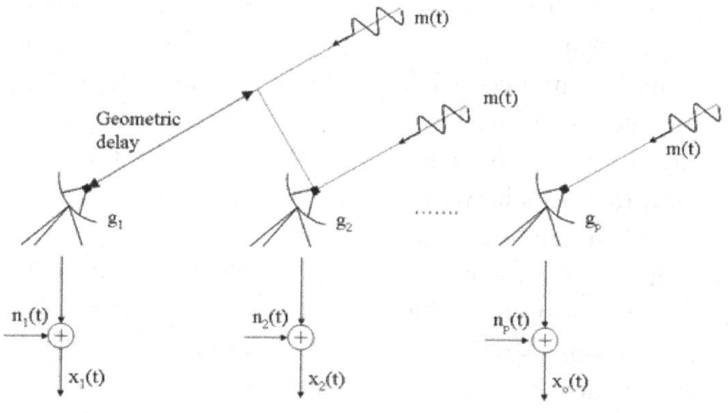

and is regularly expressed in Jansky's (Jy). Jy is a unit of flux density 10^{-26} Wm^{-2}Hz^{-1}. The sources of interest in the sky are mJy or sometimes even μJy sources.

In this chapter we will first describe the signal model for array antenna systems. This model will then be mapped onto the LOFAR array radio telescope. One of the major issues in radio astronomy and for LOFAR in particularly is radio interference. LOFAR is operating in a frequency range with many other users. As a passive user, signals from all other sources are considered RFI. How to deal with RFI will be discussed. We will conclude with a list of issues for further research.

BACKGROUND

Modern radio telescopes, such as LOFAR, use interferometric techniques to create large array telescopes to increase both the spatial resolution as well as the sensitivity. An interferometer measures the correlation between two antenna signals spaced at a distance d, which is called the baseline. Placing the antennas (or stations as we will call them in LOFAR) on various positions, a number of different baselines can be obtained. Using the rotation of the Earth a sequence of correlations is obtained with varying baseline distances and

baseline orientations with respect to the observed sky image field.

Each interferometer measures discrete components of the spatial spectrum of the observed object's brightness distribution on the sky. This spatial spectrum is often called the "visibility function". The measured visibilities are a function of the locations of the interferometers (the antennas), the location of the source in the sky, and the wavelength of observation.

In this section we will describe these signals and see how we can obtain an estimate of the source brightness distribution, the final image of the sources of interest. We will use bold lower case letters for vectors, bold upper case letter for matrices, and non-bold letters for scalars.

In Figure 1 the reception of signals in a radio astronomy array is schematically shown. In this figure the antennas are depicted as parabolic telescope dishes, but the concept is also valid for arrays of antenna elements. The received signal of each of the antenna elements $x_i(t)$ is the sum of the sky signal $m(t)$ and telescope based noise signal $n_i(t)$. The visibilities are obtained by correlating the measured telescope signals $x_i(t)$.

Antenna Array Signal Model

In this section, we follow the signal processing formalism as described in Leshem 2000 and van der Veen 2004. Consider a radio telescope antenna array consisting of p antennas, each having an output signal $x_i(t)$ for $i=1...p$. Assume that each antenna i receives a signal $m(t)$ from a celestial source. We assume narrow-band conditions hold for the signals received by the antennas. This means that the propagation differences between array elements i and j are much smaller than the inverse of the bandwidth of the signals. We model the signals in complex base-band form as white Gaussian random variables. The signals received at each of the antenna elements are identical up to a geometric time delay and up to a (complex) gain factor. As we assume narrow-band conditions, geometric time delay can be represented by a multiplicative complex factor, usually known as spatial signature a_i. Complex gains may be included in the spatial signature or may be separate factors. In our case, we assume the latter, and denote the gain by g_i. The output signal of the i-th antenna can be expressed as

$$x_i(t) = a_i\, g_i\, m(t) + n_i(t) \qquad (1)$$

where $n_i(t)$ is the noise signal from the low-noise amplifier, spill-over, background pick-up, etc. The noise contributions $n_i(t)$ are uncorrelated, that is:

$$E\{n_i(t)\,\overline{n_j(t)}\} = \delta_{ij}\sigma^2_{n_i} \qquad (2)$$

The signals $x_i(t)$ of an antenna array with p antennas can be conveniently stacked in a vector $\mathbf{x}(t)$:

$$\mathbf{x}(t) = [x_1(t), ..., x_p(t)]^T \qquad (3)$$

where T denotes the transpose operator. The spatial signature vector $\mathbf{a}$, the gain vector $\mathbf{g}$, and the noise vector $\mathbf{n}(t)$ are defined similarly. The vectors $\mathbf{a}$ and $\mathbf{g}$ vary slowly over time. The vector $\mathbf{a}$ varies because of the Earth rotation, the vector

$\mathbf{g}$ usually varies because of gain variations due to for example temperature changes affecting the behavior of the electronics. The received signal model can be written as

$$\mathbf{x}(t) = \mathbf{G}\,\mathbf{a}m(t) + \mathbf{n}(t) \qquad (4)$$

where $\mathbf{G}$ is defined as $\mathbf{G} = \mathrm{diag}(\mathbf{g})$, where the diag operator creates a diagonal matrix $\mathbf{G}$ with the components of the vector $\mathbf{g}$ stacked on its main diagonal.

The array signatures a_i represent the spatial delay factor and can therefore be written as a complex exponential:

$$a_i = \exp\left\{-i\frac{2\pi(\mathbf{b}_i - \mathbf{b}_0)^T \mathbf{s}}{\lambda}\right\} \qquad (5)$$

where $\mathbf{b}_i$ is the position vector of the i-th antenna, $\mathbf{b}_0$ is an arbitrary reference position, $\mathbf{s}$ is the direction vector of the celestial source, and λ is the wavelength of the impinging radio wave.

Let $\Re$ denote a $(p \times 3)$ matrix containing all the relative antenna position vectors $\mathbf{b}_i - \mathbf{b}_0$, then the signature vector $\mathbf{a}$ can be expressed by

$$\mathbf{a} = \exp\left\{-i\frac{2\pi\Re\mathbf{s}}{\lambda}\right\} \qquad (6)$$

Our signal model can easily be extended to multiple astronomical point-sources. For Q sources the model becomes

$$\mathbf{x}(t) = \sum_{q=1}^{Q} \mathbf{G}\,\mathbf{a}_q m_q(t) + \mathbf{n}(t) \qquad (7)$$

with $m_q(t)$ denoting the signal from the q-th source at time t, and $\mathbf{a}_q$ denoting the signature vector corresponding to the q-th source.

Let $\mathbf{A}$ be defined by stacking Q signature vectors in a (p_xQ) matrix $\mathbf{A}$: $\mathbf{A} = [\mathbf{a}_1, \dots, \mathbf{a}_Q]$, and assume Q celestial source signals m_j are stacked in a vector $\mathbf{m} = [m_1, \dots m_Q]^T$. The signal model can then also be written in a more compact form as

$$\mathbf{x}(t) = \mathbf{GAm}(t) + \mathbf{n}(t) \tag{8}$$

Covariance Model

The covariance model corresponding to the signal model described above, is expressed by:

$$\mathbf{R} = E\left\{\mathbf{x}(t)\mathbf{x}^H(t)\right\} = \mathbf{G}\,\mathbf{A}\,E\left\{\mathbf{s}(t)\mathbf{s}^H(t)\right\}\mathbf{A}^H\,\mathbf{G}^H + E\left\{\mathbf{n}(t)\mathbf{n}^H(t)\right\} \tag{9}$$

with H the Hermitian transpose operator.

The cross correlations between the sky sources and antenna noise contributions are obviously zero. With $\mathbf{B} = E\left\{\mathbf{m}(t)\mathbf{m}^H(t)\right\}$ and $\mathbf{D} = E\left\{\mathbf{n}(t)\mathbf{n}^H(t)\right\}$ we get:

$$\mathbf{R} = \mathbf{G}\,\mathbf{A}\,\mathbf{B}\,\mathbf{A}^H\mathbf{G}^H + \mathbf{D} \tag{10}$$

The matrix $\mathbf{B}$ contains the powers of the celestial sources, it is diagonal because the sources are independent. The noise matrix $\mathbf{D}$ is diagonal in case the antenna noise contributions are independent, as is usually the case for telescope dishes. The matrix is not diagonal in case there is some cross-coupling between the antenna elements, as is the case for closely-packed aperture arrays.

Each matrix element of the covariance matrix $\mathbf{R}$ represents the correlation between signals of the corresponding antennas. The matrix $\mathbf{R}$, the vector $\mathbf{x}$, and the corresponding signal and gain components from our data model. Suppose we

have actual observations with N samples of data $\mathbf{x}_n = \mathbf{x}(nT_s)$ obtained using a sampling period of T_s. We then can create a short term integration sample covariance estimate $\hat{\mathbf{R}}$

$$\hat{\mathbf{R}} = \frac{1}{N}\sum_{n=1}^{N}\mathbf{x}_n\mathbf{x}_n^H \tag{11}$$

Assuming the model describes the reality perfectly, the following relation holds: $\mathbf{R} = E\left\{\hat{\mathbf{R}}\right\}$. This observed covariance matrix forms the basis for imaging the radio sky, as will be described in the following section. In practical radio telescopes, the integration period is in the order of 1 to 60 seconds.

Imaging

As before, assume an antenna aperture array with p antennas and consider a simplified sky model with Q point-sources with power $\sigma_{s_i}^2$. The data model for this configuration is, as explained on the previous sections:

$$\mathbf{R} = \sum_{q=1}^{Q}\sigma_{s_q}^2\,\mathbf{a}_q\mathbf{a}_q^H + \mathbf{D} \tag{12}$$

In this simplified model we have ignored the gain matrices $\mathbf{G}$ as we will focus on the sky imaging principles. In practice, the gain factor of the matrix $\mathbf{G}$ can be found by calibration procedures, for example by using dedicated measurements on known celestial sources and by injecting signals with known properties in the electronic signal chain. More information on calibration in radio astronomy can for example be found in (Wijnholds, 2010).

Each sky source q in the model above corresponds, in general, to a unique signature vector $\mathbf{a}_i$ defined as in (6). The question is now, given a measured or modeled covariance matrix, how to image the sky. The covariance matrix contains correlations between all antenna pairs and are usually defined on a grid of baseline measurement points. These measurements are, when the grid is scaled with wavelength, called (u,v,w) data. The (u,v) coordinates usually span a horizontal plane, the vertical direction usually is represented by the w coordinate. Mapping the observed (u,v,w) visibilities to the desired source brightness distribution is called imaging.

For many astronomical instruments we have a planar configuration of antennas, and usually a very small field of view. This reduces the coordinate system to a (u,v) plane. If the visibilities of a source are covering the entire (u,v) plane, the source image is perfectly reconstructed by the Fourier inversion of the visibility. This is also known as the Van Cittert-Zernike theorem. In practice often only a (small) part of the (u,v) plane is measured since only a limited set of baselines can be obtained. Rotating of the Earth fills the (u,v) plane with ellipses. The final set of samples is known as the (u,v)-coverage of the telescope.

With a limited set of observations, the Fourier inversion yields a so-called dirty map. This is the actual image convolved with an instrument point spread function, also known as the dirty beam. A dirty map pixel power value corresponding to a direction **s**, $I_D(\mathbf{s})$, can be written as

$$I_D(\mathbf{s}) = \sum_{n=1}^{N} \mathbf{a}_n(\mathbf{s})^H \mathbf{R}_n \mathbf{a}_n(\mathbf{s}) \qquad (13)$$

where N is the number of covariance matrices to be averaged. For each direction **s**, a pixel value in the sky map can be computed, and a complete sky map can be computed by combining the obtained pixel values for all directions **s**. For a linear regular grid of telescopes, the **a**-vector components form a Fourier DFT vector. For a two dimensional regular grid, a sky image is created by applying a two-dimensional Fourier transform (2D-FFT) to (u,v,w) visibility data. The 2D FFT

is an efficient implementation to create images from (u,v) data for regularly spaced antenna elements. For a non-regular grid, one can create sky maps efficiently by applying 2 2D FFT using a so-called gridding approach, cf. (Thompson 1986, Perley 1994). Given observed (u,v,w) data, one can calibrate unkown gains, and apply deconvolving algorithms to create "clean" sky maps. These techniques, for example described in (Thompson 1986, and Wijnholds 2010), are outside the scope of this chapter.

Beam-Forming

In the next generation of radio telescopes, like LOFAR, the dishes are replaced or complemented by phased array antenna systems, called stations. The stations have their own beam-former which can be implemented as a time-delay beam-former, narrow band beam-former, or as a space-time beam-former. The beam-formed data of a station will be correlated with beams of other stations. For a beam-former, consider an antenna array of p antennas, as in Figure 1. The signal $m(t)$ from celestial sources is added coherently if in summing the $x_i(t)$ signals, the geometric delays are compensated for by adding artificial time delays. In analog form this can be delay lines on a printed circuit board, in the digital domain this can be discrete delay steps. In LOFAR, as will be presented in the next section, both are implemented: Analog delays on printed circuit boards in the High Band Antennas and digital discrete delay steps in the stations. An advantage of true time delays is that they are frequency independent and thus broad band. Other implementation forms of beamfomers are shown in Figure 2. The left figure shows a narrow-band beamformer. An advantage of this beamformer is that the weight coefficients can be set accurately, in contrast to the (often) discrete steps of true time delay beamformers. The right figure shows a space-time beamformer which includes a FIR filter to make it broad-band.

Figure 2. Narrow band beamforming and broadband beamforming

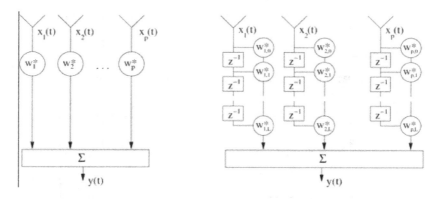

We now consider the narrow-band beam-former in some more detail. The beam-former weights corresponding to a direction $\mathbf{s}_0$ are defined by:

$$\mathbf{w} = [w_1, w_2 \dots w_p]^T \qquad (14)$$

and

$$\mathbf{w} = \exp\left\{-i\frac{2\pi\,\Re\,\mathbf{s}_0}{\lambda}\right\} \qquad (15)$$

The beam-former output is calculated as:

$$y(t) = \mathbf{w}^H\mathbf{x}(t) \qquad (16)$$

The pixel intensity in the map I of the source brightness distribution in the direction $\mathbf{s}_0$ is given by:

$$I(\mathbf{s}_0) = E\left\{\mathbf{y}(t)\,\overline{\mathbf{y}(t)}\right\} \qquad (17)$$

The most simple case is when the gains are constant. We therefore assume $\mathbf{G}$ unity or incorporated in $\mathbf{a}$. This gives:

$$I(\mathbf{s}_0) = \mathbf{w}^H\,\mathbf{A}\,E\left\{\mathbf{s}(t)\,\overline{\mathbf{s}(t)}\right\}\mathbf{A}^H\mathbf{w} + \mathbf{w}^H\,E\left\{\mathbf{n}(t)\mathbf{n}(t)^H\right\}\mathbf{w} \qquad (18)$$

$$I(\mathbf{s}_0) = \mathbf{w}^H\,\mathbf{A}\,B\,\mathbf{A}^H\mathbf{w} + \mathbf{w}^H\,\mathbf{D}\,\mathbf{w} \qquad (19)$$

For a single source

$$I(\mathbf{s}_0) = \sigma_s^2\mathbf{w}^H\,\mathbf{a}_i\mathbf{a}_i^H\mathbf{w} + \mathbf{w}^H\,\mathbf{D}\,\mathbf{w} \qquad (20)$$

If $\mathbf{s} = \mathbf{s}_0$ then $\mathbf{w}=\mathbf{a}_i$, and if the noise in uncorrelated with identical power for all antenna elements, $\mathbf{D} = \sigma_n^2\mathbf{I}$, where $\mathbf{I}$ is the identity matrix, then:

$$I(\mathbf{s}_0) = p^2\sigma_s^2 + p\,\sigma_n^2 \qquad (21)$$

Let the signal to noise ratio (SNR) of the beam-former output be defined as the ratio between the received celestial source power and the telescope noise power. From the formula above it is clear that the SNR scales linearly with the number of telescopes or antenna elements, p. This is closely related to the array-factor as defined in antenna theory, cf. (van Trees, 2002). Often, the array vectors are normalized, yielding $I(\mathbf{s}_0) = p\sigma_s^2 + \sigma_n^2$.

Although the beam-former output power was computed to demonstrate the effect the beam-former has on SNR, in practice, the beam-former

Figure 3. LOFAR system architecture

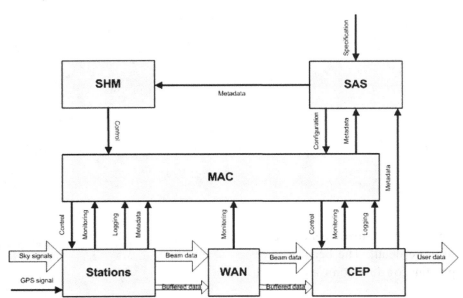

output signal (the station beam) is correlated with corresponding beams of other stations. In this way (*u,v,w*) data are obtained, which can be converted to sky maps by applying 2D Fourier transforming techniques.

Limitations

In the previous sections it looks rather straightforward to make an image: measure the visibilities, inverse Fourier transform them to a dirty image and recover this image to the final source brightness distribution. A standard method of doing this last step is CLEAN.

In practice, however, a number of limitations are present:

- **Gain variation.** The Low Noise Amplifiers are designed for the very weak astronomical signals and therefore are sensitive to environmental fluctuations. Gain and phase fluctuations are present, and as can be seen in Equation 10, this will have impact in the measured visibilities.

- **Antenna pattern.** Unwanted sidelobes makes all the individual antenna elements sensitive to strong sources in these sidelobes.
- **Ionosphere.** The outer layer of the atmosphere, the ionosphere, will cause time varying refraction and diffraction.
- **Radio Frequency Interference.** RFI can be treated as additional noise in the signal chain of the radio telescope; a (non-diagonal) extension of matrix **D** in Equation 10. RFI will be a separate topic later in the chapter.

THE LOFAR SIGNAL CHAIN

LOFAR is a wide-area sensor network for astronomy, geophysics and precision agriculture. For the scope of this chapter we only consider the astronomical application. The architecture for the astronomical applications consists of six main architectural blocks, as can be seen in Figure 3.

Figure 4. Block diagram of a LOFAR station

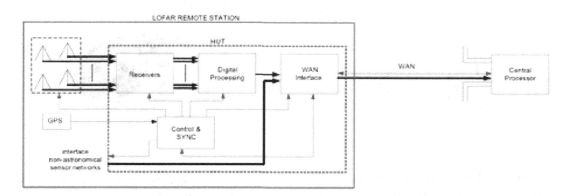

System Level

The sky signals are received by the antennas. In LOFAR two types of antennas are used, the Low Band Antenna for the frequency range from 30 to 80 MHz and the High Band Antenna for the frequency range from 120 to 240 MHz. The reason for the intermediate gap is the FM radio band which will give to much flux density. The sky signals are processed in the stations. A station selects the sky signals of interest for a particular observation out of the total sky. This process results in one or multiple beams onto the sky. This is done by means of beam-forming the signals from all of the antennas in the station as mathematically described in the previous section.

Next block in the architecture is the Wide Area Network (WAN). The WAN is responsible to transparently transport all the beam data (the beam signals on the sky) from the stations to the central processor. This is done in the Northern part of the Netherlands as well as in other countries in Europe. In the Central Processor (CEP), the data from the individual stations is combined in such a way, that user data is generated as was specified by the user.

Three additional building blocks are implemented in LOFAR for optimally control the LOFAR telescope:

1. **Scheduling And Specification (SAS).** Given the specification, the main responsibility of SAS is to schedule and configure the system in the right mode. Additionally SAS facilitates the possibility to store metadata of the system for a long term and make that information accessible for the user.

2. **Monitoring And Control (MAC).** The main responsibility of MAC is to (real-time) control the system based upon the actual configuration of that moment. Additionally MAC facilitates the (real-time) monitoring of the present state of the system.

3. **System Health Management (SHM).** SHM is identified as an autonomous block to predict and act on failures of the hardware before it actually fails. Ideally it even should pin point which system component is the cause of a failure. The reason to identify this block is because of the scale of the system and the percentage of time the system should be effectively operational.

The Station

Each LOFAR station will contain 96 Low Band Antennas (LBAs) and 48 to 96 High Band Antenna (HBA) tiles to cover the whole frequency range with sufficient sensitivity. The LBAs will cover

Figure 5. Sketch of a LOFAR low band antenna

Figure 6. Photograph of a cross-section of the LBA antenna head

the spectrum from 30 to 80 MHz and the HBA tiles will cover the spectrum from 120 to 240 MHz.

As shown in Figure 4 both antennas are connected to the receiver. In the receiver the electrical signals are amplified, filtered and converted to the digital domain within the receiver block. Subsequently, in the digital domain the data is reduced.

Finally the digital data is merged with the control data and data from the non-astronomical sensor networks, converted to the optical domain and transported via a WAN. The data path of a station is controlled and synchronized with other stations by a control and synchronization unit that is disciplined by GPS signals as reference.

THE ANTENNAS

A sketch of the LBA antenna is depicted in Figure 5.

The signals of the LBAs are amplified in the LNAs in the top of each antenna as shown in

Figure 6. Via coaxial cables the analog signals are distributed to the receiver system. These co-axial cables are not completely equal, which results in some gain and phase differences. Although these parameters can change, they are assumed stable within time scales of tens of minutes, such that they can be calibrated. In Equation 1 the output of a LBA is expressed. The gain g_i is the gain of the LBA.

Each HBA tile consists of a 4×4 array of dual polarized antenna elements mounted together in a tile. This is depicted in Figure 7.

The 16 antenna signals are amplified and analog beam-formed <u>in</u> the tile. The tile delay unit provides analog signal delays up to 15 ns in a 5-bit true time delay circuit. These 5 discrete delay lines are implemented using meandering PCB traces over the delay unit circuit board. This gives accurate and repeatable delays for in total 32 unique time delays. The delays implemented are 0.47, 0.94, 1.88, 3.76 and 7.51 ns. The beam-

Figure 7. Drawing of the high band antenna tile

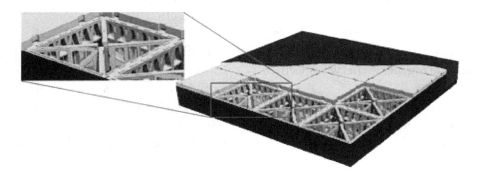

Figure 8. Observation modes in the station receiver

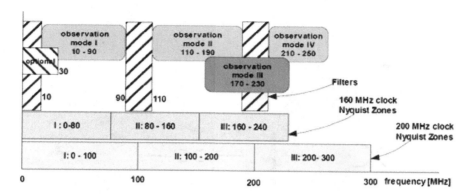

forming is done for both polarizations. Both signals are sent to the LOFAR cabinet at the field for further processing.

The Receiver System

The receiver unit selects one out of two antennas. An extra interface at the input of the receiver is provided to cope with future extensions with a third antenna. After selecting an antenna, the signal is filtered with one of the four integrated filters. These filters split the input band in four parts. After filtering, the signal is amplified and filtered again to reduce the out of band noise contribution.

For the receiver a wideband direct digital conversion architecture is adopted. This reduces the number of analogue devices used in the signal path. The A/D converter converts the analogue signal into a 12 bit digital signal at a maximum sampling rate of 200 MHz. To fill the gaps in between the Nyquist zones an alternative sample frequency of 160 MHz can be chosen as well. This results in the observation modes as depicted in Figure 8.

The Digital Processing System

After A/D conversion, the band is split into 512 equidistant sub-bands using a polyphase filterbank. The negative part of the original spectrum is omitted, i.e. the real input signal is from here on represented by complex signals. Each sub-band

signal is decimated with a factor of 1024 after filtering. Hence, the clock rate after filtering is reduced to 195 kHz and 156 kHz respectively, for a 200 MHz and 160 MHz input sampling rate. After filtering, specific sub-bands can be selected. The selection is controlled centrally at the station. The selected sub-bands have a maximum total effective bandwidth of 48 MHz per polarization.

To form beams, the antenna signals are combined. This is done with independent beam-formers for each sub-band. The weights necessary for the beam-former are calculated in the Local Control Unit (LCU) and are sent to the beam-formers each second in order to follow sources while the earth rotates. Additionally statistical measurements are performed at the station. This monitor information is forwarded to the Monitoring And Control (MAC) subsystem.

In parallel with the beam-forming a full cross correlation of all the antenna signals for one sub-band can be done as well. This is done to accommodate for online calibration of all signal paths for gain and phase differences. Furthermore this information will be used as well for RFI (Radio Frequency Interference) detection.

The raw antenna data or sub-band data can be stored in a transient buffer as well. Freezing of the buffer content can be controlled by internal or external triggers. The stored data or selections thereof can be sent to CEP for further processing.

Figure 9. Central processing facility of LOFAR

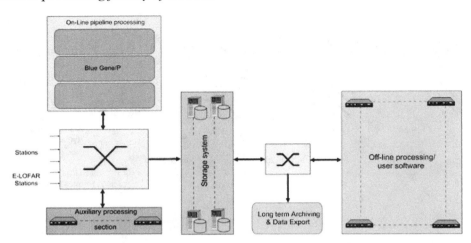

Central Processor Facility

The Central Processor hardware (CEP) system is used to combine and process the data from all the stations (see Figure 9). First the data is streamed and online processed using a BG/P supercomputer. For the imaging applications the data is correlated, while for transients applications the data of all stations is added with weights. For normal operations one rack is sufficient. Although a second rack is available as well for special and high demanding observation modes.

Storage is available for temporary storage of raw data files until these are processed off-line and products are exported to users. Additional storage is available for (limited) archiving of end-products.

A streaming processing application has been developed for the entire pipeline converting the raw station data streams into correlation products. In the first step of this pipeline the 3 Gbps of data is received, validated and synchronized (the core stations can deliver 6 Gbps). Also, additional delay is applied in order to correct for the rotation of the earth. The next step of the pipeline is a poly-phase filter bank to split the sub-bands data into 256 frequency channels. This results in channels of approximately 1 kHz. From each channel, the

correlation matrix for all stations is calculated. The filter bank and correlator operations are implemented on the Blue Gene/LPcomputer.

The output of the correlator is stored until completion of the observation. The remainder of the processing operates off-line on these datasets containing the coherent data from a complete observation. The basic tasks to be performed are typically flagging of bad samples, calibration and transformation to the image plane, resulting in an image of the sky.

CHALLENGES IN THE LOFAR SIGNAL CHAIN

Previously we mentioned the challenges in the image-mapping:

*   Gain variation.
*   Antenna pattern.
*   Ionosphere.
*   Radio Frequency Interference.

For now, we focus on the latter one, RFI. The other items are well described in the special issue of the IEEE signal processing magazine of January 2010.

Figure 10. Spectra of the LOFAR monitor campaign in February 2008

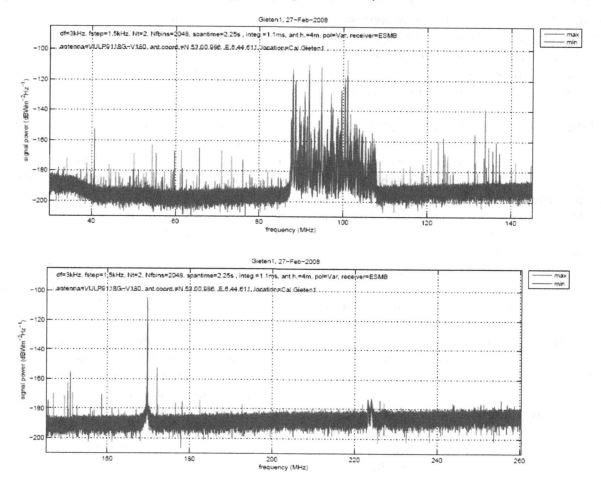

One·of the increasing issues is interference. The two antennas of LOFAR cover the frequency range from 30 MHz to 270 MHz. This spectrum is in use by many other services. In Figure 10, a typical spectrum is shown of the LOFAR band.

The spectra were obtained by a special LOFAR monitoring van. The equipment that is used for the mobile measurements is a data acquisition instrument (a Rhode & Schwarz ESMB receiver) which connects to a notebook computer for control and readout. The input is connected to either a Rhode & Schwarz HE010 active antenna or the Schwarzbeck Vulp9118G passive antenna. The FM radio band is clearly visible in the figure. Also visible are distant DAB-T signals in TV band III

channel 12 at 223-230 MHz. The strongest signal in the band is a pager signal at approximately 169 MHz.

The use of the radio spectrum in terms of signal power and time-frequency occupancy is roughly known from allocation tables, from monitoring observations as illustrated above, and for future spectrum developments from spectrum management agencies. In order to estimate the required attenuation levels, the observed spectrum power needs to be related to the LOFAR sensitivity. The calibration capabilities of LOFAR include the removal of strong sky sources such as Cas A from the data. In that respect, remaining RFI can be removed in the same way as astronomical sources are removed. This results in the requirement that RFI sources must be suppressed to levels comparable to the level of Cas A. However, as this technique reduces the number of degrees of freedom, only a very limited number of these relatively strong RFI sources can be suppressed.

Methods on Handling Interference

At the LOFAR core, the strongest observed signal (apart from the FM band which is blocked by analog filters, and apart from the spectrum below 30 MHz) was a paging signal at 169 MHz, at a level of 65 dBμV/m. This is a narrow band signal, and the LOFAR systems were designed to remain linear in the presence of transmitters generating these large field stengths. Most of the interferers (>90%) have observed power levels of less than 40 dBμV/m. That means the RFI mitigation techniques must reduce the interference for more than 40 dB. There are, however, a few additional effects which help to reduce the observed interference:

- **Spectral dilution.** With LOFAR we observe a wide band, while individual interferers are limited to smaller bands. The energy of a single narrowband RFI source is diluted by averaging all channels in the band (N). The noise power decreases with $\sqrt{N}$, whereas the RFI power decreases with N. This leads to a spectral dilution factor which scales with $\sqrt{N}$. Of course this is only valid for narrowband RFI and not for wideband modulation techniques (e.g. Spread spectrum modulation and Ultra Wide Band techniques).
- **Spatial dilution.** Observing astronomical sources requires a substantial integration time. The sources are moving with respect to the antennas, while the (terrestrial) interferers don't move. In the integration process, the RFI sources are diluted.

- **Polarization.** Most RFI sources are fixed to the horizon and polarized vertically, while LOFAR antennas are horizontally polarized (w.r.t. the horizon). This gives an addition attenuation of up to about 15 dB.

Implementing RFI mitigation techniques requires computing power. One of the constraints of the LOFAR system design is that the computing power needed for the interference mitigation should be an order of magnitude less than what is required for the astronomical signal processing. The following mitigation techniques within LOFAR can be identified:

- The observed band will be divided in sub-bands of 156/195 kHz. By choosing the *optimal location of the sub-bands*, RFI can be minimized.
- *Flagging in the time domain* (timescales from ms to 10s). Flagging is done at the central level, rather than at station level. The reason for that is the possibility to flag at kHz level, instead of the complete sub-band of 156/196 kHz.
- *Spatial RFI mitigation at the stations.* Only fixed or very slow-varying spatial nulls at the station will be applied. Fast changing interference nulling would lead to fast changes in the (sidelobe) gains which makes calibration of the instrument very difficult. Fixed spatial nulling is easily implemented in the station beam-former.
- *Spatial filtering at central level.* This will be done after correlation of the data in the central computer. Although in theory this has been researched, practical implementation will be done after the first phase of the project.
- *Reduction of the interferers in the map*, done to sky noise levels. This will be included in the map making process which involves long integration times. Interferers

Figure 11. First images of LOFAR with three station [courtesy of Sarod Yatawatta, ASTRON]

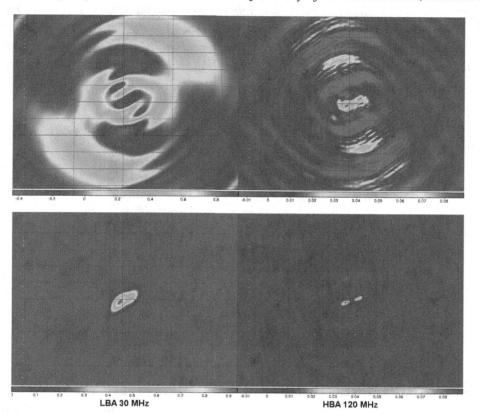

LBA 30 MHz　　　　**HBA 120 MHz**

in the map can be treated the same as strong sources in the map which should be reduced as well. Current available techniques as *Clean* and *Selfcall* will be used, as well as new techniques as spatial filtering.

RESULTS

LOFAR is being constructed at this moment. The first stations are ready. In Figure 11 one of the first results can be seen. The image is of Cygnus A. It was named by radio astronomers after the constellation in which it appears, Cygnus the Swan. Despite its great distance from us (600 million light years away), it is still by far the closest powerful radio galaxy and one of the brightest radio sources in the sky. The top two images are the dirty images of the LBA and the HBA system. Cleaned images are shown at the bottom.

The image is made using three LOFAR stations. The shortest baseline is 4.8km and the longest baseline is about 20km. Cygnus A is a double source and as can be seen, given the longest baseline of about 20 km, Cygnus A is resolved even at the lowest frequency of 30 MHz. As more stations become operational, we will have better uv-plane coverage and thus, better images.

More stations are constructed at this moment, both in The Netherlands as in other European countries. In Figure 12 a few images can be seen of the construction phase of LOFAR. In 2010 the telescope will be opened officially.

Figure 12. Construction of the LOFAR station. In the upper left image the locations of the LOFAR stations is highlighted with a green dot. In the upper right image one of the LOFAR PCB's, the receiver unit, can be seen. In the middle, a High Band Antennas tile is placed in the field. At the field 96 Low Band Antennas are placed, as can be seen in the lower left picture. Finally at the lower right image an aerial view of the superterp. The superterp is the center of LOFAR with 6 stations very close to each other.

FUTURE RESEARCH DIRECTIONS

In this book advanced signal and imaging techniques are presented, for radio astronomy in particular for this chapter. Imaging in astronomy looks rather straightforward: measure the visibilities, inverse Fourier transform them to a dirty image and recover this image to the final source brightness distribution. As we mentioned before, a number of limitations are present:

- **Gain variation.** The Low Noise Amplifiers are designed for the very weak astronomical signals and therefore are sensitive to environmental fluctuations. Gain and phase fluctuations are present, and as can be seen in Equation 10, this will have impact in the measured visibilities. Most of the gain and phase variations are relatively slow compared to the integration time. Online calibration is performed at the stations.

- **Antenna pattern.** Unwanted sidelobes makes all the individual antenna elements sensitive to strong sources in these sidelobes.
- **Ionosphere.** The outer layer of the atmosphere, the ionosphere, will cause time varying refraction and diffraction.
- **Radio Frequency Interference.** RFI can be treated as additional noise in the signal chain of the radio telescope; a (non-diagonal) extension of matrix **D** in Equation 10.

In the previous section we start with the latter one. Future research directions are towards all the limitations.

The signal processing techniques described in this chapter are not limited to the LOFAR radio astronomy application. They can be used in other applications as well. One of such applications is the seismic application, which uses seismic sensors to listen to the earth. The seismic sensors convert vibrations into electric signals which are subsequently converted to digital format. These data are sent over the LOFAR network to a central site where data-processing, storage and further data dissemination takes place.

CONCLUSION

Currently the LOFAR system is still under construction. Half of the total number of stations have been installed in the field and are operational. Furthermore the basic pipeline from antenna signal reception to astronomical imaging is in place. Since LOFAR is a complex system with lots of signal processing algorithms and hardware it will take some time to optimize the system further.

LOFAR is the first telescope of its size and architecture which will open the low frequency window for astronomers. The technology and signal processing algorithms used for LOFAR is for a significant part different from traditional telescopes. Also the data coming from the telescope needs to be post processed to reduce the data volume to acceptable levels. Nowadays an astronomer is responsible for the calibration of the data which can be done typically on a desktop computer. However this is not possible with the huge amounts of data LOFAR will generate. Hence the calibration and imaging will be done by the LOFAR system automatically which reduces the data significantly. Finally also the possibility of observing at multiple directions simultaneously will increase the efficiency of the LOFAR telescope.

REFERENCES

Bentum, M. J., van Langenvelde, H. J., Tilanus, R., & Friberg, P. (2006). *The extended sub-millimeter array–proof of concept by connecting the JCMT.* Paper presented at SPS-DARTS, Second annual IEEE Benelux/DSP Valley Signal Processing Symposium, Antwerp, Belgium.

Bentum, M. J., Verhoeven, C. J. M., Boonstra, A. J., van der Veen, A. J., & Gill, E. K. A. (2009). *A novel astronomical application for formation flying small satellites.* Paper presented at the 60th International Astronautical Congress, Daejeon, Republic of Korea, 12 – 16 October, 2009.

Bradley, R., Backer, D., Parsons, A., Parashare, C., & Gugliucci, N. E. (2005). PAPER: A Precision Array to Probe the Epoch of Reionization. *Bulletin of the American Astronomical Society, 37*, 1216.

Bregman, J. D. (2000). Concept design for a Low Frequency Array. *Proceedings of the Society for Photo-Instrumentation Engineers, 4015*, 19–33.

Gunst, A. W., & Bentum, M. J. (2007). *Signal processing aspects of the LOFAR.* Paper presented at the IEEE International Conference on Signal Processing and Communications, Dubai, United Arab Emirates.

Gunst, A. W., & Bentum, M. J. (2008). *The current design of the LOFAR instrument.* Paper presented at the URSI General Assembly, Chicago, IL.

Jansky, K. G. (1932). Directional studies of atmospherics at high frequencies. *Proceedings of IRE, 20*, 1920–1932. doi:10.1109/JRPROC.1932.227477

Jester, S., & Falcke, H. (2009). Science with a lunar low-frequency array: From the dark ages of the Universe to nearby exoplanets. *New Astronomy Reviews, 53*, 1–26. doi:10.1016/j.newar.2009.02.001

Kassim, N. E., Polisensky, E. J., Clarke, T. E., Hicks, B. C., Crane, P. C., Stewart, K. P., et al. (2005). The long wavelength array. In N. Kassim, M. Perez, W. Junor & P. Henning (Eds.), *Astronomical Society of the Pacific Conference series.* (p. 392).

Leshem, A., van der Veen, A. J., & Boonstra, A. J. (2000). Multichannel interference mitigation techniques in radio-astronomy. *The Astrophysical Journal, 131*, 355–373. doi:10.1086/317360

Morales, M. F., Bowman, J. D., Cappallo, R., Hewitt, J. N., & Lonsdale, C. J. (2006). Statistical EOR detection and the Mileura widefield array. *New Astronomy Reviews, 50*, 173–178.

Perley, R. A., Schwab, F. R., & Bridle, A. H. (1994). *Synthesis imaging in radio astronomy. Proceedings of the Third NRAO Synthesis Imaging Summer School, Astronomical Society of the Pacific Conference Series, 6.*

Peterson, J. B., Pen, U., & Wu, X. (2005). Searching for early ionization with the primeval structure telescope. In N. Kassim, M. Perez, W. Junor & P. Henning (Eds.), *Astronomical Society of the Pacific Conference series.* (p. 441).

Schilizzi, R. T. (2004). The square kilometer array. In J.M. Oschmann, Jr. (Ed.), *Ground-based telescopes. Proceedings of the SPIE, 5489*, (pp. 62-71).

Swarup, G., Kapahi, V. K., Velusamy, T., Ananthakrishnan, S., Balasubramanian, V., & Pramesh-Rao, A. (1991). Twenty-five years of radio astronomy at Tata Institute for Fundamental Research. *Current Science, 60*(2).

Van der Veen, A. J., Leshem, A., & Boonstra, A. J. (2004). Array signal processing for radio astronomy. In Hall, P. J. (Ed.), *The square kilometre array: An engineering perspective* (pp. 231–249). Dordrecht: Springer.

Weiler, K. (2000). The promise of long wavelength radio astronomy. In Stone, R. G. (Eds.), *Radio astronomy at long wavelengths* (pp. 243–256). American Geophysical Union.

Wijnholds, S. J., van der Tol, S., Nijboer, R., & van der Veen, A. J. (2010). Calibration challenges for the next generation of radio telescopes. *IEEE Signal Processing Magazine, 27*(1), 32–42.

Wootten, A., & Emerson, D. (2005). *ALMA: Imaging at the outer limits of radio astronomy.* Paper presented at the IEEE International Conference on Acoustics, Speech, and Signal Processing (ICASSP '05), Philadelphia, USA.

ADDITIONAL READING

Baars, J. W. M. DÁddario, L.R. & Thompson, A.R. (August 2009). Radio Astronomy in the Early Twenty-First Century. In *Proceedings of the IEEE*, 97(8), (pp. 1377-1381).

Boonstra, A. J., & van der Veen, A. J. (2003, January). Gain calibration methods for radio telescope arrays. *IEEE Transactions on Signal Processing, 51*(1), 25–38.

Boonstra, A. J., Wijnholts, S. J., van der Tol, S., & Jeffs, B. (March 2005). *Calibration, sensitivity and RFI mitigation requirements for LOFAR.* Paper presented at the IEEE Int. Conf. Acoust., Speech, Signal Process. (ICASSP). Philadelphia, PA, USA (pp. V-869–V-872).

Harry L. Van Trees, (March 2002). *Optimum Array Processing (Detection, Estimation, and Modulation Theory, Part IV)*, Wiley.

Högbom, J. A. (1974, June). Aperture synthesis with nonregular distribution of intereferometer Baselines. *Astronomy & Astrophysics, 15*(Suppl.), 417–426.

Kraus, J. D. (1986). *Radio Astronomy* (2nd ed.). Powell, Ohio: Cygnus-Quasar.

Leshem, A., & van der Veen, A. J. (August 2000). Radio-astronomical imaging in the presence of strong radio interference. In *IEEE Trans. Inform. Theory (Special Issue on Inform. Theoretic Imag.)*, 46, (pp. 1730–1747).

Levanda, R., & Leshem, A. Synthetic aperture radio telescopes. (2010, January). Calibration Challenges for the Next Generation of Radio Telescopes. *IEEE Signal Processing Magazine, 27*(1), 14–29. doi:10.1109/MSP.2009.934719

Nijboer, R. J., & Noordam, J. E. (2007). LOFAR calibration. In *Astron. Data Anal. Software Syst. XVI.* R. A. Shaw, F. Hill, and D. J. Bell, Eds. 376, (pp. 237–240).

Rau, U., Bhatnagar, S., Voronkov, M., & Cornwell, T. (2009, August). Advances in calibration and imaging techniques in radio interferometry. *Proceedings of the IEEE, 97,* 1472–1481. doi:10.1109/JPROC.2009.2014853

Taylor, G., Carilli, C., & Perley, R. (1999) Synthesis imaging in radio-astronomy. In *Astron. Soc. of the Pacific*, National Radio Astronomy Observatory, Charlottesville, VA. A. Thompson, J. Moran, and G. Swenson, Eds., *Interferometry and Synthesis in Radio Astronomy.* New York: Wiley, 1986.

Van der Tol, S. Jeffs, B. & van der Veen, A.J. (September 2007). Self calibration for the LOFAR radio astronomical array. In *IEEE Trans. Signal Process.,* 55, (pp. 4497–4510).

Van der Veen, A. J., Leshem, A., & Boonstra, A. J. (2004, June). Array signal processing in radio-astronomy. *Experimental Astronomy, 17,* 231–249. doi:10.1007/s10686-005-0788-y

Vos, C. M., Gunst, A. W., & Nijboer, R. (2009, August). The LOFAR Telescope System Architecture and Signal Processing. *Proceedings of the IEEE, 97*(8), 1431–1437. doi:10.1109/JPROC.2009.2020509

Wijnholds, S. J., van der Tol, S., Nijboer, R., & van der Veen, A. J. (2010, January). Calibration Challenges for the Next Generation of Radio Telescopes. *IEEE Signal Processing Magazine, 27*(1), 32–42.

Wijnholds, S. J., & van der Veen, A. J. (2009, September). Multisource Self-Calibration for Sensor Arrays. *IEEE Transactions on Signal Processing, 57*(9), 3512–3522. doi:10.1109/TSP.2009.2022894

Wilsom, T. L., Rohlfs, K., & Hüttemeister, S. (2009). *Tools of Radio Astronomy.* Springer Berlin Heidelberg.

KEY TERMS AND DEFINITIONS

Radio Astronomy: Radio astronomy is a subfield of astronomy that studies objects in space at radio frequencies.

Low-Frequency Astronomy: Radio astronomy in the frequency range from 10 MHz to 1 GHz.

Phased Array: A group of antennas in which the relative phases of the signals of the antennas are varied in such a way that the effective radiation pattern of the array is reinforced in a desired direction and suppressed in undesired directions.

Digital Signal Processing: Representation of the astronomical signals and the processing of these signals.

Beam-Forming: Beam-forming is the signal processing technique for directional signal transmission or reception.

Correlator: A device for combining two signals in the signal chain.

Section 2
Multidisciplinary Advancements in Image Processing

Chapter 10
Advances in Moving Face Recognition

Hui Fang
Swansea University, UK

Nicolas Costen
Manchester Metropolitan University, UK

Phil Grant
Swansea University, UK

Min Chen
Swansea University, UK

ABSTRACT

This chapter describes the approaches to extracting features via the motion subspace for improving face recognition from moving face sequences. Although the identity subspace analysis has achieved reasonable recognition performance in static face images, more recently there has been an interest in motion-based face recognition. This chapter reviews several state-of-the-art techniques to exploit the motion information for recognition and investigates the permuted distinctive motion similarity in the motion subspace. The motion features extracted from the motion subspaces are used to test the performance based on a verification experimental framework. Through experimental tests, the results show that the correlations between motion eigen-patterns significantly improve the performance of recognition.

INTRODUCTION

In the last couple of decades, automatic computerised face recognition techniques have been developed for the security of accessing confidential information in both the internet virtual world and the real world. Recognition systems are required

to achieve high performance in a variety of different environments such as wide camera angles, different illuminations and various expressions.

Identity subspace learning techniques have improved significantly which can be traced back to Eigen-faces (Turk, 1991) designed for face recognition. Linear Discriminate Analysis (LDA) (Belhumeur, 1997), Active Appearance Models (AAM) (Cootes, 2001), Independent

DOI: 10.4018/978-1-60960-477-6.ch010

Component Analysis (ICA) (Bartlett, 2002) and other subspace methods, including kernel-based techniques (Torrs, 2002), have been proposed to provide various face manifolds to accurately describe the range of possible facial characteristics. In addition, a number of algorithms (Plataniotis, 2003; Costen, 2002) have been developed to alleviate the major forms of extraneous variation in facial modeling, which can be summarized as PIE (pose, illumination and expression).

Although identity subspace techniques have achieved some good recognition results, motion-based face recognition is expected to improve the performance due to more information which the face sequences contain. These advanced techniques could be applied to real-world problems. A number of commercial security-based systems have been developed and used in highly confidential environments, based on techniques for face alignment and recognition. These usually require a range of constraint conditions such as constant illumination and near frontal view. These systems are liable to attacks by malicious individuals who, for example, make facial masks which resemble genuine users or otherwise simulate parameters based on a forged face model similar to the system model. With new subspace learning algorithms, largely relying upon recently developed automatic means of tracking facial distortion within sequences, the effects of these problems can be reduced by analyzing the advanced motion models.

In psychological studies, dynamic information has also been shown to make contributions to improve recognition. Lander et al. (Lander, 2005) in particular show a significant beneficial effect of non-rigid movements. It is concluded that some familiar faces have characteristic motion patterns which act as an additional clue to recognition. O'toole et al. (O'toole, 2002) also suggest the motion features called dynamic signatures can help in identification. From the psychological experimental results, we can assume that motion features help to improve the computerized face recognition.

In (O'toole, 2002), it is mentioned that facial motion modeling provides at least three factors which can be used to improve recognition; largely derived from human psychological studies, which has been a fertile area of research in recent years. The first centres on the construction of a 3D structured face representation from temporally tracked 2D face movements. The second involves the robust calculation of the mean 2D appearance of the face from multiple frames. The third aspect investigates the distinctive movements shown by an individual face by encoding the variation in appearance within a single sequence. The results of extensive experiments show that the recognition performance is significantly improved by the inclusion of these factors.

In order to simulate a similar process indicated by the human experiments, two kinds of frameworks are widely explored for achieving the best motion effects. Most algorithms, such as (Zhou, 2003), use probabilistic approaches to improve recognition. This kind of motion information is called as robust confirmation by multiple frames in psychological research. Some other algorithms (Yamaguchi, 1998; Arandjelović, 2006; Edwards, 1999) correlate the statistical distribution of the face sequences corresponding with psychological dynamic signatures.

In this chapter, to show the advantages of applying face motion, we investigate using permutated distinctive motion similarity in the motion subspace for improving face recognition as the counterpart to the psychological work by Lander et al. (Lander, 2005). The characteristic motion distributions are extracted by permuting the eigenvectors derived from the concatenation of the parameters, encoded by a statistical model followed by correlating the pairs of gallery and probe sequences. It is then possible to combine this motion feature efficiently with the identity feature to achieve a better recognition performance compared with only using the identity feature. Although AAM+LDA subspace is more suitable for encoding the identity feature, the motion similarity extracted solely from the geometric model is robust to capture the individual dynamic signature.

In section 2, we introduce the background on face recognition techniques and the extended work on video sequences. Section 3 gives the registration framework and the subspace description on how to encode the face instances in face sequences, followed by explaining the design details of the proposed algorithm. Experimental results showing the efficiency of the method and the comparison with static face recognition methods are presented in section 4. Finally, the conclusion and future work are discussed in section 5.

BACKGROUND

Static face manifold techniques have been intensively explored over the last decade. Moghaddam et al. (Moghaddam, 1997) propose a Bayesian framework to recognize the faces based not only on the distance in the subspace but also on the orthogonal residue space. Costen et al. (Costen, 2002) design a paradigm to decompose the combined shape and appearance subspace to a group of non-orthogonal subspaces representing identity, light, pose and expression variations and recognize the face based only on the identity parameters. Kang et al. (Kang, 2002) project the appearance parameters into an LDA subspace and adjust the identity parameters based on a linear relationship according to the pose and expression variations. Although these subspace techniques have been effectively designed, the extra motion information obtained from face sequences is expected to improve the recognition rate.

The direct extension from static face recognition to moving face identification is a multi-frame confirmation method. Zhou et al. (Zhou, 2003) maximize a posterior probability by accumulation through the whole sequence. Jenkins et al. (Jenkins, 2008) claim that the face recognition rate can be improved significantly by using an average shape-free face. In recent years, dynamic signatures extracted from the moving face have been exploited to further improve face recognition systems.

There are two assumptions for using the dynamic information on the face. Many algorithms made the assumption that the moving patterns of individuals are predictive, so that they can use methods such as Auto-Regressive model (ARM) and Hidden Markov Models (HMM) to calculate the probability values. Liu et al. (Liu, 2003) use a Hidden Markov Model to model the probability of face change and use the probability for identification. Conversely, others assume that the motion is un-predictive even though some familiar behavioral patterns can be used for identity. (Arandjelovic, 2006) proposes a kernel PCA subspace to analyze the similarity of two distributions from two moving face sequences. They use an appearance-based model which makes the subspace highly nonlinear and the distance equation based on the trace of mean parameters' difference and covariance of two distributions. Edwards et al. (Edwards, 1999) proposed an on-line framework to learn a linear relationship between the static facial appearance and motion information and so adjust the central appearance based on this relationship. This process improved recognition performance compared with simply taking the sequence-mean. The length of the moving sequence is an important factor when learning a robust relationship between the movement and appearance.

Our algorithm is based on the second assumption and shares a similar structure to (Arandjelovic, 2006; Edwards, 1999). The difference is highlighted on both modeling and similarity measurements. A combined shape and texture model makes a more smooth subspace projection than Kernel PCA. When the motion subspace is used for measuring the similarity, permutated eigenvectors group similar motion of subjects as pairs to calculate it.

Figure 1. Mean face built in several iteration steps

Methodology

In our work, both of the motion features, identity centre tendency and dynamic signature, are integrated under a unified structure. The system follows several steps: firstly, the non-rigid facial motion is traced and registered robustly based on the Group-wise Registration algorithm; Secondly, all the frames are encoded by the combined shape and texture model; thirdly, the motion subspace features are extracted from these images. Finally, these features are combined with the robust identity feature encoded by the AAM+LDA subspace via a parallel confidence-based structure for the verification experiment.

Experimental Database

The BANCA database (Bailly-Bailliere, 2003) used here includes 52 subjects, half male and half female, each of which (verbally) claims the identity of two individuals four times (thus each individual can be used as both a true client and an imposter). Each such claim sequence has approximately 500 frames of frontal images, of which 6 representative frames of one individual are shown in Figure 2. The database is divided into two groups, g1 and g2, intended for a leave-some-out design. The 26 subjects in g1 and an extra 10 subjects (so totally 36 subjects) form an ensemble and are used to build the model, which allows testing of the encoding of the 26 subjects in g2 and vice versa. This both avoids overfitting of the model (individuals to be recognized are not present in the ensemble) and un-representativeness

of the model (faces swap between ensemble and gallery). One of the correct claims for each test individual is used in the gallery (those frames which are encoded to give specific knowledge about particular, identified individuals), all the other claims are probes and are so used to test recognition performance.

Because of the size of the database (about 20 seconds of recording for each person, thus over 250,000 images in the ensemble), representative frames are selected for each ensemble subject by building an appearance model of that individual, encoding each fame on this AM and using k-means clustering to divide the encodings into approximately 10 groups (one for each 50 frames). The frames most representative for each group are then selected and used to build both the AAM+LDA model.

General Group-Wise Registration

In order to find the correspondences more robustly, the general group-wise registration (GGR) framework (Cootes, 2010) is adopted and improved in the proposed algorithm. The linear regression based active appearance model is suitable for finding the feature points for building the motion subspace, but the motion we intend to investigate involves subtle non-rigid deformations which are hard to synthesize using a linear combination of a limited number of basic functions, as in the AAM.

The GGR method is developed in the minimum description length (MDL) framework based on image coding theory. It iteratively optimizes the model for coding all the image instances when

Figure 2. Representative frames from a BANCA individual. Every individual in BANCA dataset is annotated with GGR. The left most image is the first frame, using the "static image" condition. The images have the landmark positions indicated on them.

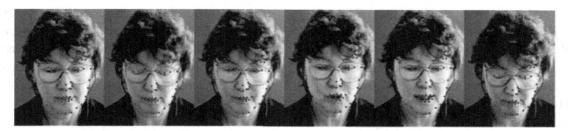

the correspondences are initialized and finds the best registration based on the updated model. These two steps represent minimizing two parts in the MDL formula: the first part is the coding length of the model and the other is the coding length of the data. The significant improvements in registration accuracy are demonstrated based on the framework, presented in (Cootes, 2010).

The implementation of the group-wise registration is summarized as follow: First, one random image is selected as a reference template and all other images are registered using a traditional template match. Next, a statistical shape model and texture model is built to represent the image set. Each image is represented in the model and the correspondences are refined by minimizing a cost function. Finally the statistical models are updated and the fitting repeated until convergence.

A coarse-to-fine scheme is adopted in the process. The model in the proposed algorithm uses a simple mean shape and texture built by warping all the faces to the mean shape using a triangular Delaunay mesh. The matching step deforms each image by minimizing a cost function. A coarse-to-fine scheme is applied to increase the number of control points and optimize their position. In the final iterations, the points are moved individually to minimize the cost. The cost function includes both shape and texture parts,

$$E = \frac{|r|}{\sigma_r} - a\sum_i (c - \frac{0.5\left\| d_i - (\Delta d_i + d_n) \right\|}{\sigma_s^2})$$

where r is the residue between the model and the current image after deformation, σ_r and σ_s are the standard deviations of the residue and shape, c is a constant, d_i is the position of the ith control point, d_n is the average position of the neighbourhood around point i and Δd_i represents the offset of the point from the average mean shape.

In Figure 1, it shows how the registration is improved by presenting the mean face built on updating the points' locations. The model represents the face structure more accurately as different subjects are progressively aligned together. In Figure 2, some samples in one sequence are shown as the registration results.

MEAN IDENTITY FEATURE FROM SEQUENCE

The faces are encoded using a combined shape and appearance model followed by the Linear Discriminate Analysis model; this takes the output of the GGR and approximates the manifold or high dimensional surface, on which the faces lie. This allows accurate coding, recognition and reproduction of previously unseen examples. Pixels defined by the GGR points as part of the face are warped to a standard shape, ensuring that the image-wise and face-wise coordinates of images are equivalent. If a rigid transformation to remove scale, location and orientation effects is performed on the point-locations, they can then be treated in the same way as the grey-levels, as again identical values for corresponding points on different faces will have the same meaning.

Redundancies between feature-point location and grey-level values can be removed and dimensionality estimated by Principal Components Analysis. Given a set of N vectors q_i (either the pixel grey-levels, or the feature-point locations) sampled from the images, the covariance matrix C of the images is calculated, and orthogonal unit Eigen-vectors Φ, and a vector of Eigen-values λ are extracted from C.

$$c = \frac{1}{N} \sum_{i=1}^{N} (q_i - \bar{q})(q_i - \bar{q})^T$$

where C is the covariance matrix reflecting the variations of the faces, represent the average vector of geometric feature or appearance feature;

Redundancies between configuration and pigmentation are removed by performing separate PCAs upon the shape and grey-levels, providing shape parameters and texture parameters (default setting to 1). These are combined to form a single vector for each image on which a second PCA is performed (Cootes, 2001). This gives a single feature vector for each image, assuming zero-mean weights and

$$x = \Phi_c^T \begin{bmatrix} W_s \Phi_s^T (q_s - \bar{q}_s) \\ \Phi_t^T (q_t - \bar{q}_t) \end{bmatrix}$$

W_s is a diagonal vector of weights compensating for the characteristic scales of the shape and texture parameters; Φ_s is the geometric eigen-vector and Φ_t is the texture eigen-vector. This appearance model allows the description of the face in terms of true, expected variation.

In the same sequence, regardless of parameter change due to different poses, lighting and expressions, the representation of the identity can be expected to be constant. Stabilization of representation by taking the mean of a number of samples is known to improve recognition performance (Zhou, 2003). However, in this case, the model will encode (even after averaging) both identity and non-identity variation. To remove the latter, a Linear Discriminate Analysis subspace (Belhumeur, 1997) is used ensure that only identity information is encoded. LDA is a linear subspace calculated from the n frames available for each of the p individuals in the ensemble, using,

$$C_B = \frac{1}{p} \sum_{i=1}^{p} (\bar{x}_i - \bar{x})(\bar{x}_i - \bar{x})^T$$

$$C_w = \frac{1}{np} \sum_{i=1}^{p} \sum_{j=1}^{n} (x_{ij} - \bar{x}_i)(x_{ij} - \bar{x}_i)^T$$

to give

$$C_D = C_B C_W^{-1}$$

where C_B is the covariance matrix between different classes, C_W is the covariance matrix within all the classes, C_D reflects the ratio between inter-class covariance and intra-class covariance, p defines the number of the classes, n defines the number in each class, x_{ij} represents the face vector and $\bar{x}_i$ represents the average vector of one class.

Eigen-vectors are then taken to provide a subspace which maximizes variation between individuals and minimizes that within them. Each frame in a sequence is projected onto this subspace to give a discriminative encoding d, before taking the mean and assessing similarity with other sequences.

$$S_c = \frac{\bar{d}_I \ \bar{d}_J}{|\bar{d}_I| \ |\bar{d}_J|}$$

where S_c is the similarity measure between identity features from two sequences.

MOTION SUBSPACE ACQUISITION AND ANALYSIS

The proposed method uses the shape model for the motion subspace when generating the identity subspace by using the AAM+LDA model. Although texture variations is the most important feature to recognize a face, the models encoding the shape variations have more benefits to analyze the face motion which is best represented by the shape changes.

When the moving images in this sequence are projected into the geometric subspace, the variation of these encoded representations reveals the movement of the face; we assume the distinctive facial motion of individuals, mainly based on the non-rigid deformations caused by talking and changing expressions after compensating the rigid similarity transform, is helpful for identity. Although this kind of feature is not robust compared with the identity subspace features, it can serve as a subsidiary feature by using an integrated framework. Based on this assumption, the characteristic motion patterns are captured by using PCA performed upon all of the encodings x of a single sequence to yield a single set of Eigen-vectors Φ. Here, each Eigen-vector represents a significant facial dynamic change. Therefore, we make the Eigen-vectors correspondent across pair of gallery sequence and test sequence before measuring similarity of the motion. The similarity of pairs of spaces can be compared as:

$$S_m = Max\{trace\left(\left(\Phi_i^{l^T}\right)\Psi_j\right) \mid \Phi_i^l$$

where S_m is the similarity value between two sequences, N defines the number of the distinctive motion patterns, Φ_i and Ψ_j are the distinctive motion patterns extracted from two sequences. Note that the ordering of the components extracted by the PCA will be determined by their associated Eigen-values and so the value of S_m is liable to be reduced by variation in the magnitude of different behavioral tendencies in different sequences of the same individual. This is overcome here by permuting the columns of the matrix describing the relationship between any two sequences before taking the trace so as to maximize Sm. In addition, Φ_i and Φ_j may be (identically) truncated to remove noisy, low-variance eigenvectors.

FRAMEWORK STRUCTURE

Although recognition may be performed on both the robust identity feature and motion patterns independently, neither shows perfect performance and it is possible to combine the two, balancing one against another. A confidence-based fusion system (Poh, 2005) is adopted, based on how definite is each type of information. The fused similarity can be calculated,

$$S_f = \frac{C_{S_c}}{C_{S_c} + aC_{S_m}}\left(\frac{S_c - \mu_{S_c}}{\sigma_{S_c}}\right) + \frac{aC_{S_m}}{C_{S_c} + aC_{S_m}}\left(\frac{S_m - \mu_{S_m}}{\sigma_{S_m}}\right)$$

Here μ and σ are the average and standard deviation of the similarity measures, allowing normalization. C is a confidence value associated with the two types of information, calculated as the absolute difference between the False Acceptance and False Rejection rates; this is dependent upon the similarity between probe and gallery for that information type and gives a greater weighting to parameters which will yield confident decisions to either accept or reject and individual. Finally α is a weight value, trading between the central tendency and movement measures. All of these

Figure 3. ROCs based on facial motion similarity (a) built by combined shape and texture model (b) built by shape model

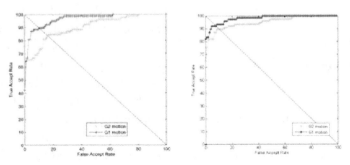

parameters can be derived from a set of calibration sequences.

EXPERIMENTS

Experiment Setup

Facial verification is an application of face recognition which seeks to determine whether an individual's real and claimed identity (as determined by some other means such as a password, identity card, or, as here, a spoken name) coincide. Thus, centring on one-to-one matches, it differs from facial familiarity or identification. Both of these involve one-to-many matches but the latter requires recording which gallery face is most similar to the probe, while the former concerns the degree of similarity to any and all of the gallery. The same features and combination methods could be used in these cases.

The proposed model yields a similarity value for each probe and its claimed identity, which will be either a True Entry or an Imposter Attack. Across the entire set of probes, each similarity can be used as a threshold, so generating a Receiver Operating Characteristic (ROC) curve, which summarizes its performance by trading between the False Acceptance (FA) and False Rejection (FR) rates as shown in Figures 3. Two performance criteria, Equal Error Rate (EER) and the

Area under the Curve (AUC) are used to present the results. In a typical authentication application, the EER is the best tradeoff, at the point of equal FA and FR. However, this parameter only concerns a single point on the ROC curve, and so is not necessarily representative of all changes (particularly those associated with the extremes of FA and FR). AUC overcomes this, measuring the total discrimination, and ranging from 1 (for perfect discrimination) to 0 (for perfect reversed discrimination). If building an actual application, an individual threshold can be chosen based on the relative risks of the two types of error.

EXPERIMENTAL RESULTS

We evaluate the performance of the static verification system by two means of selecting images. Firstly, we select the very first frame from each gallery and probe sequence and encode those alone, on the S_c measure. Secondly, we randomly select one image from each sequence and encode those parameters. The individual models which provide the motion pattern parameters are each derived from approximately 500 observations, which each has 41 dimensions. The lower of these figures limits the number of parameters which can be derived from the PCA, but some, particularly those with small Eigen-values may be spurious. There is no noticeable interaction between the means of controlling the number of parameters, but performance continues increasing until the maximum is reached. This pattern is probably an unfortunate consequence of the relatively small ensemble (36 individuals) and the high similarity between gallery and probe sequences. It is possible that with more variation between the gallery and probes some significant truncation would be required.

The advantages of the proposed work compared to the existing techniques can be summarized as:

1. The General Group-wise Registration (GGR) algorithm (Cootes, 2010) can supply more robust correspondences across all the different subjects;
2. The permutation operation can help to find the highest correlated distinguished motion pattern of the same subject;
3. The geometry based motion feature is shown to achieve a better performance and integrates with the identity feature for recognition.

The geometric based motion subspace has a better performance compared with the subspace built by combined shape and texture models presented in (Fang, 2008). Figure 3 shows the Receiver Operating Characteristic (ROC) curve, which summarizes its performance by trading between the False Acceptance (FA) and False Rejection (FR) rates, showing that the Equal Error Rate (EER) improves about 5% and 7% separately in each BANCA dataset groups.

Quantitative comparisons for the two groups with 41 eigenvectors are shown in Table 1. Although the confidence measures and means and standard deviations can be simply read from the similarity values of the calibration set (*g*1 for *g*2 verification and vice versa), the weight value α must be selected given the classification performance of the combined parameters. Note that a weight of 0 implies the use of the robust identity feature alone, and that an α value in the range 0 – 0.2 improves classification performance. This relatively low importance of movement is understandable, given the relatively low discrimination shown by the motion patterns, and corresponds to the relative difficulty humans find in learning or recognizing faces from motion alone, without grey-level information. The EER and AUC values given in Table 1 shows the combined robust identity feature and motion pattern gives a much lower EER and higher AUC than the static images.

CONCLUSION AND FUTURE WORK

We describe a framework to use the motion information in facial sequences to improve face recognition. The major correlated variations across the sequences of the same subject are captured as auxiliary information to integrate into the face recognition framework. The associated features from the moving faces achieve noticeably better performance than using the static images or other methods to use the motion information.

Table 1. Performance evaluation based on EER and AUC

Data Type	EER		AUC	
	G1	G2	G1	G2
Random Frame	4.41	3.94	0.9869	0.9965
First Frame	3.29	4.64	0.9965	0.9958
Identity Feature	3.93	1.92	0.9972	0.9991
Fusion 1	1.68	1.68	0.9995	0.9994

The results also reflect how faces are recognized by the human cognitive system when presented with video sequences. As the cognitive system accumulates the stabilized appearance information and characteristic motion from the sequence, it predominately recognizes the face based on the appearance. However, if the appearance features are indistinct but the confidence in the characteristic motion patterns is high, the decision will be made based on the latter features. This is a possible reason why face recognition can be improved using video sequences in both human and computer vision.

In future work, the geometric based motion feature will be integrated with the identity feature. Based on Figure 3, the performance of a combination of these two features is expected to achieve a better result compared with the current presented result.

REFERENCE

Arandjelovi'c, O., & Cipolla, R. (2006). An information-theoretic approach to face recognition from face motion manifolds. *Image and Vision Computing, 24*, 639–647. doi:10.1016/j.imavis.2005.08.002

Bailly-Bailliere, E., Bengio, S., Bimbot, F., et al. (2003). The banca database and evaluation protocol. In *Proceedings of ICAVBPA*, (pp. 625–638).

Bartlett, M., Movellan, J., & Sejnowski, T. (2002). Face recognition by independent component analysis. *IEEE Transactions on Neural Networks, 13*, 1450–1464. doi:10.1109/TNN.2002.804287

Belhumeur, P. N., Hespanha, J. P., & Kriegman, D. J. (1997). Eigenfaces vs. Fisherfaces: Recognition using class specific linear projection. *IEEE Transactions on Pattern Analysis and Machine Intelligence, 19*, 711–720. doi:10.1109/34.598228

Cootes, T. F., Edwards, G. J., & Taylor, C. J. (2001). Active appearance models. *IEEE Transactions on Pattern Analysis and Machine Intelligence, 23*, 681–685. doi:10.1109/34.927467

Cootes, T. F., Twining, C. J., Petrovic, V. S., Babalola, K. O., & Taylor, C. J. (in press). Computing accurate correspondences across groups of images. *IEEE Transactions on Pattern Analysis and Machine Intelligence*.

Cootes, T. F., Wheeler, G. V., Walker, K. N., & Taylor, C. J. (2002). View-based active appearance models. *Image and Vision Computing, 20*, 657–664. doi:10.1016/S0262-8856(02)00055-0

Costen, N., Cootes, T., Edwards, G., & Taylor, C. (2002). Automatic extraction of the face identity-subspace. *Image and Vision Computing, 20*, 319–329. doi:10.1016/S0262-8856(02)00004-5

Costen, N. P., Cootes, T. F., & Taylor, C. J. (2002). Compensating for ensemble-specific effects when building facial models. *Image and Vision Computing, 20*, 673–682. doi:10.1016/S0262-8856(02)00057-4

Edwards, G., Taylor, C., & Cootes, T. (1999). Improving identification performance by integrating evidence from sequences. In *Proceedings of Computer Vision and Pattern Recognition*, 1486-1491.

Fang, H., & Costen, N. (2008). Behavioral consistency extraction for face verification. *Proceedings of COST 2102 Conference*, (pp. 291-305).

Jenkins, R., & Burton, A. M. (2008). 100% accuracy in automatic face recognition. *Science, 319*, 435. doi:10.1126/science.1149656

Kang, H., Cootes, T. F., & Taylor, C. J. (2002). A comparison of face verification algorithms using appearance models. *Proceedings of British Machine Vision Conference, 2*, 477–486.

Lander, K., & Chuang, L. (2005). Why are moving faces easier to recognize? *Visual Cognition, 23*, 429–442. doi:10.1080/13506280444000382

Liu, X., & Chen, T. (2003). Video-based face recognition using adaptive Hidden Markov Models. In *Proceedings of Computer Vision and Pattern Recognition*, (pp. 26–33).

Lu, J., Plataniotis, K., & Venetsanopoulos, A. (2003). Face recognition using kernel direct discriminate analysis algorithms. *IEEE Transactions on Neural Networks*, *14*, 117–126. doi:10.1109/TNN.2002.806629

Moghaddam, B., & Pentland, A. (1997). Probabilistic visual learning for object representation. *IEEE Transactions on Pattern Analysis and Machine Intelligence*, *19*(7), 696–710. doi:10.1109/34.598227

O'Toole, A., Roark, D., & Abdi, H. (2002). Recognizing moving faces: A psychological and neural synthesis. *Trends in Cognitive Sciences*, *6*, 261–266. doi:10.1016/S1364-6613(02)01908-3

Poh, N., & Bengio, S. (2005). *Improving fusion with margin-derived confidence in biometric authentication tasks.* (LNCS 3546), (pp. 474–483).

Turk, M., & Pentland, A. (1991). Face recognition using Eigenfaces. *Proceedings of the IEEE Conference on Computer Vision and Pattern Recognition*, (pp. 586-591).

Yamaguchi, O., Fukui, K., & Maeda, K. (1998). Face recognition using temporal sequence. In. *Proceedings of IEEE International Conference on Automatic Face and Gesture Recognition*, *10*, 318–323. doi:10.1109/AFGR.1998.670968

Zhou, S., Krueger, V., & Chellappa, R. (2003). Probabilistic recognition of human faces from video. *Computer Vision and Image Understanding*, *91*, 214–245. doi:10.1016/S1077-3142(03)00080-8

ADDITIONAL READING

Aggarwal, G., Chowdhury, A., & Chellappa, R. (2004). A system identification approach for video-based face recognition. *17th International Conference on Pattern Recognition*, (pp.175-178)

Arandjelović, O., & Cipolla, R. (2006). Face recognition from video using the generic shape-illumination manifold, *In Proceedings of European Conference on Computer Vision (pp.27-40)*

Baker, S., Matthews, I., & Schneider, J. (2004). Automatic construction of active appearance models as an image coding problem. *IEEE Transactions on Pattern Analysis and Machine Intelligence*, *26*(10), 1380–1384. doi:10.1109/TPAMI.2004.77

Benedikt, L., Kajic, V., Cosker, D., Rosin, P., & Marshall, C. (2008). Facial dynamics in biomedtric identification, *Proceedings of British Machine Vision Conference*, (pp. 1065–1075)

Bettinger, F., Cootes, T., & Taylor, C. (2002). Modelling facial behaviours. *Proceedings of British Machine Vision Conference*, (pp. 797-806)

Blanz, V., & Vetter, T. (2003). Face recognition based on fitting 3D morphable model. *IEEE Transactions on Pattern Analysis and Machine Intelligence*, *25*(9), 1–12. doi:10.1109/TPAMI.2003.1227983

Cootes, T., Twining, C., Petrovic, V., Schestowitz, R., & Taylor, C. (2005). Groupwise construction of appearance model using piece-wise affine deformations. *Proceedings of British Machine Vision Conference*, (pp. 879-888)

Craw, I., Costen, N., & Kato, T. (1999). How should we represent faces for automatic recognition? *IEEE Transactions on Pattern Analysis and Machine Intelligence*, *21*(8), 725–736. doi:10.1109/34.784286

Edwards, G., Lanitis, A., Taylor, C., & Cootes, T. (1996). Modelling the variability in face images, *IEEE International Conference on Automatic Face and Gesture Recognition,* (pp. 328–333)

Edwards, G., Taylor, C., & Cootes, T. (1999). *Improving identification performance by integrating evidence from sequences* (pp. 1486–1491). IEEE Proceedings of Computer Vision and Pattern Recognition.

Gross, R., Matthews, I., & Baker, S. (2005). Generic vs. person specific active appearance models. *Image and Vision Computing, 23*(10), 1080–1093. doi:10.1016/j.imavis.2005.07.009

Hadid, A., & Pietik¨ainen, M. (2004). From still image to video-based face recognition: an experimental analysis. *Proceedings of the Sixth IEEE International Conference on Automatic Face and Gesture Recognition,* (pp.17-19)

Hadid, A., & Pietik¨ainen, M. (2009). combining appearance and motion for face and gender recognition from videos. *Pattern Recognition, 42*(11), 2818–2827. doi:10.1016/j.patcog.2009.02.011

Hanley, J., & McNeil, B. (1983). A method for comparing the areas under receiver operating characteristic curves derived from the same cases. *Radiology, 148,* 839–843.

Hill, H., & Johnston, A. (2001). Categorizing sex and identity from the biological motion of faces. *Current Biology, 11,* 880–885. doi:10.1016/ S0960-9822(01)00243-3

Jones, M., & Poggio, T. (1998). Multidimensional morphable models: a framework for representing and matching object classes. *International Journal of Computer Vision, 29,* 1–28. doi:10.1023/A:1008074226832

Kryszczuk, K., Richiardi, J., Prodanov, P., & Drygajlo, A. (2007). Reliability-based decision fusion in multimodal biometric verification systems. *EURASIP Journal on Advances in Signal Processing, 2007,* 74–82. doi:10.1155/2007/86572

Lee, K., Ho, J., Yang, M., & Kriegman, D. (2003). Video-based face recognition using probabilistic appearance manifolds, *IEEE Conference on Computer Vision and Pattern Recognition,* (pp.313)

Phillips, P., Moon, H., Rizvi, S., & Rauss, P. (2000). The FERET evaluation methodology for face recognition algorithms. *IEEE Transactions on Pattern Analysis and Machine Intelligence, 22*(10), 1090–1104. doi:10.1109/34.879790

Phillipsand, P., Flynn, P., & Scruggs, T. (2005). *Overview of the face recognition grand challenge* (pp. 947–954). IEEE Proceedings of Computer Vision and Pattern Recognition.

Su, Y., Shan, S., Chen, X., & Gao, W. (2009). Hierarchical ensemble of global and local classifiers for face recognition. *IEEE Transactions on Image Processing, 18*(8), 1885–1896. doi:10.1109/ TIP.2009.2021737

Torres, L., & Vila, J. (2002). Automatic face recognition for video indexing applications. *Pattern Recognition, 35,* 615–625. doi:10.1016/ S0031-3203(01)00064-4

Wright, J., Yang, A., Ganesh, A., Sastry, S., & Ma, Y. (2010). Robust face recognition via sparse representation. *IEEE Transactions on Pattern Analysis and Machine Intelligence, 31*(2), 210–227. doi:10.1109/TPAMI.2008.79

Zhao, W., Chellappa, R., Philips, P., & Rosenfeld, A. (2003). Face recognition: a literature survey. *ACM Computing Surveys, 35*(4), 399–458. doi:10.1145/954339.954342

Zitova, B., & Flusser, J. (2003). Image registration method: a survey. *Image and Vision Computing, 21,* 977–1000. doi:10.1016/S0262-8856(03)00137-9

Chapter 11
Facial Image Processing in Computer Vision

Moi Hoon Yap
University of Bradford, UK

Hassan Ugail
University of Bradford, UK

ABSTRACT

The application of computer vision in face processing remains an important research field. The aim of this chapter is to provide an up-to-date review of research efforts of computer vision scientist in facial image processing, especially in the areas of entertainment industry, surveillance, and other human computer interaction applications. To be more specific, this chapter reviews and demonstrates the techniques of visible facial analysis, regardless of specific application areas. First, the chapter makes a thorough survey and comparison of face detection techniques. It provides some demonstrations on the effect of computer vision algorithms and colour segmentation on face images. Then, it reviews the facial expression recognition from the psychological aspect (Facial Action Coding System, FACS) and from the computer animation aspect (MPEG-4 Standard). The chapter also discusses two popular existing facial feature detection techniques: Gabor feature based boosted classifiers and Active Appearance Models, and demonstrate the performance on our in-house dataset. Finally, the chapter concludes with the future challenges and future research direction of facial image processing.

INTRODUCTION

Face offers the most natural way of authenticating a person and has become a popular biometrics technique available for security (Zhang, 2000). By following the norm of the biometrics techniques, face recognition and facial expression recognition has emerged from manual process to automatic process. The evolving technology has motivated the computer vision scientists' involvement in facial analysis.

DOI: 10.4018/978-1-60960-477-6.ch011

The application of computer vision in face processing remains an important research field. This chapter reviews and demonstrates the computer vision techniques applied to facial image processing and analysis. In addition, we discuss the recent advances in facial image processing in computer vision.

First we outline some reliable face detection techniques and provide the best recommendation. For the last decade, Haar cascades technique is known as a popular face detection technique especially in real-time application. We illustrate the performance of Haar cascades and enhance its performance by pre-processed the images with skin colour segmentation. Additionally, we provide some demonstration on the effect of computer vision algorithms on face images, which exemplify but not typify the algorithms.

Secondly, we review the existing computer vision techniques in facial feature detection. To recognise the facial expression, it is crucial to detect the changes in facial features. Unlike face recognition, facial expression recognition aims to classify the expressions by finding a model for non-rigid patterns of facial expression and it is expected to perform better by using a set of image sequences. Besides, psychological points of view are well-known as important approach in facial expression analysis. Hence, computer animation scientists define the facial feature points based on MPEG-4 Standard, which is inspired by Facial Action Coding System (FACS). In computer vision, most of the researchers define a reasonable set of facial feature points in tracking the facial feature changes based on FACS or MPEG-4 Standard. Scale Invariant Feature Transform (SIFT) has gained popularity in image matching, but how well it works on face images is a question for which we need an answer. Two important benchmarks in facial feature point detection are also discussed in this chapter i.e. Gabor Feature based Boosted Classifiers and Active Appearance Models.

Finally, we conclude the chapter with some discussions in the future challenges of facial analysis and future direction of facial image processing in computer vision.

BACKGROUND

The study in facial expression has been conducted in last century (Darwin, 1872; P. Ekman, 1973), and within the past 12 years considerable progress has been made in automatic analysis of facial expression from digital video input (J.F Cohn, 2007; J.F. Cohn & Kanade, 2007; Fasel & Luettin, 2003; Tian, Cohn, & Kanade, 2005). In early research, Ekman et al (P. Ekman & Friesen, 1976) reported a new method of describing facial movement based on an anatomical analysis of facial action. To capture subtlety of human emotion and paralinguistic communication, automated recognition of fine-grained changes in facial expression is needed (Tian, Kanade, & Cohn, 2001).

Early researches in automatic analysis and recognition of facial actions from input video focused on the relatively tractable problem of posed facial actions acquired under well-controlled conditions. Recent work has progressed to spontaneous facial actions, subtle facial actions, variation in illumination and merging with synthesis faces (i.e. avatar separate identity from facial behaviour) (J.F Cohn, 2007). Computer facial expression analysis systems need to analyze the facial actions regardless of context, culture, gender, and so on. The accomplishments in psychological studies, human movement analysis, face detection, face tracking, and recognition motivate the automatic facial expression analysis (Tian, et al., 2005).

FACE DETECTION

Research in face detection generally means to detect human faces and putting them into ellipsoid or rectangle boxes. Face detection algorithms are quite reliable in or near real-time condition, however, there is much works that need to be done for detailed description of external and internal facial

Figure 1. Haar Cascade face detection on CMUMIT database

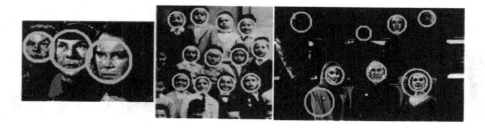

features (Ding & Martinez, 2008). Over the past years, researchers have developed various face detection techniques. These include:

- Top-down knowledge-based approach – rule-based methods
- Bottom-up feature-based approach – based on invariant face feature extraction
- Template-matching based approach – predefined template
- Appearance-based approach – learning from examples of face and non-face images to construct face characteristics, rely on statistical analysis and machine learning

Some of the representative work in face detection include Neural Network-based (Rowley, Balaju, & Kanade, 1998), Distribution-based (Sung & Poggio, 1998), Naives Bayes Classifiers(Schneiderman & Kanade, 1998), Support Vector Machine (Osuna, Freund, & Girosi, 1997), Hidden Markov Model (Nefian & Hayes, 1998), Sparse Network of Winnows (Roth, Yang, & Ahuja, 2000), Local Successive Mean Quantization Transform Features and Split up Sparse Network of Winnows Classifier (Nilsson, Nordberg, & Claesson, 2007), Information Theoritical Approach (Colmenarez & Huang, 1997) and Haar Cascades (Viola & Jones, 2001). Among these techniques, our recommendation is the Haar Cascade face detection algorithm by Viola and Jones (Viola & Jones, 2001) which is a fast and reliable algorithm in detecting faces on both image and video. Figure 1 shows the perfor-

mance of haar cascade face detection on CMUMIT database (Rowley, et al., 1998; Sung & Poggio, 1998), which produced some false positives and some false negatives. Since we are dealing with colour images/videos, we improved the accuracy of the algorithm with skin colour segmentation, which is further elaborated in next section.

It is noteworthy that the Haar Cascades implemented in this chapter is an implementation of OpenCV (Bradski, 2008) Haar Cascades algorithms, where Haar cascade for face detection is adopted from viola et al (Viola & Jones, 2001) and eyes cascade, mouth cascade and nose cascade are adopted from Castrillon-Santana et al (Castrillon-Santana, Deniz-Suarez, Anton-Canalis, & Lorenzo-Navarro, 2008).

In the following section, we will discuss about the roles of image processing techniques in improving the face detection rates.

SKIN COLOUR SEGMENTATION

Skin colour has been used widely in face detection (Hsu, Abdel-Mottaleb, & Jain, 2002; Shin, Chang, & Tsap, 2002; Storring, Andersen, & Granum, 1999). Input of colour images is typically in *RGB* format, but *RGB* components are subject to illumination. Hence, most of the techniques use colour components in the colour space, such as *HSV* or *YIQ* format (Kim, Shim, & K., 2003). Kim et al (Kim, et al., 2003) used *YCbCr* components in Matlab functions. *Y* contains luminance information, while *Cb* and *Cr* contain chrominance information. Hsu et al (Hsu, et al., 2002) used

Figure 2. (a) original image; (b) Skin colour segmentation

Figure 3. Effects of skin colour segmentation on (a) single face; (b) multiple faces

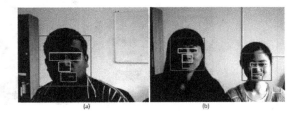

YCbCr colour space transformations to detect skin pixels in the normalized colour appearance using "reference white". Skin colour was modelled using the Gaussian distribution, then detected by Mahalanobis distance. Shin et al (Shin, et al., 2002) has done a study in determining the best colour space and they concluded that *YCbCr* is the best in their three measurements (out of four measurements). The conversion between *RGB* components and *YCbCr* components are represented by the following formulas:

$$\begin{bmatrix} Y \\ Cb \\ Cr \end{bmatrix} = \begin{bmatrix} 16 \\ 128 \\ 128 \end{bmatrix} + \begin{bmatrix} 65.481 & 128.553 & 243966 \\ -37.797 & -74.203 & 112 \\ 112 & -93.786 & -18.214 \end{bmatrix} \begin{bmatrix} R \\ G \\ B \end{bmatrix}$$

$$\begin{bmatrix} R \\ G \\ B \end{bmatrix} = \begin{bmatrix} 0.00456621 & 0 & 0.00625893 \\ 0.00456621 & -0.00153632 & -0.00318811 \\ 0.00456621 & 0.00791071 & 0 \end{bmatrix} \left(\begin{bmatrix} Y \\ Cb \\ Cr \end{bmatrix} - \begin{bmatrix} 16 \\ 128 \\ 128 \end{bmatrix} \right)$$

Each pixel was classified as skin or non-skin based on the mean and standard deviation of *Cb* and *Cr* components (Kim, et al., 2003).

To demonstrate the effectiveness of colour segmentation in separating the skin and non-skin regions, we implement skin colour segmentation by using the information from *YCrCb* colour space and *HSV* colour space. The skin region can be segmented by the presence of certain set of Chrominance values and Hue values. From our experiments, the suitable range color range works the best for the webcam model - Microsoft

Lifecam 2.0 is: *Cb<=125 or Cb>=160; Cr<=100 or Cr>=135; 26<Hue<220*. The select region is the skin region, and by using face detection algorithm, we can detect the face region, the result of skin colour segmentation is illustrated in Figure 2.

One important application in skin colour segmentation is introduced in this chapter, i.e., we apply it to reduce the false positives in Haar Cascades Face Detection. By pre-process the images with skin colour segmentation, we managed to reduce the false positives and improved the accuracy in face detection. Figure 3 illustrates the results of combining skin colour segmentation with Haar Cascades Face Detection algorithm on single face and multiple faces.

The following section is to provide a guideline of resources in facial feature point standard (MPEG-4 Standard) and introduce two existing works in facial feature point detection techniques.

FACIAL FEATURE POINTS DETECTION

There are different definitions for facial feature points. In psychology, Facial Action Coding System (FACS) (P. Ekman & Friesen, 1978) provides an objective and comprehensive language for describing facial expressions. It is a human-observer-based system designed to detect subtle changes in facial features. In computer animation, MPEG-4 Standard (Petajan, 2005) provides a detailed outline of facial feature points, where some of the feature points are too difficult to detect by computer vision algorithms. Hence, many computer vision scientists have reduced the points or define a set of feature points based on Facial Action Coding System (FACS) (P. Ekman & Friesen, 1978) into a reasonable set of feature points, which enable them to track the facial movements. For instance, Vukadinovic and Pantic (Vukadinovic & Pantic, 2005) defines facial feature points as a set of facial salient points such as the corners of the eyes, corners of the eye brows, corners and outer mid points of the lips, corners of the nostrils, tip of the nose, and the tip of the chin. Another important algorithm in localisation of facial feature points is Active Appearance Models (Matthews & Baker, 2004), which is widely used in face tracking and face alignment.

In this section, we discuss the FACS, MPEG-4 Standard, their performance in implementing the computer vision feature salient points on face images and the performance of two benchmark algorithms in facial feature point detection - Gabor Feature Based Boosted Classifiers and Active Appearance Models.

Facial Action Coding System (FACS)

The FACS is the human-observer-based system designed to detect subtle changes in facial features (P. Ekman & Friesen, 1978). It is purely descriptive and uses no emotion or other inferential labels. It is a method to measure facial expressions by identifying the muscular activity underlying transient changes in facial appearance. The System defines 46 action units, which roughly correspond to the movement of each of the individual facial muscles. FACS is the leading method for measuring facial movement in behavioural science. FACS Measurement units are AUs, not muscles because in a few appearances more than one muscle was combined into a single AU; the appearance changes produced by one muscle were sometimes separated into two or more AUs to represent relatively independent actions of different parts of the muscle. Action

Units (AU) are the smallest visually discriminable facial movement (J.F Cohn, Ambadar, & Ekman, 2006). By using FACS and viewing videotaped facial behaviour in slow motion, coders can manually code all possible facial displays. AUs may occur in a single unit or in combinations by additive (independent) or non-additive (change each other's outlook).

There are a few facial expression image databases available for research. For the ease of comparison, we chose the most representative database, i.e. Cohn-Kanade AU-Coded Face Expression Image Database (Kanade, Cohn, & Tian, 2000) and MMI database (Pantic, Valstar, Rademaker, & Maat, 2005), which are coded by certified coders and are publicly available.

MPEG-4 Standard

MPEG-4 standard is widely used in computer animation, it is inspired by FACS (P. Ekman & Friesen, 1978). Besides, it contains a comprehensive set of tools for representing and compressing content objects and the animation of those objects (Petajan, 2005). SNHC (Synthetic & Natural Hybrid Coding) in MPEG-4:

- **FDP (Face Definition Parameters)**: define different feature points on the face (Petajan, 2005) as shown in Figure 4 (labelling based on ISO/IEC 14496-2 2001 guide (ISO/IEC, 2001)
- **FAP (Face Animation Parameters)**: define the movement of different parts of the face (Puri & Eleftheriadis, 2008) as illustrated in Figure 5. FAPs are the displacements of the feature points from the neutral face definition.

With the advent of MPEG-4, there has been a move from FACS to MPEG-4 in the computer vision community. The definition is not based on the specific movement of muscles or directly

Figure 4. Face definition parameters (labelling based on ISO/IEC 14496-2 2001 guide (ISO/IEC, 2001))

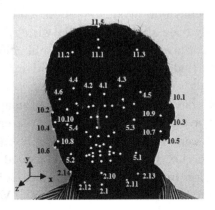

Figure 5. Face animation parameters (labelling based on ISO/IEC 14496-2 2001 guide (ISO/IEC, 2001))

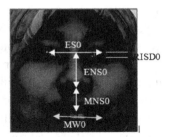

IRISD0: Iris Diameter
ES0: eye separation
ENS0: eye-nose separation
MNS0: mouth-nose separation
MW0: mouth width

related to emotions, but is a standard which is more convenient for facial animation.

Scale Invariance Features Transform (SIFT)

In this section, we would like to investigate whether or not the method for extracting distinctive invariant features from images can be used to perform reliable matching between different facial expressions. SIFT has gained popularity in solving a number of image matching problems. It transforms image data into scale-invariant coordinates relative to local features (Lowe, 2004). In staged filtering approach, the scale-invariant features are efficiently identified. First, the key

locations in scale space is identified by looking for locations that are maxima or minima of a difference-of-Gaussian function (Lowe, 1999). Then the feature vector is generated from each point to describe the local image region sampled relative to its scale-space coordinate frame, which named as "SIFT keys" by Lowe (Lowe, 1999). Lowe implemented nearest-neighbour approach in indexing on the SIFT keys to identify the candidate object models. We encourage the readers to refer to (Lowe, 1999) for the detailed discussion on SIFT.

A simple experiment implementing Lowe's algorithm (Lowe, 2004) is conducted and the result is shown in Figure 6 and Figure 7. Figure 6 shows a good match for the images with similar expression (neutral faces). On the other hand, Figure 7 illustrates a bad match for the image with different expression (a neutral face to a smiley face). This

Figure 6. Matched for images with similar facial expression

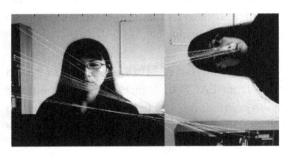

Figure 7. Unmatched for images with different facial expression

Figure 8. Facial feature point detection on image frames from MMI database and Cohn-Kanade AU-Coded Facial Expression Database (@Jeffry Cohn)

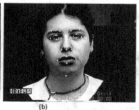

elicit that SIFT is not suitable in features tracking for facial expression analysis.

Gabor Feature Based Boosted Classifiers

Vukadinovic and Pantic (Vukadinovic & Pantic, 2005) defines 20 facial points in images of neutral faces. The defined facial feature points are: the corners of the eyes, corners of the eyebrows, corners and outer mid points of the lips, corners of the nostrils, tip of the nose, and the tip of the chin (Vukadinovic & Pantic, 2005). They implemented Viola and Jones face detector (Viola & Jones, 2001) and divided the detected face into 20 regions of interest. Then, each of the regions is examined further to predict the location of the facial feature points. They used the individual feature patch templates to detect points in the relevant region of interest. The feature models are GentleBoost templates built from the gray level intensities and Gabor wavelet features (Vukadinovic & Pantic, 2005). Donato et al (Donato, 1999) has proven that Gabor approach outperformed Principal Component Analysis (PCA), Fisher's Linear Discriminant (FLD) and Local Feature Analysis (LFA). Gabor filters remove most of the variability in image in different illumination and contrast, it also robust against small shift and deformation (Vukadinovic & Pantic, 2005).

Figure 9. Facial feature point detection on image frames from our in-house data

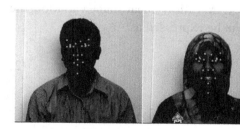

Some of the results from the implementation of this method on MMI database (Pantic, et al., 2005) and Cohn-kanade AU-Coded database (Kanade, et al., 2000) are shown in Figure 8. Figure 9 illustrates the similar algorithms on our in-house data. For further details of the Gabor Feature Based Boosted Classifiers in facial feature point detection, please refer to Vukadinovic and Pantic (Vukadinovic & Pantic, 2005).

Active Appearance Model (AAM)

Fitting an AAM to an image is to solve a non-linear optimisation problem, which consists of minimising the error between the input image and the closest model instance (Matthews & Baker, 2004). Cootes et al (Cootes, Edward, & Taylor, 1998, 2001) first proposed AAM, which is mainly used in medical segmentation and face recognition. AAM is a statistical based template matching method, where the variability of shape and texture is captured from a representative training set (Edward, Taylor, & Cootes, 1998). Principal Component Analysis (PCA) on shape and texture data allows building a parameterized face model that describe with photorealistic quality the trained face as well as unseen (Edward, et al., 1998). Due to this reason, the performance of this method is highly dependent on the training database.

Mathematically, the shape s of an AMM as the coordinates of the v vertices that make up the mesh is defined as (Matthews & Baker, 2004):

$$s=(x_1,y_1,x_2,y_2,...,x_y,y_y)^T \qquad (1)$$

Since AAMs allow linear shape variation, the shape s can be expressed as:

$$s = s_C + \sum_{t=1}^{M} p_t s_t \qquad (2)$$

where s_0 is a base shape, s_i is an n number of shape vectors (with an assumption of orthonormal), and the coefficients p_i are the shape parameters. The appearance is mathematically described by the value p_i. For further reading, please refer to (Cootes, et al., 2001; Matthews & Baker, 2004).

For further investigation, we demonstrate the algorithm by using the IMM Face Database (Stegmann, Ersboll, & Larsen, 2003) – An annotated Dataset of 240 face images as our training set. Then, we test the result with different image resources. Some of the results from the implementation of AAM on face images are illustrated in Figure 10. Figure 10(a) is an original image that exists in training database. After AAM fitting the result is shown in Figure 10(b). We found a perfect fitting as the image is part of the training database. In Figure 10(c) and Figure 10(e), the images are not in training database. The AAM fitting managed to find a closest model instance to Figure 10(c), as shown in Figure 10(d). But, the fitting is not successful for the image in Figure 10(e). Figure 10(f) shows a bad fitting. To overcome this problem, we need to expand the training database. Readers who are interested in AAM implementation can refer to AAM Library written by Yao (Yao, 2008). AAM Library source code implements the fixed Jacobian method by Tim Cootes and the Inverse Compositional method proposed by Simon Baker and Ian Matthews (Yao, 2008). It is developed under OpenCV 1.0 (Bradski, 2008) for locating facial features.

Figure 10. Implementation of AAM (a) image exists in training database (b) perfect performance of the algorithm (c) image not exist in training database (d) a reasonable result (e) image not exist in the training database (f) poor performance of the algorithm.

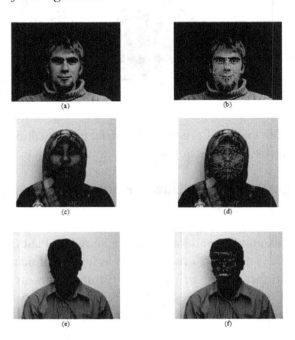

FUTURE RESEARCH DIRECTIONS

Most of the researchers (Bartlett, Hager, Ekman, & Sejnowski, 1999; Ding & Martinez, 2008; Vukadinovic & Pantic, 2005) attempt to detect the face and facial features from upright frontal view or near to upright frontal view under controlled environment. These assumptions cannot be made in a reality real-time application. Hence it is important to extend the work into different views under different condition. Consistency is always a real challenge for the computer vision community. Attempts to develop a semi-automated system in dealing with different environment are being pursued.

Another opportunity for the researchers to establish a novel idea in this area is to create a better way in facial expressions coding. The FACS and MPEG-4 Standard are the popular coding system for facial expressions coding. However, these are mainly defined based on psychological and computer animation approaches. It might be possible to do the coding on the facial feature points based on computer vision approach, or, other approaches.

It is important to understand that facial expression recognition is not emotion classification, but it covers a wide range of hidden intention and threat prediction. In this context, to automate the facial expression recognition is not only computer vision problem but it also involves psychology. One of the research issues is the accuracy of human interpretation in facial expression recognition. The hidden intention of a human can be revealed by thermal domain image/video analysis. Besides "hidden intention", human personality, emotion and mood are the factors that affect the facial expressions. The effect of personality and emotion on expressions is illustrated by Kshirsagar et al (2002) (Kshirsagar, Molet, & Magnenat-Thalmann, 2002), where they describe a system that simulates personalised facial animation with speech and expressions, modulated through mood. In 3D graphics, researchers (Egges, Kshirsagar, & Magnenat-Thalmann, 2003; Kasap & Magnenat-Thalmann, 2007) have taken these approaches in developing a synthetic talking head with personality, mood, and facial expression. The main purpose of modelling personality and mood into the synthetic models is to show "the individuality" (Egges, et al., 2003). The effect of personality and emotion on behaviour has been researched quite a lot (Johns & Silverman, 2001).

Another important issue is to obtain reliable ground truth. The subjects in the available database are the "actors". This raises the question of the reliability of the available databases for testing. These databases are basically based on the six emotions, which do not contain subtle facial changes or micro-expressions. To obtain reliable ground truth, we might need to request the subject to make some attempts to lie or to have some hidden intention during the data capture sessions.

CONCLUSION

This chapter reviewed the face detector algorithms and proposed an improved face detection algorithm based on Haar Cascade face detection technique and skin colour segmentation. Besides, we also emphasize the implementation of the computer vision techniques in facial feature point detection. Facial expression recognition remains largely unresolved, but we believe that by combining the effort from different disciplines the researchers will be able to establish promising results in predicting the human intention based on facial analysis.

In short, facial image processing remains an important research topic in computer vision. By combining the psychological factors, visual domain image/video analysis and thermal domain image/video analysis, the authors believe that it will boost up the accuracy of facial expression prediction. As for our research effort, the facial feature point tracking is underway. In near future, we will continue our research in obtaining a reliable ground truth database in order to improve the facial expression recognition in visual domain.

AKNOWLEDGMENT

This work is supported by EPSRC grant on "Facial Analysis for Real-Time Profiling" (EP/G004137/1). The authors would like to thank Jeffrey Cohn of the University of Pittsburgh for providing the Cohn-Kanade database, Pantic et al of the London Imperial College for providing the MMI database, Sung and Rowley for providing the CMUMIT database and Stegmann et. al. for providing the IMM Face Database.

REFERENCES

Bartlett, M. S., Hager, J. C., Ekman, P., & Sejnowski, T. J. (1999). Measuring facial expressions by computer image analysis. *Psychophysiology, 36*, 253–263. doi:10.1017/S0048577299971664

Bradski, G. K. (2008). *Learning OpenCV: Computer vision with the OpenCV library* (1st ed.). Sebastopol, CA: O'Reilly Media, Inc.

Castrillon-Santana, M., Deniz-Suarez, O., Anton-Canalis, L., & Lorenzo-Navarro, J. (2008). *Face and facial feature detection evaluation.* Paper presented at the Third International Conference on Computer Vision Theory and Applications, VISAPP08.

Cohn, J. F. (2007). Foundations of human computing: Facial expression and emotion. In Huang, T., Nijolt, A., Pantic, M., & Pentland, A. (Eds.), *State of the art survey. Lecture notes in artificial intelligence* (pp. 1–16). Berlin, Heidelberg: Springer.

Cohn, J. F., Ambadar, Z., & Ekman, P. (2006). Observer-based measurement of facial expressions with the facial action coding system. In J.A. Coan & A.J.B. (Eds.), *The handbook of emotion elicitation and assessment.* (pp. 203-221). Oxford, New York: Oxford University Press Series in Affective Science.

Cohn, J. F., & Kanade, T. (2007). Use of automated facial image analysis for measurement of emotion expression. In Coan, J. A., & Allen, J. J. B. (Eds.), *The handbook of emotion elicitation and assessment* (p. 483). New York, Oxford: Oxford University Press Series in Affective Science.

Colmenarez, A. J., & Huang, T. S. (1997). *Face detection with information-based maximum discrimination.* Paper presented at the IEEE International Conference Computer Vision and Pattern Recognition.

Cootes, T. F., Edward, G. E., & Taylor, C. J. (1998). *Active appearance models.* Paper presented at the European Conference on Computer Vision.

Cootes, T. F., Edward, G. E., & Taylor, C. J. (2001). Active appearance models. *IEEE Transactions on Pattern Analysis and Machine Intelligence, 23*(6), 681–685. doi:10.1109/34.927467

Darwin, C. (1872). *The expression of the emotions in man and animals.* London: J.Murray. doi:10.1037/10001-000

Ding, L., & Martinez, A. M. (2008). *Precise detailed detection of faces and facial features.* Paper presented at the IEEE Conference on Computer Vision and Pattern Recognition.

Donato, G. (1999). Classifying facial actions. *IEEE Transactions on Pattern Analysis and Machine Intelligence, 21*(10), 974–989. doi:10.1109/34.799905

Edward, G. E., Taylor, C. J., & Cootes, T. F. (1998). *Interpreting face images using active appearance models.* Paper presented at the International Conference on Automatic Face and Gesture Recognition.

Egges, A., Kshirsagar, S., & Magnenat-Thalmann, N. (2003). *A model for personality and emotion simulation.* Paper presented at the Knowledge-Based Intelligent Information & Engineering Systems (KES 2003).

Ekman, P. (Ed.). (1973). *Darwin and facial expression-a century of research in review.* New York, London: Academic Press.

Ekman, P., & Friesen, W. V. (1976). Measuring facial movement. *Environmental Psychology and Nonverbal Behavior, 1*(1), 56–75. doi:10.1007/BF01115465

Ekman, P., & Friesen, W. V. (1978). *Facial action coding system: A technique for the measurement of facial movement.* Consulting Psychologies Press.

Fasel, B., & Luettin, J. (2003). Automatic facial expression analysis: A survey. *Pattern Recognition, 36*, 259–275. doi:10.1016/S0031-3203(02)00052-3

Hsu, R., Abdel-Mottaleb, M., & Jain, A. K. (2002). Face detection in color images. *IEEE Transactions on Pattern Analysis and Machine Intelligence, 24*(5), 696–706. doi:10.1109/34.1000242

ISO/IEC. I.S. (2001). *Information technology-coding of audio-visual objects.* Geneva.

Johns, M., & Silverman, B. G. (2001). *How emotions and personality effect the utility of alternative decisions: A terrorist target selection case study.* Paper presented at the Tenth Conference on Computer Generated Forces and Behavioral Representation.

Kanade, T., Cohn, J. F., & Tian, Y. L. (2000). *Comprehensive database for facial expression analysis.* Paper presented at the Fourth IEEE International Conference on Automatic Face and Gesture Recognition (FG '00).

Kasap, Z., & Magnenat-Thalmann, N. (2007). Intelligent virtual humans with autonomy and personality: State of the art. *Intelligent Decision Technologies, 1*(1-2), 3–15.

Kim, I. S., Shim, J. H., & Yang, J. (2003). *Face detection.* Stanford University.

Kshirsagar, S., Molet, T., & Magnenat-Thalmann, N. (2002). *A multilayer personality model.* Paper presented at the 2nd International Symposium on Smart Graphics.

Lowe, D. G. (1999). *Object recognition from local scale-invariant features.* Paper presented at the International Conference on Computer Vision.

Lowe, D. G. (2004). Distinctive image features from scale-invariant keypoints. *International Journal of Computer Vision, 60*(2), 91–110. doi:10.1023/B:VISI.0000029664.99615.94

Matthews, I., & Baker, S. (2004). Active appearance models revisited. *International Journal of Computer Vision, 60*(2), 135–164. doi:10.1023/B:VISI.0000029666.37597.d3

Nefian, A. V., & Hayes, M. H. (1998). *Face detection and recognition using hidden Markov models.* Paper presented at the IEEE International Conference on Image Processing.

Nilsson, M., Nordberg, J., & Claesson, I. (2007). *Face detection using local SMQT features and split up snow classifier.* Paper presented at the IEEE Conference on Acoustics, Speech and Signal Processing, ICASSP.

Osuna, E., Freund, R., & Girosi, F. (1997). *Training support vector machines: An application to face detection.* Paper presented at the IEEE Computer Society Conference on Computer Vision and Pattern Recognition.

Pantic, M., Valstar, M. F., Rademaker, R., & Maat, L. (2005). *Web-based database for facial expression analysis.* Paper presented at the IEEE International Conference on Multimedia and Expo (ICME'05), Amsterdam, The Netherlands.

Petajan, E. (2005). MPEG-4 face and body animation coding applied to HCI. In B. Kisacanin, V. Pavlovic & T.S. Huang (Eds.), *Real-time vision for human-computer interaction.* (pp. 249-268). United States of America: Springer Science + Business Media, Inc.

Puri, A., & Eleftheriadis, A. (2008). MPEG-4: An object-based multimedia coding standard. *Mobile Networks and Applications, 3*(1), 5–32. doi:10.1023/A:1019160312366

Roth, D., Yang, M., & Ahuja, N. (2000). A snow-based face detector. *Advances in Neural Information Processing Systems, 12*, 855–861.

Rowley, H., Balaju, S., & Kanade, T. (1998). Neural betwork based face detection. *IEEE Pattern Analysis and Machine Intelligence, 20*, 22–38. doi:10.1109/34.655647

Schneiderman, H., & Kanade, T. (1998). *Probabilistic modeling of local appearance and spatial relationship for object recognition*. Paper presented at the IEEE International Conference Computer Vision and Pattern Recognition.

Shin, M. C., Chang, K. I., & Tsap, L. V. (2002). *Does colorspace transformation make any difference on skin detection?* Paper presented at the Sixth IEEE Workshop on Applications of Computer Vision. Retrieved from http://marathon.csee.usf.edu/~tsap/papers/wacv02.pdf

Stegmann, M. B., Ersboll, B. K., & Larsen, B. (2003). FAME-a flexible appearance modelling environment. *IEEE Transactions on Medical Imaging, 22*(10), 1319–1331. doi:10.1109/TMI.2003.817780

Storring, M., Andersen, H. J., & Granum, E. (1999). *Skin colour detection under changing lighting conditions*. Paper presented at the 7th Symposium on Intelligent Robotics Systems, Coimbra, Portugal.

Sung, K. K., & Poggio, T. (1998). Example-based learning for view-based human face detection. *IEEE Transactions on Pattern Analysis and Machine Intelligence, 20*(1), 39–51. doi:10.1109/34.655648

Tian, Y. L., Cohn, J. F., & Kanade, T. (2005). Facial expressions analysis. In Li, S. Z., & Jain, A. K. (Eds.), *Handbook of face recognition* (pp. 247–276). New York: Springer. doi:10.1007/0-387-27257-7_12

Tian, Y. L., Kanade, T., & Cohn, J. F. (2001). Recognizing action units for facial expression analysis. *IEEE Transactions on Pattern Analysis and Machine Intelligence, 23*(2), 97–115. doi:10.1109/34.908962

Viola, P., & Jones, M. J. (2001). *Rapid object detection using a boosted cascade of simple features*. Paper presented at the CVPR.

Vukadinovic, D., & Pantic, M. (2005). *Fully automatic facial feature point detection using Gabor feature based boosted classifiers*. Paper presented at the IEEE International Conference on Systems, Man and Cybernatics.

Yao, W. (2008). *AAM library*. Retrieved December 15, 2009, from http://code.google.com/p/aam-library/

Zhang, D. D. (2000). *Automated biometrics-technologies and systems*. Kluwer Academic Publishers.

Chapter 12
A Multispectral and Multiscale View of the Sun

Thierry Dudok de Wit
LPC2E, CNRS and University of Orléans, France

ABSTRACT

The emergence of a new discipline called space weather, which aims at understanding and predicting the impact of solar activity on the terrestrial environment and on technological systems, has led to a growing need for analysing solar images in real time. The rapidly growing volume of solar images, however, makes it increasingly impractical to process them for scientific purposes. This situation has prompted the development of novel processing techniques for doing feature recognition, image tracking, knowledge extraction, et cetera. This chapter focuses on two particular concepts and lists some of their applications. The first one is Blind Source Separation (BSS), which has great potential for condensing the information that is contained in multispectral images. The second one is multiscale (multiresolution, or wavelet) analysis, which is particularly well suited for capturing scale-invariant structures in solar images.

INTRODUCTION

The Sun is a world of paradoxes. It is our closest star and yet, distant stars and galaxies have received far more attention as far as data analysis techniques are concerned. Until the dawn of the space age, most solar images were taken in the

visible light only, since the terrestrial atmosphere absorbs most other wavelengths. Visible light, however, mostly reveals the lowest layer of the solar atmosphere, which is relatively featureless apart from the occasional presence of structures such as sunspots. Space-borne telescopes have opened the infrared, the ultraviolet and the X-ray windows, in which the Sun appears much more structured. The *vacuum ultraviolet* (VUV) range,

DOI: 10.4018/978-1-60960-477-6.ch012

whose wavelength range extends from 10 to 200 nm, has received considerable attention since it provides deep insight into the highly dynamic and energetic solar atmosphere (Aschwanden, 2005a).

The prime objective of solar image analysis is a better understanding of the complex physical processes that govern the solar atmosphere. The traditional approach consists in observing the Sun simultaneously in different wavelengths and in matching the results obtained by spectroscopic diagnostics with physical models. Indeed, key quantities such as the temperature or the density cannot be directly accessed and so a quantitative picture can only be obtained at the price of time-consuming comparisons with simulations from radiation transfer models, using strong assumptions such as local thermodynamic equilibrium. A key issue is to find new and more empirical means for rapidly inferring pertinent physical properties from such data cubes.

This situation has recently evolved with the emergence of a new discipline called *space weather*, which aims at understanding and predicting solar variability in order to mitigate its adverse effects on Earth. Manifestations of solar activity such as flares and interplanetary perturbations indeed influence the terrestrial environment and sometimes cause significant economic losses by affecting satellites, electric power grids, radio communications, satellite orbits, airborne remote sensing and also climate. This new discipline has stimulated the search for new and quicker ways of characterising solar variability. For most users of space weather, empirical quantities that are readily available are valued more than physical quantities whose computation cannot be done in real-time. The sudden need for operational services has stimulated the search for novel multidisciplinary solutions for automated data processing that rely on concepts such as feature recognition, knowledge extraction, machine learning, classification, source separation, etc. (Schrijver et al., 2007). The truly multidisciplinary character of this quest is

attested by the fact that most of these concepts are also exploited in other chapters of this book.

In most studies of the Sun, the focus has been on the identification and on the characterisation of individual solar features, such as loops (Inhester, Feng, & Wiegelmann, 2008), sunspots (Colak & Qahwaji, 2009), prominences (Labrosse, Dalla, & Marshall, 2010) and interplanetary disturbances (Robbrecht & Berghmans, 2005). As the cadence and the size of solar images increases, however, so does the need for extracting metadata and doing data mining. The human eye often remains one of the best expert systems, so tools are also needed to assist humans in visualising multiple images. The Solar Dynamics Observatory satellite, for example, which delivered its first images in April 2010, provides several times per minute 4096 x 4096 images in 7 wavelengths in the VUV simultaneously. For such purposes, it is desirable to have techniques that, in addition to extracting specific features, can display (1) multiple wavelengths simultaneously and, (2) multiple scales in a more compact way. Both tasks have not received much attention yet, but will surely become an active field of research in the next decade.

In this short overview, we shall focus on two particular concepts that are particularly appropriate for handling such tasks; they are multispectral and multiscale analysis. In both cases, potential applications will be emphasized rather than their technical aspects, which can be found in the literature. For recent reviews on feature detection algorithms, see the chapter by Pérez-Suárez et al. (2010), and also Aschwanden (2010) and Zharkova, Ipson, Benkhalil, and Zharkov (2005). The book by Starck and Murtagh (2006) is another reference in the field, although it concentrates on astronomical objects that are point-like.

A Multispectral View of the Sun

Telescopes that observe the Sun in the VUV are designed to observe preferentially one single and strong spectral line whose emission peaks in a characteristic temperature band. From the simultaneous observation of different spectral lines, one can then build a picture of the temperature layering of the solar atmosphere. More quantitative estimates require a considerable amount of modelling that is still beyond reach. Indeed, many effects such as the integration along the line of sight of optically thin and thick emissions need to be taken into account.

The solar atmosphere, however, is highly structured by an intense magnetic field. As a consequence, images taken in different wavelengths are often remarkably redundant, as are their time series. This redundancy is manifested by the same location and the similar shape of solar features, as observed in different wavelengths. Redundancy is a useful property for space weather applications because it opens the way for data reduction, i.e. for the extraction of a small set (as compared to the original one) of parameters that describe the salient features of the data. Redundancy also eases the visualisation of multiple wavelengths and helps in denoising images. A recent and promising framework for dealing with it is *blind source separation*, which aims at exploiting the coherence in the data to identify their underlying sources with the aid of the least prior information about them or about their mixing process (Choi, Cichocki, Park, & Lee, 2005).

Blind source separation has recently emerged as powerful concept in several areas such as acoustics, data compression, and hyperspectral imaging of terrestrial images. (Collet, Chanussot, & Chehdi, 2010). Given an instantaneous linear mixture of intensities that are produced by a small set of sources, blind source separation exploits the statistical properties of these sources to differentiate them in an unsupervised manner. This is hallmarked by the *cocktail party problem* (Haykin & Chen, 2005), in which the human ear attempts to isolate one single voice from a mixture of sources. In the following, we shall see how blind source separation can give new insight into solar multispectral images. Let us first focus on a case study and then discuss the implications.

A Blind Source Separation Approach

Figure 1 shows a series of 2D images of the solar limb, taken almost simultaneously by the Coronal Diagnostic Spectrometer (CDS) onboard the SoHO satellite. In this example, the spectrometer was viewing a small region of the Sun and was making 2D raster plots of the intensity of 14 spectral lines. In this particular example, the observation time is negligible as compared to the characteristic evolution time of the solar structures, which can therefore be assumed to be static. CDS offers a similar number of wavelengths as SDO and thus gives a foretaste of what SDO provides with full Sun images. The intensity of each pixel depends on various plasma parameters along the line of sight, and in particular on the temperature and on the density. A conspicuous feature of these images is their high correlation. There are two reasons for this. First, the temperature response associated with each spectral line is generally wide, and sometimes even multimodal. Second, the same line of sight may capture contributions coming from regions with different temperatures, ranging from about 10^4 to 10^6 °K. Each image therefore captures contributions associated with a mix of regions (or rather atmospheric layers) that have different temperatures.

Let us therefore assume that each pixel captures a linear mixture of a large set of pure component spectra, i.e. spectra that are associated with specific emitting regions along the line of sight. Traditionally, individual spectra have been assigned to regions such as coronal holes, the quiet Sun and active regions. The linear mixture hypothesis is reasonably well satisfied in the case of an emitting body like the Sun. For reflecting bodies such as planets, it becomes more debatable (Moussaoui et al., 2008). The shape of the spectra varies continuously along the line of sight so, a priori, it is not possible to extract the spectra in-

Figure 1. Intensity maps of the same active region at the solar limb taken on March 23, 1998 by the CDS instrument onboard the SoHO spacecraft; 8 emission lines out of 14 are shown. The characteristic temperatures of the lines increases logarithmically from top left (20,000 K) to bottom right (2.5 MK). The spectral line and its wavelength are indicated on each image. A linear scale is used for the intensity.

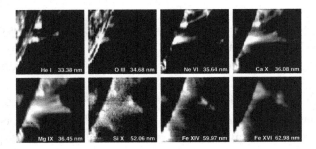

dividually. However, since they are partly redundant, one may expect all observations to be described by a small subset of them. That is, the spectral variability should only have few degrees of freedom. Many studies support this hypothesis. More than two decades ago, Lean et al. (1982) had already noticed that the solar spectral variability could adequately be described by 3 different contributions. Feldman and Landi (2008) came to the same conclusion by comparing observations with models while Amblard et al. (2008) found 3 sources by analysing solar spectra. The number 3 therefore seems to be deeply rooted in the spectral characteristics of the solar corona. The exact physical interpretation of this high redundancy has remained elusive so far. Meanwhile, it provides an ideal framework for doing data reduction by blind source separation.

The CDS spectrometer counts photons, so for each pixel the noise tends to obey a Poisson statistics. We then stabilise the variance by applying the generalised Anscombe transform, which is equivalent here to taking the square root of the intensity, see (Starck & Murtagh, 2006). In the following, we shall also normalise each image by its mean intensity.

The first step toward blind source separation is the determination of the number of sources. As a first guess we apply principal component analysis or rather the *Singular Value Decomposition* (SVD)

to the images (Golub & van Loan, 1996) and look for the dominant terms. Each image is 85 x 87 pixels in size. We fold the 85 x 87 x 14 data cube into a 7395 x 14 matrix by lexicographically ordering each image. By applying the SVD, we implicitly assume that pixel intensities $I(x,\lambda)$ can be expressed as a separable set of orthonormal components that depend either on position x or on wavelength λ

$$I(x,\lambda) = \sum_{k=1}^{14} A_k\, f_k(x) g_k(\lambda)$$

with

$$\langle f_k(x) f_l(x)\rangle = \langle g_k(\lambda) g_l(\lambda)\rangle = \delta_{kl}$$

where δ_{kl} is the Kronecker function. The weights A_k are by construction positive and they are sorted in decreasing order. The first terms in the sum have the largest weights and thus capture the salient features of the images.

Figure 2 shows the distribution of the weights A_k. Their strong ordering confirms the redundancy of the data and suggests that the salient features of the images are expressed only by 2 to 5 components. Similar results are obtained without the

Figure 2. The distribution of the weights obtained by applying the SVD to the 14 images. All weights are normalised to their largest value A_1.

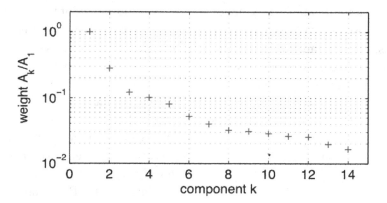

Anscombe transform, but with a weaker ordering of the weights.

The 6 spatial components $f_k(x)$ associated with the six largest weights are shown in Figure 3. These components are not realistic because intensities are non-positive and also because there is no sound justification for their orthogonality constraint.

A more careful analysis suggests that these components also mix different physical features. A more realistic prior is the statistical independence of the components, which is precisely what

Independent Component Analysis (ICA) does (Hyvärinen & Oja, 2000).

ICA has recently become a popular method for separating sources. This method often brings a significant improvement over the SVD when the probability density of the images is non-Gaussian. Most solar images indeed have a non-Gaussian probability density because they mix features with quite different intensity levels (dark corona vs. bright solar disk). For that reason it is worth incorporating this property in the source separation. ICA can be viewed as an inversion of the central

Figure 3. The 6 most significant spatial sources obtained by SVD. The linear vertical scale ranges from the lower 5% quantile to the upper 98% quantile. Only the first component (k=1) is positive.

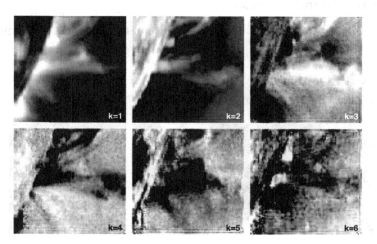

limit theorem since it uses the departure from Gaussianity as a lever to improve the discrimination of the sources by assuming that a mixture of random variables is closer to a Gaussian than the individual variables. We illustrate this here by estimating the Kullback-Leibler divergence between the probability density of the images and that of a Gaussian distribution with the same mean and variance. The Kullback-Leibler divergence reads

$$D(p\|q) = \int p(x) \ \log \frac{p(x)}{q(x)} dx$$

where $p(x)$ is the probability density of the image (estimated using a kernel histogram method) and $q(x)$ is a Gaussian density with the same mean and variance. This divergence D is positive; the larger its value, the more $p(x)$ departs from a Gaussian density. We find the divergence of most original images to be between 0.1 and 0.4, whereas that of the 3 main sources estimated by ICA and BPSS (see below) is between 0.3 and 0.8. The source images are as expected more non-Gaussian that the original images.

The sources we obtain here by ICA are closer to the expected physical picture than those found by SVD, as shown by Dudok de Wit & Auchère (2007). However, there is no sound justification for enforcing the independence of the sources; the solar atmosphere is partly transparent and so, for example, in a given active region, different

sources may have an intensity peak at the same location. An even more serious objection is the lack of positivity of the sources found.

A more realistic prior is the positivity of both the spatial components $f_k(x)$ and their mixing coefficients $g_k(\lambda)$. A natural approach for this is the recent *Bayesian Positive Source Separation* (BPSS) method, which was developed by Moussaoui, Brie, Mohammad-Djafari, and Carteret (2006). BPSS is one among of the several techniques that have recently been developed for doing BSS with positivity constraints. Our motivation for choosing it stems from the Bayesian framework that allows us to incorporate information on the noise and signal statistics. The same method has recently been compared against the ICA in the frame of hyperspectral imaging of Mars (Moussaoui et al., 2008). We assume that $f(x) = \{f_k(x)\}$ and $g(\lambda) = \{g_k(\lambda)\}$ are random matrices whose elements are independent and distributed according to Gamma probability density functions. In the following, the sources $f_k(x)$ have unit norm, as for the SVD.

We apply the BPSS to the CDS data after normalising each image to its mean value. No Anscombe transform is applied beforehand since we assume each pixel intensity to be a linear mixture of different sources. The question about the number of sources arises again. The root mean square error of the difference between the original data and the reconstructed intensities exhibits a sharp decay as the number of sources increases from one 1 to 3, and then drops more slowly. From this we expect the number of sources to be

Figure 4. The 3 sources obtained by BPSS. The linear vertical scale ranges from the lower 5% quantile to the upper 98% quantile. All sources are positive definite.

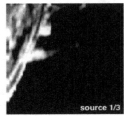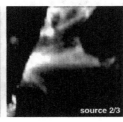

at least 3. An inspection of the sources shows that with 4 sources and more, the first 3 ones remain almost unchanged, whereas the other sources are both weaker (i.e. their corresponding mixing coefficients are smaller) and vary with their total number. In other words, the existence of 3 prevalent sources is a robust result, whereas the subsequent sources contribute much less and are not reproducible. A tentative interpretation is that 3 is the right number for obtaining a linear combination, whereas additional terms describe nonlinear effects that do not match some of the hypotheses, such as the positivity of the mixing coefficients. We therefore consider 3 sources in the following.

The 3 sources obtained by BPSS and their mixing coefficients are shown in Figure 4. These sources are by construction all positive, as are their mixing coefficients. Interestingly, they can be directly linked to specific layers of the solar atmosphere, whereas the interpretation of the sources obtained by SVD or ICA is more difficult. A result of particular interest is the clear temperature ordering in the mixing coefficients. As shown in Figure 5, each source captures emissions that correspond to a specific temperature band. The three sources respectively describe emissions originating from the coolest layers (1), from the lower solar corona (2) and from the upper hot corona (3). The cold

component is associated with the lowest layers of the solar atmosphere, in which the solar surface comes out as a bright disk. The small loops that stand out against the dark horizon are structured by the solar magnetic field. Particle acceleration processes can locally heat the plasma to several million degrees, leading to the hot diffuse structures that appear in the third source.

The key result is that all 14 spectral lines, in spite of their differences, can be classified in 3 temperature bands only, whose properties can be inferred from the data without imposing a physical model. These results are robust, in the sense that the same temperature ordering is obtained for other regions or events, provided that all three layers are properly represented in the sample. Preliminary tests with the first VUV images from SDO confirm these results.

Other Applications

Blind source separation is a concept that is much better known to the astrophysics community, which has been using it for several years, either for the analysis of multispectral images (Nuzillard & Bijaoui, 2000) or for the challenging extraction of the cosmic microwave background from images obtained by the Planck mission (Kuruoglu, 2010). The small number of sources we find in

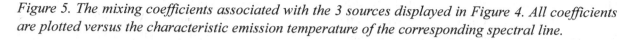

Figure 5. The mixing coefficients associated with the 3 sources displayed in Figure 4. All coefficients are plotted versus the characteristic emission temperature of the corresponding spectral line.

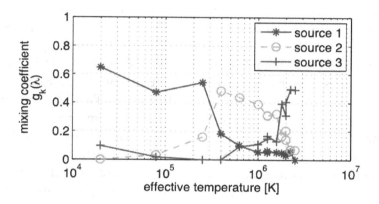

solar VUV images has several practical applications for space weather.

Temperature Maps

One of the important issues in space weather is the nowcast of the spectrally resolved solar irradiance for the specification of the upper terrestrial atmosphere. Long-term monitoring of the irradiance, however, is difficult because of instrumental constraints. An empirical approach to this problem consists in tracking solar features that emit at different temperatures (e.g. coronal holes, active regions, etc.), assign a characteristic spectrum to each of them, and adding all these contributions to obtain the total irradiance (Krivova and Solanki, 2008). Segmentation techniques are required for doing this with VUV images. By applying such a segmentation techniques on a few sources only rather than on original solar images, we reduce the computational complexity. Secondly, by using partly independent source images rather than highly redundant original images we make the mathematical problem better conditioned. Thirdly, the segmentation procedure is facilitated if the inputs have a more direct physical interpretation since this allows prior information to be incorporated more easily.

Having 3 sources only is also helpful for condensing all the pertinent information in one single image. We do so by assigning the cold, intermediate and hot sources respectively to the blue, green and red channels. This multichannel representation of the Sun is not shown here for it requires colour prints, but a variant based on ICA can be found in Dudok de Wit and Auchère (2007). This technique, which is commonly used in aerial imagery, allows us to compress all 14 images into one single image, which considerably eases their visual interpretation. We are currently developing real-time three-temperature images of the Sun, which can be used as quicklook plot for locating solar features such as coronal holes.

Denoising

Denoising is an interesting but not so frequent spinoff of blind source separation. Since the salient features of the data are well reproduced by 3 sources only, one may use the latter to reconstruct the observations while rejecting part of the incoherent noise.

One way to investigate the denoising and to qualify the fit is by inspecting the residuals, i.e. the difference between the original image and the image reconstructed from the 3 sources, see Figure 6. The vertical scale used in Figure 6 is identical to that used in Figure 1. Note that the fitted intensity matches the original one rather well. The largest outliers are observed for pixels whose intensity is large as well, so that the relative error actually remains acceptable. Its value generally remains well below 25% for such pixels.

Residuals are of course particularly interesting for detecting unsuspected features that are not properly described by the model and which may affect one or a few spectral lines locally. In a different application, we detected that way some instrumental and compression artefacts. There aren't any in this particular example.

Using other Physical Constraints

A recurrent problem with blind source separation is the lack of crisp criteria for determining the right number of sources. Although BPSS gives more realistic results than the SVD or ICA, there is still room for improvement. A better separation of the sources requires a quantitatively measurable diversity. Recently, sparsity and morphological diversity have emerged as promising criteria for further improving the separation of sources (Bobin, Starck, Fadili, & Moudden, 2008). Sparsity means that morphologically different features in an image, when projected on a suitable set of basis functions, can be characterised by a small sets of coefficients only, which eases their separation. Another advantage is the possibility to

Figure 6. Residuals obtained by removing the contribution of the 3 BPSS sources from the original images, shown in Figure 1. The vertical scale is identical to that used in Figure 1. White regions correspond to an excess of intensity as compared to that of the reconstructed image.

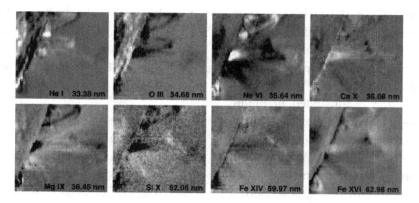

extract more sources than the number of different wavelengths. Solar images exhibit a wide range of different features with characteristic morphologies (e.g. thin loops, diffuse active regions, etc.) and our first results with this approach confirm the relevance of the sparsity concept.

A Multiscale View of the Sun

One of the most fascinating aspects of human vision is its ability to automatically adapt itself to the characteristic scale(s) of features of interest in an image. This has become one of the major challenges in artificial vision and has stimulated the development of multiscale methods for natural images (Hyvärinen, Hurri & Hoyer, 2009). Solar images, especially when taken in the VUV, provide an excellent example of natural images with a rich blend of multiple scales.

The multiscale analysis of solar images can be performed in many different ways, but wavelet (i.e. multiresolution) techniques are particularly well suited for this (Mallat, 2008). Discrete wavelet transforms are widely appreciated for their ability to decompose images into a compact set of coefficients, which can be processed for compression or for filtering and then used for rapid reconstruction of the (filtered) image. The interpretation of

these coefficients in terms of physical properties, however, is often difficult. For example, they are not translation invariant and so two images that are shifted by one pixel may have substantially different coefficients. The continuous wavelet transform, which decomposes images into a set of highly redundant coefficients, is much better suited for physical interpretation but its computational burden is considerably higher. The *à trous* (literally, with holes) is one particular algorithm that is popular in astronomical image processing since it shares some of the advantages of the previous methods. This algorithm is relatively fast and yet, the resulting wavelet coefficients have a direct interpretation (Mallat, 2008; Starck & Murtagh, 2006).

The decomposition of an image by the *à trous* algorithm is iterative: the original image, which is stored in a rectangular matrix I, is convolved with a 2D smoothing kernel S_d whose characteristic scale is d, giving a new image $I_d = I * S_d$. The scales are typically dyadic, with $d = 0, 1, 2, 4, 8, \ldots$ The largest scale must be smaller than the size of the image and I_0 is the original image. Other alternatives to the convolution with the smoothing kernel are the pyramidal median transform (Starck & Murtagh, 2006), in which the median over a $(2d+1)(2d+1)$ window centred on each pixel is computed. This is particularly appropriate for images that suffer from shot noise. Finally, we build a set of differenced images, $\{D_0=I_0-I_1, D_1=I_1-I_2, D_2=I_2-I_4, ..., D_{N/2}=I_{N/2}-I_N, D_N=I_N\}$, each of which captures structures that have a specific characteristic scale, as in the continuous wavelet transform. Note that we recover the original image simply by adding all differenced ones.

This decomposition is illustrated in Figure 7 with a solar image taken in the extreme ultraviolet by the recently launched SWAP telescope onboard the PROBA2 satellite (Berghmans et al., 2006). This image is rich in structures of all sizes. However, because of the variable optical thickness of the solar corona, many structures remain hidden in a haze that hinders their analysis. Thin features outside of the solar disk often also remain unobserved because the disk is so much brighter than the faint corona. For these reasons, it is highly desirable to enhance the image along two different directions

- For each scale, the contrast between a pixel and its local neighbourhood should be enhanced;
- For each pixel, the contrast between the different scales should be enhanced.

We process the image in several steps, using the *à trous* algorithm. First the original image is decomposed into a set of differenced images $\{D_0, D_1, ..., D_N\}$ using a Gaussian smoothing kernel. Next, each differenced image is normalised with respect to its mean absolute intensity, or root mean squared intensity. By doing so, we put all scales on equal footing. The reconstructed image, which is displayed in Figure 8b, already reveals a considerable contrast enhancement. To further

Figure 7. Sequence of differenced images $\{D_0, D_1, ... D_{128}\}$ obtained from a solar image taken in the VUV by the SWAP telescope on Jan. 24, 2010. The original image is shown in Figure 8a and is 1024 x 1024 in size. For each image, only intensities ranging from the lower 2% quantile to the upper 98% quantile are shown.

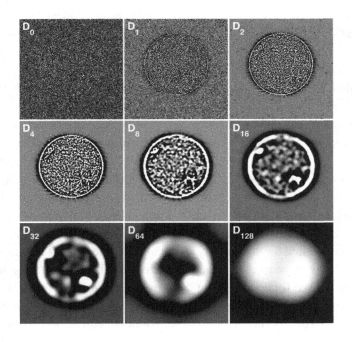

enhance the contrast, we normalise for each pixel the wavelet coefficients with respect to their mean absolute intensity, or root mean square intensity. In this last step, the enhancement is done locally only. In contrast to better-known techniques such as histogram equalization, in which the size of the neighbourhood has to be specified using criteria that are often subjective, here the size is adapted automatically to the characteristic size of the locally dominant structure, which is an important asset. This last step considerably enhances structures near the solar limb (the edge of the disk), where their identification is most difficult, see Figure 8c.

In this particular example, shot noise dominates as soon as one moves away from the limb, so that weak coronal features cannot be followed far into the corona. The image quality can be further improved by adding a filtering stage. This can be done in several ways, either by processing the wavelet coefficients, or by using the discrete wavelet transform and thresholding the wavelet coefficients, see below.

Other Applications

The multiscale image enhancement can be improved and extended in multiple ways. This is now gradually becoming an active field of investigation in solar physics.

Feature Detection

The solar feature detection and extraction problem is discussed in detail in the chapter by Pérez-Suárez et al. (2010). Here we consider this problem in the light of multiscale analysis only. The *à trous* algorithm we discussed just before has the advantage of being multi-purpose. When it comes, however, to detecting and extracting structures that have a specific shape, much better performance can be achieved by tailoring the shape of the analysing wavelets to that of the structures of interest. Typical examples are the magnetic loops that often permeate the solar atmosphere and whose conspicuous curved shape requires curved wavelets. Curvelets (Candès, Demanet, Donoho, & Ying, 2005) are ideally suited for dealing with such structures, since they have been designed to capture curved shapes. An example is shown in Figure 9, in which the differenced image D_0 from Figure 7 has been processed using the discrete curvelet transform. The wavelet coefficients have been computed and only those values exceeding a threshold determined by a preset noise level were retained. The inverse transform should then give an image in which only salient curved features are retained. We find indeed that most of the shot noise has been eliminated that way while arched-like structures in the vicinity of active regions and the solar limb now appear much more evidently. There have already been some attempts to com-

Figure 8. Solar image taken in the VUV by the SWAP telescope. The original (a) and the processed images are shown. In caption (b) only the contrast between scales has been enhanced. In caption (c) contrast enhancement with respect to the local neighbourhood has been included.

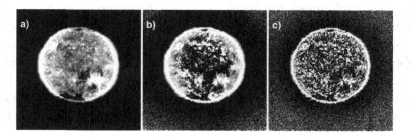

Figure 9. The differenced image D_0 from Figure 7 before (a) and after (b) noise reduction using the curvelet transform. The same vertical scale has been used for both images.

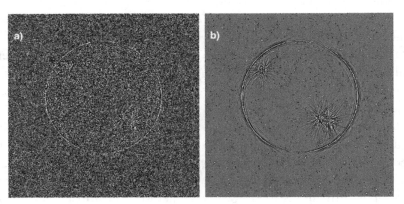

bine multiresolution techniques with more classical edge detection algorithms in order to detect features (Young & Gallagher, 2008; Ireland et al. 2008; Inhester, Feng, & Wiegelmann, 2008). An application of curvelets to the detection of coronal mass ejections has recently been reported by Gallagher, Young, Byrne, & McAteer (2010).

Denoising

The problem of denoising images is very similar to that of feature extraction, since in both cases the objective is to separate desired features from the unwanted ones. Stenborg and Cobelli (2003), for example, use a wavelet packet approach to extract features and reduce noise.

Most multiresolution methods are optimal or suboptimal for data that are affected by Gaussian noise. Solar images, however, are often based on photon counting; as a result, the noise characteristics is often a mix of Poisson and Gaussian statistics. In such cases, better results can be achieved by using the Anscombe transform (Starck & Murtagh, 2006) to stabilise the variance. In the case of an image that is affected by pure Poisson noise, this transform simply amounts to taking the square root of the pixel intensity.

An interesting issue is the onboard preprocessing of solar images since image compression is often imposed by the limited data flow between satellite and ground. Better performance can be achieved by taking into account the noise statistics in the compression scheme (Nicula, Berghmans & Hochedez, 2005). The next step would be to incorporate the multiscale nature of solar images by doing onboard denoising first. This has not yet been attempted.

Stereoscopy

The twin STEREO spacecraft that were launched in 2006 for the first time allowed solar structures to be investigated in 3D. Many tools have been developed for that purpose. Doing tomography with just two vantage points is beyond reach but the 3D stereoscopic reconstruction of contrasted features such as loops is possible (Aschwanden, 2005b; Inhester, Feng, & Wiegelmann, 2008).

Here again, multiresolution methods have the advantage of automatically adapting the size of the neighbourhood to that of the salient structures. One typically selects a feature in one image, locates it in the stereoscopic pair, and finally estimates its depth from the disparity (i.e. the displacement). This is the idea behind optical flow (Gissot & Hochedez, 2007), which has been successfully applied to solar images. A multiresolution stage can be added to this by starting the feature matching on the largest scales and refining it down to smaller scales. This not only makes the feature extraction more robust, but it also speeds up the computationally expensive matching procedure.

A powerful alternative to this, which has emerged in the field of artificial vision, is the following: in the continuous wavelet transforms, each image pixel is unfolded into a set of wavelet coefficients that uniquely describe the intensity of that pixel and its local neighbourhood properties such as texture, curvature, etc. If similar features are observed in a stereoscopic pair, then their sets of wavelet coefficients should also closely match. The procedure then goes as follows: take a pixel from one image, and correlate it with pixels in the stereoscopic pair by matching the wavelet coefficients. The two pixels that describe the same structure should be those that have the highest correlation. This has been termed the *local correlation function* (Perrin, Torresani, & Fuchs, 1999) because the correlation is truly done on a pixel per pixel basis. The advantage over the previous methods is the automatic selection of the prevalent scale and the resilience versus distortions; if the same structure does not appear identically in both images (because of the depth), the local correlation function will nevertheless manage to identify it. An application to time series has recently been reported (Souček, Dudok de Wit, Décréau, & Dunlop, 2004) and similar concepts are used in other fields, such as for character recognition.

Self-Similarity

Wavelets are ideally suited for analysing scale-invariant features in solar images because they are by construction self-similar. Turbulent cascades in the solar atmosphere and the intricate topology of the magnetic field that drives most solar dynamic processes are some among the various reasons for expecting multifractal patterns in solar images. What matters is not the spectral content of the image but rather the interplay between scales.

The multifractal structure of solar images has recently received considerable attention, for quite different reasons. One has to do with the prediction of solar flares from photospheric images and from solar magnetograms (i.e. maps of the photospheric magnetic field). Flares are associated with changes in the topology of the magnetic field, and the idea of actually quantifying the flaring probability from the multifractal structure of the magnetic field has been investigated by several authors (Criscuoli, Rast, Ermolli, & Centrone, 2007; Conlon et al., 2008). A second motivation for considering the multifractal nature of the solar atmosphere is to shed light on the mechanisms of turbulence and understand the anomalous heating of the corona (Georgoulis, 2005). This has turned out to be difficult because each line of sight integrates emissions originating from various altitudes. A third reason has to do with observational strategies. New solar telescopes have ever increasing spatial and temporal resolutions; as a consequence, the number of photons received per pixel keeps on decreasing (for fixed aperture). The question then arises as to how inhomogeneous the solar emission becomes at small scales (Delouille, Chainais, & Hochedez, 2008). For a multifractal emitting surface, for example, the inhomogeneity should not scale in the same way as for a monofractal surface because the statistical properties differ. This has implications on the definition of the dynamic range of the detector.

Another idea is to use the continuous wavelet transform to investigate the Lipschitz regularity in solar images, i.e. the local "sharpness" of discontinuities (Mallat, 2008). VUV images of the solar disk reveal many tiny bright spots, most of which are either transient bright features of the solar corona, or impacts from cosmic rays. The local Lipschitz regularity of these spots depends on their physical origin, thereby giving a means for separating them (Hochedez et al., 2002).

CONCLUSION

In this chapter, we have given a brief tour of solar imaging as seen from a blind source separation and from a multiscale perspective. The recent emergence of space weather has greatly contributed to help introducing concepts that have matured in other fields and that are now just waiting to be applied to the Sun. They are now gaining wider acceptance as it is found that empirical models can be truly complementary to better-known physical models.

Blind source separation is particularly useful for providing fast and empirical representations of the temperature distribution in the solar atmosphere. Multiscale methods have even more applications because they are deeply rooted in the scale invariant structure of the solar atmosphere. The best is yet to come as physics-based and empirical models are gradually getting closer to each other, leading to semi-empirical models in which the weaknesses of one can be complemented by the strengths of the other.

This study received funding from the European Community's Seventh Framework Programme (FP7/2007-2013) under the grant agreement nr. 218816 (SOTERIA project, www.soteria-space.eu). I thank the PROBA2/SWAP and the SoHO/CDS teams for providing respectively the SWAP and the CDS images. SWAP is a project of the Centre Spatial de Liège and the Royal Observatory of Belgium funded by the Belgian Federal Science Policy Office (BELSPO).

REFERENCES

Amblard, P. O., Moussaoui, S., Dudok de Wit, T., Aboudarham, J., Kretzschmar, M., Lilensten, J., & Auchère, F. (2008). The EUV sun as the superposition of elementary suns. *Astronomy & Astrophysics*, *487*, L13–L16. doi:10.1051/0004-6361:200809588

Aschwanden, M. J. (2005a). 2D feature recognition and 3D reconstruction in solar EUV images. *Solar Physics*, *228*, 339–358. doi:10.1007/s11207-005-2788-5

Aschwanden, M. J. (2005b). *Physics of the solar corona*. Chichester, UK: Praxis Publishing Ltd.

Aschwanden, M. J. (2010). Solar image processing techniques with automated feature recognition. *Solar Physics*, *262*, 235–275. doi:10.1007/s11207-009-9474-y

Berghmans, D., Hochedez, J.-F., Defise, J.-M., Lecat, J. H., Nicula, B., & Slemzin, V. (2006). SWAP onboard PROBA2, a new EUV imager for solar monitoring. *Advances in Space Research*, *38*, 1807–1811. doi:10.1016/j.asr.2005.03.070

Bobin, J., Starck, J.-L., Fadili, J. M., & Moudden, Y. (2008). Blind source separation: The sparsity revolution. *Advances in Imaging and Electron Physics*, *152*, 221–306. doi:10.1016/S1076-5670(08)00605-8

Candès, E., Demanet, L., Donoho, D., & Ying, L. (2005). Fast discrete curvelet transforms. *Multiscale Modeling and Simulation*, *5*, 861–899. doi:10.1137/05064182X

Choi, S., Cichocki, A., Park, H.-M., & Lee, S.-Y. (2005). Blind source separation and independent component analysis: A review. *Neural Information Processing Letters*, *6*, 1–57.

Colak, T., & Qahwaji, R. (2009). Automated solar activity prediction: A hybrid computer platform using machine learning and solar imaging for automated prediction of solar flares. *Space Weather*, *7*, 6001. doi:10.1029/2008SW000401

Collet, C., Chanussot, J., & Chehdi, K. (Eds.). (2010). *Multivariate image processing. Digital Signal and Image Processing Series*. London: Wiley.

Conlon, P. A., Gallagher, P. T., McAteer, R. T. J., Ireland, J., Young, C. A., & Kestener, P. (2008). Multifractal properties of evolving active regions. *Solar Physics, 248,* 297–309. doi:10.1007/s11207-007-9074-7

Criscuoli, S., Rast, M. P., Ermolli, I., & Centrone, M. (2007). On the reliability of the fractal dimension measure of solar magnetic features and on its variation with solar activity. *Astronomy & Astrophysics, 461,* 331–338. doi:10.1051/0004-6361:20065951

Delouille, V., Chainais, P., & Hochedez, J.-F. (2008). Quantifying and containing the curse of high resolution coronal Imaging. *Annales Geophysicae, 26,* 3169–3184. doi:10.5194/angeo-26-3169-2008

Dudok de Wit, T., & Auchère, F. (2007). Inferring temperature from morphology in solar EUV images. *Astronomy & Astrophysics, 466,* 347–355. doi:10.1051/0004-6361:20066764

Dudok de Wit, T., Lilensten, J., Aboudarham, J., Amblard, P.-O., & Kretzschmar, M. (2005). Retrieving the solar EUV spectrum from a reduced set of spectral lines. *Annales Geophysicae, 23,* 3055–3069. doi:10.5194/angeo-23-3055-2005

Feldman, U., & Landi, E. (2008). The temperature stricture of solar coronal plasmas. *Physics of Plasmas, 15,* 056501. doi:10.1063/1.2837044

Gallagher, P. T., Young, C. A., Byrne, J. P., & McAteer, R. T. J. (2010). (in press). Coronal mass ejection detection using wavelets, curvelets and ridgelets: Applications for space weather monitoring. *Advances in Space Research.* doi:10.1016/j.asr.2010.03.028

Georgoulis, M. K. (2005). Turbulence in the solar atmosphere: Manifestations and diagnostics via solar image processing. *Solar Physics, 228,* 5–27. doi:10.1007/s11207-005-2513-4

Gissot, S. F., & Hochedez, J.-F. (2007). Multiscale optical flow probing of dynamics in solar EUV images. Algorithm, calibration, and first results. *Astronomy & Astrophysics, 464,* 1107–1118. doi:10.1051/0004-6361:20065553

Golub, G. H., & van Loan, C. F. (1996). *Matrix computations.* Baltimore: The Johns Hopkins University Press.

González-Nuevo, J., Argüeso, F., López-Caniego, M., Toffolatti, L., Sanz, J. L., & Vielva, P. (2006). The Mexican hat wavelet family: Application to point-source detection in cosmic microwave background maps. *Monthly Notices of the Royal Astronomical Society, 369,* 1603–1610. doi:10.1111/j.1365-2966.2006.10442.x

Haykin, S., & Chen, Z. (2005). The cocktail party problem. *Neural Computation, 17,* 1875–1902. doi:10.1162/0899766054322964

Hochedez, J.-F., Jacques, L., Verwichte, E., Berghmans, D., Wauters, L., Clette, F., et al. (2002). Multiscale activity observed by EIT/SoHO. In H. Sawaya-Lacoste (Ed.), *Proceedings of the second solar cycle and space weather euroconference.* (pp. 115-118). Noordwijk, The Netherlands.

Hyvärinen, A., Hurri, J., & Hoyer, P. O. (2009). *Natural image statistics: A probabilistic approach to early computational vision.* Berlin: Springer.

Hyvärinen, A., & Oja, E. (2000). Independent Component Analysis: algorithms and applications. *Neural Networks, 13,* 411–430. doi:10.1016/S0893-6080(00)00026-5

Inhester, B., Feng, L., & Wiegelmann, T. (2008). Segmentation of loops from coronal EUV images. *Solar Physics, 248,* 379–393. doi:10.1007/s11207-007-9027-1

Ireland, J., Young, C. A., McAteer, R. T. J., Whelan, C., Hewett, R. J., & Gallagher, P. T. (2008). Multiresolution analysis of active region magnetic structure and its correlation with the Mount Wilson classification and flaring activity. *Solar Physics, 252,* 121–137. doi:10.1007/s11207-008-9233-5

Krivova, N. A., & Solanki, S. K. (2008). Models of solar irradiance variations: Current status. *Journal of Astrophysics and Astronomy, 29*, 151–158. doi:10.1007/s12036-008-0018-x

Kuruoglu, E. (2010). Bayesian source separation for cosmology. *IEEE Signal Processing Magazine, 27*, 43–54. doi:10.1109/MSP.2009.934718

Labrosse, N., Dalla, S., & Marshall, S. (2010). Automatic detection of limb prominences in 304 Å EUV images. *Solar Physics, 262*, 449–460. doi:10.1007/s11207-009-9492-9

Lean, J.-L., Livingston, W. C., Heath, D. F., Donnelly, R. F., Skumanich, A., & White, O. R. (1982). A three-component model of the variability of the solar ultraviolet flux 145-200 nm. *Journal of Geophysical Research, 87*, 10307–10317. doi:10.1029/JA087iA12p10307

Mallat, S. (2008). *A wavelet tour of signal processing: The sparse way.* New York: Academic Press.

Moussaoui, S., Brie, D., Mohammad-Djafari, A., & Carteret, C. (2006). Separation of non-negative mixture of non-negative sources using a Bayesian approach and MCMC sampling. *IEEE Transactions on Signal Processing, 11*, 4133–4145. doi:10.1109/TSP.2006.880310

Moussaoui, S., Hauksdóttir, H., Schmidt, F., Jutten, C., Chanussot, J., & Brie, D. (2008). On the decomposition of Mars hyperspectral data by ICA and Bayesian positive source separation. *Neurocomputing, 71*(10-12), 2194–2208. doi:10.1016/j.neucom.2007.07.034

Nicula, B., Berghmans, D., & Hochedez, J.-F. (2005). Poisson recoding of solar images for enhanced compression. *Solar Physics, 228*, 253–264. doi:10.1007/s11207-005-4998-2

Nuzillard, D., & Bijaoui, A. (2000). Blind source separation and analysis of multispectral astronomical images. *Astronomy & Astrophysics, 147*, 129–138.

Pérez-Suárez, D., Higgins, P.A., McAteer, R.T.J., Bloomfield, D.S. & Gallagher, P.T. (2010). *Solar feature detection.*

Perrin, J., Torrésani, B., & Fuchs, P. (1999). A localized correlation function for stereoscopic image matching. *Traitement du Signal, 16*, 3–14.

Robbrecht, E., & Berghmans, D. (2005). Entering the era of automated CME recognition: A review of existing tools. *Solar Physics, 228*, 239–251. doi:10.1007/s11207-005-5004-8

Schrijver, K., Hurlburt, N. E., Cheung, M. C., Title, A. M., Delouille, V., Hochedez, J.-F., et al. (2007). *Helio-informatics: Preparing for the future of heliophysics research.* Paper presented at the 210th American Astronomical Society Meeting, Honolulu, Hawaii. Retrieved January 30, 2009, from http://www.lmsal.com/helio-informatics/Welcome.html

Souček, J., Dudok de Wit, T., Décréau, P., & Dunlop, M. (2004). Local wavelet correlation: Application of timing analysis to multi-satellite CLUSTER data. *Annales Geophysicae, 22*, 4185–4196. doi:10.5194/angeo-22-4185-2004

Starck, J.-L., & Murtagh, F. (2006). *Astronomical image and data analysis.* Berlin: Springer.

Stenborg, G., & Cobelli, P. J. (2003). A wavelet packets equalization technique to reveal the multiple spatial-scale nature of coronal structures. *Astronomy & Astrophysics, 398*, 1185–1193. doi:10.1051/0004-6361:20021687

Young, C.A., & Gallagher, P. T. (2008). Multiscale edge detection in the corona. *Solar Physics, 248*, 457–469. doi:10.1007/s11207-008-9177-9

Zharkova, V. V., Ipson, S., Benkhalil, A., & Zharkov, S. (2005). Feature recognition in solar images. *Artificial Intelligence Review, 23*, 209–266. doi:10.1007/s10462-004-4104-4

Chapter 13
Automated Solar Feature Detection for Space Weather Applications

David Pérez-Suárez
Trinity College Dublin, Ireland

R.T. James McAteer
Trinity College Dublin, Ireland

Paul A. Higgins
Trinity College Dublin, Ireland

Larisza D. Krista
Trinity College Dublin, Ireland

D. Shaun Bloomfield
Trinity College Dublin, Ireland

Jason P. Byrne
Trinity College Dublin, Ireland

Peter. T. Gallagher
Trinity College Dublin, Ireland

ABSTRACT

The solar surface and atmosphere are highly dynamic plasma environments, which evolve over a wide range of temporal and spatial scales. Large-scale eruptions, such as coronal mass ejections, can be accelerated to millions of kilometers per hour in a matter of minutes, making their automated detection and characterisation challenging. Additionally, there are numerous faint solar features, such as coronal holes and coronal dimmings, which are important for space weather monitoring and forecasting, but their low intensity and sometimes transient nature makes them problematic to detect using traditional image processing techniques. These difficulties are compounded by advances in ground- and space-based instrumentation, which have increased the volume of data that solar physicists are confronted with on a minute-by-minute basis; NASA's Solar Dynamics Observatory for example is returning many thousands of images per hour (~1.5 TB/day). This chapter reviews recent advances in the application of images processing techniques to the automated detection of active regions, coronal holes, filaments, CMEs, and coronal dimmings for the purposes of space weather monitoring and prediction.

DOI: 10.4018/978-1-60960-477-6.ch013

INTRODUCTION

Astrophysics seeks to determine the physical properties of celestial bodies, primarily by studying the light they emit. This is achieved using remote observations as the distances are generally too great to allow in-situ measurements. Our Sun is the closest of all stars, by many orders of magnitude, and allows scientists to perform long-baseline synoptic studies at size scales which are impossible with other stellar objects. The Sun is also the source of life on Earth, making the study of the Sun-Earth interaction extremely important, especially in our technology-dependent society of today. The study of space weather focuses on disturbances produced by the Sun and the effects that they have on the environment near Earth, the other planets, and throughout the heliosphere. Those disturbances can affect satellites, airplane communications, long metallic oil pipe lines and electrical distribution grids, to name a few. More directly, it can affect the health of air crews and passengers on polar flights and astronauts. Accurate forecasting of those disturbances and their effects allows us to prepare for their arrival. The Sun is routinely observed by numerous ground- and space- based observatories and the study of features in real time (i.e., those currently on the solar surface) provides a better insight into what may happen at a later time elsewhere in the heliosphere.

When Galileo Galilei turned his telescope to look at the Sun in the early 17th century he became one of the first scientists to look in detail at the solar atmosphere. He pointed out that sunspots are features on the surface of the Sun and used them to study solar rotation. Since then, many observatories have studied, counted, and classified sunspots as they emerge and evolve on the Sun. These observations have been used to understand more than how the Sun rotates; historical data have made important contributions in studying the 11-year solar activity cycle and even some possibly-related Earth climate changes (e.g., the "little ice age" in Europe during the latter half of the 17th century occurred in the Maunder minimum, an almost 60-year period in which the Sun seemingly produced few sunspots, (Eddy 1976; Lockwood et al. 2010). Images were drawn by hand and classified by eye in those early observations, initially using pencil drawings before photographic plates became common use. The invention in 1969 of the Charge-Coupled Device (for which Willard S. Boyle and George E. Smith won the 2009 Nobel prize in physics) was the start of a new era for solar physics. The ability to directly digitize images at their acquisition allowed telescopes to rapidly acquire data. Shortly thereafter, space-based missions started taking observations in different wavelengths, providing a more complete view of the Sun. The many different types of instruments on-board spacecraft (i.e., imaging, spectrograph, and in-situ detectors) have also provided invaluable information for a recent field of study called "space weather". Space weather generally refers to the combined effect that all forms of solar activity have on objects within the heliosphere (including planets, their atmospheres, and satellites.) The ultimate aim of space weather research is to accurately forecast the arrival time of events which affect the heliosphere. To achieve this we need a better understanding of the different features that appear on the Sun and the resulting different forms of solar activity.

In general the solar atmosphere is stratified into temperature layers, each of which can be distinguished by the dominant type of radiative emission. The coldest, and lowest, layer emits mostly visible light (i.e., the photosphere, approximately 6,000 K) while the hottest, and highest, layer emits mostly in extreme ultraviolet (EUV) and X-ray wavelengths (i.e., the corona, more than 1 MK; Stix, 2004). Fortunately for human life, the Earth's upper atmosphere blocks most of the high energy solar radiation. However, this makes it impossible to observe the hottest layers from ground-based facilities. The observation of these layers is achieved with the use of instruments on-board high-altitude balloons, rockets, or space-craft. A variety of instruments are used to study the processes which occur on the Sun. Imaging devices are generally sensitive to restricted wavelength range, while spectrometers provide information across wavelength at the expense of losing a spatial dimension. Another important instrument is the magnetograph that measures the magnetic field (or a component of this, e.g., the longitudinal component along the line-of-sight) at some height within the solar atmosphere (typically at the photospheric surface). Finally, coronagraphs are used to study the immediate environment around the Sun. These instruments consist of an imaging unit with an occultation disk that obscures the solar disk. The reader is referred to Stix (2004) as well as the relevant documentation for each instrument (specific papers are given in the following chapter sections) for a more detailed description of solar instruments and their characteristics.

The following sections provide an overview of the main feature detection techniques that are used for space weather forecasting. The features are classified into two groups depending on the form of data necessary to detect them: spatial features observed in single images (e.g., sunspots); temporal features requiring more than one image to characterise them (e.g., transient events, such as coronal mass ejections). However there is an overlap, as some of the techniques used to segment the first kind of feature do require more than one image in order to provide a robust threshold. The identification of many other features on the Sun, not related to space weather and hence not discussed shown in this chapter, may be found in the literature reviews of image processing techniques applied to solar images. Sanchez et al. (1992) provides a substantial review on the techniques used mainly for ground-based observations, e.g., to reduce the effect of the atmospheric turbulence around the telescope. High-resolution, ground-based images have been used to detect and track the granular (and supergranular) cells on the photosphere (e.g., Rieutord et al., 2007; Potts et al.,

2004). Zharkova et al. (2005) give a very detailed description of some of the features detected as part of the European Grid of Solar Observation (EGSO) programme, while Aschwanden (2010) provides a general review of techniques used to detect a multitude of solar features.

Detection of Spatial Features

This section focuses on the detection of time-independent features. This does not preclude evolution of these features; in fact they must evolve as there is nothing truly static on the Sun. However, they may be detected without knowledge of how they evolve. The dynamics of these features can then be measured by following them across the solar disk as the sun rotates. The description of a few detection codes focusing on active regions, coronal holes, and filaments are discussed.

Active Regions

Solar active regions were historically observed as sunspot groups in the photospheric continuum. Physically, active regions are concentrations of magnetic flux that have emerged through the solar surface and are thus a manifestation of solar magnetic activity. Sunspots were the first solar feature to be cataloged and have been routinely measured by multiple observatories over the last four centuries. Observations obtained during the last 60 years have highlighted the different forms which active regions appear to take when observed in different wavelength ranges (i.e., at different temperatures). In 1972 the National Oceanic and Atmospheric Administration (NOAA1) started a sequential numbering system for active regions. Figure 1 shows an example of NOAA 10708 extracted from SolarMonitor.org (Gallagher et al., 2002). It is clear from the figure that different algorithms must be used to identify active regions at each height in the solar atmosphere. Sunspots are the signatures of active regions in the visible photosphere (Figure 1a), normally with a dark inner umbra and a surrounding penumbra. In the warmer overlying chromosphere they are observed as extended bright patches in visible and UV emission (Figures 1c and 1d). Meanwhile, in the hot higher corona they appear as high-contrast regions of EUV emission where loop-like structures can often be distinguished. The varied appearance at different temperatures (and thus heights) in the solar atmosphere makes a consistent definition of an "active region" a somewhat difficult, and occasionally controversial, task. It should be noted that NOAA only designates numbers to those regions which have a white-light signature (i.e., a sunspot in the visible continuum)2.

As is evident from Figure 1, the same detection technique cannot be used for every image across

Figure 1. These six images show NOAA active region 10708 as viewed in (a) continuum (photosphere, ~6,000 K) from MDI, (b) magnetogram from MDI, (c) H-alpha (chromosphere, ~8,000 K) from Big Bear Solar Observatory and d) 304 Å (higher chromosphere, ~10,000 K), e) 171 Å (corona, 1 MK) and f) 195 Å (corona, 1.2 MK) from the Solar and Heliospheric Observatory. Extracted from http://www. solarmonitor.org (Gallagher et al., 2002).

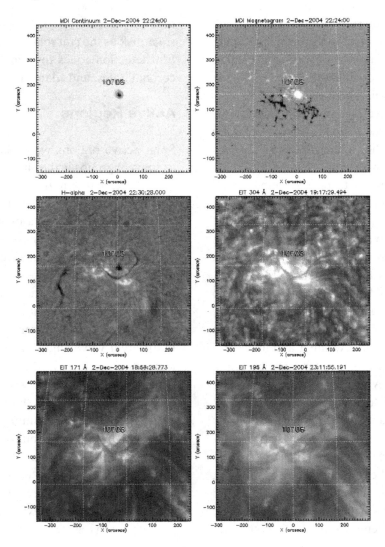

all wavelengths. However, for cataloguing purposes there is no need to detect the feature in every possible wavelength. When multiple channels are used as an input the segmentation can be done through supervised or unsupervised classification. The main difference is in how the classes are selected: the supervised method requires training values whereas classification values in the unsupervised approach are determined by the algorithm. From a space weather perspective, images of the magnetic field (i.e., magnetograms), provide the most information. The detection of active regions is essential because they are the source of several forms of solar activity, such as flares and coronal mass ejections (CMEs).

Currently, region-based flare and CME forecasting requires determination of the magnetic properties of an active region in question and its surroundings (e.g., Conlon et al., 2008; Zhang et al., 2009 and references therein). Turmon et al. (2002, 2010) describe a Bayesian technique to segment active regions in both ground- and space- based magnetogram data. The SolarMonitor Active Region Tracker (SMART, Higgins et al., 2010) is an algorithm for detecting, tracking, and cataloging active regions throughout their emergence, evolution, and subsequent decay. It extracts magnetic properties such as active region size, total magnetic flux, flux imbalance, growth or decay rate, and measurements of magnetic morphology. The SMART code operates in four main steps. First, the magnetograms are segmented into individual feature masks (Figure 2). Second, a characterization algorithm is run on each extracted region to determine its physical properties. Third, extracted regions are classified using a simple scheme. Finally, the regions are catalogued and tracked through time. Here we are interested only in the initial segmentation technique, so the reader is referred to the SMART documentation (Higgins et al., 2010) for further information on the other aspects of the algorithm.

SMART uses two consecutive magnetograms to allow for the removal of transient features as well as the extraction of time-dependent properties. The first steps applied, shown in Figure 2, are smoothing of the images (top row, middle column) with a 2D Gaussian and removal of the background using a static threshold (top row, right column). Binary masks are created from the corrected magnetograms (second and third rows, left column), setting all pixels above the threshold to one. The masks are radially dilated (second and third rows, middle column) and subtracted in order to identify and remove transient features (right column, middle row). The resulting mask is then radially dilated (bottom row, middle column). Each feature (bottom row, right column) is then characterised individually by its physical properties, which are determined from the later magnetogram.

The SMART algorithm is unique among automated active region extraction algorithms (e.g., McAteer et al. 2005a), in that it facilitates the temporal analysis of magnetic properties from first emergence of an active region through tracking the reappearance of regions over multiple solar rotations. In contrast, NOAA assign new numbers to active regions that rotate around the east limb onto the visible solar disk, irrespective of whether they are newly emerged or previously existing active regions.

Future revisions of SMART will incorporate a flare event probability. This will be determined using active region properties determined by SMART, including a measure of magnetic flux near polarity separation lines (Schrijver, 2007) and a proxy for non-potentiality (Falconer et al., 2008). The idea of using many active region properties to determine flare event probabilities was explored in Leka & Barnes (2003). The accuracy of flare prediction may increase by adapting more esoteric AR properties, such as the McAteer et al. (2005b) fractal and Conlon et al. (2009, 2010) multifractal techniques to the SMART feature characterisation. Conlon et al. (2010) propose a 2D wavelet

Figure 2. SMART steps. First image, top-left, shows the magnetogram of an active region. Next images on the right show the smoothing and thresholding at 70 G respectively. Next two rows show the masks from that image and the one obtained 96 minutes before, the growing step, and their difference. The last row shows the first mask after subtracting the difference and after dilation, which is the final segmentation mask used to plot the contours on the last image.

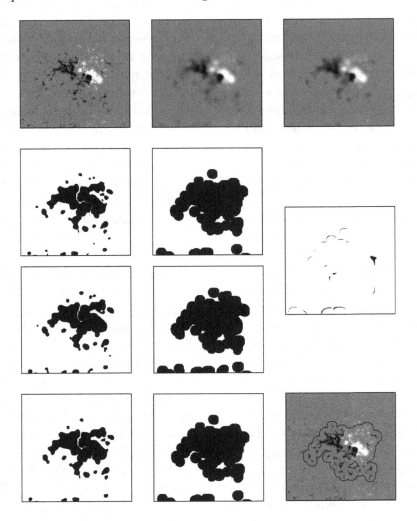

transform modulus maxima method to study the multifractal properties of active region magnetic fields whereby the segmentation of the region is provided by an adaptive space-scale partition of the fractal distribution that shows a potential link to the onset of solar flares.

Other algorithms seek to provide automated classification of sunspot groups in the same framework as human observers. Colak & Qahwaji (2008)

use white-light images and magnetograms from the Michelson Doppler Imager (MDI; Scherrer et al. 1995) onboard the *Solar and Heliospheric Observatory* (*SoHO*; Domingo et al., 1995) with neural network techniques to detect sunspots and classify active regions according to the McIntosh classification system. This system has the advantage that it achieves similar results to the NOAA active region identification scheme.

Zharkova et al. (2005) discuss two codes (developed for EGSO at Meudon observatory) to study active regions. Information about sunspot properties (i.e., size of umbra and penumbra) and intrinsic magnetic properties are obtained by comparing ground-based images of chromospheric emission (i.e., Ca II K1) and *SoHO*/MDI white-light images with *SoHO*/MDI magnetograms. These two codes are complementary: one extracts the magnetic field properties of the active region; one extracts the properties of the sunspots. The segmentation is obtained using a *Sobel* edge detection technique on the photospheric images. A global threshold segments the edges and the existing gaps are filled with the *close* and *watershed* morphological operators, followed by a new segmentation based on dynamic thresholding (constant for the MDI data; variable for ground-based images, due to Earth's unstable atmospheric conditions) to extract the sunspot umbra and penumbra. Not all active regions detected by SMART produce a photospheric sunspot signature in the continuum. Therefore, studying the output of both SMART and the Zharkova codes will help in the understanding of the production and evolution of the solar magnetic field.

Active regions can also be extracted from a number of EUV passbands. Dudok de Wit (2006) discuss a supervised clustering method for which some applications are shown in Chapter ??. Barra et al. (2009) use a fuzzy clustering technique called Spatial Possibilistic Clustering Algorithm (SPoCA). SPoCA is a multichannel, unsupervised, spatially-constrained, fuzzy clustering method that automatically segments solar EUV images into regions of interest. It has the ability to detect multiple features at once, such as active regions, coronal holes, and quiet Sun. The nature of this code allows the detection of additional features, such as filaments and coronal bright points, and the addition of images observed in other wavelengths. The algorithm attempts to find the cluster centres of the features being detected through an iterative minimization equation. Each pixel obtains a probabilistic value of belonging to one or other feature group which depends on itself, its closest neighbours, and the whole image. SPoCA includes a radial line-of-sight equalization, inclusion of an automatic evaluation of the segmentation with a *sursegmentation* method (i.e., segmenting the image into a number of classes strictly superior to the intuitively expected number of classes in the image, and then finding an aggregation criterion of the resulting partition that shows the relevant classes), and smoothing of the edges using a morphological opening with a circular isotropic element.

Coronal Holes

Coronal holes are low-density regions in the hot, high-lying solar corona that exhibit reduced EUV and X-ray emission when compared to the quiet Sun and active regions. The magnetic field distribution within coronal holes is believed to be dominated by a single polarity. This is probably due to the predominantly open nature of the magnetic field lines that extend beyond the corona and into the interplanetary medium. As a result, coronal holes give rise to the high-speed solar wind streams (Altschuler et al., 1972), causing recurring magnetic disturbances at Earth on time scales of days to months as these streams sweep past. Coronal holes can be observed in EUV and X-ray wavelengths from rocket or space-based telescopes as well as in the cooler, lower-lying chromospheric He I 10830 Å infrared absorption line from ground-based telescopes. They appear dark in EUV and X-ray because of a low emission-line strength (caused by reduced densities), while they are bright in He I because of a low absorption-line strength (caused by reduced population of the atomic state required for radiative absorption). In order to automate coronal hole detection, various teams have developed approaches mostly based on threshold segmentation. Two detection algorithms are described below: the first detects CHs from Earth using ground-based He I images, while the second uses EUV images from the Extreme ultra-violet Imaging Telescope (EIT; Delaboudinière et al., 1995) onboard *SoHO*.

Ground-based observations from the Kitt Peak Vacuum Telescope (KPVT) permitted the cataloging of coronal holes from 1974 to 2003 through manual identification. Henney & Harvey (2005) developed an algorithm motivated by the conclusion of operations of the KPVT in 2003 and the start of the synoptic observations by the SOLIS Vector Spectro-Magnetograph helium spectroheliograms and photospheric magneto-grams. The method uses a two-day averaged He I 10830 Å spectroheliogram and a two-day averaged photospheric magnetogram, weighted by an expression involving their time difference. A mask value is determined as the 10% level of the median of positive values. A morphological *closing* operation (using a square kernel func-tion as the shape operator) is applied to fill the gaps and connect nearby regions. For physical reasons, explained in Harvey & Recely (2002), areas smaller than two supergranules are removed and the mask is then multiplied by a very large value to fill the corresponding pixels on the first segmented image. The image is then smoothed to fill in small gaps and holes. This is followed by the *open* morphological operation which removes small features while preserving the size and shape of the detected regions. The magnetic properties of each candidate region are extracted from the magnetograms and the percentage of unipolarity is examined in order to disregard those with a value below a varying threshold.

More recently, Krista & Gallagher (2009) developed a coronal hole identification method that compares coronal hole properties with in-situ solar wind properties at ~1 AU. The algorithm also incorporates a space weather forecasting tool to predict the arrival of the fast solar wind streams at Earth using the Parker solar wind model (Parker, 1958). The coronal hole boundaries achieved are similar in EUV and X-ray wavelengths and agree well with the boundaries determined by eye. The

automated high-speed solar wind forecasts are also in good agreement with the observed high-speed solar wind arrival times determined from in-situ solar wind measurements. In this method, intensity histograms of a solar EUV image give a multimodal distribution, where each frequency distribution corresponds to a different form of feature on the Sun - i.e., low intensity regions, quiet Sun, and active regions. The intensity of the coronal hole boundary corresponds to the location of a local minimum between the low intensity region and quiet-Sun distributions. As Figure 3 shows, this local minimum can be enhanced us-ing a partitioning operation (1st and 2nd row in Figure 3). Through such an approach, the local histograms obtained for each sub-image have more defined minima, which aids in determining the global threshold. This method works for any time in the solar cycle regardless of the change in the overall solar intensity, as it depends solely on the mean quiet-Sun intensity within the image in question. However, the coronal hole boundaries acquired during solar maxima may be less accu-rate due to bright coronal loops intercepting the line-of-sight and obscuring parts of the coronal hole boundaries.

After segmentation, low intensity regions are classed as either coronal holes or other dark quiet-Sun features (e.g., filaments) using magne-togram data; filaments have a balanced bi-polar distribution (i.e., close to zero skewness) of mag-netic flux, whereas coronal holes have a dominant polarity (i.e., an imbalanced bi-polar distribution with a relatively large skewness). The physical properties of each coronal hole are automatically determined for forecasting purposes. For each coronal hole group the arrival time of the corre-sponding high-speed solar stream is determined at Earth and the predicted and observed arrival time is then monitored for further development of the method.

The results of the two algorithms described above have not yet been compared, however both obtain satisfying results when compared with other

Figure 3. Visualization of the coronal hole detection algorithm developed by Krista & Gallagher (2009). The 195 Å SoHO/EIT full-disk image is transformed to a Lambert equal area projection map (1st row, left). The corresponding global intensity histogram is obtained (1st row, right), as it has a unimodal distribution, no coronal hole threshold can be obtained. The map is then divided into sub-images (2nd row, left) to obtain local intensity histograms (2nd row, right) which are more likely to have a bimodal distribution. In all histograms the black and red dashed lines give the range where the threshold is searched for, and the dashed green line is the threshold found. The 3rd and 4th row images show the SoHO/EIT 195 Å full-disk image and Lambert projection map respectively with and without the low intensity region contours.

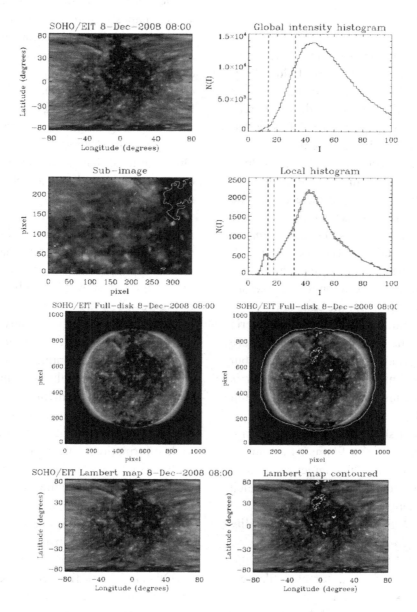

sources. Henney & Harvey (2005) compared their results with the hand-drawn coronal hole maps and found area differences of 3% or smaller. Krista & Gallagher (2009) have compared their high-speed solar wind arrival times with observed arrival times and obtained a positive correlation between the high-speed solar wind duration and the coronal hole area.

Filaments

Filaments are large volumes of very dense, cool plasma held in place by magnetic fields. They usually appear as long, dark, and thin features when observed against the solar disk, whereas they appear as bright, fuzzy arches and are called prominences at the limb. Images at chromospheric temperatures (particularly in the optical H-alpha line) provide the best outline of these features, even though filaments are also observed in the corona. H-alpha observations are routinely made from ground-based telescopes (the recent *Hinode* spacecraft does contain an H-alpha filter but, due to problems arising during launch, its use is not recommended). The images thus require pre-processing to correct for the constantly varying observing conditions caused by atmospheric seeing. Some of the corrections performed are the same for all ground-based observations, but others are instrument dependent. The reader is referred to the documentation of each algorithm for further details.

It is well established that the sudden disappearance (or eruption) of a filaments is usually associated with a CME. The characterisation of filaments can provide information with which to predict the orientation of the magnetic field associated with CMEs and hence the probability of a CME being geo-effective (i.e., its likelihood for impacting Earth). Bernasconi et al. (2005) produced a very complete, automated filament detection and characterization algorithm that is based on an existing code by Shih & Kowalski (2003). Their approach uses full-disk H-alpha im-

ages observed from Big Bear Solar Observatory (BBSO), such as the one shown in the left panel of Figure 4. The filament detection is performed by creating a mask using both threshold segmentation and an advanced morphological filtering operation. The first step removes the sunspots (their cores, or umbrae, are usually darker than filaments). This requires a filtering operation to extract only those regions with elongated shapes. These shapes are isolated by separately applying eight opening morphological operations to a filament mask with the eight linear structuring elements shown in the top-right panel of Figure 4 (Soille & Talbot, 2001). Pixels that survive at least two of these opening operations are used as seeds for a region-growing morphological filter and regions smaller than 300 pixels are deprecated. Once the mask has been created, each separate cluster is numbered and the characterisation of the detection proceeds.

In this process the position, length, area, average tilt of axis and chirality of the magnetic flux rope are extracted. The determination of the filament spine is performed using a multi-step iterative technique, shown in the bottom-right panel of Figure 4. The first iteration starts by determining the location of the two spine end points. Then it determines another vertex by adding the middle point and applying an optimization process. These steps are iterated, resulting in an array with the coordinates of the filament's spine.

The barbs of a filament are an important characteristic to take into account because they may yield information on the chirality of the flux rope within which the filament is embedded. The angle of each barb relative to the closest spine segment determines them as bear-left or bear-right. The difference between the number of bear-left and bear-right barbs establishes the chirality of the filament as left- or right- handed.

Other automatic filament detection techniques differ mainly on how the threshold is selected. In the EGSO algorithm (Fuller et al., 2005) the lower and upper thresholds used for finding seeds

Figure 4. Full disk H-alpha image where the filaments can be seen as long dark and thin structures. On the right side; top: the eight directional linear structuring elements used by advanced morphological filter. On the bottom the first step and final result of the algorithm that determine the filament's spine. The labels refer to the order in which the points are found.

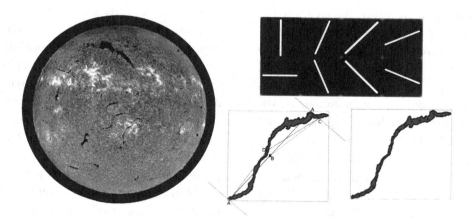

to create a mask in the segmentation of the image are calculated according to local statistics after dividing the image into smaller areas. This code uses a thinning process to obtain the skeleton of the filaments based on the *HitOrMiss* transform (Sonka et al., 1999) that removes the branches of the skeleton after iteratively computing the end points.

Detection of Temporal Features

The previous section focused on detecting features that can be localised within a single image. However, there are other forms of solar features whose detection is more complicated and requires temporal information. This section looks at two of these features - CMEs and coronal dimmings. Both may be easily identified when viewed in a sequence of images (i.e., a movie) but remain difficult to detect in a single image.

Coronal Mass Ejections

CMEs are large-scale eruptions of plasma and magnetic field from the surface of the Sun. They travel through interplanetary space with ve-

locities of up to several thousand kilometres per second, and have consequences for space-borne instruments and planetary atmospheres, often manifested as auroras on Earth and indeed other magnetic-field-protected planets (e.g., Saturn; Prangé et al., 2004). The diffuse and transient appearance of CMEs in images makes them difficult to automatically identify and track. They are best observed with the assistance of coronagraphs, a telescope attachment designed to block out the solar disk in order to better observe the surrounding corona (which is orders of magnitude fainter than the disk). They essentially create an artificial eclipse and are currently used in the Large Angle and Spectrometric Coronagraphs (LASCO; Brueckner et al., 1995) C2/3 instruments onboard *SoHO* and the Sun-Earth Connection Coronal and Heliospheric Imagers (SECCHI; Howard et al., 2008) COR1/2 instruments onboard the *Solar TErrestrial RElations Observatory* (*STEREO*; Kaiser et al., 2008). SECCHI also contains two wide-angle, visible-light imaging systems called the Heliospheric Imagers (HI).

A variety of catalogues exist that are maintained by individual instrument teams: <u>CDAW Catalog3</u> and <u>NRL LASCO CME List4</u> from *SoHO*, and the

COR1 CME Catalog5 and HI1 Event List6 from *STEREO*. These catalogues provide information on the timing and properties of a CME, including the position angle, angular width, height, velocity, and acceleration. However, the creation and population of these catalogues are time consuming and the measured parameters are subject to human bias as they include manually performed processing. The automation of CME detection is highly desirable. To date, several automated CME detection algorithms have been proposed for use with LASCO C2 data and extensible to COR1. The Computer Aided CME Tracking (CACTus) algorithm was the first attempt at automation (Berghmans et al., 2002). This was followed by the Solar Eruptive Event Detection System (SEEDS; Olmedo et al., 2008) algorithm from George Mason University and the Automatic Recognition of Transient Events and Marseille Inventory from Synoptic maps (ARTEMIS; Boursier et al., 2009) from the LASCO team at Laboratoire d'Astrophysique de Marseille. All of these automated codes rely on the use of more than one frame to detect a CME. More recently, Byrne et al. (2009) propose a method to overcome this problem for real-time detection in single images.

The pre-processing of the images followed by these groups differs from the standard methods proposed by the instrument teams. This is because the standard reduction is not optimized to detect CMEs (e.g., the presence of background stars, planets, and comets is usual within these images). It is worth noting that LASCO is the most successful comet-finder in history, having detected over *one thousand six-hundred* comets in over thirteen years of operation[7]#. The images are exposure time normalized, corrected for cosmic rays, stars or planets by different methods, and transformed to their preferred coordinate system. Figure 5 shows an example of the different transformations used by each of the methods listed previously. CACTus transforms each from the native Cartesian coordinate system to a polar coordinate system [r, position angle], where r is

the radial distance from the centre of the Sun and the position angle is the angle (anticlockwise) from a certain reference point (the ecliptic). The transformed images are stacked to produce a [r, position angle, t] data cube, which is iteratively processed to estimate the background and to remove the dust corona and rotating streamers. Following this, [r, t] slices are extracted from the cleaned datacube to proceed with the CME detection. ARTEMIS creates synoptic maps that consist of the generation of [position angle, t] images, complementary to the CACTus approach. Finally, SEEDS works in polar coordinates [r, position angle] after determination of the running-difference between two consecutive images. The method proposed in Byrne et al. (2009) does not require a coordinate transformation prior to the CME detection but, as with the other catalogues, it may be strengthened by utilising the temporal information across frames.

Figure 5 demonstrates that the same CME can presents a different signature in each of the transformations, therefore the techniques for the resulting detections are different. CMEs appear as inclined lines in CACTus, which relies on the Hough transform for detection. In ARTEMIS the CMEs appear with different morphologies, which are classified into four types: undistorted vertical streaks without temporal dispersion; quasi-symmetric arc shapes; arc shapes followed by a second structure with a dark zone in between; all remaining events with unclear signature. ARTEMIS detection involves three main steps of filtering, segmentation, and merging with high-level knowledge. Filtering is carried out line by line, removing the background with a median filter of 7-pixel width. The segmentation process is performed by a simple thresholding process with a value selected by *experience*, followed by the application of the Line Adjacency Graph (LAG; Pavlidis, 1986). This step removes small artificial "holes" by performing a morphological closure operation, identifies regions of interest, computes their geometrical and statistical parameters, and

Figure 5. Example of a CME as interpreted by the different algorithms presented here. The top left image shows a single frame of the CME as observed by SoHO/LASCO C2. The dashed line indicates the radial intensity profile as determined by ARTEMIS for multiple time steps, while the solid line indicates the fixed angle intensity profile as determined by CACTus for multiple time steps (top right images). The bottom left image illustrates the vector representation of the multiscale detection outlined in Byrne et al. (2009) whereby the curvilinear structure of the CME front is exploited. The bottom right image is a running-difference image as determined by SEEDS in order to threshold the CME intensity structure.

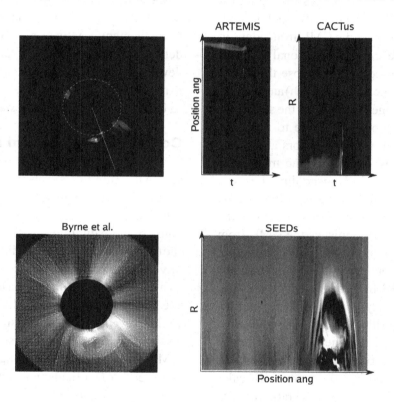

removes those that are smaller than a certain size. Finally, regions of interest are associated to the same CME if they simultaneously satisfy three empirically determined conditions that form the high-level knowledge.

A CME in SEEDs appears as a bright leading-edge enhancement (positive values) followed by a dark area deficient in brightness (negative values), with the background appearing grey (zero change). A running-difference process removes quasi-static features, such as coronal streamers. The CME is extracted by a threshold-segmentation technique and a region-growing algorithm. Positives values of the image are projected into one dimension along the angular axis, obtaining the angular intensity profile. The threshold value is obtained from this profile as a number of standard deviations (the number of standard deviations is chosen by experimental methods, and its value is often between two and four). This gives the angle of the core that is grown to cover the whole CME. The region growing algorithm connects those values between the maxima of the core-angle and a second threshold, calculated as before but just over the values outside the core-angle.

Byrne et al. (2009) apply the multiscale edge-detection method as proposed by Young & Gallagher (2008) for CME detection within a single image frame. While not presently implemented in a catalogue, the potential for automation is clear and currently under development. The *wavelet* algorithm is based upon a high and low pass filtering technique which serves to decompose the image into multiple scales. A particular scale of the decomposition best improves the signal-to-noise ratio of the CME against the background, making it easier to detect the CME front edges in a single image. The size and directional information of the filters used to decompose the image provide magnitude (i.e., edge strength) and angular information (i.e., edge normals) for the structures in the image. It becomes possible to represent the image data as a mesh of vectors across the frame (Figure 5) by combining the magnitude and angular information. Thresholding the areas of maximal edge strength corresponding to large spreads in the angular information along the curved CME front distinguishes the CME from the linear streamers in the image. Furthermore, a pixel-chaining routine may be implemented to outline the CME front edges which are then characterised with, say, an ellipse fitting routine. This detection algorithm is further strengthened by considering more than one scale in the image decomposition, effectively combining the magnitude and angular information across all scales for which the CME signal-to-noise ratio remains high. If temporal information is available (as it is when backdating a catalogue) the method may be refined by turning the single image thresholding into a spatiotemporal filter that considers the movement of the CME edges through frames as an additional constraint on the CME front detection and characterisation. Gallagher et al. (2010) further exploits the curvilinear nature of CMEs through the use of curvelets as a multiscale tool. This approach may further enhance the CME structure on scales which neglect the linear structures of streamers in the image data, which would be of great benefit to an automated detection algorithm.

The second step for any CME catalogue is the characterisation of the CME kinematics. It is not as much connected with image analysis as it is with the application of a model. However, the nature of the Hough transform used by CACTus constrains the CMEs to have constant velocity. Boursier et al. (2009) show a comparison of the catalogues produced by CACTus, SEEDs, and ARTEMIS together with the man-made CDAW showing that the automated catalogues tend to report more than twice as many CMEs as are identified by visual detection. However, the primary interest for the development of the automated algorithms is not to reproduce the human-biased results, but instead to be able to produce robust statistical analyses.

Coronal Dimmings and Bright Fronts

Coronal dimmings are usually observed as decreases in intensity in soft X-rays (Hudson et al., 1996) and EUV data (Thompson et al., 1998). The cause of these dimmings is still under debate, but the two most accepted possibilities are either a density depletion caused by an evacuation of plasma or a change in the bulk plasma temperature out of the passband of the image filter. The importance of these events and their physical cause is related to the potential of using these events to predict CMEs in the absence of coronograph images. This is currently of particular interest because NASA's recently launched *Solar Dynamics Observatory* (*SDO*) has no coronagraph instrument onboard.

The visualisation of these transient events is often optimised using base- or running- difference imaging. The base-difference method differs from the running difference method in that all images in the series have the same fixed image (usually obtained prior to the event) subtracted from them, whereas running-difference subtracts the previous image in time. Figure 6 shows a typical case of coronal dimming using running-difference imaging, where the central dark area (i.e., lower intensity than the previous image) in the middle panel expands outward over the solar surface in the right panel.

Figure 6. Coronal dimming seen in 195 Å SoHO/EIT running difference images. Courtesy of SoHO/ NASA website.

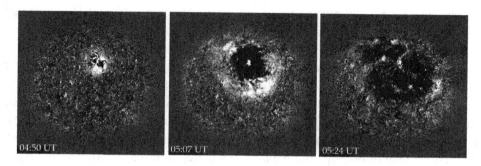

The Novel EIT wave Machine Observing (NEMO; Podladchikova & Berghmans, 2005) algorithm allows for real-time analysis of *SoHO/ EIT* data. The algorithm works in two phases - detecting an event occurrence, and extracting the information of the dimming. The detection of event occurrence is performed through statistical analysis methods based on the histogram distribution of running-difference intensities. The start of an event is characterised by a sudden increase of variance, while the sign of skewness changes and the kurtosis increases rapidly during the event. Once the start and duration of the event is known the algorithm proceeds to the extraction of the dimming. Fixed difference images are then used to extract the dimming from the background. Pixels are collected into two groups - a maximal and minimal pixel map. The maximal pixel map comprises those pixels with values that fall below -1sigma on the histogram, where sigma is the value of the variance before the event occurs. Simultaneously, the minimal pixel map is constructed by selecting the darkest 1% of all pixels from the fixed-difference image. A median filter removes all of the smaller structures considered as "noise". The final dimming region is then extracted using the minimal pixel map as seeds for a region-growing method, keeping the condition of a simply-connected region and restricting it to pixels from the maximal pixel map. After a region is extracted, the area, location, volume, mass and light curves are obtained for each event. Recently, Attrill & Wills-Davey (2009) modified NEMO to adapt it to the Atmospheric Imaging Assembly, which is the successor of *SoHO*/EIT onboard *SDO*.

OUTLOOK

Observatories from all around the world have been observing and classifying most of the features discussed here for decades. With the increase in data sizes in recent years, and due to the retirement of the experts on whom we relied to extract features manually, scientists have been turning their research toward the automation of such detections. The automation of solar feature detection gives a few advantages to the solar physicist. Firstly, and probably one of the most important advantages for statistical studies, is the creation of non-human biased catalogues. It is well known in observational science that when the observer is substituted, the records show a systematic variation. Secondly, the automation of the techniques accelerates the process of feature segmentation, which nowadays its crucial to manage the amount of data available.

The creation of a non human-biased catalogue is not an easy task. Solar features have been named after how they were observed for the first time, with their description in the hand-made catalogues depending on the observer. A comparison of how any automated algorithm performs, as compared with the previous catalogues, will always show differences. This is mainly due to the parameters used to define the feature itself. Therefore, improvements in detection are always tied with a comparison to the feature description.

The relatively sudden increase in the amount of solar data available is changing the way scientists work. There are two new space missions which will provide a huge amount of high-quality data. The first mission, launched in November 2009, is a European Space Agency mini-satellite called *Proba2*, which provides 1kx1k full-disk images of the 1 MK corona every minute and the ability to off-point from the solar disk (i.e., to track transient events such as CMEs). The second mission, launched in February 2010, is NASA's *SDO*. It will produce a 4kx4k full-disk image in 10 different filters every 10 seconds, and is also equipped with a high-resolution magnetometer that will far surpass the spatial and temporal resolution of *SoHO*/MDI. The analysis of this new data cannot be carried out in the traditional way, due to limitations in storage and bandwidth. The researcher will not be able to download a whole day of data (1.5 TB) to his personal computer. The new approach is to download only the desired feature, made possible by a pipeline of automated analysis algorithms working on each new image downloaded from the spacecraft which will feed a catalogue with feature properties. These catalogues realize the idea of virtual observatories (VO) as a collection of multiple data archives and tools to facilitate multi-instrument research. Recently the VO ideal has matured into a broader concept than simply a collection of data and tools. This is the case with the HELiophysics Integrated Observatory (HELIO[8]), a new project that provides possible links of events throughout the whole heliosphere in addition to data access.

These new instruments provide higher spatial and temporal resolution that could change feature detection either for better or worse. On the bright side, features may be better defined (i.e., boundaries will be more accurate). However, that could also make them more difficult to detect (i.e., previously contiguous bright areas may now be disconnected). This issue provides an advantage to the feature detection techniques based on fuzzy clustering (SPOCA) or supervised clustering (Dudok de Wit, 2006), where the algorithm segments the image based solely on the data provided. Relating these features to current classifications may be a difficult task. It seems clear that the definition of features cannot be linked to what is visible in only one wavelength (e.g., the detection of coronal holes can be differentiated from filaments by using the magnetic field information). More robust feature definitions will be achieved with the help of *SDO* by combining imaging of the chromosphere and corona in multiple passbands with photospheric vectormagnetograms, which provide the orientation and magnitude of the magnetic field at the solar surface.

The future perspectives for image processing applications to space weather are clear: near real-time feature extraction and analysis is crucial for forecasting. The solar physics community is well aware of this issue as is evident in the success of the Solar Image Processing Workshop series[9]. The detection of active regions, filaments, coronal holes, and tracking of CMEs is key, though a better understanding of these processes is required. An important advance would be the detection of transient events in single frames, which could, e.g., detect CMEs in single coronagraph images without the need for an image sequence. The biggest advancement could perhaps be in the search for pre-cursors for these events, and it seems that multiscale techniques may be a vital tool in this case.

REFERENCES

Altschuler, M. D., Trotter, D. E., & Orrall, F. Q. (1972). Coronal holes. *Solar Physics, 26,* 354–365. doi:10.1007/BF00165276

Aschwanden, M. J. (2010). Image processing techniques and feature recognition in solar physics. *Solar Physics, 262,* 235–275. doi:10.1007/s11207-009-9474-y

Attrill, G. D. R., & Wills-Davey, M. J. (2009). Automatic detection and extraction of coronal dimmings from *SDO*/AIA data. *Solar Physics,* 143.

Barra, V., Delouille, V., Kretzschmar, M., & Hochedez, J. (2009). Fast and robust segmentation of solar EUV images: Algorithm and results for solar cycle 23. *Astronomy & Astrophysics, 505,* 361–371. doi:10.1051/0004-6361/200811416

Berghmans, D., Foing, B. H., & Fleck, B. (2002). Automated detection of CMEs in LASCO data. In Wilson, A. (Ed.), *From solar min to max: Half a solar cycle with SOHO* (pp. 437–440).

Bernasconi, P. N., Rust, D. M., & Hakim, D. (2005). Advanced automated solar filament detection and characterization code: Description, performance, and results. *Solar Physics, 228,* 97–117. doi:10.1007/s11207-005-2766-y

Boursier, Y., Lamy, P., Llebaria, A., Goudail, F., & Robelus, S. (2009). The ARTEMIS catalog of LASCO coronal mass ejections. Automatic recognition of transient events and Marseille inventory from synoptic maps. *Solar Physics, 257,* 125–147. doi:10.1007/s11207-009-9370-5

Brueckner, G. E., Howard, R. A., & Koomen, M. J. (1995). The Large angle spectroscopic coronagraph (LASCO). *Solar Physics, 162,* 357–402. doi:10.1007/BF00733434

Byrne, J. P., Gallagher, P. T., McAteer, R. T. J., & Young, C. A. (2009). The kinematics of coronal mass ejections using multiscale methods. *Astronomy & Astrophysics, 495,* 325–334. doi:10.1051/0004-6361:200809811

Colak, T., & Qahwaji, R. (2008). Automated McIntosh-based classification of sunspot groups using MDI images. *Solar Physics, 248,* 277–296. doi:10.1007/s11207-007-9094-3

Conlon, P. A., Kestener, P., McAteer, R., & Gallagher, P. (2009). *Magnetic fields, flares & forecasts.* In AAS/Solar physics division meeting.

Delaboudinière, J. P., Artzner, G. E., Brunaud, J., & Gabriel, A. H. (1995). EIT: Extreme-Ultraviolet Imaging Telescope for the *SoHO* mission. *Solar Physics, 162,* 291–312. doi:10.1007/BF00733432

Domingo, V., Fleck, B., & Poland, A. I. (1995). *SoHO*: The Solar and Heliospheric Observatory. *Space Science Reviews, 72,* 81–84. doi:10.1007/BF00768758

Dudok de Wit, T. (2006). Fast segmentation of solar extreme ultraviolet images. *Solar Physics, 239,* 519–530. doi:10.1007/s11207-006-0140-3

Eddy, J. A. (1976). The Maunder minimum. *Science, 192,* 1189–1202. doi:10.1126/science.192.4245.1189

Falconer, D. A., Moore, R. L., & Gary, G. A. (2008). Magnetogram measures of total nonpotentiality for prediction of solar coronal mass ejections from active regions of any degree of magnetic complexity. *The Astrophysical Journal, 689,* 1433–1442. doi:10.1086/591045

Fuller, N., Aboudarham, J., & Bentley, R. D. (2005). Filament recognition and image cleaning on Meudon H-alpha spectroheliograms. *Solar Physics, 227,* 61–73. doi:10.1007/s11207-005-8364-1

Gallagher, P. T., Moon, Y. J., & Wang, H. (2002). Active-region monitoring and flare forecasting I. Data processing and first results. *Solar Physics, 209*, 171–183. doi:10.1023/A:1020950221179

Gallagher, P. T., Young, C. A., Byrne, J. P., & McAteer, R. T. J. (2010). (in press). Coronal mass ejections detection using wavelets, curvelets and ridgelets: Applications for space weather monitoring. *Advances in Space Research*. doi:10.1016/j.asr.2010.03.028

Harvey, K. L., & Recely, F. (2002). Polar coronal holes during cycles 22 and 23. *Solar Physics, 211*, 31–52. doi:10.1023/A:1022469023581

Henney, C. J., & Harvey, J. W. (2005). Automated coronal hole detection using He 1083 nm spectroheliograms and photospheric magnetograms. In K. Sankarasubramanian, M. Penn, & A. Pevtsov, (Eds.), *Large-scale structures and their role in solar activity.* (p. 261).

Higgins, P. A., Gallagher, P. T., McAteer, R. T. J., & Bloomfield, D. S. (2010). Solar magnetic feature detection and tracking for space weather monitoring. In *Advances in space research: Space weather advances* (Article in press).

Howard, R. A., Moses, J. D., Vourlidas, A., & Newmark, J. S. (2008). Sun earth connection coronal and heliospheric investigation (SECCHI). *Space Science Reviews, 136*, 67–115. doi:10.1007/s11214-008-9341-4

Hudson, H. S., Acton, L. W., & Freeland, S. L. (1996). A long-duration solar flare with mass ejection and global consequences. *The Astrophysical Journal, 470*, 629. doi:10.1086/177894

Kaiser, M. L., Kucera, T. A., & Davila, J. M. (2008). The *STEREO* mission: An introduction. *Space Science Reviews, 136*, 5–16. doi:10.1007/s11214-007-9277-0

Krista, L. D., & Gallagher, P. T. (2009). Automated coronal hole detection using local intensity thresholding techniques. *Solar Physics, 256*, 87–100. doi:10.1007/s11207-009-9357-2

Leka, K. D., & Barnes, G. (2003). Photospheric magnetic field properties of flaring versus flare-quiet active regions. I. Data, general approach, and sample results. *The Astrophysical Journal, 595*, 1277–1295. doi:10.1086/377511

Lockwood, M., Harrison, R. G., Woollings, T., & Solanki, S. H. (2010). Are cold winters in Europe associated with low solar activity? *Environmental Research Letters, 5*, 024001. doi:10.1088/1748-9326/5/2/024001

McAteer, R. T. J., Gallagher, P. T., & Ireland, J. (2005b). Statistics of active region complexity: A large-scale fractal dimension survey. *The Astrophysical Journal, 631*, 628–635. doi:10.1086/432412

McAteer, R. T. J., Gallagher, P. T., Ireland, J., & Young, C. A. (2005a). Automated boundary-extraction and region-growing techniques applied to solar magnetograms. *Solar Physics, 228*, 55–66. doi:10.1007/s11207-005-4075-x

Olmedo, O., Zhang, J., Wechsler, H., Poland, A., & Borne, K. (2008). Automatic detection and tracking of coronal mass ejections in coronagraph time series. *Solar Physics, 248*, 485–499. doi:10.1007/s11207-007-9104-5

Parker, E. N. (1958). Dynamics of the interplanetary gas and magnetic fields. *The Astrophysical Journal, 128*, 664. doi:10.1086/146579

Pavlidis, T. (1986). A vectorizer and feature extractor for document recognition. *Computer Vision Graphics and Image Processing, 35*(1), 111–127. doi:10.1016/0734-189X(86)90128-3

Podladchikova, O., & Berghmans, D. (2005). Automated detection Of Eit waves and dimmings. *Solar Physics, 228*, 265–284. doi:10.1007/s11207-005-5373-z

Potts, H. E., Barrett, R. K., & Diver, D. A. (2004). Balltracking: An highly efficient method for tracking flow fields. *Astronomy & Astrophysics, 424*, 253–262. doi:10.1051/0004-6361:20035891

Prangé, R., Pallier, L., Hansen, K. C., Howard, R., Vourlidas, A., & Courtin, R. (2004). An interplanetary shock traced by planetary auroral storms from the Sun to Saturn. *Nature, 432*, 78–81. doi:10.1038/nature02986

Rieutord, M., Roudier, T., Roques, S., & Ducottet, C. (2007). Tracking granules on the Sun's surface and reconstructing velocity fields. I. The CST algorithm. *Astronomy & Astrophysics, 471*, 687–694. doi:10.1051/0004-6361:20066491

Sanchez, F., Collados, M., & Vazquez, M. (1992). *Solar observations: Techniques and interpretation (first Canary Islands winter school of astrophysics)*. Cambridge Universtiy Press.

Scherrer, P. H. (1995). The solar oscillations investigation-the Michelson Doppler imager. *Solar Physics, 162*, 129–188. doi:10.1007/BF00733429

Schrijver, C. J. (2007). A characteristic magnetic field pattern associated with all major solar flares and its use in flare forecasting. *The Astrophysical Journal, 655*, L117–L120. doi:10.1086/511857

Shih, F. Y., & Kowalski, A. J. (2003). Automatic extraction of filaments in H solar images. *Solar Physics, 218*, 99–122. doi:10.1023/B:SOLA.0000013052.34180.58

Soille, P., & Talbot, H. (2001). Directional morphological filtering. *IEEE Transactions on Pattern Analysis and Machine Intelligence, 23*(11), 1313–1329. doi:10.1109/34.969120

Sonka, M., Hlavac, V., & Boyle, R. (1999). *Image processing analysis and machine vision*. Cole Publishing Company.

Stix, M. (2004). *The Sun: An introduction* (2nd ed.). Springer-Verlag.

Thompson, B. J., Plunkett, S. P., Gurman, J. B., Newmark, J. S., St. Cyr, O. C., & Michels, D. J. (1998). *SoHO*/EIT observations of an Earth-directed coronal mass ejection on May 12, 1997. *Geophysical Research Letters, 25*, 2465–2468. doi:10.1029/98GL50429

Turmon, M., Jone, H. P., Malanushenko, O. V., & Pap, J. M. (2010). Statistical feature recognition for multidimensional solar imagery. *Solar Physics, 262*, 277–298. doi:10.1007/s11207-009-9490-y

Turmon, M., Pap, J. M., & Mukhtar, S. (2002). Statistical pattern recognition for labeling solar active regions: Application to *SoHO*/MDI imagery. *The Astrophysical Journal, 568*, 396–407. doi:10.1086/338681

Young, C. A., & Gallagher, P. T. (2008). Multiscale edge detection in the corona. *Solar Physics, 248*, 457–469. doi:10.1007/s11207-008-9177-9

Zhang, J., Wang, Y., & Robinette, S. (2009). *Toward creating a comprehensive digital active region catalog*. In AAS/Solar Physics Division Meeting.

Zharkova, V. V., Aboudarham, J., Zharkov, S., Ipson, S. S., Benkhalil, A. K., & Fuller, N. (2005). Solar feature catalogues in EGSO. *Solar Physics, 228*, 361–375. doi:10.1007/s11207-005-5623-0

ENDNOTES

1. http://www.noaa.gov
2. http://www.ngdc.noaa.gov/stp/SOLAR/ftpsunspotnumber.htm
3. http://cdaw.gsfc.nasa.gov/CME_list/
4. http://lasco-www.nrl.navy.mil/index.php?p=content/cmelist
5. http://cor1.gsfc.nasa.gov/catalog/
6. http://www.sstd.rl.ac.uk/stereo/HIEventList.html
7. http://sungrazer.nrl.navy.mil/
8. http://www.helio-vo.eu
9. http://www.sipwork.org

Chapter 14
Image Processing Applications Based on Texture and Fractal Analysis

Radu Dobrescu
Politehnica University of Bucharest, Romania

Dan Popescu
Politehnica University of Bucharest, Romania

ABSTRACT

Texture analysis research attempts to solve two important kinds of problems: texture segmentation and texture classification. In some applications, textured image segmentation can be solved by classification of small regions obtained from image partition. Two classes of features are proposed in the decision theoretic recognition problem for textured image classification. The first class derives from the mean co-occurrence matrices: contrast, energy, entropy, homogeneity, and variance. The second class is based on fractal dimension and is derived from a box-counting algorithm. For the purpose of increasing texture classification performance, the notions "mean co-occurrence matrix" and "effective fractal dimension" are introduced and utilized. Some applications of the texture and fractal analyses are presented: road analysis for moving objective, defect detection in textured surfaces, malignant tumour detection, remote land classification, and content based image retrieval. The results confirm the efficiency of the proposed methods and algorithms.

INTRODUCTION

Image texture, defined as a function of the spatial variation in pixel intensities (grey values), is

DOI: 10.4018/978-1-60960-477-6.ch014

useful in a variety of applications and has been a subject of intense study by many researchers. It is very hard to define rigorously the texture in an image. The texture can be considered like a structure which is composed of many similar elements (patterns) named textons or texels in some regular or continual relationship. Wilson (1988) points out that textured regions are spatially extended patterns based on more or less accurate repetition of some unit cell; the origin of the term is related to the weaving craft. Gonzalez (1992) relates texture to other concepts like smoothness, fineness, coarseness, graininess and describes three approaches for texture analysis: statistical, structural and spectral.

There are two points of interest in this chapter. The first point is texture classification and segmentation based on statistical features (especially derived from the mean co-occurrence matrix). The second point of interest is the analysis and classifications of textural images based on fractal dimension with a modified box-counting algorithm. We provide some applications in more detail that illustrate the discriminating power of the selected features, the efficiency of the mean co-occurrence matrix and the modified box-counting algorithm in texture classification and segmentation. The applications refer to road following, defect detection, remote image segmentation, classification of malignant tumours, and content based image retrieval.

BACKGROUND

Texture Analysis Features

Texture analysis has been studied using various approaches, such as statistical (grey level co-occurrence matrices and the features extracted from them, autocorrelation based features and power spectrum), fractal (box counting fractal dimension) and structural (the texture is composed of primitives and is produced by the placement of these primitives according to certain placement rules). The structural approach is suitable for analyzing textures where there is much regularity in the placement of texture elements. The statistical approach utilises features to characterise the

stochastic properties of the distribution of grey levels in the image.

There are two important kinds of problems that texture analysis research attempts to solve: texture segmentation and texture classification. The process called texture segmentation involves identifying regions with similar texture and separating regions with different texture. This implies prior knowledge of the texture types and component number which exist in the analyzed image. Texture segmentation is a more difficult problem than the texture classification. The later involves deciding what texture class an observed image belongs to. Thus, one needs to have an a priori knowledge of the classes to be recognized. Because texture has many different dimensions and characteristics there is no single method of texture representation that is adequate everywhere. The features implied in the classification process can differ from one application to another. Wagner (Jähne, 1999) compares the performance of 318 textural features from 18 feature sets. On the other hand, texture classification can be used to segment multi-textured images. With this end in view, the image is divided into small textured region which are classified and indexed. Generally, the result is a coarse type of segmentation process. The segmentation fineness depends on the degree of partition of the initial images. If the degree of partition is too fine, then it is possible that the texture could disappear. Actually, for an application, a partition index is established taking into account the given image resolution and the texture fineness.

Statistical Methods of Texture Analysis

The statistical approach to texture analysis is more useful than the structural one. The simple statistical features, the mean μ and variance σ^2, can be computed indirectly in terms of the image histogram h. Thus,

$$\mu = \frac{1}{N} \sum_{i=1}^{K} x_i h(x_i)$$

$$N = \sum_{i=1}^{K} h(x_i)$$

$$\sigma^2 = \frac{1}{N} \sum_{i=1}^{K} (x_i - \mu)^2 h(x_i)$$

$N = n_1 n_2$ is the image dimension, $h(x_i)$ is the number of pixels of value x_i in an image I and K is the number of grey levels.

The shape of the image histogram provides many clues to help characterize the associated image, but for many applications it is inadequate to discriminate textures (it is not possible to indicate local intensity differences).

Another simple statistical feature is the edge density (number of edge pixels per unit area), Den_e. The density of edges, detected by a local binary edge detector, can be used to distinguish between fine and coarse textures. Den_e can be evaluated as the ratio between the number of extracted edge pixels N_e (edges must be thinned – to one pixel thickness) and image area A (number of pixels in image region):

$$Den_e = \frac{N_e}{A}$$

In order to characterize textured images, connected pixels in certain direction $d = (\Delta x, \Delta y)$ must be analyzed. For this reason, the autocorrelation function R, the difference image I_d, and co-occurrence matrices C_d, must be considered. The autocorrelation function of an image can be used to describe the fineness/coarseness of the texture (Shapiro, 2001):

$$R(\Delta x, \Delta y) = \frac{\sum_{u=0}^{n_1-1} \sum_{v=0}^{n_2-1} I(u,v) I(u+\Delta x, v+\Delta y)}{\sum_{u=0}^{n_1-1} \sum_{v=0}^{n_2-1} I^2(u,v)}$$

The histogram h_d of the difference $I_d(x,y)$ between all intensity values $I(x,y)$ and $I(x+\Delta x, y+\Delta y)$ can also be used for texture classification:

$$I_d(x,y) = I(x,y) - I(x+\Delta x, y+\Delta y)$$

From this histogram one can extract the mean μ_d and variance σ^2_d:

$$\mu_d = \frac{1}{N} \sum_{i=1}^{K} x_i h_d(x_i)$$

$$\sigma^2_d = \frac{1}{N} \sum_{i=1}^{K} \left(x_i - \mu_d \right)^2 h_d(x_i)$$

In order to reduce the computation time necessary to calculate the co-occurrence matrix, Unser suggested a method based on sum and difference histograms on the gray levels to estimate its coefficients (Unser, 1986).

The most powerful statistical method for textured image analysis is based on features extracted from the Grey-Level Co-occurrence Matrix (GLCM), introduced by Haralick (1973). GLCM is a second order statistical measure of image variation and gives the joint probability of occurrence of grey levels of two pixels, separated spatially by a fixed displacement $d=(\Delta x, \Delta y)$. The co-occurrence matrix is able to capture the spatial dependence between image pixels. Smooth texture gives a co-occurrence matrix with high values along the diagonal for small d. The range of grey level values (K) within a given image determines the dimensions of the associated co-occurrence matrix. The elements of the co-occurrence matrix C_d depend upon the displacement $d=(\Delta x, \Delta y)$:

$$C_d(i,j) = Card\{((x,y),(t,v))/I(x,y) = i, I(t,v) = j,$$
$$(x,y), (t,v) \in N_1 \times N_2, (t,v) = (x + \Delta x, y + \Delta y)\}$$

The Cartesian product $N_1 \times N_2$ is the set of all possible positions of the pixels.

There are two important variations of the co-occurrence matrix. The first is the normalized grey level co-occurrence matrix (NGLCM) N_d defined by:

$$N_d = \frac{C_d(i,j)}{\sum_{i=1}^{K}\sum_{j=1}^{K} C_d(i,j)}$$

which allows them to be considered as probabilities (Shapiro, 2001). The second is the symmetric grey level co-occurrence matrix C_s defined by:

$$C_s = C_d(i,j) - C_{-d}(i,j)$$

From the co-occurrence matrix N_d one can extract some important statistical features for texture classification. Haralick (1973) proposed 13 texture features derived from the NGLCM with high discriminating power. The following are common features used with good results in texture classification (because they have a good discriminating power): contrast Con_d, energy Ene_d, entropy Ent_d, homogeneity Omo_d, and correlation Cor_d (Shapiro, 2001).

$$Con_d = \sum_{i=1}^{K}\sum_{j=1}^{K}(i-j)^2 N_d(i,j)$$

$$Ene_d = \sum_{i=1}^{K}\sum_{j=1}^{K} N_d(i,j)^2$$

$$Ent_d = -\sum_{i=1}^{K}\sum_{j=1}^{K} N_d(i,j)\log_2(N_d(i,j))$$

$$Omo_d = \sum_{i=1}^{K}\sum_{j=1}^{K}\frac{N_d(i,j)}{1+|i-j|}$$

$$Cor_d = \frac{\sum_{i=1}^{K}\sum_{j=1}^{K}(i-\mu_x)(j-\mu_y)N_d(i,j)}{\sigma_x \sigma_y}$$

where:

$$\mu_x = \sum_{i=1}^{K}\sum_{j=1}^{K} iN_d(i,j), \quad \mu_y = \sum_{i=1}^{K}\sum_{j=1}^{K} jN_d(i,j),$$

$$\sigma^2_x = \sum_{i=1}^{K}\sum_{j=1}^{K}(i-\mu_x)^2 N_d(i,j)$$

$$\sigma^2_y = \sum_{i=1}^{K}\sum_{j=1}^{K}(j-\mu_y)^2 N_d(i,j)$$

There are a number of relations between texture aspect and statistical characteristics:

The contrast measures the coarseness of texture. The contrast is expected to be high in coarse texture, namely, when the grey levels of each pixel pair are dissimilar. Large values of contrast correspond to large local variation of the grey level.

The entropy measures the degree of disorder or non-homogeneity. Large values of entropy correspond to uniform GLCM.

The Homogeneity is expected to be large if the grey levels of each pixel pair are similar.

The Correlation shows how spread out the distribution of grey levels is.

Local Features Derived From Mean Co-Occurrence Matrix

With a view to obtaining a statistical feature relatively insensitive to texture rotation and translation, we introduce the *mean co-occurrence matrix* notion (Popescu, 2008b). For each pixel we consider $(2h+1) \times (2h+1)$ symmetric neighbourhoods, $h = 1, 2, 3, ...$ Inside each neighbourhood there are eight principal directions: 1, 2, 3, 4, 5, 6, 7, 8 (corresponding to $0°$, $45°$, ...,$315°$) and we evaluate the co-occurrence matrices $N_{d,k}$ corresponding to the displacement d determined by the central point and the neighbourhood edge point in the k direction ($k = 1,2,...,8$). For each neighbourhood type, the mean co-occurrence matrix CM_d is calculated by averaging the eight co-occurrence matrices $N_{d,k}$:

$$CM_d = (N_{h,1} + N_{h,2} + N_{h,3} + N_{h,4} + N_{h,5} + N_{h,6} + N_{h,7} + N_{h,8})/8,$$
$$h = 1, 2,...$$

Thus, for a 3×3 neighbourhood, $h = 1$; for a 5×5 neighbourhood, $h = 2$; for a 7×7 neighbourhood, $h = 3$, and so on.

The features used in the previous case, can be extracted from the mean co-occurrence matrix CM_d: contrast C, energy E, entropy Et, homogeneity O, and correlation Cor. Evidently, C, E, Et, O, and Cor depend on neighbourhood size h.

It has been observed that the addition of colour increases the classification efficiency because the colour and statistical texture features have complementary roles. For colour texture classification, colour and texture features must be extracted separately and then combined in distances evaluation. The algorithms are the same as in the grey level case. The colour texture segmentation involves four steps: image splitting, colour (RGB or HSV) separation, image classification based on colour component matrices and indexing.

Textured Images Comparison Based on Euclidian Distance

Many texture analysis applications such as texture classification, defect detection, path following, texture segmentation, interpretation of land images, etc. are based on based on decision theoretic methods, especially on Euclidian distance between two feature vectors. Unfortunately, the components and the lengths of the feature vectors are dependent on the application.

For the purpose of the evaluation of the discriminating power of the selected features we calculate the Euclidian distances between two regions: a reference one, for example $I_1(1)$ and a measured one, for example $I_1(2)$. The Euclidian distance $D\{I_1, I_2\}$ between two images I_1 and I_2, which are characterized by the feature vectors $[C_1, E_1, Et_1, O_1, Cor_1]^T$ and $[C_2, E_2, Et_2, O_2, Cor_2]^T$ is expressed by the following relation:

$$D(I_1, I_2) = \sqrt{(C_1 - C_2)^2 + (E_1 - E_2)^2 + (Et_1 - Et_2)^2 + (O_1 - O_2)^2 + (Cor_1 - Cor_2)^2}$$

Because the ranges of the initial characteristics can differ too much, for efficient Euclidian distance calculation, the characteristics must to be normalized. So, we considered that all the feature values which correspond to the reference region are equal to 1 (Popescu, 2008c).

Fractal Analysis Features

Many natural surfaces have a statistical quality of roughness and self-similarity at different scales. Fractals are very useful in modelling these properties in image processing and became popular (Mandelbrot, 1982). Self-similarity across scales in fractal geometry is the crucial concept. A deterministic fractal is defined using this concept of self-similarity as follows. Given a bounded set A in a Euclidean n-space, the set A is said to be self-similar when A is the union of N distinct (non-overlapping) copies of itself, each of which has been scaled down by a ratio r. The fractal dimension, often denoted D, is a key parameter developed in fractal geometry to measure the irregularity of complex objects. Different methods often yield significantly different D values for the same feature (Sun, 2006; Xu, 2009).

The fractal dimension D is related to the number N and the ratio r as follows (Crisan, 2005):

$$D = \frac{\log N\left(r\right)}{\log\left(1 \ / \ r\right)}$$

There are a number of methods proposed for estimating the fractal dimension D. One method is a box-counting algorithm that assumes determination of a fractal dimension like function of the object size evolution in connection with the scale factor. There are many programs for counting the fractal dimension in different forms, but box counting is the most familiar algorithm. A lot of specialized papers describe this algorithm, from the earliest ones (Keller, 1993; Sarker 1994) and till recent proposals (Bakes, 2008; Li, 2009). For the box counting basic algorithm, the image must be binary type. The method consists of splitting the image, successively, into equivalent squares with normalized side $r = 1/2, 1/4, 1/8,...$ and each time computing the number $N(r)$ of squares covered by the object image. The division process is limited by the image resolution. The fractal dimension can be obtained by plotting $lnN(r)$ for different values of $ln(1/r)$, where r is the side length of the covering boxes and calculating the slope of the resulting curve, which is approximated by a straight line. A linear regression is performed using the logarithmic coordinates $x = log(1/r)$, $y = logN(r)$. The regression slope (a) is used to determine the box counting fractal dimension FD as follows (Crisan, 2005):

$$y = a\,x + b$$

$$a = FD = \frac{n \sum\limits_{i=1}^{n} x_i y_i - \left(\sum\limits_{i=1}^{n} x_i\right)\left(\sum\limits_{i=1}^{n} y_i\right)}{n \sum\limits_{i=1}^{n} x_i^2 - \left(\sum\limits_{i=1}^{n} x_i\right)^2}$$

In the above equation, the notation significances is as follows: $x_i = \log_2(1/r_i)$, $y_i = \log_2(N(r_i))$, n is the number of partitions, $i = 1,2,3,...,n$ is the index of function point in the graphical representation.

To evaluate the fractal dimension of a grey level textured image, we apply the box-counting algorithm to contours extracted from binary images obtained using different thresholds (Popescu, 2008b). Because the binary image (and also the fractal dimension) depends on the threshold applied, we use all the significant grey levels contained in the image. The fractal dimensions computed for every grey level are plotted in a graphical representation, named the fractal dimension spectrum. Thus, the algorithm calculates a mean fractal dimension MFD from individual values FD_j of some image contours extracted from the initial grey-level image as follows:

$$MFD = \frac{1}{k} \sum\limits_{j=1}^{k} FD_j$$

The Algorithm Consists of the Following Steps

a. Converting the colour image to a 256 grey levels image;

b. Converting the 256 grey levels image to a binary image using a fixed threshold T_j;

c. Extraction the image contour using 3x3 neighbourhoods;

d. Computing the fractal dimension FD_j, from the contour image, using the box-counting algorithm;

e. Iteration of the steps b-d, for $j = 1,...,k$;

f. Determination of MFD.

In the colour image case, the natural decomposition into R, G, and B components can be considered (Popescu, 2008a). The mean fractal dimension is calculated by means of the grey level algorithm for each component: $MFDR$, $MFDG$, and $MFDB$. These characteristics are utilized as features in the texture classification process. The same procedure is applied with HSV decomposition.

Fractal-Based Techniques in Texture Analysis Applications

Fractal-based texture classification is another approach that correlates texture coarseness and fractal dimension. The fractal dimension gives a measure of the roughness of a surface. Keller (1989) uses fractal geometry to describe texture. Intuitively, the larger the fractal dimension, the rougher the texture is. Most natural surfaces (in particular textured surfaces) are not deterministic but have a statistical variation. This makes the computation of fractal dimension more difficult.

Because the analyzed regions are textured ones, their contour images (edges) are full of edge pixels, and the first points in the log-log representations give a partial *FD* equal to 2. Therefore, we proposed another fractal dimension, similar to the box-counting algorithm, which we named *effective fractal dimension - EFD* (Popescu, 2008c). Based on a modified box- counting algorithm, *EFD* is calculated by omitting the first points in the log-log representation (the points of the form $(x_i, 2x_i)$, $i = 1, 2, \ldots, k$) as follows:

$$a_E = EFD = \frac{(n-k)\sum\limits_{i=k+1}^{n} x_i y_i - \left(\sum\limits_{i=k+1}^{n} x_i\right)\left(\sum\limits_{i=k+1}^{n} y_i\right)}{(n-k)\sum\limits_{i=k+1}^{n} x_i^2 - \left(\sum\limits_{i=k+1}^{n} x_i\right)^2}$$

The threshold assessment which is used for edge extraction constitutes a problem for the fractal dimension evaluation in the grey level image case. Thus, we propose the following methods for threshold establishment:

a. The values for which the contour conserves the texture of image.

b. The value for which the highest number of contour points is obtained.

c. The values for which the contour pixel set is a nonempty one.

An example of this is presented in (Popescu, 2008c). The case study calculates two fractal dimension estimations, for 128×128 textured images I_1 (wood) and I_2 (wood with defect). From this point of view we considered the box-counting algorithm, grey level case, which utilizes the above method b. The first estimation is *FD* and the second estimation is *EFD*. The example shows that *EFD* improves texture classification process.

The full division box-counting algorithm (*FD*) considers the linear regression slope from all the points. The vector [*v*] represents the division vector, and the vector [*w*] represents the resulting box-counting vector (the numbers of squares which contain edge pixels). The vectors [*x*] and [*y*] represent the component logarithms of [*v*] and [*w*]:

$$x = \log v, \, y = \log w$$

Thus, *FD* is the fractal dimension, which is evaluated from (x, y) representation, namely log-log representation.

The effective division box counting algorithm (*EFD*) considers only the points in the sequences *x* and *y* of the basic box counting algorithm, after a start point, corresponding to first value of *x* wherefore $y \neq 2x$. Thus, shorter vector representations for box counting algorithm, [*X*] and [*Y*] are obtained, with:

$$X_i = x_{i+k}, \, Y_i = y_{i+k}, \, i = 1, 2, \ldots, n-k$$

Effective fractal dimension *EFD*, equal to a_E, is calculated. For example:

b) Image I_1
[*v*] = [2 4 8 16 32 64 128]
[*w*] = [4 16 64 253 855 2052 3587]
[*x*] = [0.301 0.602 0.903 1.204 1.505 1.806 2.107]

$[y] = [0.602\ 1.204\ 1.806\ 2.403\ 2.932\ 3.312\ 3.555]$
$FD_1 = 1.685$

Start point: (1.204, 2.403)

$[X] = [1.204\ 1.505\ 1.806\ 2.107]$
$[Y] = [2.403\ 2.932\ 3.312\ 3.555]$
$EFD_1 = 1.274$
b) Image I_2
$[v] = [2\ 4\ 8\ 16\ 32\ 64\ 128]$
$[w] = [4\ 16\ 64\ 238\ 837\ 2339\ 4459]$
$[x] = [0.301\ 0.602\ 0.903\ 1.204\ 1.505\ 1.806\ 2.107]$
$[y] = [0.602\ 1.204\ 1.806\ 2.377\ 2.923\ 3.369\ 3.649]$
$FD_2 = 1.731$

Start point: (1.204, 2.377)

$[X] = [1.204\ 1.505\ 1.806\ 2.107]$
$[Y] = [2.377\ 2.923\ 3.369\ 3.649]$
$EFD_2 = 1.417$

The graphical results for image I_1 are presented in Figure 1a (log-log diagram for *FD*) and Figure 1b (log-log diagram for *EFD*).

We can see by inspecting Figure 1a and 1b that *EFD* is smaller than *FD* and the discriminating efficiency (the absolute relative difference between fractal dimensions of two textured images) is better for *EFD* than *FD*:

$$\frac{|FD_2 - FD_1|}{FD_2} = 2.65\%,$$
$$\frac{|EFD_2 - EFD_1|}{EFD_2} = 10\%$$

APPLICATIONS BASED ON TEXTURE AND FRACTAL ANALYSIS

We will briefly present some applications of textures and fractals in road analysis for mobile object navigation, automated inspection for defect detection and localization, medical image processing for malignant tumour detection and remote sensing for land classification.

Road Analysis Based on Statistical Features Derived from the Average Co-Occurrence Matrix

The application consists of the asphalt region separation (road segmentation) in images acquired

Figure 1. Log-log diagrams (a-FD, b-EFD) for image I_1.

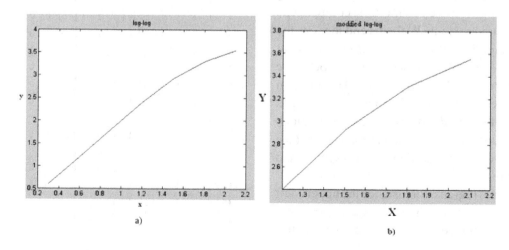

a)

b)

Figure 2. Textures for road delimitation: (a) I_1 - asphalt and pebble, (b) I_2 - asphalt and grass

I_1 I_2

from a video camera. The images from the assistant camera unit are divided into equivalent small square regions (sub-images). For illustration, Figure 2 presents two images with different texture. Image I_1 contains asphalt and pebble, and image I_2 contain asphalt and grass.

It can be easily observed from inspection of Figure 2 that the road and its neighbourhoods are characterized by different textures. The asphalt texture is considered as a reference texture. The application goal is to identify the asphalt regions and to produce an asphalt localization matrix by recognition techniques. The case study (Popescu, 2008b) is related to a method of segmentation of the road image I_1, based on multiple comparisons of the textured regions. With this end in view, the whole image is partitioned into 256 equivalent regions (16×16 block matrix). The blocks are 32×32 matrices. The values of similarity region measurement are placed in the 16×16 matrix. After the threshold process, the 16×16 block matrix becomes the segmentation matrix (see Figure 3), with 1 for carriage road (asphalt) and 0 for the rest.

One of the regions, for instance the upper – left region, is considered the reference region, $Ir = I_1(1,1)$. After image partitioning, a block matrix

(16×16) is obtained. Each block is a textured region with a well defined position. The reference region is successively compared with other regions, using Euclidian distances $D(Ir, I_1(i,j))$, $i,j = 1,2,3,..., 16$. Consider the texture features derived from the mean co-occurrence matrix CM_d. First,

Figure 3. Segmentation matrix

	1	2	3	4	5	6	7	8	9	10	11	12	13	14	15	16
1	1	1	1	1	1	1	1	1	1	1	0	0	0	0	0	0
2	1	1	1	1	1	1	1	1	1	1	0	0	0	0	0	0
3	1	1	1	1	1	1	1	1	1	1	0	0	0	0	0	0
4	1	1	1	1	1	1	1	1	1	1	0	0	0	0	0	0
5	1	1	1	1	1	1	1	1	1	1	0	0	0	0	0	0
6	1	1	1	1	1	1	1	1	1	1	0	0	0	0	0	0
7	1	1	1	1	1	1	1	1	1	1	0	0	0	0	0	0
8	1	1	1	1	1	1	1	1	1	1	0	0	0	0	0	0
9	1	1	1	1	1	1	1	1	1	1	0	0	0	0	0	0
10	1	1	1	1	1	1	1	1	1	1	0	0	0	0	0	0
11	1	1	1	1	1	1	1	1	1	1	0	0	0	0	0	0
12	1	1	1	1	1	1	1	1	1	1	0	0	0	0	0	0
13	1	1	1	1	1	1	1	1	1	1	0	0	0	0	0	0
14	1	1	1	1	1	1	1	1	1	1	0	0	0	0	0	0
15	1	1	1	1	1	1	1	1	1	1	0	0	0	0	0	0
16	1	1	1	1	1	1	1	1	1	1	0	0	0	0	0	0

the comparison result is binarized by applying a corresponding threshold. If $D(Ir, I_1(i,j)) \geq T$ (the region is not similar to Ir) then the region $I_1(i,j)$ is indexed 0; if $D(Ir, I_1(i,j)) < T$ (the region is similar to Ir) then the region $I_1(i,j)$ is indexed 1. Thus, the regions are filled with 0 or 1 depending on the region index (see Figure 3).

In order to appreciate the efficiency of the selected features, we analyzed the most unfavourable cases, namely the minimum distance between two regions coming from different textures, and the maximum distance between two regions coming from the same texture. The minimum distance value for the dissimilar textures case is greater than the maximum distance value, for the similar textures case.

Texture Defect Detection in Manufactured Surfaces

Most applications of defect detection have been in the domain of textile inspection, wood manufactured inspection, painted metal surfaces, and assessment of carpet wear.

There are two important kinds of problems that texture analysis attempts to solve in surface manufacturing production investigation: (a) Are the material textures similar in all the regions? (b) Is there a surface defect and where it is?

The major focus of this application is defect detection based on a classification process with normalized type features extracted from the mean co-occurrence matrix. Texture analysis is made on four regions which are obtained by partition of images of deal board (see Figure 4).

Figure 4. Deal board with defects (two knots).

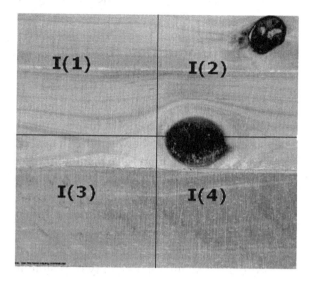

Images $I(1)$ and $I(3)$ are normal type, while images $I(2)$ and $I(4)$ contain defects (knots) - see Figure 4. Textural features: contrast (C), energy (E), entropy (Et), homogeneity (O) and variance (V) are calculated for displacement $d=10$. We consider that the texture of the region $I(1)$ is the reference. Therefore, all the feature values for $I(1)$ are equal to 1. The normalized results for the mentioned features and the distances between images are presented in Table 1.

From the table of distances, one can observe that the idea of region comparison based on statistical features extracted from average co-occurrence matrix, can be utilized to defect detection and localization. So, there are defects in regions $I(2)$, and $I(4)$, inside image I. Based on a distance

Table 1. Features and distances between regions inside image I

Image region	C	E	Et	O	V	Distances	Distance value
$I(1)$	1	1	1	1	1		
$I(2)$	0.91	0.67	5	0.90	5.67	$D(I(1),I(2))$	7.61
$I(3)$	0.96	0.98	1.44	0.99	1.58	$D(I(1),I(3))$	0.73
$I(4)$	0.89	0.42	3.22	0.86	3.28	$D(I(1),I(4))$	4.64

values comparison it is obviously that the defect of region $I(2)$ is greater than the defect of region $I(4)$.

Malignant Tumour Detection Based on Fractal Analysis

Fractal analysis is a mathematical and computer technique that quantifies complex shapes. Fractal analysis can discriminate between the shapes of benign and malignant tumours in mammography (Wang, 1998; Losa, 2005; Crisan, 2007). Thus, a method to emphasize the irregularity of the contour is to calculate and combine different forms of fractal dimension.

Experimentally, it was established that most of the information about the malignity of a tumour is contained in the contour of the tumour shape. Doubtful tumours are characterized by blurred contours which change by altering the threshold used to separate the tumour from the background (image segmentation). The outline of each image was analyzed by estimating the global fractal dimension, the local fractal dimension and local connected fractal dimension. For this purpose we used an original software package described in detail elsewhere (Crisan, 2005). In brief, the fractal dimension of each outline was measured by the box-counting algorithm and the local fractal dimension and local connected fractal dimension were estimated according to the algorithms published in (Landini, 1996; Crisan, 2005).

In order to describe the heterogeneous nature of a tumour image, we may compute, for every single point in the image, a *local fractal dimension* (box-counting dimension, for example), limited to a neighbourhood of the central pixel. Thus, instead of a single value meant to characterize the whole image, we have a set of values, one for each point in the analyzed object. The values will be represented as a histogram in order to give emphasis to the distribution of the local irregularities of the image. We may consider that the global dimension (of the whole image) is the local dimension with the highest frequency.

In some situations neither local fractal dimension approach is enough. A compact set of disconnected points (0-topological dimension) will have a higher fractal dimension. Thus, we associate with every single point in the image a *local-connected fractal dimension* in order to describe the shape structure containing the point, considering only those points inside the neighbourhood connected with the central pixel.

Although the advantages of using the local and local connected dimensions are obvious, they present an important disadvantage: the distribution of the local and local connected fractal dimensions depends on the choice of the maximum window size. For better efficiency, one can combine fractal and texture analysis. To achieve this aim, a complex image processing software system written in the programming language C++ was utilised. The product facilities divided into three categories are as follows:

- Digital image processing
  - screening of a region for processing and saving this region in a file;
  - graphical visualization of colours, brightness levels, and grey levels;
  - noise rejection using mean based filter and median filter;
  - image segmentation in binary form with a selected threshold and binary image visualization; → edge extraction with local binary filter and contour visualization;
  - selection of the image for high level processing by contour or shape;
- Fractal analysis
  - fractal dimension determination: global box counting fractal dimension for each binary threshold and mean fractal dimension, local fractal dimension, and local connected fractal dimension;

- ○ box counting fractal dimension spectrum visualization;
- ○ histogram of slope frequency visualization;
- ○ selection of slope with maximum frequency;
- Texture analysis
  - ○ co-occurrence matrices calculus;
  - ○ evaluation of texture features: energy, entropy, contrast and correlation.

This program was tested on mammographic images offered by the Department of Medical Imagistic of Fundeni Clinical Institute, Bucharest. It is known that using morphological and contour features of the mammary lesions, the American College of Radiology proposes a five classes classification named BIRADS (Breast Imaging Report Data System). The first three classes (1,2 and 3) are associated with benign lesions, having regular forms with well defined contours. Class BIRADS 5 is associated with malignant lesions. Class BIRADS 4 reports suspect anomalies; with images having a weakly defined contour, the delimitation of the tumor from the healthy tissue being very difficult to obtain. The malignacy risk of the tumors in this class is 15-50% and a simple visual diagnostic is not recommended (Taplin, 2002).

The proposed algorithm, using only the local fractal dimension as indicator, was tested on a set of 38 images (mammography BIRADS 4; from them 23 were tested as benign and 15 were tested as malignant by means of biopsy). For each case, the mean fractal dimension was calculated for different segmentation thresholds. Figure 5 shows the fractal dimension spectrum (right) of a mammogram (left).

The experiments confirmed that more than 90% of benign lesions have an average fractal dimension under the threshold 1.4, while malign lesions are characterized, by a similar percentage, with an average fractal dimension over that threshold.

Remote Land Classification

Haralick (1973) used some features extracted from grey level co-occurrence matrices to analyze aerial photograph data sets. Land classification implies that homogeneous regions with different types of terrains are to be identified (buildings, water, vegetation, etc.). At a distance, these regions seem to be textured ones. Adding colour information, the classification and segmentation processes are improved.

In Figure 6 a case study of land image containing different texture regions like vegetation, buildings, and water is presented. The investigation was made only for the grey level case and demonstrates that texture extracted from mean co-occurrence matrix are sufficient for image classification and segmentation.

Multi-textured image I (see Figure 6) is partitioned into 256 equivalent regions (16×16 block matrix): I-1, I-2,…,I-256. Because we want to

Figure 5. Mammographic image with the associated fractal dimension spectrum

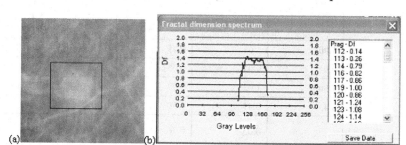

Figure 6. Land image containing buildings, vegetation, and water regions

identify three region types, a minimum of three references are necessary: one for the buildings class, one for the vegetation class (pine) and one for the water class. The references are chosen from image regions: Bd_{ref} – for buildings (I-137), V_{ref} – for vegetation (I-173), and W_{ref} – for water (I-88). Contrast and entropy are normalized at Bd_{Ref} (it has the greatest values). Similarly, energy and homogeneity take value 1.00 at W_{Ref}.

In order to evaluate the discriminating power of the selected features *C, E, Et* and *O*, the refer-

ence image Bd_{ref} is successively compared with similar regions Bd_1(I-136), Bd_2(I-152), Bd_3(I-153) and also with dissimilar ones, V_{ref} and W_{ref}. The results (Table 2) confirm that inside the class "Buildings" the distances are less than distances between classes: $D(Bd_{Ref}, V_{Ref})$ and $D(Bd_{Ref}, W_{Ref})$. In order to obtain a segmentation of the image, we introduce three thresholds: T_{Bd} for "building" texture, for "vegetation" texture T_v and T_w for "water" texture. Sequentially comparing each small region with the references Bd_{ref}, V_{ref} and W_{ref}, we obtain one of the following results:

a) $D(Bd_{ref}, x) \leq T_{Bd}$ b) $D(Bd_{ref}, x) \leq T_{Bd}$ c) $D(Bd_{ref}, x) \leq T_{Bd}$ d) Other.

Indexing the tested regions x with different indexes corresponding to the cases a, b, c and d we can obtain the segmentation of the image in Figure 6.

Adding colour information, or fractal based features, the classification and segmentation processes will be improved.

CONTENT-BASED IMAGE RETRIEVAL USING FRACTAL DIMENSION AS TEXTURAL FEATURE

Advances in modern computer and telecommunication technologies have led to huge archives

Table 2. Normalized features and distances between Bd_{Ref} and other regions.

Region/ Normalized feature	Bd_{Ref}	Bd_1	Bd_2	Bd_3	V_{Ref}	W_{Ref}
Contrast	**1.00**	1.05	0.75	0.75	0.18	0.02
Energy	0.04	0.04	0.05	0.06	0.18	**1.00**
Entropy	**1.00**	1.02	0.97	0.95	0.71	0.46
Homogeneity	0.43	0.43	0.50	0.50	0.67	**1.00**
Distance $D(Bd_{Ref}, x)$		0.05	0.26	0.26	0.92	1.85

of multimedia data in diverse application areas such as medicine, remote sensing, entertainment, education and on-line information services. To use this widely available multimedia information effectively, efficient methods for storage, browsing, indexing and retrieval must be developed. Different multimedia data types may require specific indexing and retrieval tools and methodologies. Due to the emergence of large-scale image collections, content-based image retrieval (CBIR) was proposed as a way to overcome database access difficulties. In CBIR, images are automatically indexed by summarizing their visual contents through automatically extracted quantities, or features, such as colour, texture or shape. Thus, low-level numerical features, extracted by a computer, are substituted for higher-level, text-based, manual annotations or keywords. Since the inception of CBIR, many techniques have been developed along this direction and many retrieval systems, both research and commercial, have been built. Low-level features such as colour, texture and shape of objects are widely used for CBIR. From 2000 onwards the fractal dimension was proposed as relevant indexing feature (Dobrescu, 2003; Yao, 2003; Varma, 2007). In specific applications, such as medical imaging, fractal dimension associated with low-level (textural) features play a substantial role in defining the content of the data. By adding as indicator the fractal dimension, the confidence in these textural features is improved (Calhoun, 2009; Iakovidis 2009; Abu Eid, 2010). An original method to determine the distance between two images based on fractal dimensions histograms (Dobrescu, 2010) is discussed in the following and the convergence with the classic indicators is underlined.

The Feature Integration Approach

The main limitation of feature integration in most existing CBIR systems is the heavy involvement of the user, who not only must select the features to be used for each individual query, but also must specify their relative weights. An interactive CBIR system designed to simplify this problem is discussed. This system uses the concept of integrated (or cumulative) features. The general class of accumulating features aggregate the spatial information of a partitioning irrespective of the image data. A special type of accumulative features are the global features which are calculated from the entire image, as proposed first by Smeulders *et al.* (Smeulders, 2000), which define the operator f that transpose the initial image data $i(x)$ (element of the image space x), into another data array, the resulting image field being:

$$f(x) = g \circ i(x),$$

where g is an operator on images. Accumulating features are symbolized by:

$$F_j(x) = \sum_{f(x) \in T_j} h \circ f(x)$$

where $\sum$ represents an aggregations operation (the sum in this case, but it may be a more complex operator), F_j is the set of accumulative features or a set of accumulative features ranked in a histogram. F_j is part of feature space F. T_j is the partitioning over which the value of F_j is computed. The operator h may hold relative weights, for example, to compute transform coefficients. A simple but very effective approach to accumulating features is to use the histogram, that is, the set of features $F(m)$ ordered by histogram index m. Joint histograms add local texture or local shape, directed edges and local higher-order structures.

Semantic-Based Image Retrieval Techniques

By contrast to searching based on low-level physical features, semantic feature query is based on key words, including the processing of nature language and traditional image retrieval. The goal of this kind of retrieval minimizes the semantic gap between the simple visual image feature and rich image semantics. There are two ways to reduce the semantic gap: high-level semantic features reduce low-level features and low-level features change into high-level semantics. The semantics of the image has the features of fuzziness, complexity and abstractness. It involves three semantic layers: feature semantics, object and space relation semantics, high-level semantics. Feature semantics is the colour, shape, texture and other low-level visual features of the image and it is related to the visual sense directly. Object semantics and space related semantics need to identify and extract the object characteristics, the relation between the space of the objects and other relations from the image. It involves pattern recognition, logic reasoning and other related techniques. High-level semantics mainly involves images scene semantics (such as a street or a room etc.), behavior semantics (such as performance) and emotion semantics (such as calmness, harmony, inspiration etc.). Semantic-based image retrieval commonly refers to object-based and high-level semantic image retrieval method.

From the novel solutions utilised in CBIR to improve efficieny, Latent Semantic Indexing (LSI) is the most attractive. LSI was used in document retrieval domain by finding some underlying connections among a group of documents. The most important issues implementing LSI in image retrieval systems are how to describe the image features in terms as if done in documents and how to transform the "terms" from vocabularies to features and documents to images synchronously. As LSI originates from natural language processing, applying it to source code is not straightforward (Santini, 2001). We have applied LSI to recover the architectural concepts by clustering variable names based on their similarity as opposed to clustering contexts as a whole. Next to LSI, we have therefore included a set of algorithms and methods to preprocess the source code before LSI is applied. Our contribution is that we applied them in a structured manner in order to entirely exploit their usefulness (Dobrescu, 2010a).

The Relevance Feedback Technique in the Image Retrieval

The basic idea of the image retrieval based on relevance feedback is to allow the users to evaluate the results of the search during the retrieval, pointing out the related and unrelated results and letting the computer learn the signs that users make, guiding the next-turn retrieval. A widely used relevance feedback method is to amend the search vector at one hand, and use feedback information to amend different kinds of feature vector weight in the distance formula.

In our approach an interactive system that involves the human user as a part of the retrieval process is considered. There are four main parts in this system: the image database, the feature database, the similarity metric selection part and the feature relevance computation part (Dobrescu, 2010b). At the start of processing a query, the system has no *a priori* knowledge about the query, thus all features are considered equally important and used to compute the similarity measure. The top N similar images will be returned to the user as retrieval results and the user can further refine this by sending relevance feedback to the system until he or she is satisfied. The learning-based feature relevance computation part is used to re-compute the weights of each feature based on user's feedback, and the metric selection part is used to choose the best similarity metric for the input feature weight vector based on reinforcement learning. Therefore, by considering user's feedback, the system can automatically adjust the query and provide a better approximation to the user's expectation. Moreover, by adopting relevance feedback, burdens of concept mapping and specification of weights are removed. Now, the user only needs to mark images that are relevant to the query, and the weight of each feature used for relevance computation is dynamically updated to model high level concepts and subjective perception.

Implementation of a CBIR Algorithm with Semantic Relevance Feedback

The implemented retrieval algorithm consists of the following steps (Dobrescu, 2007):

- Extraction of visual features from a database of images and provide links to the images from these features.
- Extraction of visual features from the query image using the same procedure used for feature extraction of the database images
- Comparing the features of the query image and features of images from the database and return the best matches as the retrieved images using a distance measure for comparison and a given threshold.

The feature extraction procedure was carried out in the following manner (Dobrescu, 2003):

a. Determination of the histogram peak for the local fractal dimension of the selected image – F1
b. Determination of the histogram peak for the local connected fractal dimension of the same image – F2
c. Determination of the Wavelet based sub-band coding of the image and estimation of the fractal dimension of each sub-band. The vector of fractal dimensions constituted the feature vector for each image in the database – F3

The last step in the retrieval procedure is the decision on the image relevance. Given an image, we have to decide which images in the database are relevant to it, and we have to retrieve the most relevant ones as the results of the query. In our experiments we use a likelihood ratio approach which is a Gaussian classifier, which define two classes, namely the relevance class *A* and the irrelevance class *B*. Given feature vectors of a pair of images, if these images are similar, they should be assigned to the relevance class, if not, they should be assigned to the irrelevance class. In order to evaluate the retrieval performance, two traditional measures: precision and recall were used. Precision is the percentage of retrieved images that are correct and recall is the percentage of correct images that are retrieved. Note that computation of these measures requires prior ground-truth of the database. Since our automatically generated ground-truths are not the ones required for precision and recall, these measures cannot be used directly (Dobrescu, 2008). We use modified versions of them to evaluate the performance of our algorithm. After manually grouping a smaller set of sub-images in our database, we will evaluate the performance using precision and recall too. To test the retrieval performance, we use the following procedure. Given an input query image of size $K \times K$, images are retrieved in descending order of likelihood ratio or ascending order of distance for nearest neighbour rule. If the correct image is retrieved as one of the k best matches, it is considered a success. Average rank of the correct image is also computed. This can also be stated as a nearest neighbour classification problem where the relevance class is defined to be the k best matches and the irrelevance class is the rest of the images. For this experiment, we use the non-shifted sub-images to compute the best case performance and the shifted sub-images to compute the worst case performance (Dobrescu, 2010b). We call this the worst case because the shifted sub-images overlap a sub-image in the database by only half the area. All other possible sub-images have a sub-image in the database which they overlap by more than half the area. This experimental procedure is appropriate to our problem of retrieving images which have some section in them that is like the user input image.

Solutions and Recommendations

On the Road Analysis

The majority of the texture analysis algorithms perform feature extraction in the 2-dimensional image domain, which usually renders the extraction process (e.g. Gabor filtering) computationally expensive. Our design was primarily motivated by the requirement of simplicity in feature extraction and the underlying hardware. Thus, maintaining the local neighbourhood properties is especially important in texture analysis, as texture is a visual feature associated to areas. Our scenario tests the performance of a descriptor in a texture-based image segmentation, when information about the candidate textures is available apriori. In the experiments, the datasets and methodology developed by Randen (1999) were used. It is very important to use in similarity-based retrieval experiments a large encompassing dataset. In our case we have used the Brodatz database (Brodatz, 1966), a very well known test database in the field of texture analysis. It contains 116 texture classes and 1856 images (16 images per class), and the size of the images is 128×128 pixels. The retrieval accuracy is evaluated by considering all the images as queries and counting the number of relevant matches in the top 16 retrieved images. The overall accuracy of the descriptor for this database is the average accuracy for all the queries in the database.

On the Texture Defect Detection in Manufactured Surfaces

Defect detection from images plays significant role in quality of manufactured products and its application areas continue to increase. Numerous methods have been proposed for performing this task, using sub-band domain co-occurrence matrices (Latif-Amed, 2000), wavelet transformation and vector quantization (Karras, 2001),

regularity and orientation criteria (Chetverikov, 2002), Markov Random Fields (MRF) and the auto-correlation function (Cohen 1991). The announced detection rates in these works vary between 82 and 89 percent. A better result with 4-5 per cent overall increase was reported when using general characteristics of independent components (ICs) of texture images (Sezer 2007). In our experiments we attempted to classify some of these proposed methods using the approximate resolution of employed images, *i.e.* low, medium and high, and their computational complexity. The selection of image resolution, for the textile web inspection, is largely determined from the available computational power and expected performance. High-resolution inspection images will require more computing power to inspect the entire width of a web but are desirable to detect subtle defects. On the other hand, computational requirements are low for the low-resolution images but these images cannot be expected to detect subtle defects that are lost due to the low-resolution imaging.

On the Malignant Tumour Detection Based on Fractal Analysis

It is obvious that the fractal analysis of tumours started with the seminal works of Landini (1996) and Mattfeld (1997), but after a few years the amount of research in this field became overwhelming. Fractal analysis represents a quantitative approach to parameters which until now have been rather abstract, such as heterogeneity or complexity. Our aim was to introduce this quantitative notion in the field of interpretation of histological images, which by definition leads to a certain subjectivity. There are huge differences between the fractal interpretation of the tumour shape and of the determination of the fractal dimension of the interface between the malignant epithelial tissue and the supporting tissue, i.e. stroma (Jelinek, 2010). For example, a previous study (Vasilescu, 2004) show that tumours having the mean spectral fractal dimension greater than a threshold are malignant, when to the contrary the fractal evaluation of the invasiveness of a gastric tumour shows that a decrease of fractal dimension may correlate with a higher risk of local invasion, a characteristic feature of gastric carcinoma of diffuse type. In this study the estimation of the fractal dimension was used to quantify the degree of complexity of the epithelial texture of gastric cancers, which appears statistically significantly higher than those of gastric carcinoma of diffuse type but lower than the fractal dimension of normal healthy gastric mucosa.

On the Remote Land Classification

There has been growing interest in the application of fractal geometry to observe spatial complexity of natural features at different scales. One can discuss three different fractal approaches -isarithm, triangular prism, and variogram- to characterize texture features of urban land-cover classes in high-resolution image data (Myint, 2003; Shen, 2008). Untill now, the results obtained from this type of analysis suggest that fractal-based textural discrimination methods are applicable but these methods alone may be ineffective in extracting texture features or identifying different land-use and land-cover classes in remotely sensed images. For better evaluattion of the efficiency of fractal approaches, deviation and mean of the selected features were examined in our study. Different textural features including Entropy were extracted based on the GLCM (Grey Level Co-occurrence Matrix) texture feature and used as the distinct feature value in the classification procedures. Results indicate that the proposed approach brings significant improvement of the classification rate based on the different texture feature images of various bands, allowing a better discrimination and mapping of mixed land cover types.

On the Semantic Content Base Image Retrieval

A lot of visual feature extraction algorithms, developed in the past few years, are mainly used in CBIR systems (Ko, 2005; Westerveld, 2000; Stasak, 2004). Based on different kinds of feature extraction algorithms, many famous CBIR systems have been developed, such as Photobook (Pentland, 1994) BlobWorld (Carson, 2002), etc. Some image search engines aimed at the WWW have also been developed, such as the Google image search engine, QBIC (Smeulders, 2000), Mars (Ortega, 1997), SIMPLicity (Wang, 2002) and MUVIS (Kiranyaz, 2006) but all of these systems are challenged by accuracy because their similarity matching is realized mainly by matching symbols and very little, by matching semantic content. Although the results have improved over the past years, they are still far from optimal. A real improvement can be obtained by using Latent Semanting Indexing (LSI) that solves two specific issues: the location of concepts and features in the source code, and the use of information retrieval methods to support software engineering tasks and activities. LSI is a technique in natural language processing, in particular in vectorial semantics, for analyzing relationships between a set of contexts and the terms they contain by producing a set of concepts related to the contexts and terms. For these reasons, in the original solutions discussed in this chapter, image filtering is executed before the latent semantic indexing based on multiple feature fusion to classify the images into different categories. The goal of our CBIR system is to obtain the main objects of the image through image segmentation and extracting several kinds of physical features from these objects, then fusing these features into a single vector. In this aim an "image-term" matrix is constructed, as a particular realization of a Term by Context Matrix, based on the multi-modal semantic space. On this matrix one implements a singular value decomposition operation. Then using the semantic relevance feedback one constructs strong- and weak-relevance sets through the users' choices. The "image-term" matrix is adjusted according to these strong/weak sets, and then one can get the adjusted feature vector of the query image to implement the feedback query.

FUTURE RESEARCH DIRECTIONS

1. Because, in spite of its computational efficiency, the regular partition scheme used by various box-counting methods intrinsically produces less accurate results than other methods (for example, wavelet based method), we intend to propose a novel multi-fractal estimation algorithm based on mathematical morphology to characterize the local scaling properties of textures. One estimates that the morphological multi-fractal estimation can differentiate texture images more effectively and provide more robust segmentations.

2. Despite the significant progress in the last decade, the problem of fabric defect detection still remains challenging and requires further attention. The statistical, spectral and model-based approaches give different results and therefore a combination of these approaches can give better results, than either one individually and is suggested for future research.

3. An important direction to follow in further research is to exploit the scaling property of the tumour-host borderline. This dynamics is always governed by processes of cell surface diffusion. However, more work is needed to fully determine the whole dynamical behaviour of tumour growth, focusing on the fractal aspect of the contours of the studied tumours. The aim is to demonstrate both the validity of the tumour growth model and the accuracy of the fractal dimension discriminator, especially in order to establish

the initial transition from the benign to the malignant stages of a tumour.

4. It is still difficult to conclude which combination between band and texture features like Entropy, Energy and others is the most efficient. More sample and feature classes will be tested to yield better comparisons. The preliminary results indicated that the accuracy in classifying fine resolution image data could be significantly improved using texture and fractal analysis and vegetation index.

5. The proposed framework can be easily extended to incorporate more low-level features like edges. An interesting prospect for future work refers to the development of classifiers that detect material degradation. This could lead to automatic tools to support Cultural Heritage experts in the monitoring of historical monuments, to prevent the degradation of such monuments.

CONCLUSION

1. In this chapter we present a different approach to make the descriptor sensitive to texture patterns of different sizes. This approach uses a multiresolution decomposition of the original image and has the advantage over the previous approaches of a much faster extraction process. We have shown that this approach outperforms other state-of-the art techniques in both the retrieval and segmentation applications, while being more compact and simpler to extract.

2. Our defect detection method, which allows a classification process with normalized type features extracted from the mean co-occurrence matrix and associated with the "effective" fractal dimension of the image, offers a detection rate between 90% and 92%, which is in the range of the best results reported up to now.

3. Fractal features of natural forms give to fractal analysis opportunities in various fields, medical imaging being a very important one. The proposed algorithm prove that the fractal dimension, as a way to characterize the complexity of a form, can be used for diagnosis of mammographic lesions classified BI-RADS 4, further investigations being not necessary.

4. The ability to distinguish between different types of land cover has been one of the principal aims of satellite remote sensing. In our work the effective analytical means based on GLCM were employed to extract the distinct texture feature values in image classification. The classification was conducted on a combined texture and fractal feature images using a minimum distance classifier, combined with the vegetation index (BRI). With the use of texture feature images and BRI, six types of land cover types were successfully classified. The advantage of this approach is based on the fact that the texture feature image retains sufficiently the structure and shape of the object in the image and the vegetation index has considerable potential in discriminating variables in land cover classification. The classification accuracy was improved by 80% and as high as 93% for dense vegetation when combining the texture feature with BRI although the overall accuracy is slight lower than that of object-oriented analysis method.

5. The experimental results for the retrieval of images from databases worked with excellent results, the errors in retrieval were very few or none at all. Also the processing time was small and all of this pointed to the fact that all the algorithm's functions were implemented correctly. For natural images the results were not as good and amongst the query results almost every time there were incorrect retrieval. However sorting of the results proved efficient and thus amongst the results proved efficient and thus amongst

the first 5 query results almost every time all the results were good. For the images that contain objects that contrast with the background there has been very accurate recovery; sometimes even 100% of the results were good. The processing and retrieval times were influenced a lot by the number of objects in the query item. As an indication, for the database containing 1000 images, a retrieval process took around 60 seconds, which given the complexity of the calculations involved is a good time.

REFERENCES

Abu Eid, R., & Landini, G. (2010). The complexity of the oral mucosa: A review of the use of fractal geometry. *Control Engineering and Applied Informatics, 12*(1), 10–14.

and data system assessments and management recommendations in screening mammography. *Radiology, 222*, 529–535.

Bakes, A. R., & Bruno, O. M. (2008). *A new approach to estimate fractal dimension of texture images*. IEEE International Conference on Image Processing. (LNCS 5099), (pp. 136-143).

Brodatz, P. (1966). *Textures: A photographic album for artists and designers*. Dover Publications.

Calhoun, V. D., & Adali, T. (2009). Feature-based fusion of medical imaging data. *IEEE Transactions on Information Technology in Biomedicine, 13*(5), 711–720. doi:10.1109/TITB.2008.923773

Carson, C., Belongie, S., Greenspan, H., & Malik, J. (2002). Blobworld: Image segmentation using expectation-maximization and its application to image querying. *IEEE Transactions on Pattern Analysis and Machine Intelligence, 24*(8), 1026–1038. doi:10.1109/TPAMI.2002.1023800

Chetverikov, D., & Hanbury, A. (2002). Finding defects in texture using regularity and local orientation. *Pattern Recognition, 35,* 203–218. doi:10.1016/S0031-3203(01)00188-1

Cohen, F. S., Fan, Z., & Attali, S. (1991). Automated inspection of textile fabrics using textural models. *IEEE Transactions on Pattern Analysis and Machine Intelligence, 13*(8), 803–808. doi:10.1109/34.85670

Crisan, A. D. (2005). *Image processing using fractal techniques.* Unpublished doctoral thesis, Politehnica University Bucharest.

Crisan, D., Dobrescu, R., & Planinšič, P. (2007). Mammographic lesions discrimination based on fractal dimension as an indicator. *Proceedings of the 14th International Conference on Systems, Signals and Image Processing,* IWSSIP 2007, (pp. 275-280).

Dobrescu, R. (2010b). *FractIm–a fractal application for medical diagnosis.* (Scientific Report 61-031). Retrieved from http://isis.pub.ro/proiecte/Imago

Dobrescu, R., Dobrescu, M., Mocanu, S., & Popescu, D. (2010a). Malignant skin lesions detection based on texture and fractal image analysis. *Proceedings of the International Conference of WSEAS MUSP, 10,* 21–26.

Dobrescu, R., & Ichim, L. (2008). Using fractal dimension as discriminator of infected HeLa cells from spectrophotometric images. *International Journal of Functional Informatics and Personalised Medicine, 1*(1), 53–67. doi:10.1504/IJFIPM.2008.018292

Dobrescu, R., & Ionescu, F. (2003). Fractal dimension based technique for database image retrieval. *Proceedings of the First IAFA Symposium,* (pp. 107-112).

Dobrescu, R., Vasilescu, C., & Ichim, L. (2007). Fractal and scaling analysis in tumor growth evaluation. *WSEAS Transactions on Systems, 1*(6), 102–108.

Gonzalez, R. C., & Woods, R. E. (1992). *Digital image processing.* Reading, MA: Addison Wesley.

Haralick, R. M., Shanmugam, K., & Dinstein, I. (1973). Textural features for image classification. *IEEE Transactions on Systems, Man, and Cybernetics, 3*(6), 610–621. doi:10.1109/TSMC.1973.4309314

Iakovidis, D. K. (2009). A pattern similarity scheme for medical image retrieval. *IEEE Transactions on Information Technology in Biomedicine, 13*(4), 442–450. doi:10.1109/TITB.2008.923144

independent component model for defect detection. *Pattern Recognition,* 121-133.

Jähne, B., Horst Haußecker, H., & Geißler, P. (Eds.). (1999). *Handbook of computer vision and applications, volume 2: Signal processing and pattern recognition. Ch.12: Texture analysis.* Academic Press.

Jelinek, H. F., Milosevic, N. T., & Ristanovic, D. (2010). The morphology of alpha ganglion cells in mammalian species: A fractal analysis study. *Control Engineering and Applied Informatics, 12*(1), 3–9.

Karras, D. A., & Karkanis, S. A. Iakovidis, D.K., Maroulis, D.E. & Mertzios, B.G. (2001). Improved defect detection in manufacturing using novel multidimensional wavelet feature extraction involving vector quantization and PCA techniques. *Proceedings of the 8th Panhellenic Conference on Informatics,* (pp. 139-143).

Keller, J. M., & Chen, S. (1989). Texture description and segmentation through fractal geometry. *Computer Vision Graphics and Image Processing, 45,* 150–166. doi:10.1016/0734-189X(89)90130-8

Keller, J. M., Crownover, R. M., & Chen, R. Y. (1993). On the calculation of fractal features from images. *PAMI, 10*(15), 1087–1090.

Kiranyaz, S., Ferreira, M., & Gabbouj, M. (2006). Automatic object extraction over multi-scale edge field for multimedia retrieval. *IEEE Transactions on Image Processing*, 3759–3772. doi:10.1109/TIP.2006.881966

Ko, B. C., & Byun, H. (2005). FRIP: A region-based image retrieval tool using automatic image segmentation. *IEEE Transactions on Multimedia, 7*(1), 105–113. doi:10.1109/TMM.2004.840603

Landini, G., & Rippin, J. W. (1996). How important is tumour shape? Quantification of the epithelial-connective tissue interface in oral lesions using local connected fractal dimension analysis. *The Journal of Pathology, 179*, 210–217. doi:10.1002/(SICI)1096-9896(199606)179:2<210::AID-PATH560>3.0.CO;2-T

Latif-Amet, A., Ertüzün, A. & Erçil, A. (2000). An efficient method for texture defect detection:

Li, J., Du, Q., & Sun, C. (2009). An improved box-counting method for image fractal dimension estimation. *Pattern Recognition, 42*(11), 2460–2469. doi:10.1016/j.patcog.2009.03.001

Losa, G. A., Merlini, D., Nonnemacher, T. F., & Weibel, E. R. (2005). Fractals in biology and medicine. *Birkhäuser Verlag, IV*, 1–314.

Mandelbrot, B. B. (1982). *Fractal geometry of nature*. New York: Freeman.

Mattfeldt, T. (1997). Nonlinear deterministic analysis of tissue texture: A stereological study on mastopathic and mammary cancer tissue using chaos theory. *Journal of Microscopy, 185*(1), 47–66. doi:10.1046/j.1365-2818.1997.1440701.x

Myint, S. W. (2003). Fractal approaches in texture analysis and classification of remotely sensed data: Comparisons with spatial autocorrelation techniques and simple descriptive statistics. *International Journal of Remote Sensing, 24*(9), 1925–1947. doi:10.1080/01431160210155992

Ortega, M., Rui, Y., Chakrabarti, K., Mehrotra, J., & Huang, T. S. (1997). Supporting similarity queries in MARS. *Proceedings of the 5th ACM International Multimedia Conference*, (pp. 403-413).

Pentland, A., Picard, R. W., & Sclaroff, S. (1994). Photobook: Content-based manipulation of image databases. *Proceedings of SPIE Storage and Retrieval Image and Video Databases, II*, 34–47.

Popescu, D., & Dobrescu, R. (2008b). Carriage road pursuit based on statistical and fractal analysis of the texture. *International Journal of Education and Information Technologies, 2*(11), 62–70.

Popescu, D. & Dobrescu, R. (2008c). Plastic surface similarity measurement based on textural and fractal features. *Revista de materiale plastice, 45*(2), 158-163.

Popescu, D., Dobrescu, R., & Angelescu, N. (2008a). Color textures discrimination of land images by fractal techniques. *Proceedings of the 4th WSEAS International Conference on REMOTE SENSING -REMOTE'08*, (pp. 51-56). Venice, Italy.

Randen, T., & Husøy, J. (1999). Filtering for texture classification: A comparative study. *IEEE Transactions on Pattern Analysis and Machine Intelligence, 21*, 291–310. doi:10.1109/34.761261

Santini, S., Gupta, A., & Jain, R. (2001). Emergent semantics through interaction in image databases. *IEEE Transactions on Knowledge and Data Engineering, 13*(3), 337–351. doi:10.1109/69.929893

Sarker, N., & Chaudhuri, B. B. (1994). An efficient differential box-counting approach to compute fractal dimension of image. *IEEE Transactions on Systems, Man, and Cybernetics, 24*, 115–120. doi:10.1109/21.259692

Sezer, O.G., Ercil, A., & Ertuzun, A. (2007). Using perceptual relation of regularity in the texture with

Shapiro, L., & Stockman, G. (2001). *Computer vision*. Prentice Hall.

Shen, G., & Apostolos, S. (2008). Application of texture analysis in land cover classification of high resolution image. *2008 Fifth International Conference on Fuzzy Systems and Knowledge Discovery*, vol. 3, (pp. 513-517).

Smeulders, A. W. M., Worring, M., Santini, S., Gupta, A., & Jain, R. (2000). Content-based image retrieval at the end of the early years. *IEEE Transanctions on PAMI, 22*, 1349–1380.

Stasak, J. (2004). *A contribution to image semantic analysis*. 10th Conference on Professional Information Resources.

Subband domain co-occurrence matrices. *Image and Vision Computing, 18*, 543–553.

Sun, W., Xu, G., Gong, P., & Liang, S. (2006). Fractal analysis of remotely sensed images: A review of methods and applications. *International Journal of Remote Sensing, 27*(22), 4963–4990. doi:10.1080/01431160600676695

Taplin, S.H., Ichikawa, L.E. & Kerlikowske, K. (2002). Concordance of breast imaging reporting

Unser, M. (1986). Sum and difference histograms for texture analysis. *IEEE Transactions on Pattern Analysis and Machine Intelligence, 8*, 118–125. doi:10.1109/TPAMI.1986.4767760

Varma, M., & Garg, R. (2007). Locally invariant fractal features for statistical texture classification. *Proceedings of the IEEE International Conference on Computer Vision*.

Vasilescu, C., Herlea, V., Talos, F., Ivanov, B., & Dobrescu, R. (2004). Differences between intestinal and diffuse gastric carcinoma: A fractal analysis. In Dobrescu, R., & Vasilescu, C. (Eds.), *Interdisciplinary applications of fractal and chaos theory* (pp. 144–149). Bucuresti: Editura Academiei Romane.

Wang, J., Li, J., & Wiederhold, G. (2002). SIMPLicity: Semantics-sensitive integrated matching for picture libraries. *IEEE Transactions on PAMI, 23*, 947–963.

Wang, T. C., & Karayiannis, N. B. (1998). Detection of microcalcifications in digital mammograms using wavelets. *IEEE Transactions on Medical Imaging, 17*(4), 498–509. doi:10.1109/42.730395

Westerveld, T. (2000). Image retrieval: Content versus context. *Content-Based Multimedia Information Access*, 276-284

Wilson, R., & Spann, M. (1988). *Image segmentation and uncertainty*. New York: John Wiley and Sons Inc.

Xu, Y., Ji, H., & Fermuller, C. (2009). Viewpoint invariant texture description using fractal analysis. *International Journal of Computer Vision, 83*(1), 85–100. doi:10.1007/s11263-009-0220-6

Yao, M., Yi, W., Shen, B. & Dai, H. (2003). An image retrieval system based on fractal dimension. *Journal of Zhejiang University - Science A, 4*(4), 421-425.

ADDITIONAL READING

Aksoy, S., & Haralick, R. M. (1998). Textural Features for Image Database Retrieval, *Proc. of the IEEE Workshop on Context-Based Access of Image,* Vol.CVPR'98, pp. 45-49.

Aksoy, S., & Haralick, R. M. (1998). Content-based image database retrieval using variances of gray level spatial dependencies", *Proc. of IAPR Int. Workshop on Multimedia Information Analysis and Retrieval*, 8, pp. 327-336.

Barnsley, M. (1988). *Fractals Everywhere*. Academic Press.

Bradshaw, B. (2000) Semantic Based Image Retrieval: A Probabilistic Approach", *Proc. of ACM. Multimedia*, 23(6), pp. 676—689

Chen, C. H., Pau, L. F., & Wang, P. S. P. (Eds.). (1998). *The Handbook of Pattern Recognition and Computer Vision* (2nd ed.). River Edge, NJ: World Scientific Publishing Co.

Chen, K. M., & Chen, S. Y. (2002). Color texture segmentation using feature distributions. *Pattern Recognition Letters*, 23(7), 775–771. doi:10.1016/S0167-8655(01)00150-7

Dobrescu, R., & Vasilescu, C. (2004). *Interdisciplinary Applications Of Fractal And Chaos Theory* (Romāne, A., Ed.). București.

Drimbarean, A., & Whelan, P. F. (2001). Experiments in colour texture analysis. *Pattern Recognition Letters*, 22(10), 1161–1167. doi:10.1016/S0167-8655(01)00058-7

Einstein, A., Wu, H. S., Sanchez, M., & Gil, J. (2001). Fractal characterization of chromatin appearance for diagnosis in breast cytology. *The Journal of Pathology*, 195, 366–381.

Eisenbarth, T., Koschke, R., & Simon, D. (2003). Locating features in source code. *Trans. on Soft. Eng.*, 29(3), 210–224.

Fagin, R., Kumar, R., & Sivakumar, D. (2003) Efficient similarity search and classification via rank aggregation. *Proceedings of ACM SIGMOD*, pp. 301-312

Fournier, J., Cord, M., & Philipp-Foliguet, S. (2001). Retin: A content-based image indexing and retrieval system. *Pattern Analysis and Applications Journal*, 4(2/3), 153–173. doi:10.1007/PL00014576

Gleich, D., & Zhukov, L. (2004) SVD based term suggestion and ranking system, *Proc. of the ICDM*, pp. 391–394

Gupta, A., & Jain, R. (1997). Visual Information Retrieval. *Communications of the ACM*, 40(5), 70–79. doi:10.1145/253769.253798

Jain, A. K., Murty, M. N., & Flynn, P. J. (1999). Data clustering: a review. *ACM Computing Surveys*, 31(3), 264–323. doi:10.1145/331499.331504

Jiang, W., Wan, B., Zhang, Q., & Zhou, Y. (2008) Image Search by Latent Semantic Indexing Based on Multiple Feature Fusion, *Proc. of the Congress on Image and Signal Processing*, Vol. 2, pp. 515-519

Kuhn, A., Ducasse, S., & Girba, T. (2007). Semantic clustering: Identifying topics in source code. *Information and Software Technology*, 49(3), 230–243. doi:10.1016/j.infsof.2006.10.017

Landauer, T. K., Foltz, P. W., & Laham, D. (1998). Introduction to latent semantic analysis. *Disc. Proc*, 25, 259–284.

Landini, G. (1998). Complexity in tumor growth patterns. In Losa, G. A. (Ed.), *Fractals in Biology and Medicine* (pp. 268–284). Birkhauser Verlag.

Lowe, D. G. (2004). Distinctive image features from scale-invariant keypoints. *International Journal of Computer Vision*, 60(2), 91–110. doi:10.1023/B:VISI.0000029664.99615.94

Marcus, A., & Poshyvanyk, D. (2005) The conceptual cohesion of classes, *Proc. of the ICSM*, pp.133–142

Natsev, A. P., Naphade, M. R., & Tesic, J. (2005) Learning the semantics of multimedia queries and concepts from a small number of examples, *Proc. of ACM Multimedia*, pp. 598 - 607

Nonnenmacher, T. F., Losa, G. A., & Weibel, E. R. (1993). *Fractals in biology and medicine*. Basel: Birkhauser Verlag.

Quelhas, P., & Odobez, J.-M. (2006) Natural scene image modeling using color and texture visterms. *Proc. CIVR*, pp. 411–421

Sajaniemi, J., & Prieto, R. N. (2005) An investigation into professional programmers' mental representations of variables, *Proc. of the IWPC*, pp. 55–64

Snoek, C. G., Huurnink, B., Hollink, J., de Rijke, M., Schreiber, G., & Worring, M. (2007). Adding semantics to detectors for video retrieval. *IEEE Transactions on Multimedia*, 285–288.

Wang, G., Zhang, Y., & Fei-Fei, L. (2006) Using dependent regions for object categorization in a generative framework, *Proc. CVPR*, pp. 1597–1604

Zeng, J., & Alhajj, R. (2006) Classification by multi-perspective representation method. *Proc. of SEKE*, pp.85-90

Zhai, C., & Lafferty, J. (2001) A study of smoothing methods for language models applied to ad hoc information retrieval, *Proc. SIGIR*, pp 334–342

Zheng, W., & Li, J. Si, Z., Lin, F., & Zhang, B. (2006) Using high-level semantic features in video retrieval, *Proceedings of CIVR*, pp. 370-379

Chapter 15
Real–Time Primary Image Processing

Radu Dobrescu
Politehnica University of Bucharest, Romania

Dan Popescu
Politehnica University of Bucharest, Romania

ABSTRACT

Image processing operations have been classified into three main levels, namely low (primary), intermediate, and high. In order to combine speed and flexibility, an optimum hardware/software configuration is required. For multitask primary processing, a pipeline configuration is proposed. This structure, which is an interface between the sensing element (camera) and the main processing system, achieves real time video signal preprocessing, during the image acquisition time. In order to form the working neighborhoods, the input image signal is delayed (two lines and three pixels). Thus, locally 3×3 type processing modules are created. A successive comparison median filter and a logical filter for edge detection are implemented for a pipeline configuration. On the other hand, for low level, intermediate, and high level operations, software algorithms on parallel platforms are proposed. Finally, a case study of lines detection using directional filter discusses the performance dependency on number of processors.

INTRODUCTION

Real-time image and video processing systems involve processing vast amounts of image data in a timely manner for the purpose of extracting useful information, which could mean anything from obtaining an enhanced image to intelligent scene analysis. Digital images and video are essentially multidimensional signals and are thus quite data intensive, requiring a significant amount of computation and memory resources for their processing. The amount of data increases if color is also considered. Furthermore, the time dimension of digital video demands processing massive amounts of data per second. One of the keys to real-time algorithm development is the

DOI: 10.4018/978-1-60960-477-6.ch015

exploitation of the information available in each dimension. For digital images, only the spatial information can be exploited, but for digital videos, the temporal information between image frames in a sequence can be exploited in addition to the spatial information.

The key to cope with this issue is the concept of parallel processing which deals with computations on large data sets. In fact, much of what goes into implementing an efficient image/video processing system centers on how well the implementation, both hardware and software, exploits different forms of parallelism in an algorithm, which can be data level parallelism - DLP or/ and instruction level parallelism – ILP (Hunter, 2003). DLP manifests itself in the application of the same operation on different sets of data, while ILP manifests itself in scheduling the simultaneous execution of multiple independent operations in a pipeline fashion.

Usually, in a decision theoretic based pattern recognition system for industrial applications, the classification is performed in the feature space by a distance function criterion. In applications like visual servoing, vehicle navigation, industrial inspection, multimedia, medical engineering, etc., the main requirement for the video system is the real time execution of the algorithms. In order to obtain a very high image processing speed, the primary operators (pre-processing operators) are transferred from the central computer to the sensory level.

There are two classes of digital primary image processing operators: local operators and global operators. The global operators require information from the complete image frame. They are not suitable for industrial video applications because they have two main disadvantages: long time execution and edge alteration. On the other hand, many functions like noise rejection, binary segmentation, edge extraction, erosion, dilation, area evaluation, and perimeter evaluation can be calculated with the aid of local bi-dimensional filters (Popescu, 1990).

Generally, software implementation of many image processing procedures is not compatible with on-line, real time operation requirements and with hard industrial environment conditions. Moreover, most of the required primary image processing procedures can be hardware implemented, using programmable devices. Thus, for an efficient industrial image processing system, the hardware/software co-design approach is highly recommended.

Operations like noise rejection, edge detection, binary segmentation of image, are frequently encountered. Due to the development of the integrated circuits like FPGA and DSP, these primary image processing algorithms can be implemented together with the video camera like embedded system.

There are many applications of real time primary image processing. For example, results concerning increase of image processing speed and sensor fusion for obstacle detection in robot navigation are presented in (Popescu, 2006).

Although the images can be color, for simplicity, we will consider only the grey level case. In the color image case, the results are similar; it is necessary to consider three grey level type matrices (RGB or HSV).

BACKGROUND

Definition of "Real-Time" Concept

"Real-time" is an elusive term that is often used to describe a wide variety of image/video processing systems and algorithms. Considering the need for real-time image/video processing and how this need can be met by exploiting the inherent parallelism in an algorithm, it becomes important to discuss what exactly is meant by the term "real-time". From the literature, it can be derived that there are three main interpretations of the concept of "real-time," namely real-time in the perceptual sense, real-time in the software engineering sense, and real-time in the signal processing sense.

Real-Time in Perceptual Sense

Real-time in the perceptual sense is used mainly to describe the interaction between a human and a computer device for a near instantaneous response of the device to an input by a human user.

For instance, Bovik (2005) defines the concept of "real-time" in the context of video processing, describing that *"the result of processing appears effectively 'instantaneously' (usually in a perceptual sense) once the input becomes available"*. An important item to observe here is that "real-time" involves the interaction between humans and computers in which the use of the words "appears" and "perceivable" appeals to the ability of a human to sense delays. Note that "real-time" connotes the idea of a maximum tolerable delay based on human perception of delay, which is essentially some sort of application-dependent bounded response time.

Real-Time in Software Engineering Sense

Real-time in the software engineering sense is also based on the concept of a bounded response-time as in the perceptual sense. Dougherty and Laplante (1995) point out that a *"real-time system is one that must satisfy explicit bounded response time constraints to avoid failure,"* further explaining that *"a real-time system is one whose logical correctness is based both on the correctness of the outputs and their timeliness."* Indeed, while any result of processing that is not logically correct is useless, the important distinction for "real-time" status is the all-important time constraint placed on obtaining the logically correct results. So, soft real-time refers to the case where missed real-time deadlines result in performance degradation rather than failure.

Real-Time in Signal Processing Sense

Real-time in the signal processing sense is based on the idea of completing processing in the time available between successive input samples. For example, "real-time" is defined as *"completing the processing within the allowable or available time between samples,"* and it is stated that a real-time algorithm is one whose total instruction count is *"less than the number of instructions that can be executed between two consecutive samples (Kehtarnavaz, 2005)"* An important item of note here is that one way to gauge the "real-time" status of an algorithm is to determine some measure of the amount of time it takes for the algorithm to complete all requisite transferring and processing of image data, and then making sure that it is less than the allotted time for processing.

In the following the discussion is focused on two main aspects. First, we present an example of hardware implementation of primary image processing algorithms, corresponding to real-time in the signal processing sense. Secondly, we present an example of software implementation of primary image processing algorithms, corresponding to real-time in the software engineering sense.

DIVERSITY OF OPERATIONS IN REAL TIME PRIMARY IMAGE PROCESSING

In many industrial applications of video systems, image representation must be simple and efficient. Therefore, the binary image representation and the contour representation of objects are often utilized.

In order to combine real time image processing and great flexibility, an optimal hardware and software combining solution for video systems must be adopted. For primary processing the weight is on hardware solutions.

The digital primary processing mainly consists of three stages: noise rejection, binary representation, and edge extraction. Due to the fact that noise can introduce errors in other stages (like contour detection and feature extraction) image noise rejection must be the first stage in any digital image processing application. For these algorithms local operators which act in symmetrical neighbourhoods of the considered pixels are recommended. They have the advantage of simplicity and they can be implemented to operate in real time.

The local operator calculates the resulting values in the center (i,j) of a small dimensional window (3×3, 5×5, 7×7, etc.), on the basis of all pixel values from the window. In the general symmetrical neighbourhood case, namely $(2h+1) \times (2h+1)$, the values of the input function G (grey level case) are grouped in the matrix $[G_{i,j}]$:

$$[G_{ij}] = \begin{bmatrix} G(i-h, j-h)...G(i-h, j+h) \\ ....G(i, j).............. \\ G(i+h, j-h)...G(i+h, j+h) \end{bmatrix}_{(2h+1)\times(2h+1)}$$

In the processed image G', the central window value $G(i,j)$ is replaced by the resulting value of the local filter $G'(i,j)$.

In this work we consider 3×3 windows. Therefore, for simplicity we utilized the following notations:

$a_1 = G(i-1, j-1)$, $a_2 = G(i-1, j)$, $a_3 = G(i-11, j+1)$, $b_1 = G(i, j-1)$, $b_2 = G(i, j)$, $b_3 = G(i, j+1)$, $c_1 = G(i+1, j-1)$, $c_2 = G(i+1, j)$, $c_3 = G(i+11, j+1)$

Thus,

$b_2' = F(a_1, a_2, ..., b_2, ..., c_3)$

In the preceding relation, F represents the filter function.

The neighbourhoods are not defined for pixels situated on the lines or columns near the border of the image, where the neighbourhood would extend beyond the image boundary. Different solutions exist to this problem, but we consider that the implied pixels must remain unchanging.

Noise Rejection

Noise rejection is the first step in primary image processing and it is a mandatory one. There are many local noise rejection filters, but the effective selection depends upon the application. It must take into account the following criteria: efficiency, contour preserving, on-line execution and latency. The pixel neighbourhood dimension can be established compromising between the complexity, the degree of noise rejection (good filtering), and the contour maintenance (safeguarding of contrast lines). Increasing window dimension leads to contour distortion. For this reason, we consider that a 3×3 neighbourhood can be a satisfactory compromise.

Most frequently utilized noise rejection filters are:

- Smoothing filter (mean filter, MF), based on the arithmetic mean or the weighted arithmetic mean of the values from $G_{i,j}$;
- Median filter (MeF), based on the median of values from $G_{i,j}$;
- Logical filter (binary filter BF), based on logical functions, for binary image type.

The mean filters (MF_1, MF_2 and MF_3) replace the central value b_2 by the arithmetic mean (or the weighted arithmetic mean) b_2':

MF_1 - based on arithmetic mean filter:

$$b_2' = \frac{1}{9}\left(a_1 + a_2 + a_3 + b_1 + b_2 + b_3 + c_1 + c_2 + c_3\right)$$

MF_2 - based on weighted arithmetic mean filter:

$$b_2' = \frac{1}{10}\left(a_1 + a_2 + a_3 + b_1 + 2b_2 + b_3 + c_1 + c_2 + c_3\right)$$

MF_3 - based on weighted arithmetic mean filter:

$$b_2' = \frac{1}{16}\left(a_1 + 2a_2 + a_3 + 2b_1 + 4b_2 + 2b_3 + c_1 + 2c_2 + c_3\right)$$

The mean filter can be considered as a convolution operator type. Thus, the following weight matrices are considered: H_1 - for MF_1, H_2 - for MF_2, H_3 - for MF_3:

$$[H_1] = \frac{1}{9}\begin{bmatrix} 1 & 1 & 1 \\ 1 & 1 & 1 \\ 1 & 1 & 1 \end{bmatrix},$$

$$[H_2] = \frac{1}{10}\begin{bmatrix} 1 & 1 & 1 \\ 1 & 2 & 1 \\ 1 & 1 & 1 \end{bmatrix},$$

$$[H_3] = \frac{1}{16}\begin{bmatrix} 1 & 2 & 1 \\ 2 & 4 & 2 \\ 1 & 2 & 1 \end{bmatrix}$$

Then, the filter relations will be:

$$b_2' = \left[G_{i,j}\right] \otimes [H_n], n = 1,2,3$$

For noise rejection, the median filter, which is nonlinear, has a better action with respect to contrast lines compared with the mean value based linear filters. Concerning the one dimensional median filter, it is known that a local monotonic sequence is invariant to passing through filter. Also, m-local monotonic sequences are invariant to passing through the $2k+1$ dimension median filter, if $m > 2k$.

In the 2-D case, for symmetrical neighbourhoods, these observations apply too. As a consequence, an edge will be invariant to crossing through a median filter. For the special case of the cross neighbourhood, the relation which describes the median filter function is the following:

$$b_2' = Me\ \{a_2, b_1, b_2, b_3, c_2\}$$

The above median filter does not affect the image details larger than 3×3 pixels in size and rejects noise of dimension two pixels.

In the binary image case, the values $a_1, a_2, a_3, b_1, b_2, b_3, c_1, c_2, c_3,$ and b_2' are binary ones and the noise can be rejected by logical filtering. Only "salt and pepper" binary image noise can be rejected with a 3×3 local logical filter. The filter acts in a neighbourhood $V_{i,j}$ and is characterized by the logical expression:

$$b_2' = \left[b_2 \cup (a_1 \cap a_2 \cap a_3 \cap b_1 \cap b_3 \cap c_1 \cap_2 \cap c_3)\right]$$
$$\cap \left(\overline{b_2} \cup a_1 \cup a_2 \cup a_3 \cup b_1 \cup b_3 \cup c_1 \cup c_2 \cup c_3\right)$$

For high dimensional binary noise morphological operators like erosion and dilation must be utilized. On the other hand, erosion and dilation can be considered like local 3×3 cross operators. For erosion, if any of the neighbours (a_2, b_1, b_3, c_2) has the value 0, the pixel in the middle of the mask will be 0. Dilation is the opposite operation; if the center pixel is 1, then any of the neighbours will be 1. Combining erosion with dilation it is possible to eliminate the "1" noise of dimension 2×n or m×2, where n is the number of columns and m is the number of lines. The algorithm is the following:

1. Create the eroded image E from binary image B by erosion operation;
2. Obtain the dilated image D from E by dilation operation.

Edge Detection

For edge extraction, we propose a logical function based algorithm. First, the image after noise rejection filtering (median filter) is transformed into binary form by a suitable method. Usually, the binary threshold is determined in a learning phase and is set equal with the level corresponding to the deep valley after the first peak in the gray level histogram (bimodal case). The matrix representation of the binary image is analyzed in 3×3 neighbourhoods, as in the noise rejection case in order to detect a 1 in the central position and at least a 0 in the rest. For this neighbourhood form of the edge detection algorithm, two logical function expressions are possible:

$$c_{i,j} = b_{i,j}$$
$$\wedge \left(\overline{b}_{i-1,j-1} \vee \overline{b}_{i-1,j} \vee \overline{b}_{i-1,j+1} \vee \overline{b}_{i,j-1} \vee \overline{b}_{i,j+1} \vee \overline{b}_{i+1,j-1} \vee \overline{b}_{i+1,j} \vee \overline{b}_{i+1,j+1} \right) \quad (1)$$

$$c_{i,j} = b_{i,j} \wedge \left(\overline{b}_{i-1,j} \vee \overline{b}_{i,j-1} \vee \overline{b}_{i,j+1} \vee \overline{b}_{i+1,j} \right) \quad (2)$$

The first algorithm introduces errors when edges are inclined at 45°, namely they are doubled. The second algorithm introduces errors in the case of 90° concave angles, but they are insignificant from the geometrical and topological feature extraction point of view.

The marginal pixels of the whole frame would be filtered with mirrored pixels from previous calculations or would remain unmodified.

Binary Segmentation of Images

Binary images are very useful both in low level image processing algorithms (like edge extraction, erosion, and dilation) and in intermediate image processing algorithms (like geometrical feature extraction: area, perimeter, center of mass, minimum radius, maximum radius, etc.). Commonly, in binary segmented images the object silhouette pixels are logically 1 and the background pixels are logically 0. The binary image pixels $B(i,j)$ are established by a grey level image threshold $T(i,j)$:

$$B(i,j) = \begin{cases} 1, & if \ G(i,j) \geq T(i,j) \\ 0, & if \ G(i,j) < T(i,j) \end{cases}, \quad \begin{aligned} i &\in \{0,1,...,m-1\} \\ j &\in \{0,1,...,n-1\} \end{aligned}$$

The threshold value $T(i,j)$ can be determined by different information: point, neighbourhood and frame. In many applications $T(i,j)$ is a constant.

HARDWARE ARCHITECTURES OF PRIMARY IMAGE PROCESSORS

Hardware implementation of primary image processing algorithms can be used to speed up image processing applications. Both real time requirements and flexibility make necessary hardware/software codesign. Moreover, reconfigurable applications imply reconfigurable hardware devices, such as Field Programmable Gate Arrays or DSPs (Wnuk, 2006). The execution time of FPGA image processing algorithm implementations can be one to two orders of magnitude less than software implementations. On the other hand, hardware implementation of image processing sub-systems increases the system price. This extra-cost is justified especially for large images (high resolution).

For each primary image processing algorithm, a simple and efficient hardware/software implementation must be proposed. Also these modules can be integrated on a video sensor chip. Thus, the image sensor becomes a smart one, which can faster perform most of the video system functions.

Cellular Pipeline Implementation of Primary Image Processing

In order to speed up image processing applications, two architectures of primary image processing system are implemented: a parallel one and a pipeline one.

In the parallel implementation, the modules are parallel connected and all the processing results are simultaneously presented. This architecture is recommended for medium level image processing (like feature extraction) and not for primary image processing case.

Pipeline implementation utilizes the natural modality of image acquisition from the usual CCD and CMOS sensors. This acquisition type consists of line-by-line and pixel-by-pixel scanning of the image (scene) with a fixed clock.

Examples of real time pipeline architectures which are implemented on FPGAs, for real time primary processing of color images, are presented in (Popescu, 2009).

In the pipeline implementation each module is connected with a previous module, a posterior module and a master control (Figure 1a) (Popescu, 2006). The processing elements (PE modules) are dedicated to the main video system operations such as noise rejection, binary representation, edge extraction, and feature extraction. Processing elements are similar and contain a RAM, a programmable logic circuit and a timer. The first module (noise rejection PE) can be directly connected to a digital camera. The modules achieve their synchronic tasks, in parallel, based on image acquisition clock.

The module structures (Figure 1b) are similar and contain five shift registers (SR$_1$, SR$_2$, SR$_3$, SR$_4$ and SR$_5$) and the filtering circuit (F). Two shift registers, with k bits (SR$_1$ and SR$_2$), have n_2 states, where n_2 is the number of the line pixels and k is the number of bits per pixel level. They are used to delay two image lines. Three shift registers, with k bits and three states, SR$_3$, SR$_4$ and SR$_5$, have available states. Their outputs form the neighbourhood under investigation (a_1, a_2, a_3, b_1, b_2, b_3, c_1, c_2, c_3). Thus, in order to form the neighbourhoods $V_{i,j}$, the input image signal is delayed (two lines and three pixels). The filtering circuit F accomplishes the function of the local 3×3 filter and takes a particular form for every filter used. The filtering circuit inputs are the window's pixels a_1, a_2, a_3, b_1, b_2, b_3, c_1, c_2, c_3. The central value b_2 is replaced by the processed value b'_2 which is the output of the module filter. Particularly, for cross median filter, the inputs are: a_2, b_1, b_2, b_3, c_2.

Usually, the shift registers are simulated with high speed RAM, by successive write/read operations or are implemented in FPGAs.

Figure 1. (a) Pipeline configuration for primary image processing; (b) The diagram of the local 3×3 processing element.

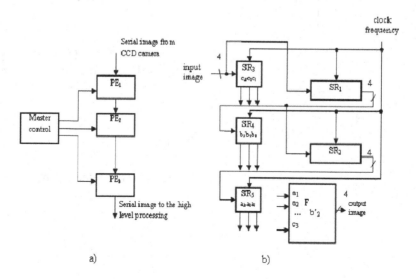

There are two proposed hardware implementations of median filters using cross neighborhoods: simultaneous comparison median filter (Popescu, 2006) and successive comparison median filter (Hiasat, 1999).

The main feature of the simultaneous comparison based median filter is the one step median computation of the data string. This filter computes the median on line, during one acquisition clock. Figure 2 represents the structure of the median filter which is based on simultaneous comparison.

The notation in Figure 2 signifies the following (it is supposed that the image is 4 bits coded): Σ – 4 bits comparators (adders whose outputs are 1 if $a_2 \geq b_1, ..., b_3 \geq c_2$); P_1, P_2, P_3, P_4, P_5 – electronic switches; M – fast EPROM. The comparison results from the adders constitute addresses of the memory M. The five bits word which is the output of a valid address contains a singular 1. Thus a singular switch (P) is selected and the output b_2' is connected to the corresponding median value (a_2, b_1, b_2, b_3 or c_2). Memory content is presented in Table 1 (valid addresses case), where X denotes 0 or 1.

The algorithm for computing the median by successive comparisons consists of determining the majority bit for each binary rank. Thus, for median evaluation of a data string, k steps are necessary, where k represents the number of bits of the gray level representation.

The implementation of logical filter for edge detection (2) is very simple and contains only two logical ports (Figure 3).

The properties of the image pre-processing algorithms presented make possible on line estimation of the main geometrical characteristics like silhouette area, perimeter (line length) and centroid coordinates (Popescu, 1990). For example, the silhouette area is evaluated by simply counting the logical 1-s obtained from the binary translation module output. Correspondingly, the perimeter is evaluated by counting the logical 1-s obtained from the contour extraction module output, with corrections for lines having a 45^0 slope:

$$P = \sum_{i,j} c_{i,j} + n_{45}(\sqrt{2} - 1)$$

Herein $c_{i,j} = b_2$' and n_{45} represents the number of the following type configurations:

1 0 , 0 1

0 1 1 0

These are detected by a simple hardware module.

Figure 2. Median filter implementation based on simultaneous comparison

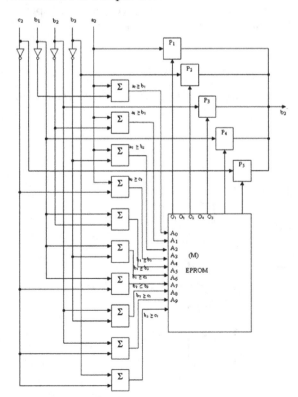

Table 1. Memory content for simultaneous comparison

ADRESSES										OUTPUTS					MEDIAN	COMMENT			
A_0	A_1	A_2	A_3	A_4	A_5	A_6	A_7	A_8	A_9	O_1	O_2	O_3	O_4	O_5	b_2'	Simultaneous comparison results			
1	1	0	0	X	0	0	0	0	X	1	0	0	0	0	a_2	$a_2 \geq b_1$	$a_2 \geq b_2$	$a_2 < b_3$	$a_2 < c_2$
1	0	1	0	0	X	0	1	X	0	1	0	0	0	0	a_2	$a_2 \geq b_1$	$a_2 \geq b_3$	$a_2 < b_2$	$a_2 < c_2$
1	0	0	1	0	0	X	X	1	1	1	0	0	0	0	a_2	$a_2 \geq b_1$	$a_2 \geq c_2$	$a_2 < b_2$	$a_2 < b_3$
0	1	1	0	1	1	X	X	0	0	1	0	0	0	0	a_2	$a_2 \geq b_2$	$a_2 \geq b_3$	$a_2 < b_1$	$a_2 < c_2$
0	1	0	1	1	X	1	0	X	1	1	0	0	0	0	a_2	$a_2 \geq b_2$	$a_2 \geq c_2$	$a_2 < b_1$	$a_2 < b_3$
0	0	1	1	X	1	1	1	1	X	1	0	0	0	0	a_2	$a_2 \geq b_3$	$a_2 \geq c_2$	$a_2 < b_1$	$a_2 < b_2$
0	X	0	0	1	0	0	0	0	X	0	1	0	0	0	b_1	$b_1 > a_2$	$b_1 \geq b_2$	$b_1 < b_3$	$b_1 < c_2$
0	0	X	0	0	1	0	1	X	0	0	1	0	0	0	b_1	$b_1 > a_2$	$b_1 \geq b_3$	$b_1 < b_2$	$b_1 < c_2$
0	0	0	X	0	0	1	X	1	1	0	1	0	0	0	b_1	$b_1 > a_2$	$b_1 \geq c_2$	$b_1 < b_2$	$b_1 < b_3$
1	X	1	1	0	1	1	1	1	X	0	1	0	0	0	b_1	$b_1 \geq b_3$	$b_1 \geq c_2$	$b_1 \leq a_2$	$b_1 < b_2$
1	1	X	1	1	0	1	0	X	1	0	1	0	0	0	b_1	$b_1 \geq b_2$	$b_1 \geq c_2$	$b_1 \leq a_2$	$b_1 < b_3$
1	1	1	X	1	1	0	X	0	0	0	1	0	0	0	b_1	$b_1 \geq b_2$	$b_1 \geq b_3$	$b_1 \leq a_2$	$b_1 < c_2$
X	0	0	0	0	0	0	0	0	X	0	0	1	0	0	b_2	$b_2 > a_2$	$b_2 > b_1$	$b_2 < b_3$	$b_2 < c_2$
1	1	1	X	0	X	0	1	0	0	0	0	1	0	0	b_2	$b_2 > b_1$	$b_2 \geq b_3$	$b_2 \leq a_2$	$b_2 < c_2$
1	1	X	1	0	0	X	0	1	1	0	0	1	0	0	b_2	$b_2 > b_1$	$b_2 \geq c_2$	$b_2 \leq a_2$	$b_2 < b_3$
0	0	X	0	1	1	X	1	0	0	0	0	1	0	0	b_2	$b_2 > a_2$	$b_2 \geq b_3$	$b_2 \leq b_1$	$b_2 < c_2$
0	0	0	X	1	X	1	0	1	1	0	0	1	0	0	b_2	$b_2 > a_2$	$b_2 \geq c_2$	$b_2 \leq b_1$	$b_2 < b_3$
X	1	1	1	1	1	1	1	1	X	0	0	1	0	0	b_2	$b_2 \geq b_3$	$b_2 \geq c_2$	$b_2 \leq a_2$	$b_2 \leq b_1$
X	0	0	0	0	0	0	1	X	0	0	0	0	1	0	b_3	$b_3 > a_2$	$b_3 > b_1$	$b_3 \leq b_2$	$b_3 < c_2$
0	X	0	0	1	1	X	0	0	0	0	0	0	1	0	b_3	$b_3 > a_2$	$b_3 > b_2$	$b_3 \leq b_1$	$b_3 < c_2$
0	0	0	X	X	1	1	1	1	1	0	0	0	1	0	b_3	$b_3 > a_2$	$b_3 \geq c_2$	$b_3 \leq b_2$	$b_3 \leq b_1$
1	1	1	X	X	0	0	0	0	0	0	0	0	1	0	b_3	$b_3 > b_1$	$b_3 > b_2$	$b_3 \leq a_2$	$b_3 < c_2$
1	X	1	1	0	0	X	1	1	1	0	0	0	1	0	b_3	$b_3 > b_1$	$b_3 \geq c_2$	$b_3 \leq a_2$	$b_3 \leq b_2$
X	1	1	1	1	1	1	0	X	1	0	0	0	1	0	b_3	$b_3 > b_2$	$b_3 \geq c_2$	$b_3 \leq a_2$	$b_3 \leq b_1$
X	0	0	0	0	0	0	X	1	1	0	0	0	0	1	c_2	$c_2 > a_2$	$c_2 > b_1$	$c_2 \leq b_2$	$c_2 \leq b_3$
0	X	0	0	1	X	1	0	0	1	0	0	0	0	1	c_2	$c_2 > a_2$	$c_2 > b_2$	$c_2 \leq b_1$	$c_2 \leq b_3$
0	0	X	0	X	1	1	1	1	0	0	0	0	0	1	c_2	$c_2 > a_2$	$c_2 > b_3$	$c_2 \leq b_1$	$c_2 \leq b_2$
1	1	X	1	X	0	0	0	0	1	0	0	0	0	1	c_2	$c_2 > b_1$	$c_2 > b_2$	$c_2 \leq a_2$	$c_2 \leq b_3$
1	X	1	1	0	X	0	1	1	0	0	0	0	0	1	c_2	$c_2 > b_1$	$c_2 > b_3$	$c_2 \leq a_2$	$c_2 \leq b_2$
X	1	1	1	1	1	1	X	0	0	0	0	0	0	1	c_2	$c_2 > b_2$	$c_2 > b_3$	$c_2 \leq a_2$	$c_2 \leq b_1$

Figure 3. Logical filter for edge detection

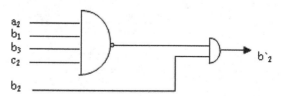

259

RUNNING SOFTWARE ALGORITHMS FOR REAL TIME IMAGE PROCESSING ON PARALLEL PLATFORMS

Classification of Software Operations Involved in Real-Time Image Processing

Traditionally, image/video processing operations have been classified into three main levels, namely low, intermediate, and high, where each successive level differs in its input/output data relationship (Kyo, 2005). Low-level operators take an image as their input and produce an image as their output, while intermediate-level operators take an image as their input and generate image attributes as their output, and finally high-level operators take image attributes as their inputs and interpret the attributes, usually producing some kind of knowledge-based control at their output.

Low-Level Operations

Low-level operations transform image data to image data. This means that such operators directly deal

with image matrix data at the pixel level. Examples of such operations include colour transformations, gamma correction, linear or nonlinear filtering, noise reduction, sharpness enhancement, frequency-domain transformations, etc. The ultimate goal of such operations is to either enhance image data, possibly to emphasize

certain key features, preparing them for viewing by humans, or extract features for processing at the intermediate-level. These operations can be further classified into point, neighbourhood (local), and global operations (Soviany, 2003). Point operations are the simplest of the low-level operations since a given input pixel is transformed into an output pixel, where the transformation does not depend on any of the pixels surrounding the input pixel. Such operations include arithmetic operations, logical operations, table lookups, threshold operations, etc. The inherent DLP in such operations is obvious, as depicted in Figure 4a, where the point operation on the pixel shown in black needs to be performed across all the pixels in the input image. Local neighbourhood operations are more complex than point operations in that the transformation from an input pixel to an output pixel depends on a neighbourhood of the input pixel. Such operations include two-dimensional spatial convolution and filtering, smoothing, sharpening, image enhancement, etc. Since each output pixel is some function of the input pixel and its neighbours, these operations require a large amount of computations. The inherent parallelism in such operations is illustrated in Figure 4.b, where the local neighbourhood operation on the pixel shown in black needs to be performed across all the pixels in the input image. Finally, global operations build upon neighbourhood operations in which a single output pixel depends on every pixel in the input image (Figure 4c). A prominent example of such an operation is the discrete Fou-

Figure 4. Parallelism in low-level image/video processing: (a) point; (b) neighborhood; (c) global

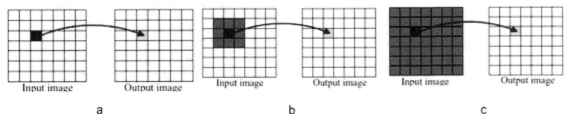

a   b   c

rier transform which depends on the entire image. These operations are quite data intensive as well.

All low-level operations involve nested looping through all the pixels in an input image with the innermost loop applying a point, neighbourhood, or global operator to obtain the pixels forming an output image. For this reason low-level operations are excellent candidates for exploiting DLP.

Intermediate-Level Operations

Intermediate-level operations transform image data to a slightly more abstract form of information by extracting certain attributes or features of interest from an image. This means that such operations also deal with the image at the pixel level, but a key difference is that the transformations involved cause a reduction in the amount of data from input to output. Intermediate operations primarily include segmenting an image into regions/objects of interest, extracting edges, lines, contours, or other image attributes of interest such as statistical features. The goal by carrying out these operations is to reduce the amount of data to form a set of features suitable for further high-level processing. Some intermediate-level operations are also data intensive with a regular processing structure, thus making them suitable candidates for exploiting DLP.

High-Level Operations

High-level operations interpret the abstract data from the intermediate-level, performing high level knowledge-based scene analysis on a reduced amount of data. Such operations include classification (recognition) of objects or a control decision based on some extracted features. These types of operations are usually characterized by control or branch-intensive operations. Thus, they are less data intensive and more inherently sequential rather than parallel. Due to their irregular structure and low-bandwidth requirements, such operations

are suitable candidates for exploiting ILP, although their data-intensive portions usually include some form of matrix–vector operations that are suitable for exploiting DLP.

Essential Architecture Features of Parallel Platforms

Hardware Architecture Design

A great deal of the present growth in the field of image/video processing is primarily due to the ever-increasing performance available on standard desktop PCs, which has allowed rapid development and prototyping of image/video processing algorithms. The desktop PC development environment has provided a flexible platform in terms of computation resources including memory and processing power. In many cases, this platform performs quite satisfactorily for algorithm development.

As discussed in the previous section, practical image/video processing systems include a diverse set of operations from structured, high-bandwidth, data-intensive, low-level and intermediate-level operations such as filtering and feature extraction, to irregular, low-bandwidth, control-intensive, high-level operations such as classification. Since the most resource demanding operations in terms of required computations and memory bandwidth involve low-level and intermediate level operations, considerable research has been devoted to developing hardware architectural features for eliminating bottlenecks within the image/video processing chain, freeing up more time for performing high-level interpretation operations. While the major focus has been on speeding up low-level and intermediate level operations, there have also been architectural developments to speed up high-level operations.

From the literature, one can see there are three major architectural features that are essential to any image/video processing system, namely single instruction multiple data (SIMD), very long instruction word (VLIW), and an efficient memory subsystem. The concept of SIMD processing is a key architectural feature found in one way or another in most modern real-time image/video processing systems (Davies, 2005). It embodies broadcasting a single instruction to multiple processors, which simultaneously execute the instruction on different portions of data in parallel, thus allowing more computations to be performed in a shorter time.

While SIMD can be used for exploiting DLP, VLIW can be used for exploiting instruction level parallelism (ILP) and thus for speeding up high-level operations (Broers, 2005). VLIW furnishes the ability to execute multiple instructions within one processor clock cycle, all running in parallel, hence allowing software-oriented pipelining of instructions by the programmer. Besides the fact that for VLIW to work properly there must be no dependencies among the data being operated on, the ability to execute more than one instruction per clock cycle is essential for image/video processing applications that require operations in the order of Giga operations per second.

An efficient memory subsystem is considered a crucial component of a real-time image/video processing system, especially for low-level and intermediate-level operations that require massive amounts of data transfer bandwidth as well as high-performance computation power. Concepts such as direct memory access (DMA) and internal versus external memory are important. DMA allows transferring of data within a system without burdening the CPU with data transfers, so it is a well-known tool for hiding memory access latencies, especially for image data.

Software Architecture Design

While translating source code from a research development environment to a real-time environment is an involved task, it would be beneficial if the entire software system is well thought out ahead of time. Considering that real-time image/video

processing systems usually consist of thousands of lines of code, proper design principles should be practiced from the start in order to ensure maintainability, extensibility, and flexibility in response to changes in the hardware or the algorithm (Sangwan, 2005). One key method of dealing with this problem is to make the software design modular from the start, which involves abstracting out algorithmic details and creating standard interfaces or application programming interfaces (APIs) to provide easy switching among different specific implementations of an algorithm. Also beneficial is to create a hierarchical, layered architecture where standard interfaces exist between the upper layers and the hardware layer to allow ease in switching out different types of hardware so that if a hardware component is changed, only minor modifications to the upper layers will be needed.

Real-Time Operating System

In a real-time image/video processing system, certain tasks or procedures have strict real time deadlines, while other tasks have firm or soft real-time deadlines. In order to be able to manage the deadlines and ensure a smoothly running system, it is useful to utilize a real time operating system. Real-time operating systems (RTOS) allow the assignment of different levels of priorities to different tasks. With such an assignment capability, it becomes possible to assign higher priorities to hard real-time deadline tasks and lower priorities to other firm or soft real-time tasks. For portable embedded devices such as digital cameras, a real-time operating system can be used to free the upper layer application from managing the timing and scheduling of tasks, and handling file input/output operations (Dobrescu, 2005). Therefore, a real-time operating system is an important key component of the software of any practical real-time image/video processing system since it can be used to guarantee meeting real-time deadlines and thus ensuring deterministic behaviour to a certain extent. To design efficient RTOS, it is necessary to assure performance related to benchmarks as: latency time, predictability, dependability, average task switch time, average pre-emption time, semaphore shuffle time, deadlock break time (Halang, 2000; Stewart, 2006; Xu, 2008)

A Parallel Platform Support for Real-Time Image Processing

Parallel Platform Model and Scheduling Principles

Our system model consists of P processor units. Each processor p_i has capacity c_i, with $c_i > 0$, $i = 1, 2,..., P$. The capacity of a processor is defined as its speed relative to a reference processor with unit-capacity. We assume for the general case that $c_1 \leq c_2 \leq \cdots \leq c_P$. The *total capacity C* of the system is defined as $C = \sum_{i=1}^{P} c_i$. A system is called *homogeneous* when $c_1 = c_2 \cdots = c_P$. The platform is conceived as a distributed system (Epema, 1999). Each machine is equipped with a single processor. In other words, we do not consider interconnections of multiprocessors. The main difference with multiprocessor systems is that in a distributed system, information about the system state is spread across the different processors. In many cases, migrating a job from one processor to another is very costly in terms of network bandwidth and service delay (Agosta, 2004), and that is the reason that we have considered from the beginning only the case of data parallelism for a homogenous system. The intention was to test the general case of image processing with both data and task parallelism, by developing a scheduling policy with two components. The *global* scheduling policy decides to which processor an arriving job must be sent, and when to migrate some jobs. At each processor, the *local* scheduling policy decides when the processor serves which of the jobs present in its queue.

Jobs arrive at the system according to one or more inter-arrival time processes. These processes determine the time between the arrivals of two consecutive jobs. The *arrival time* of job j is denoted by A_j. Once a job j is completed, it leaves the system at its *departure time* D_j. The *response time* R_j of job j is defined as $R_j = D_j - A_j$. The *service time* S_j of job j is its response time on a unit-capacity processor serving no other jobs; by definition, the response time of a job with service time s on a processor with capacity c' is s/c'. We define the *job set J(t) at time t* as the set of jobs present in the system at time t: $J(t)=\{j|A_j \leq t < D_j\}$

For each job $j \in J(t)$, we define the *remaining work* $W_j^r(t)$ at time t as the time it would take to serve the job to completion on a unit-capacity processor. The *service rate* $\sigma_j^r(t)$ of job j at time t $(A_j \leq t < D_j)$ is defined as: $\sigma_j^r(t) = \lim_{\tau \to t} \dfrac{dW_j^r(\tau)}{d\tau}$

The *obtained share* $\omega_j^s(t)$ of job j at time t $(A_j \leq t < D_j)$ is defined as: $\omega_j^s(t) = \sigma_j^r(t) / C$

So, $\omega_j^s(t)$ is the fraction of the total system capacity C used to serve job j, but only if we assume that $W_j^r(t)$ is always a piecewise-linear, continuous function of t.

Considering $W_j^r(A_j) = S_j$ and $W_j^r(D_j) = 0$ we have $\int_{A_j}^{D_j} \omega_j^s(t)dt = \int_{A_j}^{D_j} \sigma_j^r(t)dt = S_j / C$. One can define an upper bound on the sum of the obtained job shares of any set of jobs $\{1,..., J\}$:

$\sum_{j=1}^{J} \omega_j^s(t) \leq C^{-1} \sum_{i=1}^{\min(J,P)} c_i$ and furthermore the maximum obtainable total share $\omega_{max}(t)$ at time is:

$$\omega_{max}(t) = C^{-1} \sum_{i=1}^{\min(J,P)} c_i .$$

A CASE STUDY: LINES DETECTION USING DIRECTIONAL FILTERING

Theoretical Principle

Usually the problem of detecting lines and linear structures in images is solved by considering the second order directional derivative in the gradient direction, for each possible line direction (Geusebroek, 2001). Theoretically, in two-dimensions, line points are detected by considering the second order directional derivative in the gradient direction. For a line point, the second order directional derivative perpendicular to the line is a measure of line contrast, given by $\lambda = f_{ww}(x,y)$ where $f(x,y)$ is the grey-value function and the indices w denote differentiation in the gradient direction. Bright lines are observed when $\lambda < 0$ and dark lines when $\lambda > 0$. In practice, one can only measure differential expressions at a certain observation scale. By considering Gaussian weighted differential quotients in the gradient direction, $f_{ww}^{\sigma} = G_{ww}(\sigma) * f(x,y)$ a measure of line contrast is given by $r(x,y,\sigma) = \sigma^2 \left| f_{ww}^{\sigma} \right| \dfrac{1}{b^{\sigma}}$ where σ, the Gaussian standard deviation, denotes the scale for observing the line structure, and where line brightness b is given by:

$$b^{\sigma} = \begin{cases} f^{\sigma} if \ldots f_{ww}^{\sigma} \leq 0 \\ W - f^{\sigma} otherwise \end{cases}$$

Line brightness is measured relative to black for bright lines, and relative to white level W (255 for an 8-bit camera) for dark lines. The response of the second order directional derivate λ does not only depend on the image data, but it is also affected by the Gaussian smoothing scale σ. Because a line has a large spatial extent along the line direction, and only a small spatial extent (i.e., the line width) perpendicular to the line, the Gaussian filter should be tuned to optimally ac-

cumulate line evidence. For directional filtering anisotropic Gaussian filters may be used of scale σ_v and σ_w for longest and shortest axis, respectively. Line contrast is given by:

$$r'(x,y,\sigma_v,\sigma_w) = \sigma_v \sigma_w \left| f_{ww}^{\sigma_v,\sigma_w} \right| \dfrac{1}{b^{\sigma_v,\sigma_w}}$$

The optimal filter orientation may be different for each position in the image plane, depending on line evidence at the particular image point under consideration. The final line detection filter, parameterized by orientation θ, smoothing scale σ_v in the line direction, and differentiation scale σ_w perpendicular to the line, is given by

$$r''(x,y,\sigma_v,\sigma_w,\theta) = \sigma_v \sigma_w \left| f_{ww}^{\sigma_v,\sigma_w,\theta} \right| \dfrac{1}{b^{\sigma_v,\sigma_w,\theta}}$$

where $f_{ww}^{\sigma_v,\sigma_w,\theta} = G_{ww}(\sigma_v,\sigma_w,\theta) * f(x,y)$

When the filter is correctly aligned with the line, and σ_v, σ_w are optimally tuned to capture the line, filter response is maximal. Hence, the maximum per pixel line contrast over the filter parameters yields line detection:

$$R(x,y) = \arg\max_{\sigma_v,\sigma_w,\theta} r''(x,y,\sigma_v,\sigma_w,\theta)$$

The final result is obtained by considering the maximum response per pixel over all filter results. This yields the optimal orientation θ, an estimate of line thickness σ_w, the best smoothing size σ_v, and the line contrast $R(x,y)$.

Implementation of the Directional Filtering Algorithm

There are many different ways to implement a directional filtering algorithm. For example, one can create for each orientation a new filter based on σ_v and σ_w. This yields a rotation of the filters, while the orientation of the input image remains fixed. Another possibility is to keep the orientation of the filters fixed, and to rotate the input image instead. Yet another solution is to integrate the notion of orientation in the filter operation itself. In this case image pixels are accessed not only according to the size of the neighbourhood of the filter, but also on the basis of the given orientation. From these solutions, the second, which applies fixed filters to rotated image data, seems to be more suitable for parallelization. In order to stress the possibility of executing parallel operations, let us consider first the main steps of a sequential implementation.

The first step consists of rotating the original input image for a given orientation θ. This operation is made by a dedicated routine *Rotate_Image*. Then, for all combinations (σ_v, σ_w) the filtering is performed by six operations executed in sequence by six dedicated routines, as follows: (1) *Filter 1* to compute $f_{ww}^{\sigma_v,\sigma_w,\theta}$; (2) *Filter 2* to compute $b^{\sigma_v,\sigma_w,\theta}$ (both filltering operations are generalized Gaussian convolutions performed by applying two 1-dimensional filters; (3) *Binary_Op1*, a binary pixel operation having an image as argument; (4) *Binary_Op2*, a binary pixel operation having an constant value as argument; (5) *Back_Rotate_Image* to match the orientation of the original input image; (6) *Contrast* to obtain the maximum response.

It is to be noted that on a state-of-the-art sequential machine the program may take from tens of seconds up to minutes to complete, depending on the size of the input image and the extent of the chosen parameter subspace. Consequently, for the directional filtering program parallel execution is highly desired.

The program described above may be processed in parallel in two different schedules. In the first schedule all dedicated routines are forced to run in parallel, using all available processing units. The second schedule differs from the first in that the last two operations in the innermost loop of the program are run on one node only. In both schedules the *Original_Image* structure must be broadcast to all nodes. This is because the structure is applied in the initial rotation operation. In addition, in both schedules the first four operations in the innermost loop can be executed locally on partial image data structures. The only need for communication is in the exchange of image borders (shadow regions) in the two Gaussian convolutions.

In the first schedule the last two operations in the innermost loop are run in parallel as well. This requires the distributed image *Binary_Op1* to be available in full at each node, because it has an access pattern of type 'other' in the back-rotation operation. This can be achieved by executing a gather-to-all operation, which is logically equivalent to a gather operation followed by a broadcast. Finally, a partial maximum response image *Contrast* is calculated on each node, which requires a final gather operation to be executed just before termination of the program. In the second schedule the last two operations are not executed in parallel. As a result, the intermediate result image after *Binary_Op2* that produces both the back-rotated image needs to be gathered to the single node, as well as the complete maximum response image.

Performance Evaluation

The test was performed on a reduced set of operations. For each instruction utilized in the directional filtering algorithm two measurements were executed, for images having 200^2 or 1000^2 elements. The test network was configured as a cluster with 2, 4 or 8 nodes, each node being a processor working at 1 Ghz with 128 MByte RAM. Table 2 offers the predicted and measured results for the processing of an image with 1024x1024 pixels (see Figure 5. a-original, b-after processing) with 12 orientations and 4 combinations (σ_v, σ_w).

Table 2. Evaluation results

Number of processors	Schedule 1		Schdule 2	
	Predicted duration [s]	**Measured duration [s]**	**Predicted duration [s]**	**Measured duration [s]**
1	5.43	5.56	5.43	5.56
2	2.81	2.90	3.91	4.01
4	1.54	1.60	3.15	3.22
6	1.16	1.21	2.91	2.95
8	0.93	0.98	2.77	2.82

A schedule is preferred if the set of operations unique to that schedule is faster than the set of operations unique to another schedule (i.e., not in the set of operations common to both schedules). Hence, for the directional filtering program the schedule in which *all* operations are run in parallel is preferred if:

$$\theta_\sigma(P_{rotate}(size/N)+P_{max}(size/N)+P_{bcast}(size/N)+P_{gather}(size/N)) < \theta_\sigma(P_{rotate}(size)+P_{max}(size))$$

where N denotes the number of processing units and θ_σ denotes the size of the parameter subspace. For the first schedule the large number of broadcast operations is expected to have the most significant impact on performance. For the second schedule, on the other hand, the many rotations of non-partitioned image data are expected to be costly.

FUTURE RESEARCH DIRECTIONS

For the future, our work concerning hardware implementations for real time image processing will focus on FPGAs, due to their flexibility in implementing custom hardware solutions and because they provide a low-cost, flexible development of high-performance, custom parallel processors, suitable for transferring almost any kind of image/video processing algorithm from a development environment to a real-time

Figure 5. Test image for line detection: (a) original; (b) after processing

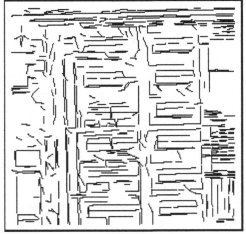

implementation. The main directions for further implementations will be: *image filtering operations* (for example 2D nonlinear image filtering for mammogram contrast enhancement in real-time, 3D image filtering operations for the removal of speckle noise in ultrasonic images, median and convolution filtering for 3D medical image processing); *low-level operations* (for example, the problem of controlling the exposure time of a charge-coupled device camera in real-time using a low-level histogram-based measure or the implementation of a box-counting algorithm), and finally *standard image processing operations* such as edge detection, moment calculation, and Hough transform.

Regarding the potential of the parallel platform for image processing, in the near future we will focus our attention on the improvement of the scheduling component, by using processor units with different processing capacities and also other service policies for the queue of jobs. We will continue implementing example programs to investigate the implication of parallelization of typical applications in the area of real-time image processing, trying to improve the performances by supporting the execution of a sequence of algorithms on the same block and by dynamic reconstruction of the post processed image.

CONCLUSION

Hardware implementation of primary image processing algorithms makes possible the on-line execution of the processing tasks, directly from video camera. The median filter and edge extraction filter based on binary image utilization are simple and efficient for further feature extraction. If a pipeline structure with k processing modules, 3×3 neighbourhood types, is utilized, then the latency is about 2k lines. The structure flexibility can be improved by FPGA implementation. The recommendation is to introduce such processing pipelines in the video camera structure.

The aim of the discussed parallel processing platform was to validate a software architecture that allows an image processing researcher to develop parallel applications. The challenge is that algorithms for processing digital images and video are developed and prototyped on desktop PCs or workstations, which are considered to be resource unlimited platforms. Adding to this the fact that the vast majority of algorithms developed to process digital images and video are quite computationally intensive, one needs to resort to specialized processors, judiciously trade-off decisions to reach an accepted solution, or even abandon a complex algorithm for a simpler, less computationally complex algorithm. The experiments show how to use parallelizable patterns, obtained for typical low level image processing operations. At the same time this architecture offers the possibility to test several modes for tasks management and scheduling and to try an optimization of the load balancing between the workstations. Each operation is implemented such that its execution can be adapted to obtain higher performance. Experiments show that our performance models are highly accurate for parallel processing using convolution functions. Given the results we are confident that the proposed architecture forms a powerful basis for automatic parallelization and optimization of a wide range of image processing applications.

REFERENCES

Agosta, G., Crespi Reghizzi, S., Falauto, G., & Sykora, M. (2004). JIST: Just-in-time scheduling translation for parallel processors. *Proceedings of the Third International Symposium on Parallel and Distributed Computing/Third International Workshop on Algorithms, Models and Tools for Parallel Computing on Heterogeneous Networks (ISPDC/HeteroPar '04)*, (pp. 122-132).

Bovik, A. (2005). Introduction to digital image and video processing. In Bovik, A. C. (Ed.), *Handbook of image & video processing*. Amsterdam: Elsevier Academic Press. doi:10.1016/B978-012119792-6/50065-6

Broers, H., Caarls, W., Jonker, P., & Kleihorst, R. (2005). Architecture study for smart cameras. *Proceedings of the European Optical Society Conference on Industrial Imaging and Machine Vision*, (pp. 39–49).

Davies, E. (2005). *Machine vision: Theory, algorithms, practicalities*. San Francisco: Morgan Kauffmann Publishers.

Dobrescu, M. (2005). *Distributed image processing techniques for multimedia applications*. Unpublished doctoral dissertation, Politehnica Univ. of Bucharest.

Dougherty, E., & Laplante, P. (1995). *Introduction to real-time imaging. Bellingham, WA/Piscataway*. NJ: SPIE Press/IEEE Press.

Epema, D. H. J., & de Jongh, F. C. M. (1999). Proportional share-scheduling in single-server and multiple-server computing systems. *Performance Evaluation Review*, *27*(3), 7–10. doi:10.1145/340242.340295

Geusebroek, J. M., Smeulders, A. W. M., & Geerts, H. (2001). A minimum cost approach for segmenting networks of lines. *International Journal of Computer Vision*, *43*(2), 99–111. doi:10.1023/A:1011118718821

Halang, W. A., Gumzej, R., Colnaric, M., & Druzovec, M. (2000). Measuring the performance of real-time systems. *The International Journal of Time-Critical Computing Systems*, *18*, 59–68.

Hiasat, A. A., Al-Ibrahim, M. M., & Gharaibeh, K. M. (1999). Design and implementation of a new efficient median filtering algorithm. *IEEE Proceedings on Vision. Image and Signal Processing*, *146*(5), 273–278. doi:10.1049/ip-vis:19990444

Hunter, H., & Moreno, J. A. (2003). New look at exploiting data parallelism in embedded systems. *Proceedings of the International Conference on Compilers, Architectures, and Synthesis for Embedded Systems*, (pp. 159–169).

Kehtarnavaz, N. (2004). *Real-time digital signal processing based on the TMS320C6000*. Amsterdam: Elsevier.

Kyo, S., Okazaki, S., & Arai, T. (2005). An integrated memory array processor architecture for embedded image recognition systems. *Proceedings of the 32nd International Symposium on Computer Architecture*, (pp. 134–145).

Popescu, D. (1990). Industrial image processing. *Buletinul IPB. Seria Inginerie Electrica*, *52*(2), 91–96.

Popescu, D., Dobrescu, R., Avram, V., & Mocanu, S. (2006). Local processors for on line processing with applications in intelligent vehicle technologies. *Proceedings of the 10th WSEAS International Conference on SYSTEMS*, (pp. 579-584). Athens, Greece.

Popescu, D., Patarniche, D., Dobrescu, R., Nicolae, M., & Dobrescu, M. (2009). Real time mobile object tracking based on chromatic information. *Proceedings of the 5th WSEAS International Conference on Remote Sensing -REMOTE'09*, (pp. 13-18). Genova, Italy.

Sangwan, R., Ludwig, R., Laplante, P., & Neill, C. (2005). Performance tuning of imaging applications through pattern-based code transformation. *Proceedings of SPIE-IS&T Electronic Imaging Conference on Real-Time Imaging*, (pp. 1–7).

Seinstra, F. J., Koelma, D., & Geusebroek, J. M. (2002). A software architecture for user transparent parallel image processing. *Parallel Computing*, *28*(7-8), 967–993. doi:10.1016/S0167-8191(02)00103-5

Soviany, C. (2003). *Embedding data and task parallelism in image processing applications.* Unpublished doctoral dissertation, Delft University of Technology.

Stewart, D. B. (2006). Measuring execution time and real–time performance. *Proceedings of Embedded System Conference.*

Wnuk, M. (2008). Remarks on hardware implementation of image processing algorithms. *International Journal of Applied Mathematics and Computer Science, 18*(1), 105–110. doi:10.2478/v10006-008-0010-2

Xu, T. (2008). *Performance benchmarking of FreeRTOS and its hardware abstraction.* Unpublished doctoral thesis, Technical University of Eindhoven.

ADDITIONAL READING

Ackenhusen, J. (1999). *Real-Time Signal Processing: Design and Implementation of Signal Processing Systems.* Englewood Cliffs, NJ: Prentice-Hall.

Akil, M. M. (2004). Architecture for Hardware Thinning and Crest Restoration in Gray-level Images, *Proceedings of SPIE-IS&T Electronic Imaging Conference on Real-Time Imaging,* SPIE, 5297, 242–253.

Amsterdam Elsevier.

Architecture for Real-Time Image Processing. *Journal of Real-Time Imaging, 8*(5), 345–356.

Arias-Estrada, M. & Xicotencatl, J. (2001). Real-Time FPGA-Based Architecture for Stereo

Batlle, J., Marti, J., Ridao, P., & Amat, J. (2002). A New FPGA/DSP-Based Parallel

Bertozzi, M., & Broggi, A. (1998). GOLD: A Parallel Real-Time Stereo Vision System for Generic Obstacle and Lane Detection. *IEEE Transactions on Image Processing, 7*(1). doi:10.1109/83.650851

Canny, J. F. (1986). A computational approach to edge detection. *IEEE Transactions on Pattern Analysis and Machine Intelligence, 8*(6), 679–698. doi:10.1109/TPAMI.1986.4767851

Chen, T., & Chung, K. (2001). A New Randomized Algorithm for Detecting Lines. *Journal of Real-Time Imaging, 7*(6), 473–481. doi:10.1006/rtim.2001.0233

Davies, E. (2005). *Machine Vision: Theory, Algorithms, Practicalities.* San Francisco, CA: Morgan Kauffmann Publishers.

Dougherty, E., & Laplante, P. (1995). *Introduction to Real-time Imaging.* Bellingham.

Gonzales, R., & Woods, R. (2002). *Digital Image Processing.* Englewood Cliffs, NJ: Prentice-Hall.

Hussmann, S., & Ho, T. (2003). A High-Speed Subpixel Edge Detector Implementation Inside a FPGA. *Journal of Real-Time Imaging, 9*(5), 361–368. doi:10.1016/j.rti.2003.09.013

Iannizzotto, G., & Vita, L. (2002). On-Line Object Tracking for Colour Video Analysis. *Journal of Real-Time Imaging, 8*(2), 145–155. doi:10.1006/rtim.2001.0267

Kao, W., Sun, T., & Lin, S. (2005). A Robust Embedded Software Platform for Versatile Camera Systems. *Proceedings of the IEEE International Symposium on Circuits and*

Kehtarnavaz, N. (2004). *Real-Time Digital Signal Processing Based on the TMS320C6000.*

Kehtarnavaz, N., & Oh, H. (2003). Development and Real-Time Implementation of a Rule-Based Auto-Focus Algorithm. *Journal of Real-Time Imaging, 9*(3), 197–203. doi:10.1016/S1077-2014(03)00037-8

Kessal, L., Abel, N., & Demigny, D. (2003). Real-Time Image Processing with Dynamically Reconfigurable Architecture. *Journal of Real-Time Imaging, 9*(5), 297–313. doi:10.1016/j.rti.2003.07.001

Kotoulas, L., & Andreadis, I. (2004). Efficient Hardware Architectures for Computation of Image Moments. *Journal of Real-Time Imaging, 10*(6), 371–378. doi:10.1016/j.rti.2004.09.002

Lee, J., Ko, J., & Kim, E. (2004). A Real-Time Face Detection and Tracking for Surveillance System using Pan/Tilt Controlled Stereo Camera. *Proceedings of SPIE-IS&T Electronic Imaging Conference on Real-Time Imaging*, SPIE, 5297, 152–162.

Lins, A., & Williston, K. (2004). *Processors for Consumer Video Applications*. Berkeley Design.

Mahlknecht, S., Oberhammer, R., & Novak, G. (2004). A Real-Time Image Recognition System for Tiny Autonomous Mobile Robots. *Proceedings of the 10th IEEE Real-Time and Embedded Technology and Applications Symposium,* 324–330.

Meribout, M., Nakanishi, M., & Ogura, T. (2002). Accurate and Real-Time Image Processing on a New PC-Compatible Board. *Journal of Real-Time Imaging, 8*(1), 35–51. doi:10.1006/rtim.2001.0269

Nishikawa, Y., Kawahito, S., & Inoue, T. (2005). Parallel Image Compression Circuit for High-Speed Camera. *Proceedings of SPIE-IS&T Electronic Imaging Conference on Real-Time Imaging*, SPIE, 5671, 111–122.

Paschalakis, S., & Bober, M. (2004). Real-Time Face Detection and Tracking for Mobile Videoconferencing. *Journal of Real-Time Imaging, 10*(2), 81–94. doi:10.1016/j.rti.2004.02.004

Plaza, A., & Chang, C.-I. (2007). *High performance computing in remote sensing*. Boca Raton: CRC Press.

Popescu, D., Ionescu, G., & Dobrescu, R. (1995). Sensor Fusion for Mobile Robot Control, *ISMCR'95 Proceedings*, Slovakia, 415-418.

Sangwan, R., Ludwig, R., & Neill, C. (2005). Software Visualization Techniques for Real-Time Imaging Applications. *Proceedings of SPIE-IS&T Electronic Imaging Conference on Real-Time Imaging*, SPIE, 5671, 30–35.

Schmidt, D. C. (2002). Middleware for real-time and embedded systems. *Communications of the ACM, 45*(6), 43–48. doi:10.1145/508448.508472

Shapiro, L., & Stockman, G. (2000). *Computer Vision*. Prentice Hall.

Sicilliano, B., & Katib, O. (2008). *Handbook of Robotics*. Springer Verlag. doi:10.1007/978-3-540-30301-5

SPIE, 4303, 59–66.

Systems, 5, 5015–5018.

Technology. http://www.bdti.com.

Tsai, P., Chang, C., & Hu, Y. (2002). An Adaptive Two-Stage Edge Detection Scheme for Digital Color Images. *Journal of Real-Time Imaging, 8*(4), 329–343. doi:10.1006/rtim.2001.0286

Venugopal, S., Castro-Pareja, C., & Dandekar, O. (2005). An FPGA-Based 3D Image Processor with Median and Convolution Filters for Real-Time Applications. *Proceedings of SPIE-IS&T Electronic Imaging Conference on Real-Time Imaging*, SPIE, 5671, 174–182.

Vision. *Proceedings of SPIE-IS&T Electronic Imaging Conference on Real-Time Imaging,*

Wakerly, J. (2000). *Digital Design: Principals and Practices*. Englewood Cliffs, NJ: Prentice Hall.

WA/Piscataway, NJ: SPIE Press/IEEE Press.

Wilson, R., & Spann, M. (1988). *Image Segmentation and Uncertainty*. New York: John Wiley and Sons Inc.

Chapter 16
Recent Advances in Corneal Imaging

Abdulhakim Elbita
Bradford University, UK

Rami Qahwaji
Bradford University, UK

Stanley Ipson
Bradford University, UK

Taha Y. Ahmed
Bradford University, UK

K. Ramaesh
Bradford University, UK

T. Colak
Bradford University, UK

ABSTRACT

The cornea is the convex and transparent covering in the front of the eye. It is responsible for most of the focusing power required to create an image on the retina. Injuries and various pathologies such as (keratoconus, lattice dystrophy, dry eye, conjunctivitis, etc.) of the cornea compromises vision. Due to loss of corneal transparency, hence scattering light rays passing through the cornea, in severe cases, total sight loss can occur

A confocal microscope can be used to provide a sequence of images (of variable quality) at different depths from the front surface of the eye, showing the various corneal layers and structures. From these images, Ophthalmologists can extract clinical information on the state of health of a patient's cornea. Currently, analyses of these images are mainly based on visual interpretation of these images or on semi-automatic methods, which might contribute to making erroneous diagnoses.

DOI: 10.4018/978-1-60960-477-6.ch016

This chapter details work with sequences of corneal images from a confocal microscope to develop enhancement methods to improve the visual quality of the images. Due to involuntary movements of the subject's eye during image capture, the images suffer both lateral and longitudinal translations, and work is ongoing to attempt to register adjacent images in the sequence. Currently this registration uses an approach based on the Scale Invariant Feature Transforms (SIFT) algorithm. Registration is a necessary stage in the construction of a 3D model of the subject's cornea for use as a diagnostic aid. The algorithms, results, progress and suggestions for future work are presented in this chapter.

INTRODUCTION

The Structure of the Cornea

The cornea is a collection of cells and proteins that constitute a very highly organized structure. It is the clear outer layer, covering the front of the eye. The cornea must remain transparent to transmit and refract light. The cornea does not contain blood vessels to feed or protect it from infection, However it receives its nourishment from tears and the aqueous humour (fluid filling the Anterior chamber which is the space between the lens and cornea). Figure 1 shows the anatomy of the eye, and inspecting it we can recognize that the anterior surface of the cornea, which

is not uniformly curved (H.E. Kaufman 2000; MedicineNet 2010).

The dimensions of the cornea are, on average, 12.6 mm in the horizontal direction median and 11.7 mm in the vertical median. Its thickness is not uniform, with the central cornea thinner than the peripheral cornea (520 μm =< thickness <= 650 μm) (H.E.Kaufman 2000). The cornea has a tear film on its front surface and three main layers separated by two thin membranes. The Epithelial layer is the outermost layer of the cornea and is separated by the Bowman's membrane from the central Stroma layer, which in turn is separated by the Descemet's membrane from the innermost Endothelium layer as shown in Figure 2. Approximate thicknesses of these layers in the

Figure 1. The anatomy of the Eye (MedicineNet 2010)

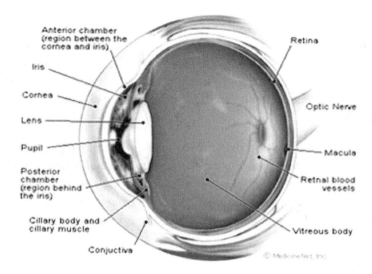

Figure 2. The corneal layers

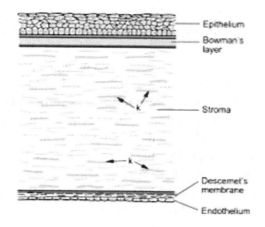

normal cornea are given in Table 1 (H.E. Kaufman 2000; Reinstein, Archer et al. 2008).

The Functions of the Cornea

The cornea, has two functions (Institute 2010) (Institute 2010):

- It acts as a shield which helps to protect the eye from germs, dust and other harmful environmental factors.
- It refracts and transmits light into the eye, providing most of the focusing power needed to form an image on the retina

Some Diseases and Disorders Affecting the Cornea

Corneal pathology is diverse, some of which are very briefly outlined below (Institute 2010) (Institute 2010):

- **Allergic response:** Mainly immune mediated, symptoms like itching, burning, stinging and watery discharge. It's a common condition, usually a multitude of factors are incorporated.
- **Conjunctivitis (Pink Eye):** Caused by bacteria or viral infections Causing red, uncomfortable eye with different types of discharge depending on the aetiology.
- **Dry Eye:** Reduction in tears production compromises integrity of the corneal surface and the epithelial layer in particular, the tear /corneal interface is the first and most crucial refractive plane, this process suffers considerably in dry eyes, not to mention damaged epithelial cells are more prone to infections.
- **Fuchs'Dystrophy:** This disease affects both eyes and is slightly more common in women than in men. Whereby there is a malfunction of the endothelial (innermost layer), this layer acts as an osmotic pump system, i.e it pumps water out of the cornea, when it becomes dysfunctional the cornea cells

Table 1. The constituent layers of the cornea and their approximate thicknesses

order	Layer Name	Constituents	The Layer thickness
1	Epithelium layer	Squamous cell layer	About 52 to 65 µm ≈ 10% of the cornea thickness
		Wing cells	
		Basal cells	
2	Bowman's Layer		About 8 to 14 µm ≈ 2% of the cornea thickness
3	Stroma Layer		About 442 to 552 µm ≈ 85% of the cornea thickness
4	Descemet's membrane	Anterior banded layer	About 10 to 12 µm≈ 2% of the cornea thickness
		Non-banded layer	
5	Endothelium layer		About 4 to 6About 4 - 6 µm ≈ 1% of the cornea thickness

swell up with water and invariably reduces vision.

- **Herpes Zoster (Shingles):** This disease is caused by the virus that causes chickenpox. later in life the dormant virus is reactivated and may affect the corneal nerves, in chronic cases scarring can take place.

- **Keratoconus:** The cornea becomes conical rather than spherical, it progresses with age and causes irregular astigmatism and short-sightedness, its due to poor alignment of the corneal collagen lamellae. This disease usually affects both eyes.

- **Lattice Dystrophy:** This disease is caused by an accumulation of amyloidal deposits (abnormal protein fibres) in the cornea in a linear pattern and may reduce vision.

- **Map-Dot-Fingerprint Dystrophy:** This disease usually affects adults between the ages of 40 and 70 and occurs in both eyes. However it may develop earlier in life. This disease usually affects the epithelium causing it to have map-like appearance (e.g., large, slightly gray patches similar to continents on a map that could also have opaque dots near the patches)

- **Stevens - Johnson Syndrome:** This disease has many names and is a disorder of the skin that can also affect the eyes. When present on the face, especially around the mouth, it causes a lot of pain. In some cases this disease leads to severe vision loss. The underlying pathophysiological process is not fully understood.

The Confocal Microscope

A confocal microscope can be used for examining thick transparent objects such as the cornea by illuminating them with a focused spot of light. Its operation is dependent on integrating two techniques; point-by-point illumination of the specimen and the rejection of out-of-focus light (Semwogerere, 2005). The confocal microscope,

whose principle of operation is shown in Figure 3, solves the problem of the presence of light from out-of-focus objects when using conventional light microscopes.

In a confocal microscope, a pinhole aperture acts as a point source of light, which is focused by a 45° partially (50%) reflecting mirror and an objective lens at a point in the specimen. The light scattered back from the focal point in the specimen is collected by the objective lens and after passing through the partially reflecting mirror is focused onto a duplicate pinhole aperture. A detector behind this aperture measures the light passing through the aperture. Only a small fraction of any light scattered back from other more weakly (un-focused) illuminated points in the specimen falls on the detector because it is out of focus in the plane of the pin hole. An objective lens with a high-numerical-aperture is used. By scanning the focused spot systematically through the volume of the specimen, a picture of the specimen can be built up.

Comparing Confocal Microscopes with conventional microscopes, the former can provide two very important advantages for corneal imaging: reduced depth of field and improved lateral resolution. Today there are many types of confocal

Figure 3. The principles of confocal microscopy. Light reflected from the specimen illuminated at the focal point is "co-focused" at the confocal detection aperture (JAY C. ERIE 2009).

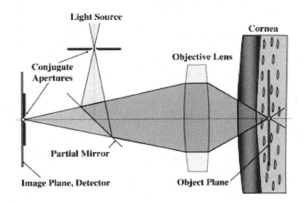

Figure 4. The Confoscan 4 Microscope (NI-DEKtechnologies 2010)

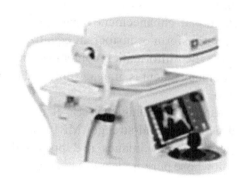

microscopes available, but the three most commonly used in practice are (Jay C. Erie 2009):

- The Tandem Scanning Confocal microscope (TSCM)
- The ConfoScan 4 Slit-Scanning Confocal Microscope (ConfoScan 4-SSCM)
- The Heidelberg Retina Tomograph Rostock Corneal Module laser scanning confocal microscope (HRT or HRT3)

The Confoscan4 Microscope can be used with two Lenses operating at different magnifications of 20x and 40x. The specifications for both probes are shown in Table 2.

Hence, the ConfoScan 4 Slit-Scanning Confocal Microscope (SSCM) enables images of the cornea to be scanned at different depths (default layer separation 5 μm, minimum separation 1 μm)

and immediately viewed for diagnostic purposes. During a scan, the instrument locates the rear of the cornea (no signal back from the vitreous humour) and then steps forward in 5 μm steps until it reaches the front surface of the cornea (no signal back from the tear layer). This cycle is repeated 3 times during a 20 s scan which provides about 350 images.

This chapter is organised as follows: the second section covers existing work on corneal image enhancement and image registration. Also, we describe the challenges associated with processing corneal images. In the third section we describe our methods for image enhancement and image registration. Some concluding remarks and suggestions for future work are presented in the last section.

BACKGROUND AND CHALLENGES

Image Enhancement

Researchers frequently use image enhancement to improve the characterics of an image and its visual appearence. For example in medical applications very small details can play a critical role in the diagnosis and hence effective treatment of diseases. It is also often useful to highlight particular features while displaying medical images. This section reviews some papers which are relevant to the enhancement of the corneal images obtained using confocal microscopy.

Table 2. Characteristics of the 20x probe and 40x probe lenses

Feature	20x probe	40x probe
Working distance	12 mm (through air)	2 mm (through gel)
Inspected field	460 x 690 μm	460 x 345 μm
Image size	384 x 576 pixel	768 x 576 pixel
Magnification	250 X	500 X
Minimum scan step	1 mm	1 mm

The authors of (Scarpa, Fiorin et al. 2007) used a Confoscan4 confocal microscope with 40x lenses to provide a sequences of images. To decrease the problem of bluring they used a high-pass frequency filter and a sinusoidal transformation curve to adjust the intensity.

The authors of (Tinena, Sobrevilla et al. 2009) incorporated a CCD camera in a confocal microscope to capture images of the corneal endothelium, which were displayed on a monitor for viewing. They introduced a computer system for the automatic segmentation based on the watershed algorithm. The markers needed were introduced using fuzzy techniques. Their proposed algorithm included the following steps:

- **Fuzzy Markers generation:** The authors used a fuzzy if-then rules algorithm to obtain a set of markers and their confidences (area of the marker and the greatest contrast).
- **Fuzzy Watershed-based segmentation:** This process provided cell-sized segments and a reference for each region, which consisted of the centroid of the cell and its degree of confidence.
- **Fuzzy-based cells consistency measure:** They modified the confidence degrees of the remaining cells using fuzzy concentrations and/or dilations of the initial degrees of confidence. Cells that have pentagonal, hexagonal or heptagonal shapes are kept and other cells are deleted. The cells remaining are the final results of the segmentation.

The authors concluded that fuzzy techniques are a powerful approach to deal with the problems of corneal images (noise and poor quality).

The authors of (Xujia, Shishuang et al. 2008) developed an enhancement method based on 2D Empirical Mode Decomposition. Using this method, they segregated the image into low and high frequency information content. The high frequency content was expanded and then added to the low frequency content to enhance the image. This method was shown to be more powerful than traditional enhancement techniques (e.g., histogram transform or histogram equalization), and the detailed information content of the image after enhancement is more definite and more structured.

The authors of (Haiguang, Li et al. 1994) worked with low-field-strength MRI (magnetic resonance imaging) images and proposed an algorithm whose purpose is noise smoothing at a given filtering level while introducing very little modification of the structures in the original image. The proposed algorithm generated four sub-images by low-pass filtering the image along the four main directions, and the final filtered image is a weighted (through the differences between these sub-images and the original image) combination of them. This algorithm was implemented using array processors and was found to be effective in reducing the image noise and enhancing edges.

The authors of (Yang, Su et al.) proposed a method of image enhancement applying a wavelet transform (WT) (of unspecied type) to the original image and then the Haar transform (HT) to the detail sub-images. The authors' algorithm included five steps: decomposition of the images with WT; decomposition of all high frequency sub-images with HT; enhancement of the high frequency fields using a soft thresold function; enhancement of high frequency coefficients by applying different weight values in different sub-images. The final step applied the inverse HT and inverse WT to obtain the enhanced image.

The authors of (Khan, Khan et al. 2007) proposed an ehancement method using a non-subsampled wavelet pyramid decomposition of low-pass region. They preprocessed images using a low-pass filter to remove illumination pattern and an adaptive median filter to remove thermal noise. The authors concluded that this method was affective, especially in low constrast and non-uniformly illuminated images.

Image Registration

In general, image registration is the process of overlaying two or more images of a scene which are not aligned (in position, orientation or scale) because they were taken from different viewpoints, at different times or using different sensors. The process of registering two images is based on finding common features in the images which will enable a transformation to be found to properly align the images. The image registration procedure involves four basic steps: detecting prominent features, matching features between images, generating the mapping function, and transforming and re-sampling one of the images.

The authors of (Scarpa, Fiorin et al. 2007) applied an automatic registration procedure on pairs of images, which were obtianed from a Confoscan4 confocal microscope with 40x lens. The process applied included the following steps:

- The first image is processed to select the two regions of interest (ROIs) with the highest brightness and contrast.
- For each ROI the normalized correlation method is applied to compute the shift along the x and y axes which gives the best match in the second image. Assuming that the shift that occurs between two successive images is small, this begins by matching the ROI within only a small patch of the second image which is enlarged until a good match is found.
- The registration is assumed successful, if the shifts in the x and y directions for the first ROI are equal in magnitude to the shifts in the x and y directions for the second ROI, with an error tolerance of ± 1 pixel.

The registration procedure failed for 3% of the images and the average difference between automatic and manual registrations was 1.5 pixels.

The authors of (Ito, Aoki et al. 2008) registered enhanced dental radiograph images, using a Phase-Only Correlation (POC) method and corrected nonlinear distortions based on a Thin-Plate Spline model. The authors' fully automatic algorithm to determine matching points between two images includes the folowing steps:

- Images are enhanced using local area contrast enhancement and morphological filtering.
- The displacement and rotation between the two enhanced images are estimated using POC techniques and then normalised in a new registered image.
- Harris corner detection and POC are used to find landmark points in the reference and registered images which are used to estimate the parameters of a thin plate spline model which corrects the non-linear distortion in the registered image.

The authors compared their proposed algorithm with three other algorithms and found it more useful for matching poor and blurred dental radiographs.

Challenges Associated with Processing Corneal Image

There are significant practical difficulties in achieving useful models of the human eye from a confocal microscope. There are two main problems areas.

The first is the quality of the individual images which are subject to blurring, non uniform illumination and noise. Many of the images are very dark, making the characteristics of the cornea's cells unclear, because it is not possible to adjust the illumination level during a single scan of the cornea so it is optimum for all the layers. The spherical shapes of the corneal layers, causes non uniform reflection of illumination light from the different areas of the cornea, and the different attenuation of light along the different paths of illumination.

The second problem is caused by movements of the eye during the scanning process. Respiration, cardiac pulse and other factors cause images of adjacent layers to be displaced laterally and may cause images in the capture sequence to be out of sequence in terms of depth. This also means that the difference in depth between captured layers is not necessarily uniform or the same as the instrument setting. The amount of movement varies from patient to patient and from scan to scan.

The confoscan4 microscope can optionally incorporate a Z-Ring detachable contact element, which increases image stability to give a more precise location along the z-axis (NIDEKtechnologies 2010). The resulting images are still misaligned because of shifts in the x-y plane.

SUMMARY

From the previous survey it becomes obvious that corneal imaging is still an emerging science, with limited algorithms being applied for enhancement and registration. The techniques discussed in this section have been applied to different images but not to a bench mark dataset. In addition, different groups seem to be focusing on different aspects of corneal imaging (enhancement only or registration only) and no systematic or modular approach to tackle the general challenges in corneal imaging is presented. Some of these issues will be addressed in the next section.

PROCESSING CORNEAL IMAGES

Cornea Image Data

In this work, we have used data from (BioImLab 2007). This data consists of 3 folders (subject1, subject2 and subject3). Each folder contains a

Figure 5. Proposed algorithm for image enhancement based on DCT

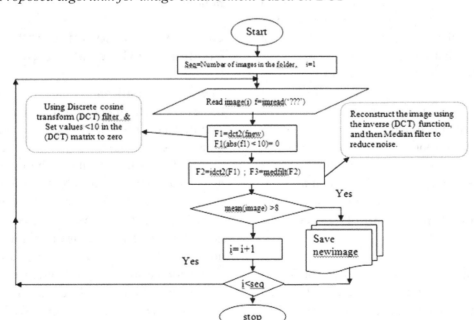

full sequence of images from the epithelium to the endothelium layers (144, 85 and 127 respectively). These images were obtained using a ConfoScan 4 confocal microscope and image field of 460×350μm at 40X magnification. Every image is of size 768×576 and is saved in JPEG format.

Enhancement of the Images

Image enhancement aims to enhance the visual characteristics of the image with respect to the viewers and/or the imaging application. Initially, we worked on the BioImLab data to reduce the blur present in the images using several general methods of enhancement (high-pass filter, low-pass filter, Gaussian filter, dilation, Erosion), but without especially good results.

Two other techniques were therefore investigated, the first applying Discrete Cosine Transform (DCT) based filtering to reduce illumination variation followed by median filtering to reduce noise. The steps of this proposed enhancement algorithm are displayed in Figure 5.

The second technique was fuzzy image enhancement, which is one of the well-established tools used for the contrast enhancement of digital images. For this work, we have implemented the image J plugin (Alestra 2008). Three 3 membership sets are defined: Dark, Gray and Bright. These sets and the membership functions are defined based on the Fuzzy set theory, which is a modification of the conventional (crisp) set theory. It allows every pixel to be either a full member of a set or a partial member of two or more sets with varying degrees of membership for every set. There are different methods for determining the membership values such as: Subjective evaluation and elicitation, Ad-hoc forms, Converted frequencies or probabilities, Physical measurements or Learning and adaption (e.g. using neural networks). For more information please refer to (Daley 1999). For this work, the membership values were determined interactively using ad-hoc observations.

Figure 6 shows some representative examples of the image enhancement results using both

Figure 6. Shows some representative examples of the image enhancement results using DCT and fuzzy methods

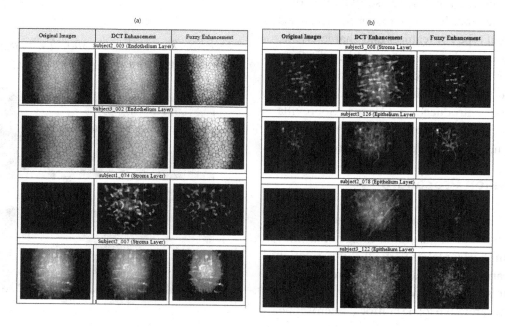

Figure 7. Two samples showing correct and incorrect matching points found when using the SIFT algorithm

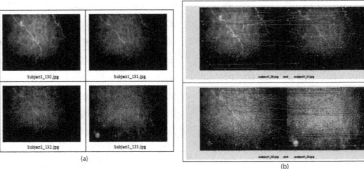

DCT and Fuzzy techniques. From Figure 6, it can be seen that most of the corneal images are blurred, have non uniform luminosity, are low in contrast and have peripheral regions darker than the central regions. Also, the layers of the cornea consist of different types of cells with different features and visual textures. This would make it seem unlikely that a single algorithm will achieve optimum improvement of the observations and clarify the appearance of all the layers of the cornea. A preliminary enhancement step can be useful, because the accuracy of the classification and registration may depend on this operation, and the enhanced images are useful in their own right. On the other hand each layer of the cornea has different types of features and characteristic, whose further enhancement could benefit from algorithms that use differents types of methods specific to the features in each layer. The fuzzy if-then rules introduced in (Tinena, Sobrevilla et al. 2009) could be useful for the Endothelium layer, but not be as efficient for the Stroma layer.

Registration Procedure

In the general sense, image registration is a process that aims to align digital images that may suffer from horizontal or vertical shifts, rotation or scale variations. There are factors (respiration, cardiac pulse, … etc.) which cause movements of the cornea relative to the confocal microscope.

This means that image alignment may change and shifts in the x and y directions are necessary to form registered images. In order to begin to compensate for these movements, we have used the Scale Invariant Feature Transforms (SIFT) algorithm[1] (A. Vedaldi ; Lowe January 5, 2004) to obtain matching points in pair of images and hence to estimate the shifts along the x and y axes. In the results of this work two particular sources of error were present: spuriously matched points and duplicate matched points. These are present in the samples of results shown in Figure 7.

Figure 8. Two examples showing results from applying algorithm_1

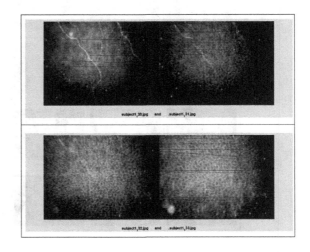

To remove these sources of error, two different approaches were investigated and these are summarized as follows:

1. Algorithm_1, operates on the array of descriptors (128 × number of point matches) between image A (descriptor_a) and image B (descriptor_b), which the SIFT procedure provides.

    ◦ *Normalize the descriptor_a and descriptor_b, to provide descriptor_a1(128,N) and descriptor_b1(128,M).*

    ◦ *Calculate the squared distances between two sets of points (point matches), which is an array of dimensions (N, M).*

    ◦ *Extract the number of the smallest values in the resulting distance matrix, by finding the smallest values in each column.*

    ◦ *Calculate the shift values along x- and y-axis (Δx and Δy) for each extracted matching point.*

    ◦ *Calculate the median values of Δx and Δy to estimate the magnitude of the shift between the pair of images.*

    This algorithm provides a satisfactory elimination of the inaccurate point matches, but has

Figure 10. Three examples showing results from applying Algorithm_2

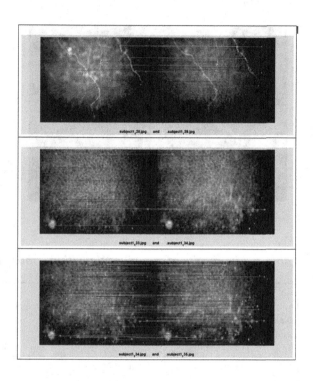

the disadvantage that it takes from 47 s to 52 s to register a pair of images, such as those shown in Figure 8, on a 2.93 GHz Intel® Core™ Duo PC with 4 GB of memory with Windows Vista Enterprise operating system.

Figure 9. Algorithm_2 for registering corneal images

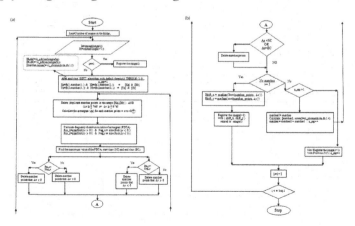

Figure 11. Superimposed set of registered adjacent images in the original sequence from the confocal microscope.

2. Algorithm_2 is summarised in the flowchart shown in Figure 9. This algorithm provides more accurate results than Algorithm_1, and also take less time (from 9 s to 11 s) to register a pair of images, as shown in Figure 10.

The algorithms attempt to register pairs of adjacent images in the given sequence, providing the shifts along x and y axes. The matching process sometimes fails because movement may have caused unrelated images to be adjacent in the sequence and in this case the shifts along the x and y axes are set to zero. An example of the

Table 3. The results of applying Algorithm_2 based on the SIFT algorithm

Name of first image	Name of second image	Shifts between adjacent images		Name of Image	Shifts between first image and subsequent images		
		Shift x	Shift y		Shift x	Shift y	
subject1_124.jpg	subject1_125.jpg	1	1	subject1_124.jpg	0	0	
subject1_125.jpg	subject1_126.jpg	-1	1	subject1_125.jpg	1	1	
subject1_126.jpg	subject1_127.jpg	-7	1	subject1_126.jpg	0	2	
subject1_127.jpg	subject1_128.jpg	-3	2	subject1_127.jpg	-7	3	
subject1_128.jpg	subject1_129.jpg	-4	2	subject1_128.jpg	-10	5	
subject1_129.jpg	subject1_130.jpg	-2	-1	subject1_129.jpg	-14	7	
subject1_130.jpg	subject1_131.jpg	-3	-4	subject1_130.jpg	-16	6	
subject1_131.jpg	subject1_132.jpg	-2	-2	subject1_131.jpg	-19	2	
subject1_132.jpg	subject1_133.jpg	2	-1	subject1_132.jpg	-21	0	
subject1_133.jpg	subject1_134.jpg	-2	-1	subject1_133.jpg	-19	-1	
subject1_134.jpg	subject1_135.jpg	-4	2	subject1_134.jpg	-21	-2	
subject1_135.jpg	subject1_136.jpg	-2	-2	subject1_135.jpg	-25	0	
subject1_136.jpg	subject1_137.jpg	1	-3	subject1_136.jpg	-27	-2	
subject1_137jpg	subject1_138.jpg	-1	1	subject1_137jpg	-26	-5	
subject1_138.jpg	subject1_139.jpg	-8	5	subject1_138.jpg	-27	-4	
subject1_139.jpg	subject1_140.jpg	-5	2	subject1_139.jpg	-35	1	
subject1_140.jpg	subject1_141.jpg	-2	-1	subject1_140.jpg	-40	3	
subject1_141.jpg	subject1_142.jpg	-3	-1	subject1_141.jpg	-42	2	
subject1_142.jpg	subject1_143.jpg	-4	2	subject1_142.jpg	-45	1	
subject1_143.jpg	subject1_144.jpg	-8	-5	subject1_143.jpg	-49	3	
				subject1_144.jpg	-57	-2	

registration of a short sequence of images is shown in Figure 11, which shows the superimposed images and Table 3, which shows the shifts in x and y between pairs of adjacent images and the shifts in x and y between the first image in the sequence and the subsequent images, which is used to form the image shown in Figure 11.

Validation

Verification of Algorithm_2 was carried out by taking four different types of corneal images and then adding noise and known random shifts along the x and y axes for testing registration. Two dif-

ferent amounts of Gaussian noise (10% and 20% of maximum image intensity) were added to the images with random shifts along the x and y axes. Sample results are shown in Table 4 and Figure 12 for 10% added noise and in Table 5 and Figure 13 for 20% added noise.

The results are encouraging and show that the method works well when the two images are from the same layer. False matching between images from different layers would be avoided by performing a prior classification of the images. Discrepancies between the estimated and true shifts are indicated in Tables 4 and 5 by shading and amount to 2 out of 21 cases with 10% noise

Table 4. Comparison of estimated shifts, using Algorithm_2 applied to 21 simulated images, with the actual shifts applied. Images in the second column are first column enhanced images with 10% of Gaussian added noise and with random shifts along the x and y axes.

First Image	Second Image	Estimated shifts		Actual shifts	
		Shift x	Shift y	Shift x	Shift y
'subject1_124a.jpg'	'subject1_124b.jpg'	-11	-6	-11	-6
'subject1_125a.jpg'	'subject1_125b.jpg'	10	-20	10	-20
'subject1_126a.jpg'	'subject1_126b.jpg'	5	26	5	26
'subject1_127a.jpg'	'subject1_127b.jpg'	-2	30	-2	30
'subject1_128a.jpg'	'subject1_128b.jpg'	-14	-42	-14	-41
'subject1_129a.jpg'	'subject1_129b.jpg'	34	11	34	11
'subject1_130a.jpg'	'subject1_130b.jpg'	-24	2	-24	2
'subject1_131a.jpg'	'subject1_131b.jpg'	-5	14	-5	14
'subject1_132a.jpg'	'subject1_132b.jpg'	32	8	32	8
'subject1_133a.jpg'	'subject1_133b.jpg'	-35	-7	-35	-7
'subject1_134a.jpg'	'subject1_134b.jpg'	-8	5	-8	6
'subject1_135a.jpg'	'subject1_135b.jpg'	9	3	9	3
'subject1_136a.jpg'	'subject1_136b.jpg'	-1	-16	-1	-16
'subject1_137a.jpg'	'subject1_137b.jpg'	-9	-5	-9	-5
'subject1_138a.jpg'	'subject1_138b.jpg'	-11	4	-11	4
'subject1_139a.jpg'	'subject1_139b.jpg'	-27	-26	-27	-26
'subject1_140a.jpg'	'subject1_140b.jpg'	15	-3	15	-3
'subject1_141a.jpg'	'subject1_141b.jpg'	-43	6	-43	6
'subject1_142a.jpg'	'subject1_142b.jpg'	-4	-14	-4	-14
'subject1_143a.jpg'	'subject1_143b.jpg'	-6	13	-6	13
'subject1_144a.jpg'	'subject1_144b.jpg'	7	17	7	17

Figure 12. Examples of image registration with Gaussian noise at 10% of maximum intensity

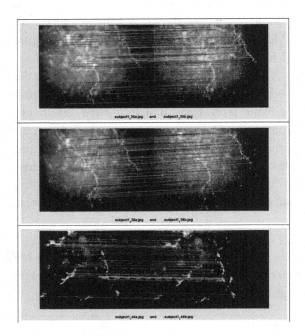

Figure 13. Examples of image registration with Gaussian noise at 20% of maximum intensity

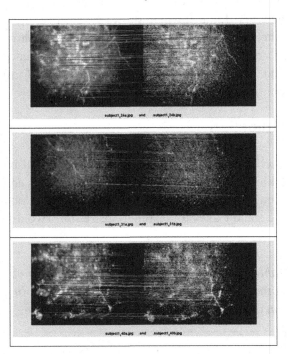

and 7 out of 21 cases with 20% noise. In all cases the discrepancies are no more than ±1 pixel which is still an acceptable level since the objects to be matched in practical cases may change shape slightly between images. These results indicate that the registration method can tolerate the presence of significant amounts of noise, which is important because the images of some layers will be darker and contain a higher percentage of noise than others. Enhancement of images during pre-processing may be able to reduce the noise but is unlikely to eliminate it.

CONCLUSION AND FUTURE WORK

In this work, we have investigated novel applications in the fields of corneal image enhancement and registration. We hope this work will lead to more interesting developments in corneal imaging. Our wider aim is to work towards creating

automated computer systems for tracing nerves and to providing 3D models from corneal images. This is an emerging area of research, as indicated by (Scarpa, Fiorin et al. 2007),(A Zhivov 2009). Nerve tracing will potentially provide clinical information relating to changes of nerves in the cornea. From inspection of the epithelium images, the nerves in the cornea have some similarities with retinal nerves. For this reason, we are going to investigate vessel tracking algorithms applied to retinal images, and apply them to corneal nerve tracing. Systems such as these require more than one stage before they can be accomplished, such as layer identification, efficient enhancement and efficient registration.

The aim of layer identification is to automatically process the corneal image to identify its layer (Epithelium, Stroma or endothelium). Texture analysis methods, neural networks and other artificial intelligence approaches will be

Table 5. Comparison of estimated shifts, using Algorithm_2 applied to 21 simulated images, with the actual shifts applied. Images in the second column are first column enhanced images with 20% of Gaussian added noise and with random shifts along the x and y axes.

Name of first Image	Name of second Image	The shifts applied to the first image to give the second		The calculated shifts	
		Shift X	Shift Y	Shift X	Shift Y
'subject1_124a.jpg'	'subject1_124b.jpg'	-33	-9	-33	-9
'subject1_125a.jpg'	'subject1_125b.jpg'	-16	-28	-16	-28
'subject1_126a.jpg'	'subject1_126b.jpg'	5	15	5	15
'subject1_127a.jpg'	'subject1_127b.jpg'	-14	-11	-14	-11
'subject1_128a.jpg'	'subject1_128b.jpg'	14	13	14	14
'subject1_129a.jpg'	'subject1_129b.jpg'	51	1	52	1
'subject1_130a.jpg'	'subject1_130b.jpg'	-28	-15	-28	-14
'subject1_131a.jpg'	'subject1_131b.jpg'	-14	27	-14	27
'subject1_132a.jpg'	'subject1_132b.jpg'	-28	-28	-28	-28
'subject1_133a.jpg'	'subject1_133b.jpg'	59	-3	59	-3
'subject1_134a.jpg'	'subject1_134b.jpg'	24	38	24	39
'subject1_135a.jpg'	'subject1_135b.jpg'	-1	8	-1	8
'subject1_136a.jpg'	'subject1_136b.jpg'	11	12	11	11
'subject1_137a.jpg'	'subject1_137b.jpg'	-9	7	-9	7
'subject1_138a.jpg'	'subject1_138b.jpg'	28	27	28	27
'subject1_139a.jpg'	'subject1_139b.jpg'	-11	1	-11	1
'subject1_140a.jpg'	'subject1_140b.jpg'	-5	18	-5	18
'subject1_141a.jpg'	'subject1_141b.jpg'	2	-41	2	-42
'subject1_142a.jpg'	'subject1_142b.jpg'	-12	8	-12	8
'subject1_143a.jpg'	'subject1_143b.jpg'	6	-20	6	-19
'subject1_144a.jpg'	'subject1_144b.jpg'	-19	-28	-19	-28

investigated to study their possible uses in the classification of the corneal images.

For the efficient enhancement we are suggesting to use two different models of enhancement. The first enhancement model is more general and will be used to clarify the images to allow automatic classification of the original images to identify their layers. After classification, we will use other methods of enhancement related to the natures of the layers because in each layer we have different image features. The output of enhancement and smoothing algorithms should not change the structure of the images, or introduce artefacts that could affect diagnosis. Fuzzy

image enhancement could be used to enhance the images of endothelium layer. A logarithm based high-pass and low-pass filter, or algorithms based on the Discrete Cosine Transform could be used for enhancing other layers. However, staff from Glasgow Eye Hospital will be asked to confirm which enhancement is the most appropriate to use with each layer.

The aim of efficient registration is to obtain the best set of descriptors with minimum processing time to register the images. The SIFT and other algorithms (PCA-SIFT "Principal Components Analysis applied to SIFT descriptors", GLOH "Gradient Location and Orientation Histogram",

Harris affine region detector, Harris-Laplace detector, KLT (Kanade-Lucas-Tomasi) ... etc.) will be investigated for this stage.

ACKNOWLEDGMENT

F. Scarpa, D. Fiorin, A. Ruggeri. "In Vivo Three-Dimensional Reconstruction of the Cornea from Confocal Microscopy Images". Proc. 29th IEEE EMBS Annual International Conference, City Internationale, Lyon, France, Aug 23-26, pp. 747-750, 2007.

REFERENCES

Alestra, S. (2008). *Fuzzy contrast enhancement.* Retrieved from http://svg.dmi.unict.it/iplab/imagej/Plugins/Fuzzy%20Image%20Processing/Fuzzy%20Contrast%20Enhancement/FuzzyContrastEnhancement.htm

BioImLab. (2007). *Biomedical imaging laboratory (BioImLab) is part of the Department of Information Engineering (DEI) of the University of Padova, Italy.* Retrieved from http://bioimlab.dei.unipd.it/index.html

Daley, W.D. (1999). *Computational approaches for decision support in real time broiler processing.* Georgia Institute of Technology: A Unit of the University System of Georgia.

Haiguang, C., & Li, A. (1994). A fast filtering algorithm for image enhancement. *IEEE Transactions on Medical Imaging, 13*(3), 557–564. doi:10.1109/42.310887

Institute, N. E. (2010). Facts about the cornea and corneal disease. Retrieved from http://www.nei.nih.gov/health/cornealdisease/

Ito, K., Aoki, T., et al. (2008). *Medical image registration using phase-only correlation for distorted dental radiographs.* 19th International Conference on Pattern Recognition, 2008. ICPR 2008. Erie, J.C., McLaren, A.W. & Patel, S.V. (2009). Perspectives confocal microscopy in ophthalmology. Elsevier.

Kaufman, H. E., Barron, B. A., McDonald, M. B., & Kaufman, S. C. (Eds.). (2000). *Companion handbook to the Cornea.*

Khan, M. A., & Khan, M. K. (2007). *Endothelial cell image enhancement using non-subsampled image pyramid* (pp. 1057–1062).

Lowe, D.G. (2004). Distinctive image features from scale-invariant keypoints. *International Journal of Computer Vision.*

MedicineNet. (2010, March 9, 2010). *Definition of cornea.* Retrieved from http://www.medterms.com/script/main/art.asp?articlekey=7248

Reinstein, D. Z., & Archer, T. J. (2008). Epithelial thickness in the normal cornea: Three-dimensional display with very high frequency ultrasound. *NIH Public Access, 24,* 571.

Scarpa, F., Fiorin, D., et al. (2007). *In vivo three-dimensional reconstruction of the cornea from confocal microscopy images.* 29th Annual International Conference of the IEEE Engineering in Medicine and Biology Society, 2007. EMBS 2007.

Semwogerere, D.W. (2005). *Confocal microscopy.*

Tinena, F., Sobrevilla, P., et al. (2009). *On quality assessment of corneal endothelium and its possibility to be used for surgical corneal transplantation.* IEEE International Conference on Fuzzy Systems, 2009. FUZZ-IEEE 2009.

Xujia, Q., Shishuang, L., et al. (2008). *Medical image enhancement method based on 2D empirical mode decomposition.* The 2nd International Conference on Bioinformatics and Biomedical Engineering, 2008. ICBBE 2008.

Yang, Y., & Su, Z. (2010). Medical image enhancement algorithm based on wavelet transform. *Electronics Letters*, *46*(2), 120–121. doi:10.1049/el.2010.2063

Zhivov, A., Stachs, O. S., Stave, J., & Guthoff, R. F. (2009). In vivo three-dimensional confocal laser scanning microscopy of corneal surface and epithelium. *The British Journal of Ophthalmology*, *93*, 667–672. doi:10.1136/bjo.2008.137430

ENDNOTE

[1]   Implemented by Andrea Vedaldi, Brian Fulkerson

Chapter 17
Parameter Based Multi–Objective Optimization of Video CODECs

F. Al-Abri
Loughborough University, UK

E.A. Edirisinghe
Loughborough University, UK

C. Grecos
University of the West of Scotland, UK

ABSTRACT

This chapter presents a generalised framework for multi-objective optimisation of video CODECs for use in off-line, on-demand applications. In particular, an optimization scheme is proposed to determine the optimum coding parameters for a H.264 AVC video codec in a memory and bandwidth constrained environment, which minimises codec complexity and video distortion. The encoding/decoding parameters that have a significant impact on the performance of the codec are initially obtained through experimental analysis. A mathematical formulation by means of regression is subsequently used to associate these parameters with the relevant objectives and define a Multi-Objective Optimization (MOO) problem. Solutions to the optimization problem are reached through a Non-dominated Sorting Genetic Algorithm (NSGA-II). It is shown that the proposed framework is flexible on the number of objectives that can jointly be optimized. Furthermore, any of the objectives can be included as constraints depending on the requirements of the services to be supported. Practical use of the proposed framework is described using a case study that involves video content transmission to a mobile hand.

INTRODUCTION

Recently a significant amount of research effort has been focused on optimizing video CODECs, especially within the standardization activities of MPEG/JVT. This is due to the high demand of applications requiring efficient on-demand and real-time video coding, supported by the effective usage of capture, processing and display devices

DOI: 10.4018/978-1-60960-477-6.ch017

and transmissions mediums, which are practically operational under varying constraints.

To this date many optimization methods have been proposed in literature. They can be broadly classified into two categories, namely,

- Algorithm-based optimizations
- Parameter-based optimizations

The algorithm-based optimization methods focus on the direct performance optimization of a given algorithm. Alternatively, parameter-based optimization methods optimize given objectives through the optimal selection of coding parameters. Whilst the first approach focuses on the optimisation of a CODEC when the algorithms are being developed, the latter enables the optimal operation of a standardized algorithm by the selection of optimal sets of parameters.

With the standardization of H.264, the video optimization research has mainly been focused on this standard. Due to the vast amount of effort that has been put into the algorithmic optimization of H.264 during it's international standardization, the need at present is a unified framework that is capable of selecting the vast number of coding parameters one can set and specify to obtain the CODEC's optimum performance under given constraints. Therefore the focus of optimisation subsequent to the standardization activity of any video CODEC is parameter based optimisation.

This chapter presents a multi-objective optimisation strategy for the parameter-based optimisation of H.264 AVC (Advanced Video Coding) standard. Amongst a detailed experimental analysis of the proposed framework the chapter provides a case study that demonstrates the practical use of the approach.

Background

H.264/AVC is the latest block-oriented motion-compensation-based coding standard developed by the ITU-T Video Coding Experts Group

(VCEG) in collaboration with the ISO/IEC Moving Picture Experts Group (MPEG), a partnership known as the Joint Video Team (JVT). The standard defines the syntax of an encoded video bitstream together with the method of decoding this bitstream, rather than explicitly defining a CODEC as an encoder-decoder pair. H.264/AVC represents a significant leap in video compression technology, with typically a 50% reduction of the average bit rate for a given video quality, compared to MPEG–2 and about a 30% reduction compared with MPEG–4 Part 2 (Ostermann, J., et al 2004) Besides the classical application areas of videoconferencing and broadcasting of TV content (satellite, cable, and terrestrial), the improved compression capability of H.264/MPEG4-AVC has enabled new services and opened new markets and opportunities for the industry, such as Mobile TV and transmission and storage of HD content (Marpe, D., et al 2006).

H.264 AVC (Wiegand et al., 2003 & I.T.U.T.S, 2005) provides high coding efficiency through added features and functionality. However, such features and functionality also require additional resource consumption, i.e., computational complexity and memory usage. Specifically the cost effectiveness of a H.264 CODEC is shaped directly by the computational complexity of the coding algorithms (such as motion estimation and compensation, transform and entropy coding) and their memory requirements. The efficiency of a coding algorithm is further dependent on various parameters used in defining the operational status of the CODEC at a given time. A significant number of coding options are available through the selection of various combinations of a large number of coding parameters. Therefore an important problem that needs solving can be hypothesized as follows:

"Given a video sequence, what combination of coding parameters should be used so as to achieve the optimum performance of the CODEC?"

The choice of the right parameter set is of utmost importance. The parameter-based optimization of a H.264 video CODEC can provide a solution for this problem. In the literature of video coding there have been several attempts to utilise joint or multi-objective optimization to enhance the performance of a codec in different aspects related either to the final outcome of the compression or to the process of coding or decoding itself. However these have been limited by their scope, i.e. the optimisation strategies have been limited to optimising either two or maximum three objectives, following approaches specific to the objectives. These approaches are reviewed below.

- In (Vanam, R., et al, 2007) the best coding parameters were found using two fast algorithms, GBFOS-basic and the GBFOS-iterative algorithms for finding the H.264 parameter settings that take about 1% and 8%, respectively, of the number of tests required by an exhaustive search. Both algorithms performed within a maximum PSNR difference of 0.71 dB when using the same training and test data set, and were proved to be robust to changes in both test and training data sets. The paper considered encoding time and mean squared error (MSE) as measures for complexity and distortion. The GBFOS algorithm is used for obtaining parameter settings that result in D-C points that are close to optimal. An improvement to the above was proposed in (Vanam, R., et al, 2009) where two algorithms for finding additional parameter settings for the GBFOS-basic algorithm were presented. The two algorithms have a tradeoff between the number of encodings and the number of additional parameter settings generated. The additional parameter settings in the low-encode time region are useful for example an H.264 encoder on a cell phone, since it allows the encoder to choose a parameter setting that yields high-

er PSNR, while satisfying the encoding speed constraint. The algorithms for ASL videos were tested on both PocketPCs and Linux platforms. It was demonstrated that the additional parameter settings generated by CLSA and DPSPA improve the PSNR by up to 0.71 dB and 0.43 dB, respectively

- The research in (He, Z., Liang, Y et al, 2005) developed a power-rate-distortion (P-R-D) analysis framework for mobile video encoding and communication under energy constraints. Specifically the paper analyzed the encoding mechanism of typical video coding systems, and developed a parametric video encoding architecture which is scalable in computational complexity. Using the P-R-D model, given a power supply level and a bit rate, the power-scalable video encoder is able to find the best configuration of complexity control parameters to maximize the video quality. The P-R-D analysis established a theoretical basis and provided a practical guideline in system design and performance optimization for wireless video communication under energy constraints.

- A joint complexity-distortion optimization approach was proposed in (Li, S. et al 2009) for real-time H.264 video encoding under a power-constrained environment. The power constraints were translated to the encoding computational costs measured by the number of scaled CUs. The solved problems include the allocation and utilization of the computation resources. A computation allocation model (CAM) with virtual computation buffers was proposed to optimally allocate the computational resources to each video frame. By referring to the coding feature of previous frames, the temporal and spatial relationships were utilized to help decide coding parameters efficiently, while avoiding the time-consuming content analysis algorithms.

- For receiver aware optimization, a generic Rate-Distortion-Complexity model that can generate metadata necessary for MPEG-21 DIA based on the available receiver resources was proposed in (Van der Schaar, M. & Andreopoulos, Y., 2005). The complexity model was based on predicting the computational complexity of motion-compensated wavelet video coders by explicitly modelling the complexity involved in decoding a bitstream by a generic receiver. It was shown that the receivers can then negotiate with the media server/proxy, the transmission of a bitstream, having a desired complexity level based on their resource constraints. The network and receiver R-D-C adaptation were validated using a real-time multimedia streaming test-bed.

- A rate-distortion optimization framework was proposed in (Chakareski, Frossard 2006) which enables multiple senders to coordinate their packet transmission schedules, such that the average quality over all video clients is maximized. The target of the optimization function is to find the best transmission schedules for the video packets of each stream for a given available bandwidth on the shared channel. The performance of this framework was examined for two canonical problems in video streaming: bandwidth adaptation and packet loss adaptation. Significant gains in performance on the order of several dBs, both jointly over all the videos and also across the individual streams, were registered in each of the two scenarios under examination over a conventional system for distributed streaming which does not take into account the distortion information associated with the video packets.

- The research conducted by (Zhang, Y., et al. 2007) proposed a new end-to-end distortion model for R-D optimized coding mode selection. The model takes the end-to-end distortion as the sum of several separable distortion items but it keeps track of the error propagated distortion through recursive calculation. The algorithm claimed to achieve a number of advantages. Firstly, the physical explanation to the distortion items in the distortion model is obvious, e.g., the distortion caused by error concealment, which is helpful in the derivation of a proper Lagrange multiplier for R-D optimized mode selection in packet-loss environment. Secondly, the algorithm tackled the problem of subpixel MCP, because the separated distortion items can suppress the estimation errors caused by pixel averaging operations. Thirdly, the algorithm handled the deblocking filtering, as each distortion item in the algorithm need be calculated only once either before or after coding mode selection

- A linear programming (LP) technique that performs rate-distortion based optimization for constrained video coding was proposed in (Sermadevi, Hemami 2003). LP has been used to generate optimal solutions for the minmax problem under VBR buffer constraints and is applied to MPEG2. The results showed an improvement in average PSNR of approximately 0.5 dB - 2.0 dB over the initial TM5 solution, while maintaining an almost constant PSNR level.

- (Pu, W., Lu, Y., & Wu, F., 2006) analysed the power consumption and video quality separately and then posed a uniform optimization framework for the two objectives based on the investigation of the power consumption caused by the encoding and transmission. The study also proposed a complexity-configurable H.264 encoder based on the study of the FME algorithm in H.264 reference software. Using this model, given any quality level, an estimate of the appropriate encoding parameters can be created, so that the total power consumption is minimized. Or in the contrary, given any available power level, the proper encoding parameters are also calculated, so that the video quality is optimized

- In the performance and complexity joint optimization method used by (Jianning Zhang et al. 2003), the modules of the H.264 AVC codec were optimized respectively in the order of their computation proportion. Adaptive hexagon-based search (AHBS) and memory motion compensation (MMC) have been proposed for encoder and decoder. The aim of the optimization model in this research was to maximize the performance of coding system subject to some complexity constraint, or to maximize the complexity performance ratio without constraints. However AHBS achieved the best coding performance: 0.4dB improvement in low motion sequences, and 0.2dB improvement in high motion sequences compared with full search. The decoding process is accelerated up to 5 times totally with proposed MMC and other optimized techniques.

- Another study has tackled the problem of Multi-Objective Optimization for Video Streaming. The study solved three different aspects of the optimization problem. The first part introduced a pre-roll delay-distortion optimization (DDO) for uninterrupted content-adaptive video streaming over low capacity, constant bitrate (CBR) channels using MOO (Ozcelebi, Tekalp & Civanlar 2007). The second phase dealt with cross-layer optimized video rate adaptation and scheduling scheme to achieve maximum application layer" Quality-of-Service (QoS), maximum video throughput (video seconds per transmission slot), and QoS fairness for wireless video streaming (Ozcelebi et al. 2007). In the third part an end-to-end streaming system where the

end-users can view the video in mono or stereo mode depending on their display capabilities was implemented and MOO formulations were proposed. Simulations conducted using the IS-856 numerology over ITU Pedestrian A and Vehicular B channels showed that the proposed system without video rate adaptation achieves significant improvements over the state-of-the-art wireless schedulers in terms of user QoS and application-layer QoS fairness. These gains were achieved without sacrificing the overall system throughput; on the contrary, the proposed framework provided gains on the throughput as well when compared to the schedulers that are considered.

The above efforts prove the practical demand for a generalised framework for multi objective optimisation of video CODEC where optimised performance can be achieved under any number of specified objectives and constraints.

THEORY OF MULTI-OBJECTIVE OPTIMIZATION

As the name suggest, a multi-objective optimization problem (MOOP) has a number of objective functions which are to be minimized or maximized. Moreover, the problem usually has a number of constraints which any feasible solution must satisfy. The general form of the MOOP may be stated as appears in (Deb, Kalyanmoy 2001) as follows:

Equation 1

Minimize/Maximize $f_m(X)$, $m=1,2,\ldots,M$;

$$\text{subject to } g_j(X) \geq 0, \quad j=1,2,\ldots,J;$$
$$g_i\left(X\right) \geq 0, \quad j=1,2,\ldots,J;$$

$h_k(X)=0$, $k=1,2,\ldots,K$;

$$x_i^{(L)} \leq x_i \leq x_i^{(U)}, i=1,2,\ldots,n.$$

There are M objective functions $f(X) = (f_1(X), f_2(X), \ldots, f_M(X))^T$ considered in Equation 1. A solution X is a vector of n decision variables: $X = (x_1, x_2, \ldots, x_n)^T$. The terms $g_j(X)$ and $h_k(X)$ are called inequality and equality constraint functions respectively. The last set of constraints is called variable bounds, the value of each decision variable x_i is restricted within a range of lower $x_i^{(L)}$ and an upper $x_i^{(U)}$ bound. These bounds form a *decision variable space*, or simply the decision space. Each feasible solution is subjected to J inequality and K equality constraints. If any solution X does not meet all of $(J+K)$ constraints and all of the variable bounds, the solution is called an *infeasible solution*. On the other hand, if a solution X satisfies all constraints and variable bounds, it is known as a *feasible solution*. It is noted that the entire decision variable space need not be feasible. The set of all feasible solutions is called the *feasible region*.

It is important to note that not all feasible solutions in the feasible region are optimal. In other words, the feasible region not only contains optimal solutions, but also solutions that are not optimal. Figure 1 illustrates the feasible region and a number of feasible solutions of a two-objective optimization problem with two conflicting objectives. Assume that the target of this optimization problem is to maximize both objectives. It is obvious that solution 'a' is better than any one of 'b', 'c' and 'f' in terms of both objectives. Therefore 'b', 'c' and 'f' are feasible solutions but not optimal. On the other hand, solution 'd' has a smaller value for objective 1, but has a larger value for objective 2 than the point 'a'. When both objectives are equally important, none of these two solutions can be said to be better than the other with respect to both objectives. When this relationship exists between two solutions, they are called *non-dominated* solutions. There exist many such solutions (highlighted with green) in

the feasible region. For clarity, these solutions are joined with a dashed curve in the figure. All solutions lying on this curve are called *Pareto-optimal* solutions or *Pareto-optimal* set. The curve formed by joining these solutions is known as a *Pareto-optimal front (curve)*. In fact, the task in multi-objective optimization (MOO) is to find a set of *Pareto-optimal* solutions in the feasible region (Deb, Kalyanmoy 2001)

MULTI-OBJECTIVE OPTIMIZATIOPN FRAMEWORK FOR VIDEO CODECS

Following the theory of multi-objective optimisation presented above the framework for optimization of video CODECs can be described by the following three steps:

- **Step 1:** Comprehensive profiling experiments are carried out, both in the encoder and decoder, to determine the coding parameters that have a significant impact on each of the objectives/constraints, i.e. rate, distortion, memory utilization and computational cost.

Figure 1. Feasible region of a two objective optimization problem (Adapted from Zitzler, 1999)

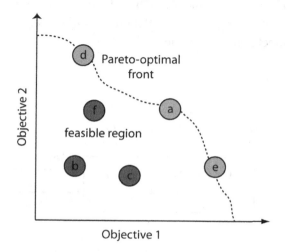

Figure 2. Proposed framework- A conceptual illustration

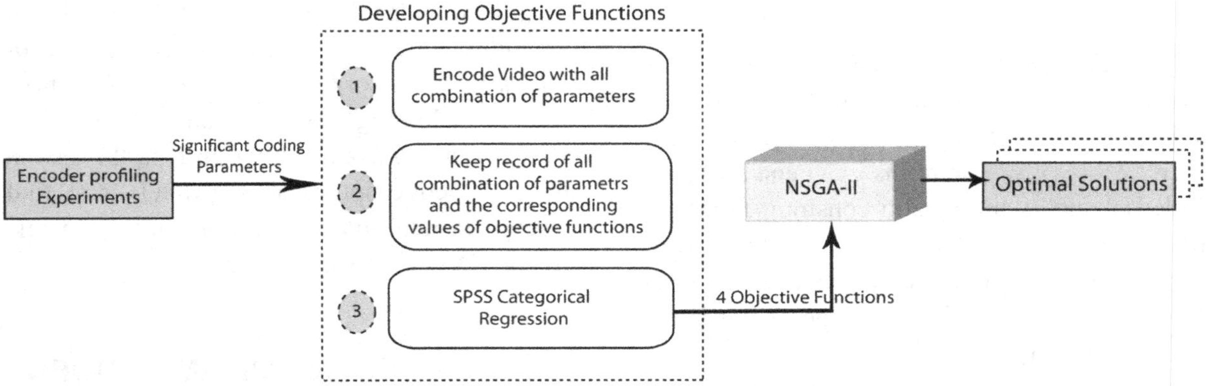

Developing Objective Functions

Encoder profiling Experiments → Significant Coding Parameters →

1. Encode Video with all combination of parameters
2. Keep record of all combination of parametrs and the corresponding values of objective functions
3. SPSS Categorical Regression

→ 4 Objective Functions → NSGA-II → Optimal Solutions

- **Step 2:** The objective function for each objective/constraint is developed using a suitable regression procedure.
- **Step 3:** The objective functions are used within a multi-objective optimization strategy that uses a genetic algorithm (GA) to produce optimal solutions.

These steps are explained in detail in section 4. Figure 2 illustrates a block diagram of the proposed framework.

Note that in a practical application the proposed framework assumes that a backward channel is available that can be used to notify the encoder about the decoder constraints, such as, limitations in decoder memory and computational power (see Figure 3). A typical web based video on-demand application using TCP/IP communication protocol meets this requirement.

PROFILING: DETERMINING THE SIGNIFICANT CODING PARAMETERS

To determine the coding parameters that have a significant influence on a H.264 CODEC's bit-rate, distortion, memory utilization and computational complexity a comprehensive analysis (i.e. CODEC profiling) is required for each of the four objectives in both the encoder and decoder sides. Since the computational complexity, bit rate and video quality performance greatly depend on the content of the source video, in this analysis 8 test sequences were chosen with distinct content and motion characteristics so that the results could reflect generality. The picture formats of the selected video clips include QCIF, CIF and SIF. Each sequence is in 4:2:0 sampling format and has 112 frames. "Claire" and "Paris" are conversational video sequences with simple motion of the foreground and fixed background. Moderate movements in

Figure 3. Application scenario

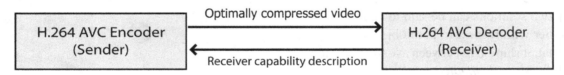

Optimally compressed video

H.264 AVC Encoder (Sender) — Receiver capability description — H.264 AVC Decoder (Receiver)

the foreground and slight movements in the background characterize the video sequences, "Foreman" and "Mobile". The sequences "Garden" and "Coastguard" show fast motion on both foreground and background regions. The most complicated motion characteristics are represented in "Football" and "Tennis" sequences. The above eight test sequences represent a wide range of videos with different properties and behaviours; from low to high detailed scenes, from moderate to high movement, from fixed to changing background. Therefore it is possible to systematically assess the performance of the proposed video CODEC through these test sequences.

In each profiling experiment a video sequence is coded with an exhaustive list of different combinations of coding parameters, shown in Table 1, leading to observing the affect each parameter has on a specific objective. The profiling experiments for the encoder and decoder are carried out separately, under four different categories as shown in Table 1.

Encoder Profiling Experiments

Computational Complexity analysis in the encoder is achieved by identifying coding parameters that directly affect the computational complexity of important sections/stages of the encoder such as, intra prediction, transform coding, quantization, motion estimation and compensation. The number of central processing unit (CPU) cycles required to

Table 1. The initial list of parameters used in profiling experiments

Parameter	Meaning
Resolution	Image width and height in luminance samples.
NumberReferenceFrames	Sets maximum number of references stored in buffer for motion estimation and compensation.
UseFME	Enable Fast motion estimation algorithms (0 disable, 1-3 enable).
SearchRange	Sets allowable search range for motion estimation.
RDOptimization	Enable rate distortion optimized mode decision.
Slicegroup	Number of slice group to be used.
IntraPeroid	Period of I-frames, i.e. frame will be coded using intra slices every IntraPeriod frames.
QP	Sets quantization parameter value.
InterSearch4x4	Enable 4 x 4 inter prediction & motion compensation.
InterSearch4x8	Enable 4 x 8 inter prediction & motion compensation.
InterSearch8x4	Enable 8 x 4 inter prediction & motion compensation.
InterSearch8x8	Enable 8 x 8 inter prediction & motion compensation.
InterSearch8x16	Enable 8 x 16 inter prediction & motion compensation.
InterSearch16x8	Enable 16 x 8 inter prediction & motion compensation.
InterSearch16x16	Enable 16 x 16 inter prediction & motion compensation.
Intra4x4ParDisable	Disable intra 4 x 4 vertical & horizontal prediction modes.
Intra4x4DiagDisable	Disable intra 4 x 4 diagonal down-Left and diagonal down-right prediction modes.
Intra4x4DirDisable	Disable intra 4 x 4 vertical right, vertical left, horizontal down, and horizontal up prediction modes.
Intra16x16ParDisable	Disable intra 16 x 16 vertical & horizontal prediction modes
Intra16x16PlaneDisable	Disable intra 16 x 16 plane prediction mode
DisableThresholding	Disable threshold of quantized coefficients that is used for discarding expensive coefficients. If after quantization there are only few small nonzero coefficients in a macroblock and the cost of coding these coefficients exceeds a fixed threshold, these coefficients are forced to zero

perform key encoding functions is estimated using Intel's VTune Performance Analyzer (Intel(R) VTune(TM) Performance Analyzer 9.0., 2008), which enables the collection of run-time data indicating the number of cycles consumed by each function of the H.264 CODEC.

Memory Utilization in the encoder can be analysed through the study of dynamic memory allocated to variable encoding parameters such as the number of reference frames, picture resolution etc. The "malloc" and "calloc" functions within the 'C' program code can be used for identifying encoding parameters that significantly impact dynamic memory allocation (e.g., those that results in over 10% share).

For a given video sequence, *rate* (i.e., bit rate, measured in Kbits/s) and *distortion* (i.e., quality, PSNR measured in dB) are sensitive to coding parameters. Modifying coding parameters such as using optional coding modes or choosing different R-D optimization algorithms will affect the output of the encoder, resulting in different levels of bit rate and visual quality. Experiments can be performed in order to find out those coding parameters that can significantly influence the bit rate ($\geq$ 10%) and quality ($\geq$ 0.2 dB).

Decoder Profiling Analysis

Analysis of the decoder is limited to finding out the decoder parameters that have a significant effect on only the decoder's computational complexity and memory utilization. Note that the decoder parameters have no impact on rate and distortion.

Computational Complexity of the decoder is analysed using the same method used at the encoder end. The major functionality of the decoder includes the de-blocking filter, motion compensation and entropy decoding.

Memory utilization at the decoder can be analysed in a manner similar to that described for use within the encoder.

The results of the above profiling experiments lead to the identification of coding parameters that have significant impact on computational complexity, memory utilisation, rate and distortion, both at encoder and decoder ends (Li, 2007). These are tabulated in Table 2.

In this analysis an ideal network is assumed where the network does not introduce any data loss or delay. Based on this observation, we assume that the quality and bit rate of the video received by the decoder are the same as that at the encoder output. In other words, we deduce that the coding parameters affecting the bit rate

Table 2. Significant parameters (Li, 2007)

	Encoder			Decoder	
	Computational Complexity	*Memory Utilisation*	*Rate/Distortion*	*Computational Complexity*	*Memory Utilisation*
Resolution	X	X	X	X	X
NumberReferenceFrames	X	X			X
Use FME	X	X			
SearchRange	X	X			
RDOptimisation	X				
SliceGroup		X			X
IntraPeriod			X	X	
QP			X		
DisableThresholding			X		

and reconstructed video quality on both encoder and decoder are identical.

After the identification of significant coding parameters objective functions are determined for complexity, memory utilisation, rate and distortion.

Equation 2

$$\min F(X) = (F_{complexity}(X), \; F_{memory}(X), \; F_{rate}(X), \; F_{distortion}(X))$$

subject to $G_{rate}(X) \leq R,$

$$G_{memory}(X) \leq M,$$

$$x_i^L \leq x_i \leq x_i^U, \; i=1,2,\ldots,n$$

DETERMINING THE OBJECTIVE FUNCTIONS

The constraint multi-optimization problem was formulated in Equation (2). We add to that the objective analysis results from the previous section, which reveals the decision variables to be used at the encoder (n=9) and decoder (n=6). Each decision variable x_i is restricted within a range $\left[x_i^L, x_i^U\right]$ inclusive.

The aim of this section is to obtain the four objective functions $F_{complexity}, F_{memory}, F_{rate}$ and $F_{distortion},$ and two constraint functions G_{memory} and $G_{rate}.$

Note that

$$G_{memory}(X) = F_{memory}(X) \; and$$

$$G_{rate}(X) = F_{rate}(X)$$

To obtain each objective function, an analysis of the mathematical relationship between the decision variables (coding parameters) and each objective is carried out in the following order:

1. For each objective a large number of experiments are carried out at the encoder and decoder based on all possible combinations of settings of the aforementioned coding parameters.

2. The values obtained for the objective and the corresponding parameter settings are used to form a data set for linear categorical regression.

3. The linear equations terms are defined to include all significant decision variables (i.e. the coding parameters).

4. Finally the categorical regression function of SPSS is used to fit the data set of (II) and to determine the coefficients of each significant polynomial term of the objective.

SPSS Categorical Regression

Categorical regression (SPSS Categories, 2007) in SPSS (originally, Statistical Package for the Social Sciences) quantifies categorical data by assigning numerical values to the categories, resulting in an optimal linear regression equation for the transformed variables. Categorical regression is also known by the acronym CATREG, for *cat*egorical *reg*ression. Standard linear regression analysis involves minimizing the sum of squared differences between a response (dependent) variable and a weighted combination of predictor (independent) variables. Variables are typically quantitative, with (nominal) categorical data recoded to binary or contrast variables. As a result, categorical variables serve to separate groups of cases, and the technique estimates separate sets of parameters for each group. The estimated coefficients reflect how changes in the predictors affect the response. Prediction of the response is possible for any combination of predictor values.

An alternative approach involves regressing the response on the categorical predictor values themselves. Consequently, one coefficient is estimated for each variable. However, for categorical variables, the category values are arbitrary. Coding the categories in different ways yield different coefficients, making comparisons across analyses of the same variables difficult.

CATREG extends the standard approach by simultaneously scaling nominal, ordinal, and numerical variables. The procedure quantifies categorical variables so that the quantifications reflect characteristics of the original categories. The procedure treats quantified categorical variables in the same way as numerical variables. Using nonlinear transformations allow variables to be analyzed at a variety of levels to find the best-fitting model. CATREG operates on category indicator variables. The category indicators should be positive integers. Discretization is used to convert fractional-value variables and string variables into positive integers.

Multi-Objective Optimization using Genetic Algorithms

The objective functions obtained from the SPSS regression are then fed into the NSGA-II multi-objective optimization tool (NSGA-II, 2009). Two reasons have contributed to proposing to use a GA based approach as against a Lagrangian Multiplier based approach for the above. The first reason is that a GA has the ability to find multiple optimal solutions in a single simulation run. The second reason is the presence of a well established, popular, public domain, GA software tool, Non-dominated Sorting Genetic Algorithm NSGA-II (Deb et al., 2002) that can effectively be utilized in the proposed work.

Non-dominated sorting genetic algorithm II (NSGA) was proposed in (Deb et al., 2000) which is at present one of the popular evolutionary algorithms (EAs) used in multi-objective optimization research. Since NSGA II works with a population of solutions, it can be extended to maintain a multiple set of solutions. With an emphasis for moving toward the Pareto-optimal region, the NSGA-II can find multiple Pareto-optimal solutions in one single simulation run (Deb et al., 2000). These multiple solutions will provide a set of coding parameters that will produce optimum solutions to the constraint problem.

Experiments, Results & Analysis

The practical use of the proposed multi-objective optimisation framework is best demonstrated by the use of experiments specifically designed to illustrate the steps that require to be taken to achieve optimum performance of a video CODEC. In this section we demonstrate the above using a H.264/AVC codec implementation (i.e. JM 15.1) publicly made available by the Joint Video Team (JVT) (H.264/AVC Reference Software, 2009). As the video sequence to be encoded and decoded under a full analysis of the intended optimal performance of the CODEC, the 'Foreman' sequence is being used. However eight video sequences of different nature are used in the analysis stage to determine the most significant parameters. The parameter sets that are chosen for experimentation is tabulated in Table 3.

Note the presence of parameters that are specific to the encoder only. The values within braces in the 'Range of values' column illustrates the step size of parameter increments, when used.

Determining the Objective Functions of the H.264 AVC Encoder

The computational complexity of the encoder is measured using the Intel VTune Performance Analyser tool (Intel(R) VTune(TM) Performance Analyzer 9.0., 2008). The experiments for obtaining computational complexity data are being performed on a Pentium-4 2.8GHz computer with the selected coding parameter setting ranges as tabulated in Table 2. The number of reference frames (x_1) is varied from 1 to 5 in increments of two. The search window size (x_2) is assumed to take either of the two values 16 or 32. The control variable FME, i.e., x_3, that represents Fast Motion Estimation search mode can take values from 0 to 3 corresponding to four motion estimation modes.

Table 3. Encoder and decoder coding parameters and their values used in the experiments

Coding Parameter	Variable name	Encoder or Decoder	Range of values
NumberRefrences	X_1	Both	1 - 5 (+2)
SearchRange	X_2	Encoder	16 – 32
UseFME	X_3	Encoder	0 – 3
RDOptimization	X_4	Encoder	0 – 2
SliceGroup	X_5	Both	1 – 7
QP	X_6	Both	17 - 49 (+4)
IntraPeriod	X_7	Both	0 – 16 (+3)
DisableThreshold	X_8	Both	0 – 1
Resolution	X_9	Both	1=qcif, 2=cif

The control variable RDOptimisation, i.e. x_4, which represent the rate-distortion optimization (RDO) mode is assumed to take one of four possible values represented by a number in the range 0 to 2. The resolutions of the video sequences used in the experiments is represented by the variable x_9 and is assumed to take one of two possible values, 1 or 2, corresponding to QCIF and CIF resolutions, respectively. Thus a total of 144 combinations of the five control variables are tested.

Table 4 illustrates the use of a selected set of parameter combinations and the corresponding complexity result obtained. Note that only the coding parameters that were determined to be significant in the prior profiling experiments (Li, 2007) have been considered.

Similarly Table 5 illustrates a selected set of significant [21] parameter combinations used to analyse memory utilization.

Since rate and distortion have the same control parameters, they can use identical data sets as shown in Table 6. The quantization parameter (x_6) is varied from 17 to 49 with assumed increments of four for experimental purposes. The control variable (x_7), IntraPeriod, can take values: 0 (means that the first frame is coded as an I-frame and subsequent frames are coded as P-frames), 1 (all frames coded as I-frames), 4, 7, 10, 13 or 16. The

Table 4. The average computational complexity per frame measured in seconds for test video sequence 'Foreman'

x_1	x_2	x_3	x_4	x_9	Complexity (seconds)
1	16	1	1	1	60532
3	16	1	1	1	160064
5	16	1	1	1	258714
1	32	1	1	1	171762
3	32	1	1	1	489053
5	32	1	1	1	802448
1	16	2	1	1	27163
3	16	2	1	1	60604
5	16	2	1	1	95224

Table 5. The average memory (MBytes) utilisation for foreman sequence for a selected set of parameter combinations

x_1	x_2	x_3	x_5	x_9	Memory
1	16	1	1	1	4.84
2	16	1	1	1	6.47
3	16	1	1	1	8.1
4	16	1	1	1	9.73
5	16	1	1	1	11.36
1	32	1	1	1	7.93
2	32	1	1	1	12.62
3	32	1	1	1	17.31
4	32	1	1	1	22
5	32	1	1	1	26.69
1	48	1	1	1	13.03

control variables (x_8) and (x_9) represents DisableThreshold and resolution of video respectively, and can take only one of two possible values.

All experimental data sets of which samples were shown in Tables 4-6 are subsequently fed into the SPSS categorical regression tool to estimate the objective function for each objective. The outcomes of the results are as follows:

Equation 3

$$F_{Complexity} = 0.284 * x_1 + 0.19 * x_2 - 0.506 * x_3 + 0.118 * x_4 + 0.444 * x_9$$
$$F_{Memory} = 0.131 * x_1 + 0.36 * x_2 - 0.086 * x_3 + 0.079 * x_5 + 0.964 * x_9$$
$$F_{Rate} = -0.63 * x_6 - 0.249 * x_7 + 0.026 * x_8 + 0.334 * x_9$$
$$F_{Distortion} = -0.992 * x_6 - 0.2 * x_7 + 0.033 * x_8 + 0.093 * x_9$$

The SPSS CATREG also produces the quantified values for each decision variable and objec-

Table 6. The average PSNR (dB units) and Rate (kbits) for foreman sequence for selected combinations of coding parameters.

x_6	x_7	x_8	x_9	PSNR	Rate
17	1	1	1	44.85	68.011643
21	1	1	1	41.74	49.544
25	1	1	1	38.86	34.671286
29	1	1	1	35.84	22.854214
33	1	1	1	33.2	15.495
37	1	1	1	30.68	10.6615
41	1	1	1	28.06	7.191857
45	1	1	1	25.72	5.064143
49	1	1	1	23.22	3.279071
17	4	1	1	44.13	34.475786

Figure 4. Rate vs. Distortion in different stages of the optimization

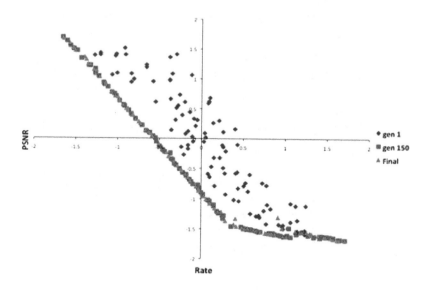

tives. Depending on the data set each category of a decision variable is assigned a numerical value that also reflects the characteristics of the original categories. For example the variable UseFME or x_3 originally has 4 categories (0,1,2,3) each representing one Fast Motion Estimation mode. The quantified version of the same variable will have 4 different numerical values (-1.729, 0.537, 0.679,

0.512). The estimated values reflect how changes in the decision variable affect the response or the objective function in this case.

Optimising the Encoder Performance

These functions are then fed to the NSGA-II optimization tool (NSGA-II, 2009) along with

Figure 5. Optimal results for constrained rate vs. PSNR

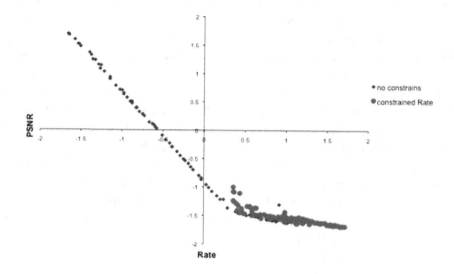

Figure 6. The affect of adding memory control to the optimization framework

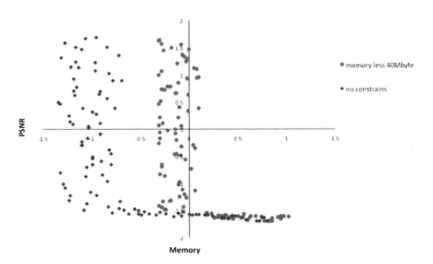

the quantified values of decision variables. The NSGA-II provides all sets of optimal results that jointly minimize complexity, memory, bit-rate and maximizes quality. Since it is complex to visualize the optimality of the results in a single 4D graph, we plot pair wise graphs as illustrates below.

Figure 4 shows the PSNR (averaged over all frames of the video), that measures objective video quality, versus bit rate values in different stages (i.e. generations of the GA being used) of the optimization process. The value of the PSNR and bitrate are based on the quantified values of the decision variables. It is shown that the spread of points converge to the Pareto-optimal solutions at the final generation of the optimization.

When constraints are added to the optimization model the same number of optimal results are produced but are concentrated within the constrained range. Figure 5, illustrates this effect. When the rate was constrained to the range 30kb/s to 100kbs, the black scatter of points results, as compared to the long blue line of points when no constraints are applied. Similarly figure 6 illustrate the result when memory is constrained to be less than 40MByte.

It is noted that the optimisation procedure described above results in a number of optimal

solutions. An example set of coding parameter values that lead to an optimal coding performance is tabulated in Table 7.

Determining the Objective Functions of H.64 AVC Decoder

It was proved that there are a total of six coding parameters (x_1 and x_4 to x_9 as listed in Table 2) that significantly affect the decoder's computational complexity, memory utilization, received bit rate and distortion. The proposed multi-objective optimization framework assumes a lossless network, which means that the decoder can receive all the data transmitted by the encoder at the specific rate. Under such a situation, the reconstructed video quality at the decoder depends entirely on the coding parameters used by the encoder. Thus, the objective functions for video quality (distortion) and rate at the decoder can use the functions derived at the encoder for the said objectives.

As in the case of the encoder analysis, the SPSS categorical regression tool was used to estimate the objective function for each objective. The results are formulated in equation set (4) below.

Table 7. An example of a parameter combination that results in an optimal solution

Parameter	Value
NumReferences	1
RDOptimisation	1
SearchRange	16
SliceGroup	1
UseFME	3
QP	17
IntraPeriod	0
DisableThreshold	1
Resolution	1 (QCIF)

Equation 4

$$F_{Complexity} = 0.998 * x_6 - 0.046 * x_2$$
$$F_{Memory} = 0.325 * x_1 + 0.925 * x_9$$

The rate and distortion objective function are the same as for the encoder, as explained above. We noticed that the regression has discarded the variable SliceGroup from all the decoder functions as it does not affect on any of them.

Optimising the Decoder Performance

The NSGA-II tool is subsequently used to obtain optimal parameter combinations, as described above in optimising the encoder performance. Figures 7 and 8 illustrate pair-wise graphs between rate vs. distortion and memory utilisation vs. complexity. Figure 7 illustrates that the results diverge to the pareto-optimal curve towards the final generation. For memory vs. complexity optimization, illustrated in figure 8, there was no clear pareto-optimal line, as compared to that

Figure 7. Optimal results for rate against distortion

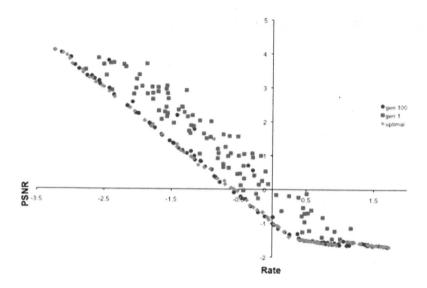

Figure 8. Optimal results for memory against complexity

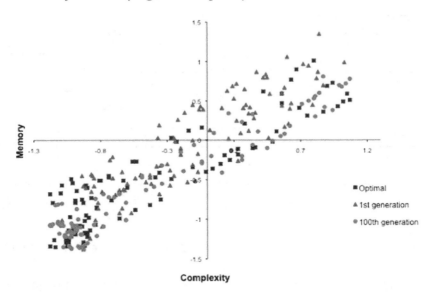

obtained for rate vs. distortion. The final optimal result was scattered in some ranges of the graph and lined up in other ranges. Figure 9 demonstrate the affect of constraining the rate within a specified range. The optimal solution is restricted to that range of the rate.

It is noted that the optimisation procedure described above results in a number of optimal solutions. An example set of coding parameter

values that lead to an optimal coding performance for the decoder is tabulated in Table 8.

Case Study

A general application setting for the multi-objective optimization framework can be explained in the context of an on-demand video streaming service provider where a feedback channel exits

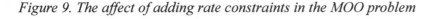

Figure 9. The affect of adding rate constraints in the MOO problem

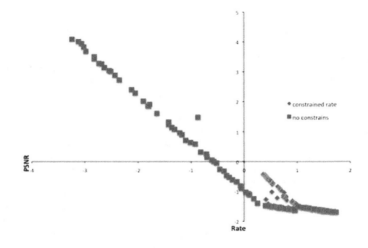

Table 8. An example of a parameter combination that results in an optimal solution

Parameter	Value
NumReferences	5
QP	49
IntraPeriod	17
DisableThreshold	0
Resolution	1 (QCIF)

between the client and the server. The recorded video streams are available for clients upon request. The framework can be applied in the following manner:

1. The service provider receives the raw video from the camcorders.
2. According to the video characteristics a set of objective functions is selected to represent it in the MOO.
3. The MOO is applied with different sets of constraints that match different client specifications.
4. The constraints and the results of different sets of optimal parameters are recorded in a lookup-table along with an encoded bitstream of the original video.
5. When a client request a specific video along with some information on its capabilities, the

lookup table will be used to decide on which version of the bitstream is best suitable for the client.

The practical use of the Multi-objective optimization framework can be demonstrated through providing a solution to support the following scenario.

"A user with a smart mobile phone with maximum internal dynamic memory of 110 MB is requesting to watch the video sequence Foreman. The mobile phone is connected to the internet through a GPRS class A network with maximum download speed of 100kbit/s. What are the best possible set of parameters to encode the requested sequence in the best quality, lowest rate and using minimum encoding resources of CPU and memory?"

Table 9. Optimal solutions for the mobile scenario

Parameters	Optimal Solutions			
	Set 1	Set 2	Set 3	Set 4
NumberRefrences	1	1	1	1
SearchRange	32	32	32	32
UseFME	1	1	1	1
RDOptimization	0	0	0	0
SliceGroup	1	2	1	1
QP	17	21	17	17
IntraPeriod	3	6	3	6
DisableThreshold	1	1	1	1
Resolution	2	1	1	1

From the given scenario we could extract 2 constraints; one is restricting the maximum bitrate and the other controlling the allowed memory usage. Thus the general set of constraints on equation set 2 can be written as:

Equation 5

$$\min F(X) = \left(F_{Complexity}(X), F_{Memory}(X), F_{Rate}(X), F_{Dsitortion}(X) \right)$$
$$subject \text{ to } \ G_{Rate}(X) \leq 100,$$
$$G_{Memory}(X) \leq 110,$$
$$x_i^L \leq x_i \leq x_i^U, \qquad\qquad i = 1,2,\ldots\ldots9$$

The NSGA-II tool (NSGA-II, 2009) can now be used to obtain the optimal solutions, given the above constraints. Parameters values that result on the CODEC's optimal performance are finally obtained as tabulated in Table 9.

In a practical application that consists of a number of video clips/files, for each video an off-line multi-objective optimisation can be carried out using the proposed framework. The parameter sets that result in optimal performance of the CODEC under known constraints (bandwidth determined by the transmission channel/medium, memory utilisation and complexity determined by the device limitations and quality determined by subjective/objective requirements) can be determined and stored in a look-up table. When a need arises and the encoder receives the constraints, it can use the look-up table to pick up the optimal parameters. Thus the operation will be real-time. This support scenario is especially beneficial in the optimal delivery of streaming video over wired networks.

CONCLUSION

This chapter presented a multi-objective optimization framework for video CODECs that is capable of determining sets of coding parameters that can be used for obtaining optimum performance of a CODEC operating under multiple constraints.

In particular we have used the proposed general framework to demonstrate the determination of coding parameters that can minimize the computational complexity, memory utilization and bit rate, while achieving the maximum visual quality. According to the simulation results, the framework can yield Pareto-optimal or near Pareto optimal solutions. In other words, it can produce coding parameter sets for the optimum performance of a given video CODEC under multiple constraints. We have demonstrated the practical use of the proposed framework with the help of an example case study, from the area of video delivery to a mobile phone.

The delay and packet loss are further practical constraints under which a typical CODEC will have to operate. Although the proposed framework is general and can thus be readily extended to cover any number of different constraints, further research is required to evaluate practical feasibility of such an evaluation.

REFERENCES

Chakareski, J., & Frossard, P. (2006). Rate-distortion optimized distributed packet scheduling of multiple video streams over shared communication resources. *IEEE Transactions on Multimedia*, *8*(2), 207–218. doi:10.1109/TMM.2005.864284

Deb, K., Agrawal, S., Pratap, A., & Meyarivan, T. (2000). A fast elitist non-dominated sorting genetic algorithm for multi-objective optimization: NSGA-II. *Proceedings of the Parallel Problem Solving from Nature VI Conference*, (pp. 849-858).

Deb, K., & Kalyanmoy, D. (2001). *Multi-objective optimization using evolutionary algorithms*. John Wiley & Sons.

Deb, K., Pratap, A., Agarwal, S., & Meyarivan, T. (2002). A fast and elitist multiobjective genetic algorithm: NSGA-II. *IEEE Transactions on Evolutionary Computation, 6*(2), 182–197. doi:10.1109/4235.996017

H.264/AVC Reference Software. (2009). *JM 15.1.* Retrieved from http://iphome.hhi.de/suehring/tml

He, Z., Liang, Y., Chen, L., Ahmad, I., & Wu, D. (2005). Power-rate-distortion analysis for wireless video communication under energy constraints. *IEEE Transactions on Circuits and Systems for Video Technology, 15*(5), 645–658. doi:10.1109/TCSVT.2005.846433

Intel Corporation. (2008). *Intel(R) VTune(TM) Performance Analyzer 9.0.*

Li, S., Yan, L., Feng, W., Shipeng, L., & Wen, G. (2009). Complexity-constrained H.264 video encoding. *IEEE Transactions on Circuits and Systems for Video Technology, 19*(4), 477–490. doi:10.1109/TCSVT.2009.2014017

Li, X. (2007). *Enhancements & optimizations to H.264/AVC video coding.* Unpublished doctoral thesis, Loughborough University.

Marpe, D., Wiegand, T., & Sullivan, G. (2006). The H.264/MPEG4 advanced video coding standard and its applications. *IEEE Communications Magazine, 44*(8), 134–143. doi:10.1109/MCOM.2006.1678121

NSGA-II. (2009). *Kanpur Genetic Algorithms Laboratory.* Retrieved March 2009, from http://www.iitk.ac.in/kangal/codes.shtml

Ostermann, J., Bormans, J., List, P., Marpe, D., Narroschke, M., & Pereira, F. (2004). Video coding with H.264/AVC: Tools, performance and complexity. *IEEE Circuits and Systems Magazine, 4*(1), 7–28. doi:10.1109/MCAS.2004.1286980

Özçelebi, T., Sunay, M. O., Tekalp, A. M., & Civanlar, M. R. (2007). Cross-layer optimized rate adaptation and scheduling for multiple-user wireless video streaming. *IEEE Journal on Selected Areas in Communications, 25*(4), 760–769. doi:10.1109/JSAC.2007.070512

Özçelebi, T., Tekalp, A. M., & Civanlar, M. R. (2007). Delay-distortion optimization for content-adaptive video streaming. *IEEE Transactions on Multimedia, 9*(4), 826–836. doi:10.1109/TMM.2007.895670

Pu, W., Lu, Y., & Wu, F. (2006). Joint power-distortion optimization on devices with MPEG-4 AVC/H.264 codec. *IEEE International Conference on Communications, 1*, 441-446.

I.T.U.T.S. Sector (2005). *Advanced video coding for generic audiovisual services.* ITU-T Recommendation H.264. (ITU-T Rec.H, 14496-14410).

Sermadevi, Y., & Hemami, S. S. (2003). Linear programming optimization for video coding under multiple constraints. In *Proceedings of the IEEE Data Compression Conference,* (pp. 53–62).

SPSS Inc. (2007). *SPSS Categories™ 16.0 manual guide.*

Van der Schaar, M., & Andreopoulos, Y. (2005). Rate-distortion-complexity modeling for network and receiver aware adaptation. *IEEE Transactions on Multimedia, 7*(3), 471–479. doi:10.1109/TMM.2005.846790

Vanam, R., Riskin, E. A., Hemami, S. S., & Ladner, R. E. (2007). Distortion-complexity optimization of the h.264/mpeg-4 avc encoder using the gbfos algorithm. *Proceedings of the 2007 Data Compression Conference,* Washington, DC, USA, (pp. 303–312).

Vanam, R., Riskin, E. A., Hemami, S. S., & Ladner, R. E. (2009). H.264/MPEG-4 AVC encoder parameter selection algorithms for complexity distortion tradeoff. *Proceedings of the Data Compression Conference.*

Wiegand, T., Sullivan, G., Bjntegaard, G., & Luthra, A. (2003). Overview of the H.264/AVC video coding standard. *IEEE Transactions on Circuits and Systems for Video Technology, 13,* 560–576. doi:10.1109/TCSVT.2003.815165

Zhang, J., He, W., Yang, S., & Zhong, Y. (2003). Performance and complexity joint optimization for H.264 video coding. *Proceedings of the 2003 International Symposium on Circuits and Systems, 2,* II-888-II-891.

Zhang, Y., Gao, W., Lu, Y., Huang, O., & Zhao, D. (2007). Joint source-channel rate-distortion optimization for H.264 video coding over error-prone networks. *IEEE Transactions on Multimedia, 9*(3), 445–454. doi:10.1109/TMM.2006.887989

Zitzler, E. (1999). *Evolutionary algorithms for multiobjective optimization: Methods and applications.* Master's thesis, Swiss Federal Institute of technology (ETH), Zurich, Switzerland.

ADDITIONAL READING

Coello, C. A. C. (2002). *Van Veldhuizen, D. A. Lamont, G. B.* Evolutionary Algorithms For Solving Multi-Objective Problems, Kluwer Academic.

H.264/14496-10 AVC Reference Software Manual. (2009, Jan), Retrieved from http://iphome.hhi.de/suehring/tml/JM%20Reference%20Software%20Manual%20 (JVT-AE010).pdf

Richardson, I. (2003). *H.264 and MPEG-4 video compression: Video Coding for Next-generation Multimedia.* John Wiley & Sons. doi:10.1002/0470869615

Sullivan, G. J., & Wiegand, T. (2005, Jan). Video compression – from concepts to the H.264/AVC standard. *Proceedings of the IEEE, 93*(1), 18–31. doi:10.1109/JPROC.2004.839617

Chapter 18
Towards Rapid 3D Reconstruction using Conventional X–Ray for Intraoperative Orthopaedic Applications

Simant Prakoonwit
University of Reading, UK

ABSTRACT

A rapid 3D reconstruction of bones and other structures during an operation is an important issue. However, most of existing technologies are not feasible to be implemented in an intraoperative environment. Normally, a 3D reconstruction has to be done by a CT or an MRI pre operation or post operation. Due to some physical constraints, it is not feasible to utilise such machine intraoperatively. A special type of MRI has been developed to overcome the problem. However, all normal surgical tools and instruments cannot be employed. This chapter discusses a possible method to use a small number, e.g. 5, of conventional 2D X-ray images to reconstruct 3D bone and other structures intraoperatively. A statistical shape model is used to fit a set of optimal landmarks vertices, which are automatically created from the 2D images, to reconstruct a full surface. The reconstructed surfaces can then be visualised and manipulated by surgeons or used by surgical robotic systems.

1. INTRODUCTION

The ability to be able to access and update 3D patient-specific surface models of the bones and other structures during an operation is very valuable. It can augment the surgeon's view of the objects of interest and help with navigation,

DOI: 10.4018/978-1-60960-477-6.ch018

measurement and surgical planning. The 3D model can also be used with a surgical robot to accurately carry out surgical procedures on an up-to-date models of the structure of interest.

A common approach to obtain 3D models of bones or other structures is to make use of some imaging techniques such as CT or MRI. The main disadvantage of those techniques is that, due to the scanning geometry of the machines, they cannot be employed directly to obtain and update 3D models during an operation. Normally, the models have to be reconstructed before or after an operation. If the model is required to be updated during an operation, the patient has to be moved from the operating table to a CT or MRI machine before returning to the operating table again. Some MRI machines are designed to be used intraoperatively. However, all the equipment must only be non-metallic. This seriously limits the use in orthopaedic applications. Also any metal surgical robots cannot be utilised. In addition, they are expensive and/or induce high radiation doses (Kumar T. Rajamani, et al., 2007).

The main aim of this chapter is to propose a new possible method towards rapidly reconstructing and updating patient-specific 3D surface models of bones using a small number of 2D conventional X-ray images feasible for intraoperative applications. The technique should be capable of reconstructing 3D bone surface models without having to move patient from the operating table.

2. BACKGROUND

In an intraoperative environment where the scanning geometry of a CT or MRI is not suitable, a C-arm conventional X-ray system can be used to acquire a number of 2D images to reconstruct a full 3D volumetric description, in terms of voxels, of an object of interest, e.g. (Atesok, et al., 2007; Ritter, Orman, Schmidgunst, & Graumann, 2007; Zbijewski & Stayman, 2007). However, to reconstruct at a reasonable resolution, the number of 2D images required is very high, e.g. 40 to 180 images, and to extract the surface of an object from the reconstructed voxels is very computationally expensive. Moreover, due to the large number of 2D X-ray images required, the patient is inevitably subjected to high dose of radiation.

Another approach is to use statistical shape analysis and modelling, e.g. (Cootes, Taylor, Cooper, & Graham, 1995; Dryden & Mardia,

1998; Kumar T. Rajamani, et al., 2007; Guoyan Zheng, Dong, Rajamani, Zhang, & Styner, 2007), which has been an important tool in 3D model reconstruction from incomplete data. In this approach, only a small number of sparse landmark vertices on the surface of an object, e.g. a bone, are needed to be determined. Those sparse landmark vertices alone contain inadequate information for the complete 3D surface reconstruction of an object. Hence, *a priori* knowledge is required. A statistical model can be reconstructed from a set of training surfaces representing reasonable variations of the surfaces of an object of interest. In intraoperative applications, the statistical model is then used as prior knowledge in the reconstruction process to fit to the patient anatomy using intraoperatively acquired sparse landmark vertices. Thus, in conclusion, the aim of statistical shape model fitting is to extrapolate from an extremely sparse and incomplete set of 3D landmark vertices to a complete and reasonably accurate 3D anatomical surface. The fitting process aligns and deforms the statistical shape model to fit the sparse landmark vertices. Therefore the model-based approach is widely accepted due to their ability to effectively represent objects.

Many researchers have presented methods for fitting statistical models to sparse vertices to create full patient-specific surfaces. The input data can be acquired from a small number of 2D X-ray images (Benameur, Mignotte, Labelle, & guise, 2005; benameur, et al., 2003; M. Fleute & Lavallee, 1999; Lamecker, Wenckeback, & Hege, 2006) or digitised sparse vertices (M. Fleute & Lavallee, 1998; Markus Fleute, Lavallee, & Julliard, 1999; K. T. Rajamani, Ballester, Nolte, & Styner, 2005; K. T. Rajamani, Styner, & Joshi, 2004). Principle Component Analysis (PCA) based statistical shape models have been widely used in surface extrapolation. In (Markus Fleute, et al., 1999), a joint optimisation technique is applied to fit the statistical model to intraoperatively digitised landmark vertices. In this method both pose and deformation are concurrently optimized. Later,

Chan et al (Chan, Edwards, & Hawkes, 2003), used a different approach by optimising deformation and pose separately and implementing the iterative closest point (ICP) method. Rajamani, et al (Kumar T. Rajamani, et al., 2007), developed a new technique using a Mahalanobis distance weighted least square fit of the deformable model to the landmark vertices. The surface model of a proximal femur was reconstructed in (Guoyan Zheng, et al., 2007) by using a Dense-Point Distribution Model (DPDM) and employing two-stage deformation processes: statistical instantiation, which stably instatiates a surface model from the DPDM using a statistical approach; kernel-based deformation, which further refines the surface model.

The existing methods mentioned above are capable of producing 3D surface models. However, the performance of the algorithms and the accuracy of the resulting models very much depend on how well the input sparse landmark vertices are selected and how well they can represent the important features of the surface. The existing methods of creating the input sparse landmark vertices are not fully automatic and the landmark vertices are not optimally distributed on the surfaces resulting in some important features may not be well captured or represented. In this chapter, a method is proposed to solve this problem. The main contributions of this work, therefore, are: a new scanning geometry; a novel automatic, without user intervention, method for determining 3D landmark vertices which are optimally distributed on a surface. Some proof of concept computational experiments have been conducted to reconstruct a femur bone. The results show that the proposed method is capable of reconstructing an acceptable 3D surface.

3. PROPOSED METHOD

In our proposed method, a small number, e.g. 5, of X-ray images are taken at appropriate projection angles. Edges are then extracted from the boundary of an object of interest. The 3D landmark vertices are determined by paring all the edges in all images. Once the landmark vertices are found, a statistical shape model is employed to fit the landmarks to create a full 3D surface model.

3.1 Edge Detection and Representation

Each X-ray image is enhanced prior to edge detection to reduce the noise level, sharpen the edges and to enhance the contrast of the edges. Techniques such as adaptive smoothing and adaptive histogram equalization can be used to enhance the image. In this research, the grayscale enhancement method described in (Abidi, Mitckes, Abidi, & Liang, 2003) has been adopted. First, linear regression is applied to stretch the pixel range within the given image so that the maximum and minimum pixel values cover wider range. Then gamma intensity adjustment is performed followed by histogram equalization. Those procedures require manual parameter adjustment. However, a set of common parameters can be use for most of the X-ray images in the application.

The edge contours are extracted by using Gradient edge detectors with relaxation labeling to reinforce the underlying boundaries whilst suppressing spurious edges due to noise. To reconstruct landmark vertices, the most effective way is to represent edge contours analytically. Therefore, those edge contours are fitted with cubic B-spline curves. Each edge contour c is represented by a sequence of 3rd order parametric curve segments. To detect edge contours and conduct B-spline curve fitting, we adopt the techniques described in (Matalas, 1996).

3.2 Landmark Point Reconstruction

In this section, we consider how multiple 3D landmark points on a 3D surface can be found by pairing 2D X-ray images, using the characteristics of epipolar geometry. The method is based on our previous work (Prakoonwit & Benjamin, 2007). In contrast to current methods, e.g. (K. T. Rajamani, et al., 2005; Kumar T. Rajamani, et al., 2007; G. Zheng, Dong, & Nolte, 2006; Guoyan Zheng, et al., 2007), where landmark points have to be manually determined based on some specific features of the structures of interest, the proposed method automatically reconstructs landmark points and automatically captures the salient features of surfaces.

As each point on the edge contours is generated by a ray from the relevant X-ray source, the epipolar plane contains the rays linking the X-ray sources to the epipolar tangents points on the image planes. The landmark point is therefore the intersection of a pair of rays from the two distinct X-ray sources. For example, consider a bone-shaped object in Figure 1. Landmark point f_1, and the corresponding epipolar tangent points t_1 and t_2 can be found from the known epipoles, derived from the intersections between the epipolar baseline and the image planes, and edge contours c_1, c_2. Two rays, r_1, r_2 from the X-ray sources are defined by linking $X_1 t_1$ and $X_2 t_2$ respectively. Then landmark point f_1 is the intersection of r_1 and r_2.

Note that in practice, it is unlikely that r_1 and r_2 will intersect exactly. This problem is solved by finding the shortest segment between the two rays and choosing the middle point on the line as the frontier point.

While the plane rotates about the epipolar baseline and touches other parts of the surface, additional landmark points, e.g. f_1, f_2, f_3, can be determined. The more complex the surface is, the more additional information is contained in the associated edge contours and therefore the more additional local epipolar tangent points are generated, thus providing the 3D locations of furthur landmark points.

So far, we have discussed pairing of two X-ray images to reconstructed 3D landmark points. The same fundamentals can further be applied to a set of more than two image. Consider a set of N_{pl} X-ray sources, irradiating surface S from distinct directions, where no potential epipolar baseline passes through S. From this set, $N_{pl}(N_{pl}-1)/2$ image pairs can be created. If a convex object is projected on to the image planes, then $N_{pl}(N_{pl}-1)$ landmark points will be determined. If the viewing angles are uniformly distributed in solid angle, smooth

Figure 1. Epipolar plane Π rotating about epipolar baseline creating epipolar tangent lines, epipolar tangent points and landmark points

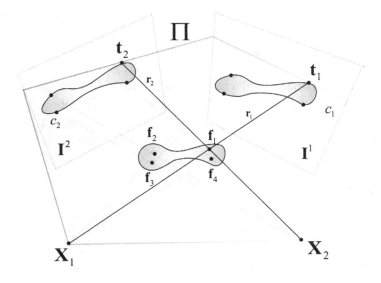

shapes, with monotonic change of gradient, result in epipolar tangent points at approximately equal increments of tangent angle on the corresponding edge contour. If N_{pl}= 10, on each 2D edge contour, interaction with nine other views will establish 18 epipolar tangent points and 18 landmark points, each centered on an arc of ±10°, for a total of 360°. A nonconvex object will generate more epipolar tangent points and landmark points from the pairing.

Figure 2 shows an example, when N_{pl}=3 and the 2D X-ray images are I^1, I^2 and I^3. Three image pairs (I^1, I^2), (I^1, I^3) and (I^2, I^3) can be created. For simplicity, we assume that all the image planes are at right angle to one another. The X-ray sources are at ∞ and all X-rays are parallel to the associated image plane's normal. If the object is a sphere, the corresponding edge contours c_1, c_2 and c_3 are all circles, Figure 2(a). Six landmark points f_i, $i = 1,2,3, ..., 6$, can then be determined by the intersections of the relevant rays in the three image pairs as illustrated in Figure 2(b). Image

pairs (I^1, I^2), (I^1, I^3) and (I^2, I^3) generate landmark point pairs (f_1, f_2), (f_3, f_4) and (f_5, f_6) respectively. The related epipolar tangent lines and the epipolar tangents points t_j, $j = 1, 2, 3, ..., 12$, on the image planes are shown in Figure 2(c).

3.3 Data Acquisition Geometry

In our previous work (Prakoonwit & Benjamin, 2007), Benjamin and the author presented a full surface reconstruction from a set of 2D conventional X-ray images taken from 10 to 15 projection directions. However, the number of projections is too high for intraoperative applications. Therefore, we propose to use only about 5 projections. This would create a set of surface points, but the information acquired from this limited set of images is not adequate to fully reconstruct a surface model using the method described in (Prakoonwit & Benjamin, 2007). Instead, the reconstructed surface points will be use as a set of optimal landmark points which will be extrapolated to create

Figure 2. A simple three X-ray source system (a) a sphere is projected onto three image planes I^1, I^2 and I^3 containing edge contours c_1, c_2 and c_3 (b) frontier points f_i, $i = 1,2,3, ..., 6$, are reconstructed from the corresponding edge contours (c) distribution of epipolar tangent points t_j, $j = 1, 2, 3, ..., 12$, on the corresponding edge contours on the image planes.

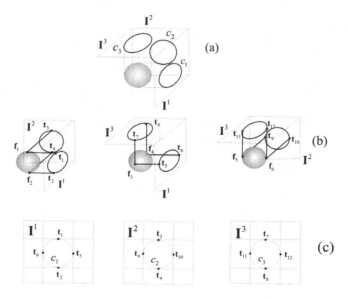

a full surface model using a statistical model as prior knowledge.

3.3.1 X-Ray Projection System

To make the method feasible for intraoperative applications, we propose the X-ray scanning geometry shown in Figure 3. A single 2D X-ray array detector is located underneath the operating table covering the area of interest. The X-ray base can be installed on the ceiling of the operating theatre, or mounted on a mobile arm. It contains 5 X-ray sources with different projection directions as required. The X-ray sources are turn on in turn to sequentially create 2D X-ray images from different projection angles, defined by α and β. In some applications where multiple X-ray sources may not be economical, a single X-ray source can be used with a mechanism to adjust the X-ray source's position and direction to sequentially create the X-ray images from different projection directions.

3.3.2 X-Ray Projection Angle Determination

In this section, the aim is to create a set of surface points optimally distributed on the object's surface. The optimal distribution means the surface points approximately cluster closely on highly curved parts of the surface and are widely spread on smooth or flat parts. To approximately create the optimal distribution of the landmark points, the following condition must be satisfied: $\delta s \propto 1 / \bar{\kappa}$, where $\bar{\kappa}$ = mean curvature and δs is a length of an arc on an edge contour c between any two consecutive epipolar tangent points. Thus the length of an arc must be inversely proportional to the local mean curvature. In other words, the higher curvature the shorter distance between two epipolar tangent points. Thus the epipolar tangent points cluster closely in highly curved parts (high $\bar{\kappa}$) and widely spread in smooth flat parts (low $\bar{\kappa}$) of the edge contour. If the normal vectors at epipolar tangent points are considered instead of the tangents, the *gap angle* between any two consecutive normal vectors must be a constant.

Figure 3. Proposed scanning geometry. An X-ray projection direction is defined by angles α and β in Xw, Yw, Zw coordinates system.

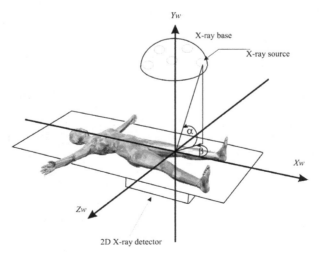

Thus the gap angle $\delta\theta$ between normal vectors $\mathbf{n}_1$ and $\mathbf{n}_2$ of epipolar tangent points $\mathbf{t}_1$ and $\mathbf{t}_2$ must be equal to the angle between any other two consecutive normal vectors on a curve segment. Provided that this condition is satisfied, the reconstructed 3D frontier points on the object itself will also approximately gather closely on highly curved parts of the surface and the surface and broadly spread on smooth flat parts.

In *n*-image system, assuming that we select the positions of the X-ray sources optimally, $\delta\theta$, the gap angle for any arc between consecutive epipolar tangent points, on any edge contour in any image plane, must be the same, i.e. $\delta\theta=\pi/(N_{pl}-1)$, where N_{pl} is the number of X-ray sources.

As the distribution of the reconstructed landmark points depends on the positions of the X-ray image planes and the associated X-ray sources, the problem is how to find a set of X-ray projection directions that create such optimal or near optimal distribution. Especially, when the positions and directions of the X-ray sources are restricted by the geometry of the scanning system such as the one shown in Figure 3.

This problem is similar to the widely studied problems of camera network design and the next best view (Mason, 1997; Mason & Grun, 1995; Olague & Mohr, 2002; Pito, 1999). The optimization problem in that case is similar to our problem. The complexity of camera positions is in the class of NP-complete problems (Mason, 1997; Olague & Mohr, 2002). Many approaches have been developed to find a reasonable optimization method (Mason, 1997; Mason & Grun, 1995; Olague & Mohr, 2002). One needs to search for the set that produces the optimal distribution condition on all image planes. The search space is enormous, non-linear, complex and poorly understood. The problem cannot be easily formulated in terms of mathematical equations in order to use traditional optimization techniques (Olague & Mohr, 2002).

Olague and Mohr (Olague & Mohr, 2002) successfully used the multi-cellular genetic algorithm, a specialized variant of the standard genetic algorithm. It follows the tree-based genetic programming representation (Kinner, 1994; Koza, 1992), which is assumed to be known and fixed. In this method, each camera variable is partitioned into a cell. The evolutionary process is applied to each camera separately allowing the appearance of subpopulations with good quality individuals. Binary representation is used in order to allow the algorithm to randomly search the whole space. The multi-cellular genetic algorithm evaluates the fitness of the whole system of multiple cameras, rather than the fitness of each cell individually. Since the camera placement problem and our X-ray source placement problem are similar, we adopt the same technique.

The objective is to minimise the error function for the multi-cellular genetic algorithm is defined by

$$\varepsilon = \overline{|\Delta\omega|}_{\delta\theta} = \frac{1}{N}\sum_{i=1}^{N}|\omega_i - \delta\theta| \qquad (1)$$

where ω_i is the angle between the normal vectors at the two ends of the ith arc between two consecutive epipolar tangent points, and N is a total number of the arcs on all edge contours on all image planes in the set. In ideal case, ε is equal to zero. The value of ω_i depends on the projection angles α and β. It can be calculated using the method explained in the previous section. The ranges of α and β are $a\leq\alpha\leq90°$, $0°\leq\beta\leq360°$ where a is a constant depending on the shape and size of the X-ray scanning geometry. Assuming that the projection distance is defined by the fix geometry of the scanning system and the corresponding projection angles, the parameters to be encoded in the multi-cellular algorithm are α and β. The detailed method on how to implement the multi-cellular algorithm can be found in (Olague & Mohr, 2002).

Figure 4. Problem: how to extrapolate sparse surface points to a complete bone surface?

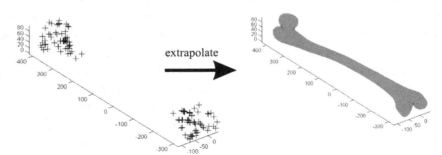

3.4 Surface Reconstruction

From the previous section, we now have a set of landmark points distributed on the surface. The next stage is to extrapolate those points to a complete 3D surface model of a bone. The problem statement is shown in Figure 4. However, reconstructing an acceptable 3D surface model from such sparse data is a challenging task as reviewed in section 1. Normally, without prior knowledge, it is rather impossible or it is not computationally viable to accurately reconstruct a full surface description for our applications. Therefore, a statistically based shape model is considered to be used as prior knowledge. Based on such model, the reconstructed landmark points can be applied to infer the full anatomical information in a robust way.

There are many approaches for constructing statistical shape models. Staib and Duncan (Staib & Duncan, 1992) and Szekelely, et al (Szelely, Lelemen, Brechbuler, & Gerig, 1996) applied Fourier representations for statistical models. Modal analysis can also be applied to create a decomposition of shapes into a basis of fundamental deformations, e.g. (Pentland & Scalroff, 1991). It is also feasible to use surface features such as crest-lines and to perform modal analysis on the features (Subsol, Thirion, & Ayache, 1996). Another approach is to consider a statistical model with modal representation based on Principal Component Analysis (PCA). This method

was first proposed by Cootes, et al (Cootes, et al., 1995). Since then, it has been widely used in model reconstruction and object recognition. The method has been proven to be suitable for many orthopaedic applications, e.g. (Markus Fleute, et al., 1999; Kumar T. Rajamani, et al., 2007; Guoyan Zheng, et al., 2007). This chapter is, therefore, based on the statistical shape model approach (Cootes, et al., 1995; Guoyan Zheng, et al., 2007).

3.4.1 Dense-Point Distribution Model Construction

In this work, we use a dense-point distribution model (DPDM) (Guoyan Zheng, et al., 2007), which is a type of statistical shape models, in our surface reconstruction. Building the statistical dense-point distribution model requires the following steps: acquiring, aligning, matching the training shapes; and Dense-Point Distribution model construction using PCA. The two steps are detailed as follows.

(1) *Acquiring, aligning and matching the training shapes.* A set of 30 computer-generated surface models of femurs with randomly varied shapes and sizes are used as the training shapes for building the statistical shape model representing femurs. Each individual computer-generated surface model is represented by a set of triangle mesh containing 4484 triangular facets and 2244 vertices. The shapes of the femurs are randomly

Figure 5. Some computer generated femurs in the training set

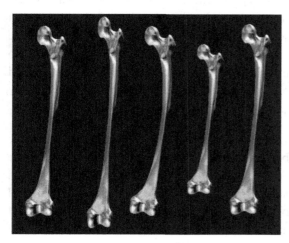

deformed to represent the possible variations in the real population. Some of the femurs are shown in Figure 5.

Let $\mathbf{vr}_i=[xr_{i,0}, yr_{i,0}, zr_{i,0}, xr_{i,1}, yr_{i,1}, zr_{i,1},...,xr_{i,Nr-1}, yr_{i,Nr-1}, zr_{i,Nr-1}]^T, i=0,1,...,M\text{-}1$, be a vector representing Nr vertices of the ith surface model in the set of M training models where, in this case, $M = 30$ and $Nr = 2244$. The 3D position of nth vertex of surface model i^{th} is defined by $[xr_{i,n}, yr_{i,n}, zr_{i,n}]$.

However, DPDM requires more vertices to create a dense cloud of vertices on the surface. This can be obtained by iteratively refining the training computer-generated surface models to create the DPDM. The refining method proposed in (Guoyan Zheng, et al., 2007) is adopted in this work. It uses a simple subdivision scheme call *loop scheme*, (Loop, 1987), to guarantee that the surface mesh is smooth. The level of subdivision and the number of dense vertices required depend on the application. Zhen, et al, (Guoyan Zheng, et al., 2007) recommends that, for this type of applications, the maximum edge length of all triangles should be less than 1.5mm. Therefore, approximately one-level or two-level subdivision is found to be adequate.

Having been subdivided, each surface model is now represented by a set of dense vertices

$$\mathbf{v}'_i = \left[x_{i,0}, y_{i,0}, z_{i,0}, x_{i,0}, y_{i,0}, z_{i,0},..., x_{i,N-1}, y_{i,N-1}, z_{i,N-1} \right]^T$$

where $i=0,1,...,M\text{-}1$ and $N =$ number of vertices after the subdivision process, $N>>N_r$.

The next stage is to align all the femur surfaces in the training set. The alignment method used in (Markus Fleute, et al., 1999; Szelely, et al., 1996) is implemented in this work. The method performs least-squares minimisation of the distances between a sparse and unorganised set of vertices and a dense set of vertices. Therefore, only the template model is required to be dense-point model, the rest can be the original models. This would reduce the computational workload in the alignment process. Therefore, suppose surface $\mathbf{v}'_0$ is selected to be a template, then surfaces $\mathbf{vr}_j, j = 1,2,...,M\text{-}1$ can be aligned with $\mathbf{v}'_0$, one by one. The alignment parameters, can then be used to align all the dense-point shapes $\mathbf{v}'_i, i = 1,2,...,M\text{-}1$, creating a new set of training 3D shapes represented by $\mathbf{v}_i, i = 0,1,2,...,M\text{-}1$ where $\mathbf{v}_0 = \mathbf{v}'_0$. The details of the alignment algorithm can be found in (Markus Fleute, et al., 1999).

(2) *Dense-point distribution model construction using PCA*. Once all the surface vertices of the models in the training set have been aligned, the statistics of the set of aligned 3D models can be captured by a dense-point distribution model. The DPDM is created by applying PCA on aligned $\mathbf{v}_i$ as described in (Cootes, et al., 1995; Markus Fleute, et al., 1999; Guoyan Zheng, et al., 2007). The covariance matrix, $\mathbf{S}$, can be calculated using

$$\mathbf{S} = \frac{1}{(M-1)} \sum_{i=0}^{M-1} d\mathbf{v}_i \bullet d\mathbf{v}_i^T \qquad (2)$$

where $d\mathbf{v}_i = \mathbf{v}_i - \bar{\mathbf{v}}$ and $\bar{\mathbf{v}} = \dfrac{1}{M} \sum_{i=0}^{(M-1)} \mathbf{v}_i$. Following the method in (Guoyan Zheng, et al., 2007), the principle axes are described by the unit eigenvectors, $\mathbf{p}_i$, of $\mathbf{S}$ such that

$$\mathbf{S} \cdot \mathbf{p}_i = \lambda_i \cdot \mathbf{p}_i, \; i = 0, ..., M' - 1 \qquad (3)$$

where λ_i is the ith eigenvectors of $\mathbf{S}$, $\lambda_i \geq \lambda_{i+1}$ and $M' \leq M - 1$.

A 3D model in the training set can be approximated using the mean model and a linear combinations with weights $\mathbf{w} = \left[w_0, w_1, ..., w_{m'-1} \right]^T$, $-3\sqrt{\lambda_i} \leq w_i \leq 3\sqrt{\lambda_i}$ (Cootes, et al., 1995), obtained from the first M' modes:

$$\mathbf{v} = \bar{\mathbf{v}} + \sum_{i=0}^{M'-1} \left(w_i \cdot \mathbf{p}_i \right). \qquad (4)$$

3.4.2 Model Fitting

Given reconstructed landmark points $\mathbf{f}_j = [x_j \; y_j \; z_j]^T$, $j=0,1,...Nf-1$ and $Nf<<N$, the whole surface can be reconstructed by fitting the DPDM represented by Equation (4) to the landmark points $\mathbf{f}$. However, before model fitting can be done, the affine registration has to be employed to align the reconstructed frontier points to the mean surface model of the DPDM.

This kind of registration is a well-known problem. There are many different methods available. In this work, the ICP algorithm (Besl & McKay, 1992; Chen & Medioni, 1992; G. Zheng, et al., 2006; Guoyan Zheng, et al., 2007), which is one of the most widely used methods, is implemented. The ICP algorithm relies on the search of pairs of closest vertices, and the computation of a paired-point matching transformation. The resulting transformation is then applied to one set

of points. The process is progress iteratively until convergence. To prevent the ICP algorithm from converging to a local minimum, we use a set of landmark points on the mean surface model and the reconstructed landmark points, to initialise the registration procedure. The aligned reconstructed frontier points are represented by

$$\mathbf{f}'_j = \begin{bmatrix} x_j & y_j & z_j \end{bmatrix}^T, j=0,1,...,Nf-1.$$

After the reconstructed frontier points have been aligned with the mean surface model, the homologous points of the aligned reconstructed landmark points on the dense mean surface model of the DPDM can be determined. Hence the DPDM can be use to represent the reconstructed surface by minimising the following error function (K. T. Rajamani, et al., 2004; Guoyan Zheng, et al., 2007).

$$E_w = \rho \cdot \sum_{j=0}^{Nf-1} \left\{ \left\| \mathbf{f}'_j - \left[\left(\bar{\mathbf{v}}_k \right)_j + \sum_{i=0}^{M'-1} \left(w_i \cdot \mathbf{p}_i(k) \right) \right] \right\|^2 \right\} + \sum_{i=0}^{M'-1} \left(\frac{w_i^2}{\lambda_i} \right) \qquad (5)$$

where $\rho =$ a factor that controls the relative weighting between the two terms, $\left(\bar{\mathbf{v}}_k \right)_j$ represents the kth point $\bar{\mathbf{v}}_k$ on the mean surface model $\bar{\mathbf{v}}$ of the DPDM is the closest point to the jth aligned reconstructed frontier point $\mathbf{f}'_j$. The details on how to determine ρ and optimise Eq D to estimate the full surface model which fits the reconstructed frontier points can be found in (Guoyan Zheng, et al., 2007).

4. EXPERIMENTAL RESULTS AND DISCUSSION

Computational experiments have been performed to show the capability of the proposed method. Femur bones have been selected as a test object. A set of 30 computer generated femurs are used as a training set as described earlier. A computer

Figure 6. (a) Test femur to be reconstructed; (b) Corresponding edge contours on the five 2D X-ray images

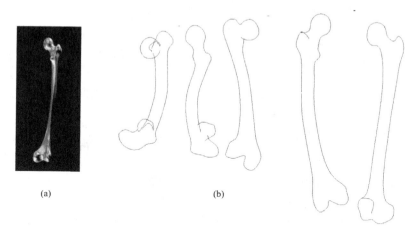

(a)   (b)

generated test femur to be reconstructed and its corresponding edge contours in the five 2D X-ray images are shown in Figure 6 (a) and (b) respectively. Note that, the test femur is different from those in the training set.

2D X-ray images of the test femur taken at 5 different projection angles determined by the method described in section 3.3.2 are simulated. The ranges of projection angle used are $50° \leq \alpha \leq 90°$ and $0° \leq \beta \leq 360°$. The edge contours of the femur on all images are shown in Figure 6 (b). The edge contours are used for the landmark vertex reconstruction as explained in section 3.2. The reconstructed landmark vertices are displayed in Figure 7(a). The 3D surface reconstruction using the DPDM to fit the landmark vertices can be found in Figure 7(b). It demonstrates that the landmark vertices are approximately well distributed on the surface. The error distance distribution of the reconstructed 3D surface. The error distance is defined as the shortest distance from the all vertices of the reconstructed femur to the original test femur.

The results show that the proposed method could produce a reasonable 3D surface of the test object. The landmark vertices are automatically generated and adapted to effectively capture the salient features of the femur. There are more landmark vertices on the detailed or highly curved parts of the surface and less landmark vertices on the less detailed or flat parts of the surface. This is to guarantee that all the important features of the surface are captured and represented for the model fitting process. However, the distribution of the landmark vertices depends on the ranges of the projection angles α and β. Ideally, the ranges should be $0 \leq \alpha \leq 90°$ and $0 \leq \beta \leq 360°$. Due to the scanning geometry, the full range of α would not be feasible for intraoperative applications. The results show that the rage of $50° \leq \alpha \leq 90°$ is still feasible.

5. FUTURE RESEARCH DIRECTION

In this chapter, a novel approach towards 3D reconstruction using conventional X-ray images for intraoperative orthopaedic applications is presented. The method still has some main limitations which require some future research in the following aspects.

5.1 Object Identification and Segmentation

In real circumstances, there are many bones and other structures in the volume. To automatically extract and reconstruct only the object of interest,

Figure 7. (a) The reconstructed landmark vertices; (b) 3D reconstructed surface and the reconstructed landmark vertices

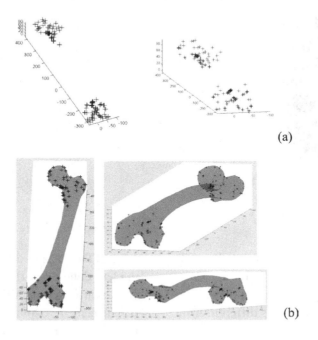

(a)

(b)

a robust object identification and segmentation has to be developed. This is not a trivial task, because objects in an X-ray image are overlapped or superimposed. If the type of object, e.g. femur, is known in advance, then some prior knowledge can be employed in the segmentation processes to simplify the problem. The method based on attributed Relational Graph (ARG) developed by Wang, et al (Wang, Li, Ding, & Ying, 2005), for security applications can be adopted and modify to extract a particular bone for the reconstruction. Note that, the method also relies heavily on the

Figure 8. Error distribution of the reconstructed femur compared with the original test femur

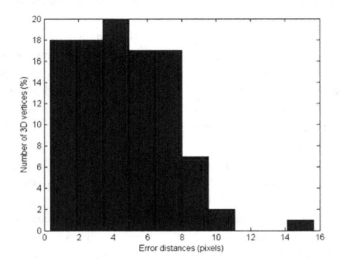

edge contours of objects. Therefore, it should work well only on high density objects or tissues such as bones where edges clearly appear on a 2D X-ray image.

5.2 Statistical Shape Model Fitting

Even though Statistical shape modelling is a robust tool for 3D surface reconstruction from sparse vertices, there is a major limitation. Instances of the model can only be deformed in ways found in the training set. That means the method cannot reconstruct any pathological conditions or deformations which are not included in the training set. Any substantial changes to a bone during an operation may also cause the reconstruction process to fail. To solve this problem, it is suggested that the edges contours from 2D X-ray images can be used to adjust the statistical shape model before fitting it to landmark vertices.

REFERENCES

Abidi, B., Mitckes, M., Abidi, M., & Liang, J. (2003). *Grayscale enhancement techniques of X-ray images for carry-on luggage*. Paper presented at the SPIE 6th International Conference on Quality Control by Artificial Vision, Gatlinburg, TN, USA.

Atesok, K., Finkelstein, J., Khoury, A., Peyser, A., Weil, Y., & Liebergall, M. (2007). The use of intraoperative three-dimensional imaging (ISO-C-3D) in fixation of intraarticular fractures. *Injury. International Journal of the Care of the Injured, 38*, 1163–1169.

Benameur, S., Mignotte, M., Labelle, H., & Guise, J. A. d. (2005). A herarchical statistical modeling approach for the unsupervised 3D biplanar reconstruction of the scoliotic spine. *IEEE Transactions on Bio-Medical Engineering, 52*(12), 2041–2057. doi:10.1109/TBME.2005.857665

Benameur, S., Mignotte, M., Parent, S., Labelle, H., Skalli, W., & Guise, J. A. d. (2003). 3D/2D registration and segmentation of scoliotic vertebrae using statistical models. *Computerized Medical Imaging and Graphics, 27*, 321–337. doi:10.1016/S0895-6111(03)00019-3

Besl, P., & McKay, N. D. (1992). A method for registration of 3D shapes. *IEEE Transactions on Pattern Analysis and Machine Intelligence, 14*(2), 239–256. doi:10.1109/34.121791

Chan, C. S., Edwards, P. J., & Hawkes, D. J. (2003). Integration of ultrasound based registration with statistical shape models for computer-assisted orthopaedic surgery. *Proceedings of SPIE Medical Imaging*, 414-424.

Chen, Y., & Medioni, G. (1992). Object modeling by registratoin of multiple range images. *Image and Vision Computing, 10*(3), 145–155. doi:10.1016/0262-8856(92)90066-C

Cootes, T. F., Taylor, C. J., Cooper, D. H., & Graham, J. (1995). Active shape models-their training and application. *Computer Vision and Image Understanding, 61*(1), 38–59. doi:10.1006/cviu.1995.1004

Dryden, I., & Mardia, K. (1998). *Statistical shape analysis*. John Wiley and Sons.

Fleute, M., & Lavallee, S. (1998). *Building a complete surface model from sparse data using statistical shape models: Application to computer assisted knee surgery system*. Paper presented at the 1st International Conference of Medical Image Computing and Computer-Assisted Intervention.

Fleute, M., & Lavallee, S. (1999). *Nonrigid 3D/2D registration of images using statistical models*. Paper presented at the 2nd International Conference on Medical Image Computing and Computer-Assisted Intervention.

Fleute, M., Lavallee, S., & Julliard, R. (1999). Incorporating a statistically based shape model into a system for computer-assisted anterior cruciate ligament surgery. *Medical Image Analysis, 3*(3), 209–222. doi:10.1016/S1361-8415(99)80020-6

Kinner, K. E. (1994). *Advances in genetic programming.* Cambridge, MA: MIT Press.

Koza, J. R. (1992). *Genetic programming, on the programming of computers by means of natural selection.* Cambridge, MA: MIT Press.

Lamecker, H., Wenckeback, T. H., & Hege, H. C. (2006). *Atlas-based 3D-shape reconstruction from X-ray images.* Paper presented at the Proceedings of the 18th International Conference on Pattern Recognition.

Loop, C. T. (1987). *Smooth subdivision surfaces based on triangles.* Master of Science thesis, University of Utha, Salt Lake.

Mason, S. (1997). Heuristic reasoning strategy for automated sensor placement. *Photogrammetric Engineering and Remote Sensing, 63*(9), 1093–1102.

Mason, S., & Grun, A. (1995). Automatic sensor placement for accurate dimensional inspection. *Computer Vision and Image Understanding, 61*(3), 454–467. doi:10.1006/cviu.1995.1034

Matalas, I. (1996). *Segmentation techniques suitable for medical images.* Unpublished doctoral thesis, Imperial College, University of London, London.

Olague, G., & Mohr, R. (2002). Optimal camera placement for accurate reconstruction. *Pattern Recognition, 35,* 927–944. doi:10.1016/S0031-3203(01)00076-0

Pentland, A., & Scalroff, S. (1991). Closed-form soultions for physically based shape modeling and recognition. *IEEE Transactions on Pattern Analysis and Machine Intelligence, 13,* 715–729. doi:10.1109/34.85660

Pito, R. (1999). A solution to the next best view problem for automated surface acquisition. *IEEE Transactions on Pattern Analysis and Machine Intelligence, 21*(10), 1016–1030. doi:10.1109/34.799908

Prakoonwit, S., & Benjamin, R. (2007). Optimal 3D surface reconstruction from a small number of conventional 2D X-ray images. *Journal of X-Ray Science and Technology, 15*(4), 197–222.

Rajamani, K. T., Ballester, M. A. G., Nolte, L. P., & Styner, M. (2005). *A novel and stable approach to anatomical structure morphing for enhanced intraoperative 3D visualization.* Paper presented at the SPIE Medical Imaging: Visualization, Image-guided Procedures, and Display.

Rajamani, K. T., Styner, M., & Joshi, S. C. (2004). *Bone model morphing for enhanced surgical visualization.* Paper presented at the 2004 IEEE Int. Symp. Biomedical Imaging: From Nano to Macro.

Rajamani, K. T., Styner, M. A., Talib, H., Zheng, G., Nolte, L. P., & Ballester, M. A. G. (2007). Statistical deformable bone models for robust 3D surface extrapolation from sparse data. *Medical Image Analysis, 11,* 99–109. doi:10.1016/j.media.2006.05.001

Ritter, D., Orman, J., Schmidgunst, C., & Graumann, R. (2007). 3D soft tissue imaging with a mobile C-arm. *Computerized Medical Imaging and Graphics, 31,* 91–102. doi:10.1016/j.compmedimag.2006.11.003

Staib, L. H., & Duncan, J. S. (1992). Boundary finding with parametrically deformable models. *IEEE Transactions on Pattern Analysis and Machine Intelligence, 17,* 1061–1075. doi:10.1109/34.166621

Subsol, G., Thirion, J. P., & Ayache, N. (1996). *Application of an automatically built 3D morphometric brain atlas: Study of cerebral ventricle shape.* Paper presented at the Visualization in Biomedical Computing.

Szelely, R., Lelemen, A., Brechbuler, C., & Gerig, G. (1996). Segmentation of 2D and 3D objects from MRI volume data using constrained elastic deformations of flexible Fourier surface models. *Medical Image Analysis, 1,* 19–34.

Wang, L.-L., Li, Y.-X., Ding, J.-L., & Ying, W.-Q. (2005). *ARG-based segmentation of overlapping objects in multi-energy X-ray image of passenger accompanied baggage.* Paper presented at the MIPPR 2005: Image Analysis Techniques.

Zbijewski, W., & Stayman, J. W. (2007). *xCAT: A mobile, flat-panel volumetric X-ray CT for head and neck imaging.* Paper presented at the IEEE Nuclear Science Symposium Conference Record.

Zheng, G., Dong, X., & Nolte, L. P. (2006). *Robust and accurate reconstruction of patient-specific 3D surface models from sparse point sets: a sequential three-stage trimmed optimization approach.* Paper presented at the 3rd Int. Workshop Medical Imaging and Augmented Reality.

Zheng, G., Dong, X., Rajamani, K. T., Zhang, X., & Styner, M. (2007). Accurate and robust reconstruction of a surface model of the proximal femur from sparse-point data and dense-point distribution model for surgical navigation. *IEEE Transactions on Bio-Medical Engineering, 54*(12), 2109–2122. doi:10.1109/TBME.2007.895736

Chapter 19
Arabic Optical Character Recognition:
Recent Trends and Future Directions

Husni Al-Muhtaseb
King Fahd University of Petroleum and Minerals, Saudi Arabia

Rami Qahwaji
University of Bradford, UK

ABSTRACT

Arabic text recognition is receiving more attentions from both Arabic and non-Arabic-speaking researchers. This chapter provides a general overview of the state-of-the-art in Arabic Optical Character Recognition (OCR) and the associated text recognition technology. It also investigates the characteristics of the Arabic language with respect to OCR and discusses related research on the different phases of text recognition including: pre-processing and text segmentation, common feature extraction techniques, classification methods and post-processing techniques. Moreover, the chapter discusses the available databases for Arabic OCR research and lists the available commercial Software. Finally, it explores the challenges related to Arabic OCR and discusses possible future trends.

INTRODUCTION

Arabic is the first language for more than 400 million people in the world. It is also used by more than 1 billion Muslims all over the world as a second language, for it is the language in which the Holy Qur'an was revealed. Arabic was added to the official languages of the United Nations in 1973 as the sixth language. The other five official languages (Chinese, English, French, Russian and Spanish) were chosen when the United Nations was founded. Also as has been reported by National Geographic (National Geographic, 2004), Arabic is expected to be one of the 5 major languages by 2050. Its importance is expected to rise, as English declines.

DOI: 10.4018/978-1-60960-477-6.ch019

Arabic is one of the Semitic languages. The Arabic script is being used/ had been used in other languages. Some of which are Hausa, Kashmiri, Kazak, Kurdish, Kyrghyz, Malay, Morisco, Pashto, Persian/Farsi, Punjabi, Sindhi, Tatar, Turkish, Uyghur, and Urdu (United Nations, 2006).

Arabic Optical Character Recognition (OCR) is an important and emerging application and research area. An OCR tool could be used to avoid retyping a scanned document or to convert the text images in the scanned document to an editable text. Such tool takes the scanned document as a picture and recognizes the text in the picture to make it available in text format.

Optical Arabic text recognition has received renewed extensive research after the recent successes in optical character recognition. Arabic text recognition, which was not researched as thoroughly as Latin, Chinese, or Japanese, is receiving more attentions from both Arabic and non-Arabic-speaking researchers.

Irrespective of the language under consideration, some typical applications of text recognition include: cheque verification, office automation, reading postal address, writer identification, and signature verification. Searching scanned documents available on the internet and searching Arabic historical manuscripts are also emerging applications. When Arabic is considered, there is real need to advance these applications.

This chapter provides a general overview of the state-of-the-art in Arabic text recognition technology. Section 2 presents the characteristics of Arabic text with respect to OCR. Section 3 introduces a typical general model for Arabic OCR. Related research on the pre-processing of text images is discussed in Section 4. Section 5 addresses the literature on segmentation of Arabic Text. Common feature extraction techniques are presented in Section 6. Section 7 discusses the used classification methods in Arabic text recognition. The post-processing related research work is addressed in Section 8. Section 9 discusses the available databases for Arabic OCR research. Section 10 lists available Commercial Arabic OCR Software. Section 11 discuses the challenges related to Arabic OCR and discusses possible future trends.

CHARACTERISTICS OF ARABIC TEXT

Arabic is a cursive language written from right to left. It has 28 basic alphabets. An Arabic letter might have up to four different shapes depending on the position of the letter in the word: whether it is a standalone letter, connected only from right (initial form), connected only from left (terminal form), or connected from both sides (medial form). Letters of a word may overlap vertically (even without touching).

Arabic letters do not have fixed size (height and width). Letters in a word can have diacritics (short vowels) such as *Fat-hah, Dhammah, Shaddah, sukoon* and *Kasrah*. Moreover, *Tanween* may be formed by having double *Fat-hah*, double *Dhammah*, or double *Kasrah*. Figure 1 lists these diacritics. These diacritics are written as strokes, placed either on top of, or below, the letters. A different diacritic on a letter may change the meaning of a word. Readers of Arabic are used to

Figure 1. Arabic short vowels (diacritics)

Fat-hah ً	Dhammah ً	Shaddah ً
Kasrah ً	Sukoon ً	TanweenFat-h ً
Tanween Dhamm ً		Tanween Kasr ً

Figure 2. An example of an Arabic sentence indicating some characteristics of Arabic text

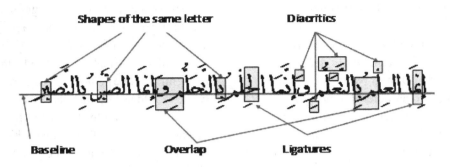

reading un-vocalized text by deducing the meaning from context.

Figure 2 shows some of the characteristics of Arabic text. It shows a base line, overlapping letters, diacritics, and two shapes of *Noon* character (initial and medial).

As stated earlier, Arabic has 28 main letters, which are shown in Figure 3. Based on the guidelines of the Arab Standardization and Metrology Organization (ASMO), some of the main letters had been extended into separate letters for easiness of presentations and usability. The standard Arabic codepages (character sets) ASMO-449, ASMO-708 and ISO 8859-6 define 36 Arabic letters (See Figure 4). When OCR is considered, it is needed to add *Lam-Alef* in its 4 different forms. Although *Lam-Alef* is a sequence of two alphabets, they are written as one set. This sequence should be treated as one set. So, four more sets should be added to the alphabets; one with bare *Alef*, the second with *Alef-Maddah*, the third with *Alef-up-Hamza* and the fourth with *Alef-down-Hamza* as shown in Figure 5. This expands the number of Arabic letters to 40. Each alphabet can take different number of shapes (from 1 to 4). Hence, the total number of shapes is 125 (one letter has only one shape, others have two, and most have four shapes).

Table 1 shows the basic Arabic letters with their categories. They are grouped onto 3 different classes according to the number of shapes a letter takes. The first Class (class 1) consists of a single shape of the *Hamza* which comes in stand-alone state (Number 1 in Table 1). *Hamza* does

Figure 3. Basic Arabic 28 letters

ا ب ت ث ج ح خ د ذ ر ز س ش ص ض ط ظ ع غ ف ق ك ل م ن ه و ي

Figure 4. Extend Arabic letters by ASMO

ء آ أ ؤ إ ئ ا ب ة ت ث ج ح خ د ذ ر ز س ش ص ض ط ظ ع غ ف ق ك ل م ن ه و ى ي

Figure 5. Expanded Arabic alphabets by adding different versions of Lam-Alef sequence

ء أ أ ؤ إ ئ ا ب ة ت ث ج ح خ د ذ ر ز س ش ص ض ط ظ ع غ ف ق ك ل م ن ه و لآ لأ لإ لا ى ي

not connect with any other letter. The second class (class 2) presents the letters that can come either as standalone or connected only from right (medial category). This class consists of *Alef Madda, Alef up Hamza, Waw Hamza, Alef down Hamza, Alef, Tah Marboutah, Dal, Dhal, Ra, Zain, Waw, Lam Alef Madda, Lam Alef Hamza up, Lam Alef Hamza down,* and *Lam Alef* (numbers 2-5, 7, 9, 15-18, and 35-39 in Table 1). The third class (class 3) consists of the letters that can be connected from either side or both sides and/or can appear as standalone. This class consists of *Hamza Kursi, Baa, Taa, Thaa, Jeem Haa, Khaa, Seen, Sheen, Sad, Dhad, Dhaa, THaa, Ain, Gain, Faa, Qaaf, Kaaf, Laam, Meem, Noun, Haa, Yaa* (numbers 6, 8, 10-14, 19-33, and 40 in Table 1). Table 2 shows a summary of these classes.

Although an Arabic letter might have up to 4 different shapes, each letter is saved using only one code. It is the duty of a built-in driver to make contextual analysis to decide the right shape to display, depending on the previous and next characters if available. When it is needed to consider different shapes of Arabic letters for a given Arabic text file, a contextual analysis algorithm is needed. Such algorithm takes the letter, its predecessor, and its successor and identifies the right shape depending on the classes of the letters.

A GENERAL MODEL FOR AN AUTOMATIC ARABIC TEXT RECOGNITION SYSTEM

Early reviews covering Arabic text recognition can be found in (Amin, 1998). More recent reviews could also be found in Lorigo and Govindaraju (Lorigo & Govindaraju, Off-line Arabic Handwriting Recognition: A survey, 2006), Haraty and Ghaddar (Haraty & Ghaddar, 2004), Ball (Ball, 2007), Burrow (Burrow, 2004), AL-Shatnawi, and Omar (AL-Shatnawi & Omar, 2008), Aburas and Gumah (Aburas & Gumah, Arabic handwriting recognition: Challenges and solutions, 2008) and Nikkhou and Choukri (Nikkhou & Choukri, 2005).

Arabic text recognition systems could be divided into two categories: Handwritten text recognition systems and printed text recognition systems. An example of handwritten text is shown in Figure 6. The handwritten recognition systems could be categorized into online recognition and offline recognition. The on-line recognition aims to recognize the characters while the writer is writing on a tablet using a stylus (see Manfredi et al. (Manfredi, Cha, Yoon, & Tappert, 2005)). Printed Arabic text recognition systems could be designed to recognize one single type of fonts (Mono-font), a set of specified fonts (Multi-font), and/ or any type of fonts (Omni-fonts). Figure 7 shows these categories. Arabic recognition systems could also address special purpose data such as numerals only, isolated character only, postal addresses, or literals numbers. The systems could also address cursive open vocabulary text such as cursive letters and letters, numerals and punctuations. Figure 8 shows these data types.

A typical generic model for an automatic Arabic text recognition system is shown in Figure 9. The automatic Arabic text recognition process starts by scanning the image containing the Arabic text. The scanned image is analyzed in the pre-processing phase to improve its condition. The pre-processing phase could include noise

Table 1. Basic shapes of Arabic alphabets

no	Stand-alone	Term.	Medial	Initial	Shapes	Class
1	ء	ء	ء	ء	1	1
2	آ	ـآ	ـآ	آ	2	2
3	أ	ـأ	ـأ	أ	2	2
4	ؤ	ـؤ	ـؤ	ؤ	2	2
5	إ	ـإ	ـإ	إ	2	2
6	ئ	ـئ	ـئـ	ئـ	4	3
7	ا	ـا	ـا	ا	2	2
8	ب	ـب	ـبـ	بـ	4	3
9	ة	ـة	ـة	ة	2	2
10	ت	ـت	ـتـ	تـ	4	3
11	ث	ـث	ـثـ	ثـ	4	3
12	ج	ـج	ـجـ	جـ	4	3
13	ح	ـح	ـحـ	حـ	4	3
14	خ	ـخ	ـخـ	خـ	4	3
15	د	ـد	ـد	د	2	2
16	ذ	ـذ	ـذ	ذ	2	2
17	ر	ـر	ـر	ر	2	2
18	ز	ـز	ـز	ز	2	2
19	س	ـس	ـسـ	سـ	4	3
20	ش	ـش	ـشـ	شـ	4	3
21	ص	ـص	ـصـ	صـ	4	3
22	ض	ـض	ـضـ	ضـ	4	3
23	ط	ـط	ـطـ	طـ	4	3
24	ظ	ـظ	ـظـ	ظـ	4	3
25	ع	ـع	ـعـ	عـ	4	3
26	غ	ـغ	ـغـ	غـ	4	3
27	ف	ـف	ـفـ	فـ	4	3
28	ق	ـق	ـقـ	قـ	4	3
29	ك	ـك	ـكـ	كـ	4	3
30	ل	ـل	ـلـ	لـ	4	3
31	م	ـم	ـمـ	مـ	4	3
32	ن	ـن	ـنـ	نـ	4	3
33	ه	ـه	ـهـ	هـ	4	3
34	و	ـو	ـو	و	2	2
35	آل	آل	ـآل	آل	2	2
36	أل	أل	ـأل	أل	2	2
37	إل	إل	ـإل	إل	2	2
38	ال	ال	ـال	ال	2	2
39	ى	ـى	ـى	ى	2	2
40	ي	ـي	ـيـ	يـ	4	3

Table 2. Classes of Arabic alphabets depending on number of possible basic shapes

Class	# of possible shapes	Alphabets
1	1	ء
2	2	ى ال إل أل آل و ز ر ذ د ة ا إ ؤ آ أ
3	4	ي ه ن م ل ك ق ف غ ع ظ ط ض ص ش س خ ح ج ث ت ب ئ

Figure 6. An example of Arabic handwritten text

Figure 7. Arabic text recognition categories

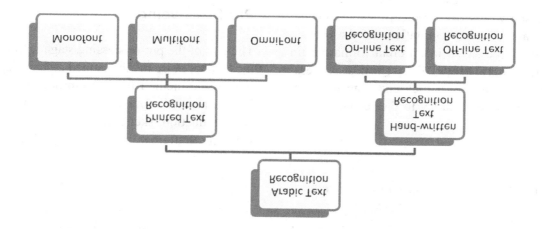

removal, skew/ slant detection and correction and normalization.

Usually, the text image is segmented into smaller images containing the lines. Depending on the used feature extraction and classification techniques, a character-based segmentation phase may or may not be required. Since Arabic text is cursive, some techniques require the segmentation of Arabic text before the feature extraction phase. During segmentation, the Arabic text image is segmented into lines. Furthermore, the line im-

ages could be segmented into words/ sub-words and then to characters or even sub-characters based on the used technique. If the image under consideration contains tables and figures, then their text is extracted for recognition. Figure 10 shows character based OCR. Figure 11 shows segmentation free OCR.

The feature extraction phase is applied to a line, a word, a sub-word, a character, or sub-character based on the method used. The features are extracted from basic units (a word, a sub-word,

Figure 8. OCR addressed data types

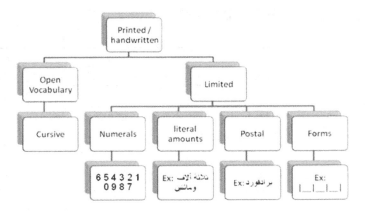

a character, or sub-character) and used in the classification and recognition. The actual recognition is done through the classification phase that produces text representation of sequences of words, sub-words, or characters that represent the text image. The representations of these basic units could be saved in different format (plain Unicode text, HTML, PDF ...). The post-processing phase is usually a spell-checking tool that could add more accuracy to the recognized resulted text.

The following sections highlight the advances and the major stages of an Arabic text recognition system.

Pre-Processing

The aim of the pre-processing stage is to reduce the level of noise in the scanned image. Depending on the method used, pre-processing might include tasks like image binarization (threshold), noise reduction, skeltonization, and component labelling. Different pre-processing classes have

Figure 9. Optical text recognition architecture

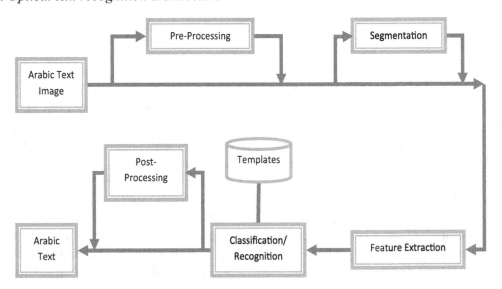

Figure 10. Character-based segmentation OCR

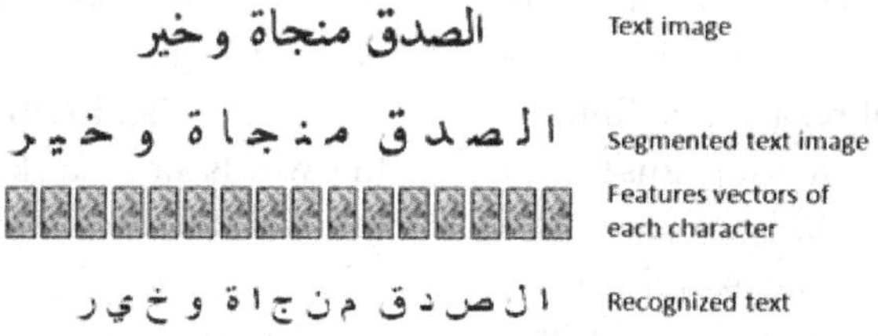

الصدق منجاة وخير Text image

ا ل ص د ق م ن ج ا ة و خ ي ر Segmented text image

Features vectors of
each character

ا ل ص د ق م ن ج ا ة و خ ي ر Recognized text

Figure 11. Segmentation-free OCR

الصدق منجاة وخير Text line image

Features vector of the
whole text line

ا ل ص د ق م ن ج ا ة و خ ي ر Recognized text

been proposed for different tasks including normalization, slope correction, slant correction and thinning, see for example Al-Ma'adeed et al. work (Al-Ma'adeed, Elliman, & Higgins, 2004). Sari et al., in (Sari & Sellami, Cursive Arabic Script Segmentation and Recognition System, 2005), used a statistical based smoothing algorithm for smoothing and noise reduction. Sarfraz et al. (Sarfraz, Nawaz, & Al-Khuraidly, Offline Arabic Text Recognition system, 2003) have introduced pre-processing techniques for the removal of isolated pixels, skew detection and correction.

A baseline estimation of handwritten words was described by Pechwitz and Margner in (M & Margner, 2002) where features related to the baseline were examined. Khorsheed and Clocksin (Khorsheed & Clocksin, 1999) used Stentiford's algorithm for thinning. Al-Khatib and Mahamud (Al-Khatib & Mahamud, 2006) addressed removing curvature effects, tilt/skew correction, and noise filtering. Another scheme for tilt correction was introduced by Sarfraz and Shahab in (Sarfraz & Shahab, An Efficient Scheme for Tilt Correction

in Arabic OCR System, 2005) which was based on finding the character *Alef* in the image and detecting the skew angle.

A transform technique (Hough Transform) known for its ability to handle distortions and noise was used by Touj et al for the recognition of Arabic printed characters in (Touj, Ben, & Amiri, 2003). Mahmoud (Mahmoud, Arabic Character-Recognition Using Fourier Descriptors and Character Contour Encoding, 1994) used normalized Fourier descriptors for Arabic OCR along with contour analysis. The contour of the primary part of the character, the dot, and the hole were extracted. Then the Fourier descriptors were computed and used for training. The normalized Fourier descriptors technique is invariant to scale, rotation, and translation. However, there is a trade-off between the gained accuracy and the processing speed. Zahour et al. introduced another contour based method to extract text-lines (Zahour, Taconet, Mercy, & Ramdane, 2001). This method was based on a partial contour tracing algorithm. It was known to be slant sensitive.

A thinning algorithm based on clustering the data image using neural network was used by Altuwaijri and Bayoumi in (Altuwaijri & Bayoumi, 1998). M. Shirali-Shahreza and S. Shirali-Shahreza pointed that when removing noise from Arabic text images, care should be taken not to remove dots that are part of the Arabic script (Shirali-Shahreza & Shirali-Shahreza, 2006). A thinning algorithm for handling poor quality Arabic text images was introduced by Cowell and Hussain in (Cowell & Hussain, 2001).

Segmentation

Zidouri et al. presented a printed Arabic character segmentation based on adaptive dissection. Their system showed promising results with some problems related to character overlapping and ligatures (Zidouri, Sarfraz, Shahab, & Jafri, 2005). Zheng et al. performed line segmentation as well as word and sub-word segmentation (Zheng, Hassin, & Tang, 2004) using horizontal histograms. However, character segmentation was based on the analysis of the upper contour of the sub-word under consideration.

Several research techniques bypass the error-prone segmentation phase by applying Hidden Markov Models (HMMs). See for examples Khorsheed (Khorsheed M. S., 2007). However, bypassing segmentation could solve the segmentation challenge only but not the remaining Arabic OCR challenges.

Sari and Sellami reported a handwritten character segmentation algorithm for isolated words. The reported algorithm was based on topological rules, which were constructed during the feature extraction phase (Sari & Sellami, Cursive Arabic Script Segmentation and Recognition System, 2005).

Some segmentation techniques divide the word into several segments where each segment could be a character, part of a character, or a group of more than one character. This might be done using morphological operations such as closing followed by opening. Similar technique was used by Lorigo and Govindaraju (Lorigo & Govindaraju, Segmentation and Pre-Recognition of Arabic Handwriting, 2005) to over-segment the words into strokes and glyphs, then reduce the possible breakpoints using prior knowledge of letter shapes. Elgammal and Ismail suggested a similar graph-based segmentation technique (Elgammal & Ismail, 2001). The suggested technique was based on the topological relation between the baseline and the line adjacency graph representation of the text, where the text is segmented onto graph units representing sub-characters. Finally, a grammar-based tool is used to construct the characters from these units.

Kandil and El-Bialy (Kandil & El-Bialy, 2004) suggested a centreline independent segmentation technique based on upwards spikes that segment an image onto isolated characters, diacritics, Hamzas, and sub-words or words.

Hadjar and Ingold presented a technique for extracting homogenous regions of complex structure Arabic documents such as newspapers (Hadjar & Ingold, 2003). In their research work, they discussed other several related segmentation algorithms including thread extraction, frame extraction, image text separation and text line extraction. A wavelet transform based segmentation algorithm was introduced by Broumandnia et al. in (Broumandnia, Shanbehzadeh, & Nourani, 2007) where segmentation points were detected by the projection of horizontal edges and their locations on baseline. Syiam et al. (Syiam, Nazmy, Fahmy, Fathi, & Ali, 2006) described an Arabic OCR system that uses histogram clustering method for the segmentation of the Arabic words.

Features Extraction

The main objective of feature selection in recognition systems is to provide minimal and efficient representation for the original input data to maximize both the effectiveness and the efficiency of the recognition process, while minimizing the processing time and complexity. According to Cheriet et al. (Cheriet, Kharma, Liu, & Suen, 2007), feature extraction methods could be classified into three categories: geometric features, structural features, and feature space transformations methods. Examples of popular geometric features include moments, histograms, and direction features. Examples of structural features include registration, line element features, Fourier descriptors, and topological features. Examples of the transformation methods include principal component analysis (PCA) and linear discriminant analysis (LDA).

Khorsheed and Clocksin used structural features for cursive Arabic words to recognize Arabic text using HMM (Khorsheed & Clocksin, 1999). The used features were the curvatures of word segments. The length of these segments was relative to other word segment lengths, while the position was relative to the baseline and the description of curved word segments. The results of this method were used to train a HMM Model to perform the recognition. Jianying et al. features included loops, cusp distance and crossing distance (Iwayama & Ishigaki, 2000). Aburas et al. used different types of features which included structural features and statistical features. Some of these features were loops, endpoints, dots, branch-points, relative locations, height, sizes, pixel densities, histograms of chain code directions, moments and Fourier descriptors (Aburas & Rehiel, Off-line Omni-style Handwriting Arabic Character Recognition System Based on Wavelet Compression, 2007). Ebrahimi et al. (Ebrahimi & Kabir, 2008) used characteristic loci as part of their features. Al-Taani (Al-Taani, 2005) suggested a feature extraction algorithm based on primary and secondary primitive features. Mahmoud in his digits recognition system (Mahmoud, Recognition of writer-independent off-line handwritten Arabic (Indian) numerals using hidden Markov models, 2008) used unit features based on the digits. The extracted features were based on angle-, distance-, horizontal-, and vertical-span

features. Majumdar developed a feature extraction scheme based on the digital curvelet transform. The features included the curvelet coefficients of the image and its morphologically altered versions (Majumdar, 2007). Ball used the character-based Word Model Recognizer (WMR) features (Ball, 2007). The model consisted of 74 features. Gagne and Parizeau suggested sub-character based features based on the orientation and curvature of the strokes (Gagne & Parizeau, 2006). A feature fusion was proposed by Sun et al. (Sun, Zeng, Liu, Heng, & Xia, 2005). They extracted two groups of feature vectors with the same sample and established the correlation criterion function between the two groups of feature vectors.

Al-Muhtaseb and Qahwaji (Al-Muhtaseb & Qahwaji, A Single Feature Extraction Algorithm for Text Recognition of Different Families of Languages, 2009) developed a language-independent single feature extraction algorithm for the recognition of Arabic, English and Bangla languages. The algorithm depends on a single type of features, which is the density distribution of the text images.

Classification

Researchers are using different techniques to recognize Arabic text. These techniques include statistical pattern recognition (Jain et al. (Jain, Duin, & Mao, 2000)), structural pattern recognition (Gupta (Gupta, Nagendraprasad, Liu, P., & Ayyadurai, 1993)), machine learning techniques including artificial neural networks (ANN) (Al-Omaria and Al-Jarrah (Al-Omaria & Al-Jarrah, 2004)), support vector machines (SVM) (Pat and Ramakrishnan (Pat & Ramakrishnan, 2008)), and multiple classifier methods (Gazzah and Ben Amara(Gazzah & Ben Amara, 2008) Chang et al. (Chang, Chen, Zhang, & Yang, 2009)).

Al Aghbari and Brook(Al Aghbari & Brook, 2009) used a 3-layer generalized neural network (NN) with 210 input nodes and 1100 output nodes to recognize Arabic words from historical manuscripts. The recognition rate was reported to be around 80%. Shaaban (Shaaban, 2008) used a two-phase classifier to recognize segmented printed Arabic characters. The classifier was based on parallel neural networks. Phase one of the classifier categorizes every Arabic character into one of eight categories. The eight categories where designed based on similarities between characters. Phase two recognizes the characters. The reported classification rate was 98%. Saeed and AlBakoor (Saeed & Albakoor, 2009) used a NN consisting of 81 input nodes, 200 hidden nodes and 28 output nodes. Three Arabic fonts were used: Arial, Arabic Transparent and Simplified Arabic. The reported recognition rate for printed Arabic was 93%.

Some other prototype systems for Arabic text/ character recognition have been reported. ORAN prototype system reported by Zidouri et al. (Zidouri A., 2004) was applied to the Naskh font and a recognition rate of 97.5% was reported.

RECAM reported by Sari and Sellami (Sari & Sellami, Cursive Arabic Script Segmentation and Recognition System, 2005) is a cursive Arabic handwritten script recognition system using word segmentation. A multi-font recognition system of printed Arabic text using the BYBLOS speech recognition system was reported by LaPre et al. in (LaPre, Zhao, Raphael, Schwartz, & Makhoul, 1996). Hamami and Berkani (Hamami & Berkani, 2002) introduced a multi-font multi-size recognition system for printed Arabic characters. The system is based on the detection of holes and concavities. Trenkle et al. (Trenkle, Gillies, Erlandson, Schlosser, & Cavin, 2001) presented a printed Arabic text recognition system with recognition rate of 93% for high quality documents and 89% for less quality documents.

Cheung et al. presented an Arabic single-font recognition system with 90% accuracy in (Cheung, Bennamoun, & Bergmann, A Recognition-Based Arabic Optical Character Recognition System, 1998). An online Arabic handwritten recognition system was presented by the same group of researchers in (Cheung, Bennamoun, & Bergmann,

An Arabic optical character recognition system using recognition-based segmentation, 2001). Aburas and Rehiel (Aburas & Rehiel, Off-line Omni-style Handwriting Arabic Character Recognition System Based on Wavelet Compression, 2007) introduced a Wavelet Compression based system for Off-line Omni-style Handwriting Arabic Character Recognition with a recognition rate of 97% in some cases.

An Arabic OCR system that uses histogram clustering method for the segmentation of Arabic words has been reported with a recognition accuracy of 91.5% by Syiam et al. (Syiam, Nazmy, Fahmy, Fathi, & Ali, 2006). Feature extraction in the reported system was based on a combination of the principle component analysis (PCA) network and characters geometric features. The classifier was designed using a decision tree induction algorithm and Multi-layered Perceptron network (MLP).

A multi-font Arabic OCR system using Hough transform for feature extraction and HMMs for classifications with 96.8% recognition rate, on some cases, was reported by Ben Amor and Ben Amra in (Ben Amor & Ben Amra, 2005). Bazzi et al. (Bazzi, LaPre, Makhoul, Raphael, & Schwartz, 1997) reported an earlier system that could be used for the recognition of English and Arabic printed text. They reported an accuracy rate of 95% for specific DARPA data.

Most of the above recognition/ classification techniques were developed to recognize isolated characters. When cursive text is considered, as a complete word or a complete string/ line, a segmentation phase is needed to segment the image into isolated characters before using one of the above techniques. The segmentation process is believed to be error-prone (see Cheriet et al. (Cheriet, Kharma, Liu, & Suen, 2007)). This is one of the motives supporting the use of HMM for the recognition of cursive Arabic script. No segmentation is needed for most of these cases, except to segment the page image into line images).

Touj et al. proposed an approach for multi-writers Arabic handwritten recognition in (Touj, Ben Amara, & Amiri, 2005). The technique uses a hybrid planar Markov model to follow the horizontal and vertical variations of writing. The model is based on different segmentation levels: horizontal, natural and vertical. Experiments using planar Markov models for Arabic handwriting have shown promising results as reported in (Miled & Ben Amara, 2001). Their results varied from 47% to 67% for different fonts. However, when they considered a selected 100 sub-words they reported an accuracy of more than 99%.

LaPre et al. used HMM based on BBN BY-BLOS Speech Recognition System to recognize multi-font printed Arabic by modifying the feature extraction phase (LaPre, Zhao, Raphael, Schwartz, & Makhoul, 1996). Khorsheed (Khorsheed M. S., 2007) used the HTK speech tool in Omni-font Arabic text recognition. The HTK is based on HMM. A multi-font Arabic OCR system using Hough-transform for feature extraction and HMMs for classifications with 96.8% accuracy in some cases was reported by Ben Amor and Ben Amra in (Ben Amor & Ben Amra, 2005). Bazzi et al. reported a HMM system that could be used for the recognition of English and Arabic printed text with accuracy reaching 95% for specific DARPA data (Bazzi, LaPre, Makhoul, Raphael, & Schwartz, 1997).

Al-Muhtaseb et al. (Al-Muhtaseb, Mahmoud, & Qahwaji, Recognition of Off-line printed Arabic text Using Hidden Markov Models, 2008) used HMM to recognize printed Arabic text of open-vocabulary of eight fonts. The reported recognition rates for single font recognition, were between 97.86% and 99.9%. The same technique was used to recognize multi-fonts with recognition rates varying from 95.61% for the 8 fonts together, to 99.2% for a category of 2 fonts (Al-Muhtaseb H., Arabic Text Recognition of Printed Manuscripts, 2010). The block diagram of the used system is shown in Figure 12.

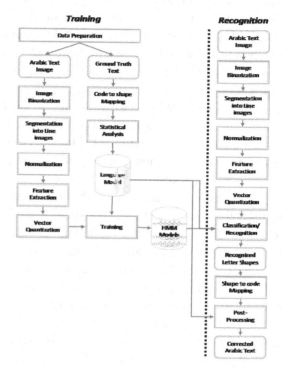

Figure 12. Printed Arabic text recognition block diagram (Al-Muhtaseb H., Arabic Text Recognition of Printed Manuscripts, 2010)

El-Mahallawy (El-Mahallawy, 2008) reported the classification of Arabic text using a HMM-based system with word error rates (WER) in the range of 0.84%-1.35% for mono-font and character error rates (CER) in the range 0.18%-0.28%. For multi-font classifications the reported word error rates were in the range 1.0%-4.4% and the reported character error rates were in the range 0.21%-0.91%.

Post-Processing and Statistical Analysis

Post-processing is the task of correcting a recognized text produced by an OCR system. Several researchers reported that post-processing could increase the recognition rates noticeably. Sari and Sellami presented a contextual-based technique for correcting Arabic words generated by OCR systems in (Sari & Sellami, Morpho-Lexical analysis for correcting OCR-generated arabic words (MOLEX), 2002). A rule-based system for correcting Arabic words operating only at the morpho-lexical level was used. An OCR system that uses linguistic information including affixes was proposed by Kanoun et al. in (Kanoun, Ennaji, Lecourtier, & Alimi, 2002). Borovikov et al. built a filter based post-OCR accuracy boost system (Borovikov, Zavorin, & Turner, 2004). The system combines different post-OCR correction filters, including a commercial spell-checker to improve the OCR results.

Statistical information of Arabic text could be used for post-processing. Statistical analysis related to the syllables of written Arabic text is still receiving little research interest. Statistical results were published in (Khedher & Abandah, 2002) for written Arabic syllables of length 1 to 8 letters. It also showed the percentages of these syllables. The analyzed text consisted of 252647 words and 1126420 characters. A second research work which aimed to prepare Arabic syllable dictionary for written Arabic to be used in OCR was introduced in (Elarian, 2006). The used text was taken from an Arabic newspaper. Al-Sulaiti (Al-Sulaiti, 2004) used a text of 842684 words

for a similar purpose. The study provided syllables of length 1 to 17. The long syllables in the study were mainly due to typos in the used text. Al-Muhtaseb et al. (Al-Muhtaseb, Mahmoud, & Qahwaji, Statistical Analysis for the support of Arabic Text Recognition, 2007) pursued statistical analysis for two books of standard classic Arabic. The statistical analysis included characters distribution in addition to written syllables of Arabic words with N-grams statistics. Several researchers have used the statistical probabilities of Arabic letters in HMM-based OCR systems.

Al-Muhtaseb (Al-Muhtaseb H., Arabic Text Recognition of Printed Manuscripts, 2010) proposed an Arabic OCR post-processing technique to correct the errors of an OCR system. The proposed technique included character-level post-processing and word level post-processing. Character level post-processing depends on encoding the shapes of letters into their letter codes. In word level post-processing, the incorrect words were figured out using a domain dictionary. Then, trials to correct each incorrect word through single substitution, deletion, or insertion were carried out. The post-processing module used the learned knowledge from the OCR system to prioritize the correcting operations between characters. The

Figure 13. Block diagram of post-processing stage. (Al-Muhtaseb H., Arabic Text Recognition of Printed Manuscripts, 2010)

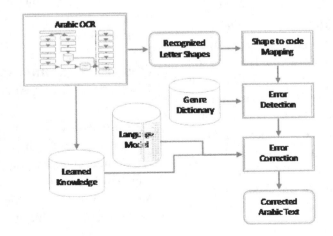

post-processing phase at the character level led to improving the recognition rates for single font classifications as the improvement in recognition reached 1%. The post-processing phase at word level proved to have positive improvement on the multi-font classifications as the improvement reached 0.8%. The block diagram presenting this technique is shown in Figure 13.

Databases

Few Arabic databases with limited content are available for research in Arabic text recognition. Some of them have been prepared for specific domains and applications such as cheques, numerals contents, and postal addresses. Farah et al. have used Arabic literal amounts (words representing numbers) of 4800 words (Farah, Souici, & Sellami, 2006). A database consisting of 26,459 Arabic names, presenting 937 Tunisian town/village names, handwritten by 411 different writers was presented by Pechwitz and Maergner in (Pechwitz & Maergner, HMM Based Approach for Handwritten Arabic Word Recognition Using the IFN/ENIT- Database, 2003) and was used in several research experiments. A database prepared from text involving 100 persons, where each person wrote 67 literal numbers, 29 of the most popular words in Arabic, three sentences representing numbers and quantities used in checks, and a free subject chosen by the writer (around 4700 handwritten words) was reported by Al-Ma'adeed et al. in (Al-Ma'adeed, Elliman, & Higgins, 2004). Alotaibi presented a small database consisting of digits. This database involved 17 persons each wrote 10 digits for 10 times (Alotaibi, 2003). An Arabic and Persian database of isolated characters consisting of 220,000 handwritten forms filled by more than 50,000 writers was presented by Soleymani and Razzazi in (Soleymani & Razzazi, 2003). The databases by Al-Ohali et al. in (Al-Ohali, Cheriet, & Suen, 2003) presented 29,498 images of subwords, 15,175 images of Indian-Arabic digits and image samples of both legal and courtesy amounts

taken from 3000 real-life bank cheques. Another database for bank-cheques included 70 words of Arabic literal amounts extracted from 5000 cheques written by 100 persons was introduced by Maddouri et al. in (Maddouri, Amiri, Belaïd, & Choisy, 2002). An automatically generated printed database of 946 Tunisian town names is discussed by Margner and Pechwitz in (Margner & Pechwitz, Synthetic Data for Arabic OCR System Development, 2001). Hamid and Haraty used 360 handwritten addresses of around 4000 words (Hamid & Haraty, 2001). The addresses were collected from students and faculty members at the Lebanese American University. Dehghan et al. (Dehghan, Faez, Ahmadi, & Shridhar, 2001) Presented a database consisting of more than 17820 names of 198 cities of Iran. Kharma et al. presented a general database with signatures which has 37,000 words, 10,000 digits, 2,500 signatures, and 500 free-form Arabic sentences (Kharma, Ahmed, & Ward, 1999). Each person wrote the 28-character alphabet ten times. DARPA Arabic Corpus consists of 345 scanned pages of printed text in 4 different fonts (Makhoul, Schwartz, Lapre, & Bazzi, 1998). The research presented by Trenkle et al. in (Trenkle, Erlandson, Gillies, & Schlosser, 1995) used 700 digitized pages from 45 printed documents. The segmentation work by Melhi in (Melhi, 2001) was based on around 240 digitized pages written by 178 persons, where each person wrote one or two pages of 10 previously prepared pages consisting of 13 lines each.

A technique to automatically generate a database for OCR systems was presented in (Margner & Pechwitz, Synthetic Data for Arabic OCR System Development, 2001). The technique which was designed to generate English databases for OCR systems was modified and used to generate Arabic Tunisian town names. Generating printed text databases automatically ensures 100% correctness of the ground truth information and allows constructing large databases. A database for the Arabic OCR of printed and handwriting text was introduced by Ben Amara et al. in (Ben

Figure 14. A proposed minimal Arabic script (Al-Muhtaseb, Mahmoud, & Qahwaji, A Novel Minimal Script for Arabic Text Recognition Databases and Benchmarks, 2009)

عزة كأب جنة طغق ميس غضيى كاف ضنط أمن فظل حنف نسكه خصنهم صنك إظ رؤى ضب
ك آح ملأ سطوع يقرء تضدح لاغ لاٱت لإض هج حت جح صخ فلي يجىئ نضص فق عض نحظ
يلغ مس ظمان ڤن طاقي تلات لأح لأه يؤس نم اٱت الڤورز ڤ ط ڤ ل ئ

Amara, Mazhoud, Bouzrara, & Ellouze, 2005). The database includes images of text phrases, words/ sub-words, isolated characters, digits, and signatures.

Benchmark databases are important to allow comparisons of efficient methods proposed by researchers. The Tunisian town/village names database (Pechwitz, Maddouri, Märgner, Ellouze, & Amiri, 2002) might be an example of a special purpose Arabic benchmark handwritten database for postal town/village names. Recently, new open-vocabulary databases that can be used as benchmarks, have been reported. Al-Muhtaseb (Al-Muhtaseb H., Arabic OCR, 2009) developed two datasets: PATS-A01 and PATS-A02. Each dataset contains eight different fonts of printed Arabic text. The PATS-A01 dataset has 2766 line images representing 65062 words, while the PATS-A02 dataset represents 5771 words and 318 line images. These datasets are designed so they both contain enough samples of basic shapes of Arabic alphabets. Slimane et al. (Slimane, Ingold, Kanoun, Alimi, & Hennebert, 2009) reported an open-vocabulary, multi-font, multi-size and multi-style Arabic printed text image database (APTI). APTI database was generated using 113284 words. The database was reported to have 10 Arabic fonts, 10 font sizes and 4 font styles. Al-Muhtaseb et al. (Al-Muhtaseb, Mahmoud, & Qahwaji, A Novel Minimal Script for Arabic Text Recognition Databases and Benchmarks, 2009) have proposed a minimal Arabic script that could be used to collect handwritten Arabic text. Although the proposed script has only three lines of Arabic text, it provides coverage of all basic

Arabic shapes of the alphabets. Figure 14 shows the proposed minimal Arabic script.

El-Mahallawy (El-Mahallawy, 2008) used 9 Arabic fonts of different sizes. For each training set he used 25 pages of text selected from Arabic websites. Each page contains around 200 words.

COMMERCIAL ARABIC OCR SOFTWARE

Several OCR software products with Arabic text recognition capabilities are available in the market. The following is a listing of some of these products:

- Readiris™ Pro from I. R. I. S. is an OCR solution for converting paper documents to digital files. The software work for different languages. A Middle East version is available for Arabic Farsi and Hebrew (IRIS, 2009).
- VERUS™ Middle East Standard from NovoDynamics is designed to recognize Arabic, Farsi, Dari, and Pashto languages, including embedded English and French (NovoDynamics, 2009).
- Sakhr™ Automatic Reader from Sakhr is an OCR solution that addresses the Arabic language. It supports Arabic, Farsi, Pashto, Jawi, and Urdu. (Sakhr, 2009).
- OmniPage from Nuance Communications is an optical character recognition application that supports more than 25 languages including Arabic (Nuance-Communications, 2009).

The data sheets of these software products claim a recognition rate reaching above 99%. However, no standard benchmarks were used to support such claims.

Independent researchers have evaluated earlier versions of some of these products by using different types of documents. The evaluation resulted in different percentages of recognition ranging from 10% to almost 100%. Several factors affected the recognition rates. Some of these factors were document quality, used fonts, and pre-trained fonts. Examples of OCR software evaluations could be found in Marton et al. (Marton, Bulbul, & Kanungo, 1999).

CHALLENGES AND FUTURE TRENDS

In this chapter we have presented the advancements and recent findings in Arabic OCR. We have shown the research directions in different phases of Arabic OCR (pre-processing, segmentation, feature extraction, classifications, and post-processing). Most of the research work in the field is being stimulated by the Arabic characteristics related to OCR. Some of these challenging characteristics are the connectivity of Arabic script, the use of dots as part of letters, the use of diacritics, having different shapes of letters depending on the context, the ligatures in handwriting and some printed fonts, and character overlapping.

Two main approaches are used to automatically recognize cursive Arabic scripts: segmentation-based and segmentation-free (holistic) approaches. The first approach is still being used by the majority of researchers where the Arabic text image is segmented into images of characters and then each character image is separately recognized. When learning algorithms are used in this approach, SVM, ANN or combination of both are used. In the second approach, each line image or word image is treated as a whole object. In this approach, HMMs have been used as the main recognition engine. As segmentation is error-prone, HMMs approaches perform better than SVM and ANN approaches. However, the number of classes used is higher than that used by SVM or ANN. It also needs a language model to be used with the classifications. So, when segmentation is considered, the trend is to avoid explicit segmentation as it is error-prone. Using HMM techniques have proven to help in the recognition process.

When looking at suggested prototypes and systems, the first question that might arise is why not to compare the performance of some of these systems? The reported systems used different datasets for different purposes and different applications. Systems developed to recognize only numerical digits consisting of ten isolated shapes cannot be compared to systems developed to recognize isolated or cursive letters consisting of hundreds of shapes. However, comparisons among systems addressing the same datasets exist. The most famous one is the comparisons between the systems developed to recognize Tunisian towns (El Abed & Margner, 2008). A competition of Arabic handwriting recognition systems based on the public databases (IFN/ENIT and ADAB) were carried out in ICDAR 2005, 2007 and 2009 conferences. The algorithms were compared based on their recognition rate. Other features such as the recognition speed, word length, writing style, and character connectivity were also considered.

The previous competitions are carried out for handwritten Arabic text. No competition currently exists for printed Arabic text because of the lack of a public and trusted database that can be used as a benchmark. A trusted database benchmark should have a very accurate transcription (ground truth information). The minimal Arabic script proposed in (Al-Muhtaseb, Mahmoud, & Qahwaji, A Novel Minimal Script for Arabic Text Recognition Databases and Benchmarks, 2009) could be used to build a printed Arabic database.

Despite the fact that several Arabic databases are used in Arabic OCR, It is not clear if the databases reported in literature contain accurate statistical distribution for the different shapes of Arabic characters or not. In some cases, some characters are appearing by more than 50 times compared to other characters (see (Melhi, 2001) as an example). Moreover, in the case of handwritten, if a database is targeted, it will be very hard to require writers to write long text, say one page or more. Even if we are able to collect two handwritten pages or more per writer, we need to ensure that all characters have adequate frequency of occurrence. We have noticed that none of the available handwritten databases claims to cover all basic shapes of Arabic letters except what has been proposed by Al-Muhtaseb (Al-Muhtaseb H., Arabic OCR, 2009).

Two types of text can be recognised by OCR systems: special purpose data and open vocabulary data. Special purpose data could be classified into several categories: numerals, postal addresses, literal amounts, and isolated letters in forms. Each of which has its own applications. These categories require specialised datasets for testing and training. As for the open vocabulary databases, not many accurate databases exist, as explained previously. Hence, researchers need to focus on providing open vocabulary Arabic datasets that has the potential to be used for different OCR purposes.

A wide range of different feature extraction schemes were used in the literature. In some research works tens of features of different types were extracted. We have noticed that the recent trends seem to focus on different types of features to represent the addressed characters. In addition to the complexity and the over-head of using different types of features, the reported accuracies in most publications did not meet the expectations. Efficient feature extraction schemes could still be proposed and developed.

Arabic multi-fonts recognition is still receiving less research work. This research gap needs to be thoroughly tackled.

More research work is needed for the post-processing of Arabic OCR. The room for re-searchers to contribute to this area and propose new techniques is still open. One future direction is to expand the post-processing techniques to include more OCR learning-based and knowledge extraction techniques. It could also be enhanced by adding morphology and syntax stages to the post-processing process.

Finally, one important application of Arabic OCR is writer identification. This emerging application is still an open area of research.

REFERENCES

Aburas, A. A., & Gumah, M. E. (2008). Arabic handwriting recognition: Challenges and solutions. (pp. 1-6).

Aburas, A. A., & Rehiel, S. M. (2007). Off-line omni-style handwriting Arabic character recognition system based on wavelet compression. *Arab Research Institute For Science & Engineering*, *3*, 123–135.

Al Aghbari, Z. A., & Brook, S. (2009). HAH manuscripts: A holistic paradigm for classifying and retrieving historical Arabic handwritten documents. *Expert Systems with Applications*, *36*, 10942–10951. doi:10.1016/j.eswa.2009.02.024

Al-Khatib, W., & Mahamud, S. (2006). *Toward content-based indexing and retrieval of Arabic manuscripts*. King Fahd University of Petroleum & Minerals.

Al-Ma'adeed, S., Elliman, D. & Higgins, C. (2004). A data base for Arabic handwritten text recognition research. *International Arab Journal on Information Technology, 1*.

Al-Muhtaseb, H. (2009). *Arabic OCR online*. Retrieved from http://faculty.kfupm.edu.sa/ics/muhtaseb/ArabicOCR

Al-Muhtaseb, H. (2010). *Arabic text recognition of printed manuscripts*. Unpublished doctoral thesis, University of Bradford, School of Computing, Informatics, and Media.

Al-Muhtaseb, H., Mahmoud, S., & Qahwaji, R. (2008). Recognition of off-line printed Arabic text using hidden Markov models. *Signal Processing, 88*(12), 2902–2912. doi:10.1016/j.sigpro.2008.06.013

Al-Muhtaseb, H., & Qahwaji, R. (2009). *A single feature extraction algorithm for text recognition of different families of languages.* Third Mosharaka International Conference on Communications, Computers and Applications (MIC-CCA2009). Amman, Jordan.

Al-Muhtaseb, H. A., Mahmoud, S., & Qahwaji, R. S. (2007). *Statistical analysis for the support of Arabic text recognition.* International Symposium on Computer and Arabic Language. Riyadh, Saudi Arabia.

Al-Muhtaseb, H.A., Mahmoud, S.A. & Qahwaji, R.S. (2009). A novel minimal script for Arabic text recognition databases and benchmarks. *International Journal of Circuits, Systems and Signal Processing*, 145-153.

Al-Ohali, Y., Cheriet, M., & Suen, C. (2003). Databases for recognition of handwritten Arabic cheques. *Pattern Recognition*, 111–121. doi:10.1016/S0031-3203(02)00064-X

Al-Omaria, F., & Al-Jarrah, O. (2004). Handwritten Indian numerals recognition system using probabilistic neural networks. *Advanced Engineering Informatics, 18*, 9–16. doi:10.1016/j.aei.2004.02.001

AL-Shatnawi, A. & Omar, K. (2008). Methods of Arabic language baseline detection-the state of art. [IJCSNS]. *International Journal of Computer Science and Network Security, 8*(10), 137–143.

Al-Sulaiti, L. (2004). *Designing and developing a corpus of contemporary Arabic. The University of Leeds, State University of New York at Buffalo.* Leeds, UK: The University of Leeds.

Al-Taani, A. T. (2005). An efficient feature extraction algorithm for the recognition of handwritten Arabic digits. *International Journal of Computational Intelligence, 2*, 107–111.

Alotaibi, Y. A. (2003). *High performance Arabic digits recognizer using neural networks.* The International Joint Conference On Neural Networks, (pp. 670-674).

Altuwaijri, M., & Bayoumi, M. (1998). A thinning algorithm for Arabic characters using ART2 neural network. *IEEE Transactions On Circuits And Systems-II: Analog And Digital Signal Processing, 45*, 260–264. doi:10.1109/82.661669

Amin, A. (1998). Off-line Arabic character recognition: The state of the art. *Pattern Recognition, 31*, 517–530. doi:10.1016/S0031-3203(97)00084-8

Ball, G. R. (2007). *Arabic handwriting recognition using machine learning approaches.* Computer Science and Engineering, State University of New York at Buffalo. State University of New York at Buffalo.

Bazzi, I., LaPre, C., Makhoul, J., Raphael, C., & Schwartz, R. (1997). *Omnifont and unlimited-vocabulary OCR for English and Arabic.* 4th International Conference Document Analysis and Recognition (ICDAR '97), (pp. 842-846).

Ben Amara, N. E., Mazhoud, O., Bouzrara, N., & Ellouze, N. (2005). ARABASE: A Relational Database for Arabic OCR Systems. *International Arab Journal of Information Technology, 2*, 259–266.

Ben Amor, N., & Ben Amra, N. E. (2005). *Multifont Arabic character recognition using Hough transform and hidden Markov models.* 4th International Symposium on Image and Signal Processing and Analysis, (pp. 285-288).

Borovikov, E., Zavorin, I., & Turner, M. (2004). *A filter based post-OCR accuracy boost system.* 1st ACM Workshop on Hardcopy Document Processing, (pp. 23-28).

Broumandnia, A., Shanbehzadeh, M., & Nourani, M. (2007). *Segmentation of printed Farsi/Arabic words*. IEEE/ACS International Conference on Computer Systems and Applications, (pp. 761-766).

Burrow, P. (2004). *Arabic handwriting recognition*. Edinburgh, UK: The University of Edinburgh.

Chang, Y., Chen, D., Zhang, Y., & Yang, J. (2009). An image-based automatic Arabic translation system. *Pattern Recognition*, *42*, 2127–2134. doi:10.1016/j.patcog.2008.10.031

Cheriet, M., Kharma, N., Liu, C. L., & Suen, C. (2007). *Character recognition systems: A guide for students and practitioners*. Wiley-Interscience.

Cheung, A., Bennamoun, M., & Bergmann, N. W. (1998). *A recognition-based Arabic optical character recognition system*. IEEE International Conference on Systems, Man, and Cybernetics, (pp. 4189-4194).

Cheung, A., Bennamoun, M., & Bergmann, N. W. (2001). An Arabic optical character recognition system using recognition-based segmentation. *Pattern Recognition*, *34*, 215–233. doi:10.1016/S0031-3203(99)00227-7

Cowell, J., & Hussain, F. (2001). *Thinning Arabic characters for feature fxtraction*. Fifth International Conference on Information Visualization (IV'01), (pp. 181-185).

Dehghan, M., Faez, K., Ahmadi, M., & Shridhar, M. (2001). Handwritten Farsi (Arabic) word recognition: A holistic approach using discrete HMM. *Pattern Recognition*, 1057–1065. doi:10.1016/S0031-3203(00)00051-0

Ebrahimi, A., & Kabir, E. (2008). A pictorial dictionary for printed Farsi subwords. *Pattern Recognition Letters*, *29*, 656–663. doi:10.1016/j.patrec.2007.11.008

El Abed, H., & Margner, V. (2008). *Arabic text recognition systems-state of the art and future trends*. International Conference Innovations in Information Technology (IIT 2008), (pp. 692-696). Al-Ain, Qatar.

El-Mahallawy, M. (2008). *A large scale HMM-based omni font-written OCR system for cursive scripts*. Cairo University, Faculty of Engineering.

Elarian, Y. S. (2006). *A lexicon of connected components for Arabic optical text recognition*. Amman, Jordan: Jordan University of Science and Technology.

Elgammal, A. M., & Ismail, M. A. (2001). *A graph-based segmentation and feature extraction framework for Arabic text recognition*. The 6th International Conference on Document Analysis and Recognition (ICDAR01), (pp. 622-626).

Farah, N., Souici, L., & Sellami, M. (2006). Classifiers combination and syntax analysis for Arabic literal amount recognition. *Engineering Applications of Artificial Intelligence*, 29–39. doi:10.1016/j.engappai.2005.05.005

Gagne, C., & Parizeau, M. (2006). Genetic engineering of hierarchical fuzzy regional representations for handwritten character recognition. [IJDAR]. *International Journal on Document Analysis and Recognition*, *8*, 223–231. doi:10.1007/s10032-005-0005-6

Gazzah, S., & Ben Amara, N. (2008). Neural networks and support vector machines classifiers for writer identification using Arabic script. *The International Arab Journal of Information Technology*, *5*(1), 92–101.

Gupta, A., Nagendraprasad, M. V., Liu, A. P., Wang, P., & Ayyadurai, S. (1993). An integrated architecture for recognition of totally unconstrained handwritten numerals. [IJDAR]. *International Journal on Document Analysis and Recognition*, *7*, 753–773.

Hadjar, K., & Ingold, R. (2003). Arabic newspaper page segmentation. *Seventh International Conference on Document Analysis and Recognition (ICDAR 2003), 2*, 895-899.

Hamami, L., & Berkani, D. (2002). Recognition system for printed multi-font and multi-size Arabic characters. *The Arabian Journal for Science and Engineering, 27*, 57–72.

Hamid, A., & Haraty, R. (2001). *A neuro-heuristic approach for segmenting handwritten Arabic text.* ACS/IEEE International Conference on Computer Systems and Applications (AICCSA'01), (pp. 110-113). Beirut, Lebanon.

Haraty, R. A., & Ghaddar, C. (2004). Arabic text recognition. *International Arab Journal of Information Technology, 1*, 156–163.

IRIS. (2009). *I.R.I.S.-OCR software and document management solutions.* Retrieved from http://www.irislink.com/

Iwayama, N., & Ishigaki, K. (2000). *Adaptive context processing in online handwritten character recognition* (pp. 469–474).

Jain, A. K., Duin, R. P., & Mao, J. (2000). Statistical pattern recognition: A review. *IEEE Transactions on Pattern Analysis and Machine Intelligence, 22*, 4–37. doi:10.1109/34.824819

Kandil, A., & El-Bialy, A. (2004). *Arabic OCR: A centerline independent segmentation technique.* The 2004 International Conference on Electrical, Electronic, and Computer Engineering (ICEEC '04), (pp. 412-415).

Kanoun, S., Ennaji, A., Lecourtier, Y., & Alimi, A. M. (2002). *Linguistic integration information in the AABATAS arabic text analysis system.* Eighth International Workshop on Frontiers in Handwriting Recognition (IWFHR'02), (pp. 389-394).

Kharma, N., Ahmed, M., & Ward, R. (1999). A new comprehensive database of handwritten Arabic words, numbers, and signatures used for OCR testing. *Proceedings of the 1999 IEEE Canadian Conference on Electrical and Computer Engineering,* (pp. 766-768). IEEE.

Khedher, M. Z., & Abandah, G. (2002). *Arabic character recognition using approximate stroke sequence.* Third International Conference on Language Resources and Evaluation (LREC2002).

Khorsheed, M. S. (2007). Offline recognition of omnifont Arabic text using the HMM ToolKit (HTK). *Pattern Recognition Letters, 28*, 1563–1571. doi:10.1016/j.patrec.2007.03.014

Khorsheed, M. S., & Clocksin, W. F. (1999). Structural features of cursive Arabic script. *10th British Machine Vision Conference, 2*, 422-431.

LaPre, C., Zhao, Y., Raphael, C., Schwartz, R., & Makhoul, J. (1996). *Multi-font recognition of printed Arabic using the BBN BYBLOS speech recognition system.* IEEE International Conference On Acoustics, Speech And Signal Processing, (pp. 2136-2139).

Lorigo, L., & Govindaraju, V. (2005). *Segmentation and pre-recognition of Arabic handwriting.* Eighth International Conference on Document Analysis and Recognition (ICDAR'05), (pp. 605-609).

Lorigo, L., & Govindaraju, V. (2006). Off-line Arabic handwriting recognition: A survey. *IEEE Transactions on Pattern Analysis and Machine Intelligence, 28*, 712–724. doi:10.1109/TPAMI.2006.102

Maddouri, S. S., Amiri, H., Belaïd, A., & Choisy, C. (2002). *Combination of local and global vision modelling for Arabic handwritten words recognition.* Eighth International Workshop on Frontiers in Handwriting Recognition (IWFHR'02), (pp. 128-135). IEEE.

Mahmoud, S. (1994). Arabic character-recognition using Fourier descriptors and character contour encoding. *Pattern Recognition, 27*, 815–824. doi:10.1016/0031-3203(94)90166-X

Mahmoud, S. (2008). Recognition of writer-independent off-line handwritten Arabic (Indian) numerals using hidden Markov models. *Signal Processing, 88*, 844–857. doi:10.1016/j.sigpro.2007.10.002

Majumdar, A. (2007). Bangla basic character recognition using digital curvelet transform. *Journal of Pattern Recognition Research, 2*, 17–26.

Makhoul, J., Schwartz, R., Lapre, C., & Bazzi, I. (1998). A script-independent methodology for optical character recognition. *Pattern Recognition, 31*, 1285–1294. doi:10.1016/S0031-3203(97)00152-0

Manfredi, M. L., Cha, S.-H., Yoon, S., & Tappert, C. (2005). Handwriting copybook style analysis of pseudo-online data. (pp. D2.1--D2.5).

Margner, V., & Pechwitz, M. (2001). *Synthetic data for Arabic OCR system development.* The 6th International Conference on Document Analysis and Recognition, ICDAR'01, (pp. 1159-1163).

Margner, V., Pechwitz, M., & El Abed, H. (2005). *ICDAR 2005 Arabic handwriting recognition competition.* International Conference on Document Analysis and Recognition, (pp. 70-74).

Marton, G. A., Bulbul, O., & Kanungo, T. (1999). *OmniPage vs. Sakhr: Paired model evaluation of two Arabic OCR products.* SPIE Conference on Document Recognition and Retrieval VI. San Jose, CA.

Melhi, M. (2001). *Off-line Arabic cursive handwriting recognition using artificial neural networks.* Bradford, UK: Bradford University.

Miled, H., & Ben Amara, N. E. (2001). *Planar Markov modeling for Arabic writing recognition: Advancement state.* Sixth International Conference on Document Analysis and Recognition, (pp. 69-73).

National Geographic. (2004, 2). *English in decline as a first language, study says.* Retrieved December 12, 2009, from http://news.nationalgeographic.com/news/2004/02/0226_040226_language.html

Nikkhou, M., & Choukri, K. (2005). *Survey on Arabic language resources and tools in the Mediterranean countries.* NEMLAR, Center for Sprogteknologi, University of Copenhagen.

NovoDynamics. (2009). *NovoDynamics VERUS™ standard.* Retrieved from http://www.novodynamics.com/

Nuance-Communications. (2009). *OmniPage OCR software.* Retrieved from http://www.nuance.com/omnipage/

Pat, P. B., & Ramakrishnan, A. (2008). Word level multi-script identification. *Pattern Recognition Letters, 29*, 1218–1229. doi:10.1016/j.patrec.2008.01.027

Pechwitz, M., Maddouri, S.S., Märgner, V., Ellouze, N. & Amiri, H. (2002). IFN/ENIT - database of handwritten Arabic words. *CIFED*, 129-136.

Pechwitz, M., & Maergner, V. (2003). *HMM based approach for handwritten Arabic word recognition using the IFN/ENIT database.* The Seventh International Conference on Document Analysis and Recognition (ICDAR 2003), (pp. 890-894).

Pechwitz, M., & Margner, V. (2002). *Baseline estimation for Arabic handwritten words.* Eighth International Workshop on Frontiers in Handwriting Recognition (IWFHR'02), (pp. 479-484).

Saeed, K., & Albakoor, M. (2009). Region growing based segmentation algorithm for typewritten and handwritten text recognition. *Applied Soft Computing, 9*, 608–617. doi:10.1016/j.asoc.2008.08.006

Sakhr. (2009). *Sakhr software*. Retrieved from http://www.sakhr.com/

Sarfraz, M., Nawaz, S. N., & Al-Khuraidly, A. (2003). *Offline Arabic text recognition system*. International Conference on Geometric Modeling and Graphics (GMAG'03), (pp. 30-36). London.

Sarfraz, M., & Shahab, S. A. (2005). *An efficient scheme for tilt correction in Arabic OCR system*. International Conference on Computer Graphics, Imaging and Vision: New Trends, (pp. 379-384).

Sari, T. & Sellami, M. (2002). Morpho-lexical analysis for correcting OCR-generated Arabic words (MOLEX). *Frontiers in Handwriting Recognition*, 461-466.

Sari, T., & Sellami, M. (2005). Cursive Arabic script segmentation and recognition system. *International Journal of Computers and Applications, 27*, 161–168. doi:10.2316/Journal.202.2005.3.202-1518

Shaaban, Z. (2008). *A new recognition scheme for machine-printed Arabic texts based on neural networks* (pp. 707–710).

Shirali-Shahreza, M. H., & Shirali-Shahreza, S. (2006). *Persian/Arabic text font estimation using dots*. Sixth IEEE International Symposium on Signal Processing and Information Technology, (pp. 420-425).

Slimane, F., Ingold, R., Kanoun, S., Alimi, A. M., & Hennebert, J. (2009). *A new Arabic printed text image database and evaluation protocols* (pp. 946–950).

Soleymani, M., & Razzazi, F. (2003). *An efficient front-end system for isolated Persian/Arabic character recognition of handwritten data-entry forms*. WSEAS Multiconference (p. 6). WSEAS.

Sun, Q.-S., Zeng, S.-G., Liu, Y., Heng, P.-A., & Xia, D.-S. (2005). A new method of feature fusion and its application in image recognition. *Pattern Recognition*, 2437–2448. doi:10.1016/j.patcog.2004.12.013

Syiam, M., Nazmy, T. M., Fahmy, A. E., Fathi, H., & Ali, K. (2006). *Histogram clustering and hybrid classifier for handwritten Arabic characters recognition. The 24th IASTED international conference on Signal processing, pattern recognition, and applications* (pp. 44–49). ACTA Press.

Touj, S., Ben, N. E., & Amiri, H. (2003). *Generalized Hough transform for Arabic optical character recognition*. International Conference on Document Analysis and Recognition (ICDAR), (pp. 1242-1246).

Touj, S., Ben Amara, N., & Amiri, H. (2005). Arabic handwritten words recognition based on a planar hidden Markov model. *International Arab Journal of Information Technology, 2*, 318–325.

Trenkle, J., Erlandson, E., Gillies, A., & Schlosser, S. (1995). *Arabic character recognition*. Symposium on Document Image, (pp. 191-195). Bowie, Maryland.

Trenkle, J., Gillies, A., Erlandson, E., Schlosser, S., & Cavin, S. (2001). *Advances In Arabic text recognition*. Symposium on Document Image Understanding Technology (SDIUT 2001), (pp. 159-168). Columbia, Maryland.

United Nations. (2006). *United Nations Arabic Language Programme*. Retrieved December 14, 2009, from http://www.un.org/depts/OHRM/sds/lcp/Arabic/

Zahour, A., Taconet, B., Mercy, P., & Ramdane, S. (2001). *Arabic hand-written text-line extraction*. Sixth International Conference on Document Analysis and Recognition (ICDAR '01), (pp. 281-285). IEEE.

Zheng, L., Hassin, A., & Tang, X. (2004). A new algorithm for machine printed Arabic character segmentation. *Pattern Recognition Letters, 25*, 1723–1729. doi:10.1016/j.patrec.2004.06.015

Zidouri, A. (2004). *ORAN: A basis for an Arabic OCR system*. International Symposium on Intelligent Multimedia, Video and Speech Processing, (pp. 703-706). Hong Kong.

Zidouri, A., Sarfraz, M., Shahab, S. A., & Jafri, S. M. (2005). *Adaptive dissection based subword segmentation of printed Arabic text*. Ninth International Conference on Information Visualization (IV'05), (pp. 239-243). IEEE Computer Society.

Compilation of References

Abbate, A., Koay, J., Frankel, J., Schroeder, S., & Das, P. (1997). Signal detection and noise suppression using a Wavelet transform signal processor: Application to ultrasonic flaw detection. *IEEE Transactions on Ultrasonics, Ferroelectrics, and Frequency Control, 44*(1), 14–26. doi:10.1109/58.585186

Abed-Meraim, K., Moulines, E., & Loubaton, P. (1997). Prediction error method for second-order blind identification. *IEEE Transactions on Signal Processing, 45*(3), 694–705. doi:10.1109/78.558487

Abidi, B., Mitckes, M., Abidi, M., & Liang, J. (2003). *Grayscale enhancement techniques of X-ray images for carry-on luggage.* Paper presented at the SPIE 6th International Conference on Quality Control by Artificial Vision, Gatlinburg, TN, USA.

Abrar, S., & Nandi, A. K. (2010). An adaptive constant modulus blind equalization algorithm and its stochastic stability analysis. *IEEE Signal Processing Letters, 17*(1), 55–58. doi:10.1109/LSP.2009.2031765

Abu Eid, R., & Landini, G. (2010). The complexity of the oral mucosa: A review of the use of fractal geometry. *Control Engineering and Applied Informatics, 12*(1), 10–14.

Aburas, A. A., & Rehiel, S. M. (2007). Off-line omni-style handwriting Arabic character recognition system based on wavelet compression. *Arab Research Institute For Science & Engineering, 3*, 123–135.

Aburas, A. A., & Gumah, M. E. (2008). Arabic handwriting recognition: Challenges and solutions. (pp. 1-6).

Acharya, S., Alonso, R., Franklin, M., & Zdonik, S. (1995). Broadcast disks: Data management for asymmetric communication environments. In *Proceedings of ACM Sigmod*, (pp. 199-210).

Acharya, S., Alonso, R., Franklin, M., & Zdonik, S. (1996). Prefetching from a broadcast disk. In *Proceedings of the International Conference on Data Engineering* (ICDE), (pp. 276-285).

Acharya, S., Franklin, M., & Zdonik, S. (1997). Balancing push and pull for data broadcast. In *Proceedings of ACM Sigmod Conference*, (pp. 183-194).

Agosta, G., Crespi Reghizzi, S., Falauto, G., & Sykora, M. (2004). JIST: Just-in-time scheduling translation for parallel processors. *Proceedings of the Third International Symposium on Parallel and Distributed Computing/Third International Workshop on Algorithms, Models and Tools for Parallel Computing on Heterogeneous Networks (ISPDC/HeteroPar '04)*, (pp. 122-132).

Akhtman, J., Bobrovsky, B. Z., & Hanzo, L. (2003). Peak-to-average power ratio reduction for OFDM modems. In. *Proceedings of IEEE GLOBECOM, 2003*, 1188–1192.

Aksoy, D., & Franklin, M. "Scheduling for Large-Scale On-Demand Data Broadcasting", In Proceedings of IEEE Info COM Conference, pp.651-659, 1998.

Al Aghbari, Z. A., & Brook, S. (2009). HAH manuscripts: A holistic paradigm for classifying and retrieving historical Arabic handwritten documents. *Expert Systems with Applications, 36*, 10942–10951. doi:10.1016/j.eswa.2009.02.024

Al-Ataby, A., Al-Nuaimy, W., Brett, C. R., & Zahran, O. (2009). *Automatic detection and classification of weld flaws in TOFD data using Wavelet transform and support vector machines.* Non-Destructive Testing 2009 conference. Blackpool, UK: BINDT.

Alberge, F., Duhamel, P., & Nikolova, M. (2002). Adaptive solution for blind identification/equalization using deterministic maximum likelihood. *IEEE Transactions on Signal Processing, 50*(4), 923–936. doi:10.1109/78.992140

Alestra, S. (2008). *Fuzzy contrast enhancement.* Retrieved from http://svg.dmi.unict.it/iplab/imagej/Plugins/Fuzzy%20Image%20Processing/Fuzzy%20Contrast%20Enhancement/FuzzyContrastEnhancement.htm

Alhaghagi, H., Green, R. J., & Hines, E. L. (2010). Double heterodyne photoparametric amplification techniques for optical wireless communications and sensing applications. *Proceedings of ICTON 10*, session We.D3.2, Munich.

Al-Khatib, W., & Mahamud, S. (2006). *Toward content-based indexing and retrieval of Arabic manuscripts.* King Fahd University of Petroleum & Minerals.

Allard, R. J., Werner, D. H., & Werner, P. L. (2003). Radiation pattern synthesis for arrays of conformal antennas mounted on arbitrarily-shaped three-dimensional platforms using genetic algorithms. *IEEE Transactions on Antennas and Propagation, 51*(5), 1054–1062. doi:10.1109/TAP.2003.811510

Al-Ma'adeed, S., Elliman, D. & Higgins, C. (2004). A data base for Arabic handwritten text recognition research. *International Arab Journal on Information Technology, 1*.

Al-Muhtaseb, H., Mahmoud, S., & Qahwaji, R. (2008). Recognition of off-line printed Arabic text using hidden Markov models. *Signal Processing, 88*(12), 2902–2912. doi:10.1016/j.sigpro.2008.06.013

Al-Muhtaseb, H. (2009). *Arabic OCR online.* Retrieved from http://faculty.kfupm.edu.sa/ics/muhtaseb/ArabicOCR

Al-Muhtaseb, H. (2010). *Arabic text recognition of printed manuscripts.* Unpublished doctoral thesis, University of Bradford, School of Computing, Informatics, and Media.

Al-Muhtaseb, H. A., Mahmoud, S., & Qahwaji, R. S. (2007). *Statistical analysis for the support of Arabic text recognition.* International Symposium on Computer and Arabic Language. Riyadh, Saudi Arabia.

Al-Muhtaseb, H., & Qahwaji, R. (2009). *A single feature extraction algorithm for text recognition of different families of languages.* Third Mosharaka International Conference on Communications, Computers and Applications (MIC-CCA2009). Amman, Jordan.

Al-Muhtaseb, H.A., Mahmoud, S.A. & Qahwaji, R.S. (2009). A novel minimal script for Arabic text recognition databases and benchmarks. *International Journal of Circuits, Systems and Signal Processing*, 145-153.

Al-Ohali, Y., Cheriet, M., & Suen, C. (2003). Databases for recognition of handwritten Arabic cheques. *Pattern Recognition*, 111–121. doi:10.1016/S0031-3203(02)00064-X

Al-Omaria, F., & Al-Jarrah, O. (2004). Handwritten Indian numerals recognition system using probabilistic neural networks. *Advanced Engineering Informatics, 18*, 9–16. doi:10.1016/j.aei.2004.02.001

Alotaibi, Y. A. (2003). *High performance Arabic digits recognizer using neural networks.* The International Joint Conference On Neural Networks, (pp. 670-674).

AL-Shatnawi, A. & Omar, K. (2008). Methods of Arabic language baseline detection-the state of art. [IJCSNS]. *International Journal of Computer Science and Network Security, 8*(10), 137–143.

Al-Sulaiti, L. (2004). *Designing and developing a corpus of contemporary Arabic. The University of Leeds, State University of New York at Buffalo.* Leeds, UK: The University of Leeds.

Al-Taani, A. T. (2005). An efficient feature extraction algorithm for the recognition of handwritten Arabic digits. *International Journal of Computational Intelligence, 2*, 107–111.

Altschuler, M. D., Trotter, D. E., & Orrall, F. Q. (1972). Coronal holes. *Solar Physics, 26*, 354–365. doi:10.1007/BF00165276

Altuwaijri, M., & Bayoumi, M. (1998). A thinning algorithm for Arabic characters using ART2 neural network. *IEEE Transactions On Circuits And Systems-II: Analog And Digital Signal Processing, 45*, 260–264. doi:10.1109/82.661669

Amblard, P. O., Moussaoui, S., Dudok de Wit, T., Aboudarham, J., Kretzschmar, M., Lilensten, J., & Auchère, F. (2008). The EUV sun as the superposition of elementary suns. *Astronomy & Astrophysics, 487*, L13–L16. doi:10.1051/0004-6361:200809588

Amermend, D., & Aristugi, M. (2006). An index allocation method for data access over multiple wireless broadcast channel. *IPSJ Digital Courier, 2*, 852–862. doi:10.2197/ipsjdc.2.852

Amin, A. (1998). Off-line Arabic character recognition: The state of the art. *Pattern Recognition, 31*, 517–530. doi:10.1016/S0031-3203(97)00084-8

Anastasi, R., Madaras, E., Seebo, J., & Winfree, W. (2007). Terahertz NDE for aerospace applications. In Chen, C. H. (Ed.), *Ultrasonic and advanced methods for nondestructive testing and material characterization* (pp. 279–302). Dartmouth, RI: World Scientific. doi:10.1142/9789812770943_0012

Aschwanden, M. J. (2005a). 2D feature recognition and 3D reconstruction in solar EUV images. *Solar Physics, 228*, 339–358. doi:10.1007/s11207-005-2788-5

Aschwanden, M. J. (2005b). *Physics of the solar corona*. Chichester, UK: Praxis Publishing Ltd.

Aschwanden, M. J. (2010). Solar image processing techniques with automated feature recognition. *Solar Physics, 262*, 235–275. doi:10.1007/s11207-009-9474-y

Aschwanden, M. J. (2010). Image processing techniques and feature recognition in solar physics. *Solar Physics, 262*, 235–275. doi:10.1007/s11207-009-9474-y

Atesok, K., Finkelstein, J., Khoury, A., Peyser, A., Weil, Y., & Liebergall, M. (2007). The use of intraoperative three-dimensional imaging (ISO-C-3D) in fixation of intraarticular fractures. *Injury. International Journal of the Care of the Injured, 38*, 1163–1169.

Attrill, G. D. R., & Wills-Davey, M. J. (2009). Automatic detection and extraction of coronal dimmings from *SDO/AIA* data. *Solar Physics, 143*.

Bäck, T. (1996). *Evolutionary algorithms in theory and practice: Evolution strategies, evolutionary programming, genetic algorithms*. Oxford University Press.

Bäck, T., Hammel, U., & Schwefel, H. P. (1997). Evolutionary computation: Comments on the history and current state. *IEEE Transactions on Evolutionary Computation, 1*(1), 3–17. doi:10.1109/4235.585888

Bakes, A. R., & Bruno, O. M. (2008). *A new approach to estimate fractal dimension of texture images*. IEEE International Conference on Image Processing. (LNCS 5099), (pp. 136-143).

Ball, G. R. (2007). *Arabic handwriting recognition using machine learning approaches*. Computer Science and Engineering, State University of New York at Buffalo. State University of New York at Buffalo.

Bannister, J. A., Fratta, L., & Gerla, M. (1990). *Topological design of the wavelength-division optical network*. INFOCOM '90. Ninth Annual Joint Conference of the IEEE Computer and Communication Societies, (pp. 1005–1013).

Barbara, D. (1999). Mobile computing and databases-a survey. *IEEE Transactions on Knowledge and Data Engineering, 11*(1), 108–117. doi:10.1109/69.755619

Barra, V., Delouille, V., Kretzschmar, M., & Hochedez, J. (2009). Fast and robust segmentation of solar EUV images: Algorithm and results for solar cycle 23. *Astronomy & Astrophysics, 505*, 361–371. doi:10.1051/0004-6361/200811416

Barry, J. R., Kahn, J. M., Krause, W. J., Lee, E. A., & Messerschmitt, D. G. (1993). Simulation of multipath impulse response for indoor wireless optical channels. *IEEE Journal on Selected Areas in Communications, 11*(3), 367–379. doi:10.1109/49.219552

Barry, J. R., Kahn, J. M., Lee, E. A., & Messerschmitt, D. G. (1991). High-speed nondirective optical communication for wireless networks. *IEEE Network, 5*(6), 44–54. doi:10.1109/65.103810

Bartlett, M. S., Hager, J. C., Ekman, P., & Sejnowski, T. J. (1999). Measuring facial expressions by computer image analysis. *Psychophysiology, 36*, 253–263. doi:10.1017/S0048577299971664

Bazzi, I., LaPre, C., Makhoul, J., Raphael, C., & Schwartz, R. (1997). *Omnifont and unlimited-vocabulary OCR for English and Arabic*. 4th International Conference Document Analysis and Recognition (ICDAR '97), (pp. 842-846).

Bellon, J., Sibley, M. J. N., Wisely, D. R., & Greaves, S. D. (1999). Hub architecture for infra-red wireless networks in office environments. *IEEE Proceedings in Optoelectronics*, *146*(2), 78–82. doi:10.1049/ip-opt:19990313

Ben Amara, N. E., Mazhoud, O., Bouzrara, N., & Ellouze, N. (2005). ARABASE: A Relational Database for Arabic OCR Systems. *International Arab Journal of Information Technology*, *2*, 259–266.

Ben Amor, N., & Ben Amra, N. E. (2005). *Multifont Arabic character recognition using Hough transform and hidden Markov models*. 4th International Symposium on Image and Signal Processing and Analysis, (pp. 285-288).

Benameur, S., Mignotte, M., Labelle, H., & Guise, J. A. d. (2005). A herarchical statistical modeling approach for the unsupervised 3D biplanar reconstruction of the scoliotic spine. *IEEE Transactions on Bio-Medical Engineering*, *52*(12), 2041–2057. doi:10.1109/TBME.2005.857665

Benameur, S., Mignotte, M., Parent, S., Labelle, H., Skalli, W., & Guise, J. A. d. (2003). 3D/2D registration and segmentation of scoliotic vertebrae using statistical models. *Computerized Medical Imaging and Graphics*, *27*, 321–337. doi:10.1016/S0895-6111(03)00019-3

Bentum, M. J., van Langenvelde, H. J., Tilanus, R., & Friberg, P. (2006). *The extended sub-millimeter array–proof of concept by connecting the JCMT*. Paper presented at SPS-DARTS, Second annual IEEE Benelux/DSP Valley Signal Processing Symposium, Antwerp, Belgium.

Bentum, M. J., Verhoeven, C. J. M., Boonstra, A. J., van der Veen, A. J., & Gill, E. K. A. (2009). *A novel astronomical application for formation flying small satellites*. Paper presented at the 60th International Astronautical Congress, Daejeon, Republic of Korea, 12 – 16 October, 2009.

Benveniste, A., & Goursat, M. (1984). Blind equalizer. *IEEE Transactions on Communications*, *32*(8), 871–883. doi:10.1109/TCOM.1984.1096163

Berghmans, D., Hochedez, J.-F., Defise, J.-M., Lecat, J. H., Nicula, B., & Slemzin, V. (2006). SWAP onboard PROBA2, a new EUV imager for solar monitoring. *Advances in Space Research*, *38*, 1807–1811. doi:10.1016/j.asr.2005.03.070

Berghmans, D., Foing, B. H., & Fleck, B. (2002). Automated detection of CMEs in LASCO data. In Wilson, A. (Ed.), *From solar min to max: Half a solar cycle with SOHO* (pp. 437–440).

Bernasconi, P. N., Rust, D. M., & Hakim, D. (2005). Advanced automated solar filament detection and characterization code: Description, performance, and results. *Solar Physics*, *228*, 97–117. doi:10.1007/s11207-005-2766-y

Besl, P., & McKay, N. D. (1992). A method for registration of 3D shapes. *IEEE Transactions on Pattern Analysis and Machine Intelligence*, *14*(2), 239–256. doi:10.1109/34.121791

Bessios, A. G., & Nikias, C. L. (1995). POTEA: The power cepstrum and tricoherence equalization algorithm. *IEEE Transactions on Communications*, *43*(11), 2667–2671. doi:10.1109/26.481216

BioImLab. (2007). *Biomedical imaging laboratory (BioImLab) is part of the Department of Information Engineering (DEI) of the University of Padova, Italy*. Retrieved from http://bioimlab.dei.unipd.it/index.html

Bitmead, R. R., Kung, S. Y., Anderson, B. D. O., & Kailath, T. (1978). Greatest common divisors via generalized Sylvester and Bezout matrices. *IEEE Transactions on Automatic Control*, *23*(6), 1043–1047. doi:10.1109/TAC.1978.1101890

Bobin, J., Starck, J.-L., Fadili, J. M., & Moudden, Y. (2008). Blind source separation: The sparsity revolution. *Advances in Imaging and Electron Physics*, *152*, 221–306. doi:10.1016/S1076-5670(08)00605-8

Bogert, B., Healy, M., & Tukey, J. (1963). The Quefrency analysis of time series for echoes: Cepstrum, pseudo-autocovariance, cross cepstrum, and saphe cracking. In M. Rosenblatt (Ed.), *Proceedings of the Symposium on Time Series Analysis*. (pp. 209-243). New York: John Wiley and Sons Inc.

Borovikov, E., Zavorin, I., & Turner, M. (2004). *A filter based post-OCR accuracy boost system*. 1st ACM Workshop on Hardcopy Document Processing, (pp. 23-28).

Boursier, Y., Lamy, P., Llebaria, A., Goudail, F., & Robelus, S. (2009). The ARTEMIS catalog of LASCO coronal mass ejections. Automatic recognition of transient events and Marseille inventory from synoptic maps. *Solar Physics, 257*, 125–147. doi:10.1007/s11207-009-9370-5

Bovik, A. (2005). Introduction to digital image and video processing. In Bovik, A. C. (Ed.), *Handbook of image & video processing*. Amsterdam: Elsevier Academic Press. doi:10.1016/B978-012119792-6/50065-6

Bradley, R., Backer, D., Parsons, A., Parashare, C., & Gugliucci, N. E. (2005). PAPER: A Precision Array to Probe the Epoch of Reionization. *Bulletin of the American Astronomical Society, 37*, 1216.

Bradski, G. K. (2008). *Learning OpenCV: Computer vision with the OpenCV library* (1st ed.). Sebastopol, CA: O'Reilly Media, Inc.

Bregman, J. D. (2000). Concept design for a Low Frequency Array. *Proceedings of the Society for Photo-Instrumentation Engineers, 4015*, 19–33.

Brigham, E. (1988). *The fast fourier transform and its applications*. NJ: Prentice-Hall Inc.

Brodatz, P. (1966). *Textures: A photographic album for artists and designers*. Dover Publications.

Broers, H., Caarls, W., Jonker, P., & Kleihorst, R. (2005). Architecture study for smart cameras. *Proceedings of the European Optical Society Conference on Industrial Imaging and Machine Vision*, (pp. 39–49).

Broumandnia, A., Shanbehzadeh, M., & Nourani, M. (2007). *Segmentation of printed Farsi/Arabic words*. IEEE/ACS International Conference on Computer Systems and Applications, (pp. 761-766).

Brueckner, G. E., Howard, R. A., & Koomen, M. J. (1995). The Large angle spectroscopic coronagraph (LASCO). *Solar Physics, 162*, 357–402. doi:10.1007/BF00733434

Burrow, P. (2004). *Arabic handwriting recognition*. Edinburgh, UK: The University of Edinburgh.

Burse, K., Yadav, R. N., & Shrivastava, S. C. (2010). Channel equalization using neural networks: A review. *IEEE Transactions on Systems, Man and Cybernetics. Part C, Applications and Reviews, 40*(3), 352–357. doi:10.1109/TSMCC.2009.2038279

Byrne, J. P., Gallagher, P. T., McAteer, R. T. J., & Young, C. A. (2009). The kinematics of coronal mass ejections using multiscale methods. *Astronomy & Astrophysics, 495*, 325–334. doi:10.1051/0004-6361:200809811

Cacciola, M., Morabito, F. C., & Versaci, M. (2007). Computational intelligence methodologies for non-destructive testing/evaluation applications. In Chen, C. H. (Ed.), *Ultrasonic and advanced methods for nondestructive testing and material characterization* (pp. 493–516). Dartmouth, RI: World Scientific. doi:10.1142/9789812770943_0021

Calhoun, V. D., & Adali, T. (2009). Feature-based fusion of medical imaging data. *IEEE Transactions on Information Technology in Biomedicine, 13*(5), 711–720. doi:10.1109/TITB.2008.923773

Candès, E., Demanet, L., Donoho, D., & Ying, L. (2005). Fast discrete curvelet transforms. *Multiscale Modeling and Simulation, 5*, 861–899. doi:10.1137/05064182X

Candy, J. (1988). *Signal processing: The modern approach*. NJ: McGraw-Hill Inc.

Cao, G. (2003). A scalable low-latency cache invalidation strategy for mobile environments. *IEEE Transactions on Knowledge and Data, 15*(5), 1251. doi:10.1109/TKDE.2003.1232276

Carruthers, J. B., Carroll, S. M., & Kannan, P. (2003). Propagation modelling for indoor optical wireless communications using fast multi-receiver channel estimation. *IEE Proceedings. Optoelectronics, 150*(5), 473–481. doi:10.1049/ip-opt:20030527

Carruthers, J. B., & Kahn, J. M. (1997). Modeling of nondirected wireless infrared channels. *IEEE Transactions on Communications, 45*(10), 1260–1268. doi:10.1109/26.634690

Carruthers, J. B., & Kannan, P. (2002). Iterative site-based modelling for wireless infrared channels. *IEEE Transactions on Antennas and Propagation, 50*(5), 759–765. doi:10.1109/TAP.2002.1011244

Carson, C., Belongie, S., Greenspan, H., & Malik, J. (2002). Blobworld: Image segmentation using expectation-maximization and its application to image querying. *IEEE Transactions on Pattern Analysis and Machine Intelligence, 24*(8), 1026–1038. doi:10.1109/TPAMI.2002.1023800

Cartz, L. (1995). *Nondestructive testing: Radiography, ultrasonics, liquid penetrant, magnetic particle, eddy current*. Illinois: ASM International.

Castrillon-Santana, M., Deniz-Suarez, O., Anton-Canalis, L., & Lorenzo-Navarro, J. (2008). *Face and facial feature detection evaluation*. Paper presented at the Third International Conference on Computer Vision Theory and Applications, VISAPP08.

Chakareski, J., & Frossard, P. (2006). Rate-distortion optimized distributed packet scheduling of multiple video streams over shared communication resources. *IEEE Transactions on Multimedia*, *8*(2), 207–218. doi:10.1109/TMM.2005.864284

Chan, C. S., Edwards, P. J., & Hawkes, D. J. (2003). Integration of ultrasound based registration with statistical shape models for computer-assisted orthopaedic surgery. *Proceedings of SPIE Medical Imaging*, 414-424.

Chang, Y., Chen, D., Zhang, Y., & Yang, J. (2009). An image-based automatic Arabic translation system. *Pattern Recognition*, *42*, 2127–2134. doi:10.1016/j.patcog.2008.10.031

Charlesworth, J. P., & Temple, J. (2001). *Engineering applications of ultrasonic time-of-flight diffraction* (2nd ed.). Baldock, UK: Research Studies Press.

Chen, K., Yun, X., He, Z., & Han, C. (2007). Synthesis of sparse planar arrays using modified real genetic algorithm. *IEEE Transactions on Antennas and Propagation*, *55*(4), 1067–1073. doi:10.1109/TAP.2007.893375

Chen, H., Xiao, Y., & Shen, X. (2006). Update–based cache access and replacement in wireless data access. *IEEE Transactions on Mobile Computing*, *5*(12), 1734–1748. doi:10.1109/TMC.2006.188

Chen, C.-H., Chi, C.-Y., & Chen, W. T. (1996). New cumulant-based inverse filler criteria for deconvolution of nonminimum phase systems. *IEEE Transactions on Signal Processing*, *44*(5), 1292–1297. doi:10.1109/78.502346

Chen, Y., & Beaulieu, N. C. (2005). NDA estimation of SINR for QAM signals. *IEEE Communications Letters*, *9*(8), 688–690. doi:10.1109/LCOMM.2005.1496583

Chen, Y., & Medioni, G. (1992). Object modeling by registratoin of multiple range images. *Image and Vision Computing*, *10*(3), 145–155. doi:10.1016/0262-8856(92)90066-C

Cheng, D. K. (1989). *Field and wave electromagnetics*. Addison-Wesley Publishing.

Cheng, Q. (2009). A constant-modulus algorithm for carrier frequency offset estimation in OFDM systems. *Proceedings of the 2009 IEEE Region 10 Conference (TENCON 2009)*, Singapore.

Cheriet, M., Kharma, N., Liu, C. L., & Suen, C. (2007). *Character recognition systems: A guide for students and practitioners*. Wiley-Interscience.

Chetverikov, D., & Hanbury, A. (2002). Finding defects in texture using regularity and local orientation. *Pattern Recognition*, *35*, 203–218. doi:10.1016/S0031-3203(01)00188-1

Cheung, A., Bennamoun, M., & Bergmann, N. W. (2001). An Arabic optical character recognition system using recognition-based segmentation. *Pattern Recognition*, *34*, 215–233. doi:10.1016/S0031-3203(99)00227-7

Cheung, A., Bennamoun, M., & Bergmann, N. W. (1998). *A recognition-based Arabic optical character recognition system*. IEEE International Conference on Systems, Man, and Cybernetics, (pp. 4189-4194).

Chi, C.-Y., Chen, C.-Y., Chen, C.-H., & Feng, C.-C. (2003). Batch processing algorithms for blind equalization using higher order statistics. *IEEE Signal Processing Magazine*, *20*(1), 25–49. doi:10.1109/MSP.2003.1166627

Childers, D., Skinner, D., & Kemerait, R. (1977). The cepstrum: A guide to processing. *Proceedings of the IEEE*, *65*(10), 1428–1443. doi:10.1109/PROC.1977.10747

Choi, S., Cichocki, A., Park, H.-M., & Lee, S.-Y. (2005). Blind source separation and independent component analysis: A review. *Neural Information Processing Letters*, *6*, 1–57.

Christensen, H. C., Schüz, J., Kosteljanetz, M., Poulsen, H. S., Thomsen, J., & Johansen, C. (2004). Cellular telephone and risk of acoustic neuroma. *American Journal of Epidemiology*, *159*(3), 277–283. doi:10.1093/aje/kwh032

Chui, C. K. (1992). *An introduction to wavelets.* San Diego: Academic Press.

Chung, Y., Chen, C. C., & Lee, C. (2006). Design and performance evaluation of broadcast algorithms for time constrained data retrieval. *IEEE Transactions on Knowledge and Data Engineering, 18*(11), 1526–1543. doi:10.1109/TKDE.2006.171

Cioffi, J. M., Dudevoir, G. P., Eyuboglu, M. V., & Forney, G. D. Jr. (1995). MMSE decision feedback equalisation and coding- Part I and II. *IEEE Transactions on Communications, 43*(10), 2582–2604. doi:10.1109/26.469441

Cioffi, J. M., Jagannathan, S., Mohseni, M., & Ginis, G. (2007). CuPON: The copper alternative to PON 100 Gb/s DSL networks. *IEEE Communications Magazine, 45*(6), 132–139. doi:10.1109/MCOM.2007.374437

Cohen, F. S., Fan, Z., & Attali, S. (1991). Automated inspection of textile fabrics using textural models. *IEEE Transactions on Pattern Analysis and Machine Intelligence, 13*(8), 803–808. doi:10.1109/34.85670

Cohn, J. F., & Kanade, T. (2007). Use of automated facial image analysis for measurement of emotion expression. In Coan, J. A., & Allen, J. J. B. (Eds.), *The handbook of emotion elicitation and assessment* (p. 483). New York, Oxford: Oxford University Press Series in Affective Science.

Cohn, J. F. (2007). Foundations of human computing: Facial expression and emotion. In Huang, T., Nijolt, A., Pantic, M., & Pentland, A. (Eds.), *State of the art survey. Lecture notes in artificial intelligence* (pp. 1–16). Berlin, Heidelberg: Springer.

Cohn, J. F., Ambadar, Z., & Ekman, P. (2006). Observer-based measurement of facial expressions with the facial action coding system. In J.A. Coan & A.J.B. (Eds.), *The handbook of emotion elicitation and assessment.* (pp. 203-221). Oxford, New York: Oxford University Press Series in Affective Science.

Colak, T., & Qahwaji, R. (2009). Automated solar activity prediction: A hybrid computer platform using machine learning and solar imaging for automated prediction of solar flares. *Space Weather, 7*, 6001. doi:10.1029/2008SW000401

Colak, T., & Qahwaji, R. (2008). Automated McIntosh-based classification of sunspot groups using MDI images. *Solar Physics, 248,* 277–296. doi:10.1007/s11207-007-9094-3

Collet, C., Chanussot, J., & Chehdi, K. (Eds.). (2010). *Multivariate image processing. Digital Signal and Image Processing Series.* London: Wiley.

Colmenarez, A. J., & Huang, T. S. (1997). *Face detection with information-based maximum discrimination.* Paper presented at the IEEE International Conference Computer Vision and Pattern Recognition.

COMSOL AB. (2006a). *FEMLAB electromagnetics module (Version 3.1).*

COMSOL AB. (2006b). *FEMLAB reference manual (Version 3.1).*

Conlon, P. A., Gallagher, P. T., McAteer, R. T. J., Ireland, J., Young, C. A., & Kestener, P. (2008). Multifractal properties of evolving active regions. *Solar Physics, 248,* 297–309. doi:10.1007/s11207-007-9074-7

Conlon, P. A., Kestener, P., McAteer, R., & Gallagher, P. (2009). *Magnetic fields, flares & forecasts.* In AAS/Solar physics division meeting.

Cootes, T. F., Edward, G. E., & Taylor, C. J. (2001). Active appearance models. *IEEE Transactions on Pattern Analysis and Machine Intelligence, 23*(6), 681–685. doi:10.1109/34.927467

Cootes, T. F., Taylor, C. J., Cooper, D. H., & Graham, J. (1995). Active shape models-their training and application. *Computer Vision and Image Understanding, 61*(1), 38–59. doi:10.1006/cviu.1995.1004

Cowell, J., & Hussain, F. (2001). *Thinning Arabic characters for feature fxtraction.* Fifth International Conference on Information Visualization (IV'01), (pp. 181-185).

Crisan, A. D. (2005). *Image processing using fractal techniques.* Unpublished doctoral thesis, Politehnica University Bucharest.

Crisan, D., Dobrescu, R., & Planinšič, P. (2007). Mammographic lesions discrimination based on fractal dimension as an indicator. *Proceedings of the 14th International Conference on Systems, Signals and Image Processing, IWSSIP 2007,* (pp. 275-280).

Criscuoli, S., Rast, M. P., Ermolli, I., & Centrone, M. (2007). On the reliability of the fractal dimension measure of solar magnetic features and on its variation with solar activity. *Astronomy & Astrophysics, 461,* 331–338. doi:10.1051/0004-6361:20065951

Daley, W.D. (1999). *Computational approaches for decision support in real time broiler processing.* Georgia Institute of Technology: A Unit of the University System of Georgia.

Darwin, C. (1872). *The expression of the emotions in man and animals.* London: J.Murray. doi:10.1037/10001-000

Davies, E. (2005). *Machine vision: Theory, algorithms, practicalities.* San Francisco: Morgan Kauffmann Publishers.

Deb, K., & Kalyanmoy, D. (2001). *Multi-objective optimization using evolutionary algorithms.* John Wiley & Sons.

Deb, K., Pratap, A., Agarwal, S., & Meyarivan, T. (2002). A fast and elitist multiobjective genetic algorithm: NSGA-II. *IEEE Transactions on Evolutionary Computation, 6*(2), 182–197. doi:10.1109/4235.996017

Deb, K., Agrawal, S., Pratap, A., & Meyarivan, T. (2000). A fast elitist non-dominated sorting genetic algorithm for multi-objective optimization: NSGA-II. *Proceedings of the Parallel Problem Solving from Nature VI Conference,* (pp. 849-858).

Dehghan, M., Faez, K., Ahmadi, M., & Shridhar, M. (2001). Handwritten Farsi (Arabic) word recognition: A holistic approach using discrete HMM. *Pattern Recognition,* 1057–1065. doi:10.1016/S0031-3203(00)00051-0

Delaboudinière, J. P., Artzner, G. E., Brunaud, J., & Gabriel, A. H. (1995). EIT: Extreme-Ultraviolet Imaging Telescope for the *SoHO* mission. *Solar Physics, 162,* 291–312. doi:10.1007/BF00733432

Delmas, J.-P., Gazzah, H., Liavas, A. P., & Regalia, P. A. (2000). Statistical analysis of some second-order methods for blind channel identification/equalization with respect to channel undermodeling. *IEEE Transactions on Signal Processing, 48*(7), 1984–1998. doi:10.1109/78.847785

Delouille, V., Chainais, P., & Hochedez, J.-F. (2008). Quantifying and containing the curse of high resolution coronal Imaging. *Annales Geophysicae, 26,* 3169–3184. doi:10.5194/angeo-26-3169-2008

Desai, A., & Milner, S. (2005). Autonomous reconfiguration in free-space optical sensor networks. *IEEE Journal on Selected Areas in Communications, 23*(8), 1556–1563. doi:10.1109/JSAC.2005.852183

Dickenson, R. J., & Ghassemlooy, Z. (2003). A feature extraction and pattern recognition receiver employing wavelet analysis and artificial intelligence for signal detection in diffuse optical wireless communications. *IEEE Transactions on Wireless Communications, 10*(2), 64–72. doi:10.1109/MWC.2003.1196404

Ding, Z., Kennedy, R. A., Anderson, B. D. O., & Johnson, C. R. (1991). IllConvergence of Godard blind equalizers in data communications systems. *IEEE Transactions on Communications, 39*(9). doi:10.1109/26.99137

Ding, L., & Martinez, A. M. (2008). *Precise detailed detection of faces and facial features.* Paper presented at the IEEE Conference on Computer Vision and Pattern Recognition.

Djahani, P., & Kahn, J. M. (2000). Analysis of infrared wireless links employing multibeam transmitters and imaging diversity receivers. *IEEE Transactions on Communications, 48*(12), 2077–2088. doi:10.1109/26.891218

Dobrescu, R., Dobrescu, M., Mocanu, S., & Popescu, D. (2010a). Malignant skin lesions detection based on texture and fractal image analysis. *Proceedings of the International Conference of WSEAS MUSP, 10,* 21–26.

Dobrescu, R., & Ichim, L. (2008). Using fractal dimension as discriminator of infected HeLa cells from spectrophotometric images. *International Journal of Functional Informatics and Personalised Medicine, 1*(1), 53–67. doi:10.1504/IJFIPM.2008.018292

Dobrescu, R., Vasilescu, C., & Ichim, L. (2007). Fractal and scaling analysis in tumor growth evaluation. *WSEAS Transactions on Systems, 1*(6), 102–108.

Dobrescu, M. (2005). *Distributed image processing techniques for multimedia applications.* Unpublished doctoral dissertation, Politehnica Univ. of Bucharest.

Dobrescu, R. (2010b). *FractIm—a fractal application for medical diagnosis.* (Scientific Report 61-031). Retrieved from http://isis.pub.ro/proiecte/Imago

Dobrescu, R., & Ionescu, F. (2003). Fractal dimension based technique for database image retrieval. *Proceedings of the First IAFA Symposium*, (pp. 107-112).

Domingo, V., Fleck, B., & Poland, A. I. (1995). *SoHO*: The Solar and Heliospheric Observatory. *Space Science Reviews, 72*, 81–84. doi:10.1007/BF00768758

Donato, G. (1999). Classifying facial actions. *IEEE Transactions on Pattern Analysis and Machine Intelligence, 21*(10), 974–989. doi:10.1109/34.799905

Dougherty, E., & Laplante, P. (1995). *Introduction to real-time imaging. Bellingham, WA/Piscataway*. NJ: SPIE Press/IEEE Press.

Dryden, I., & Mardia, K. (1998). *Statistical shape analysis*. John Wiley and Sons.

Dudok de Wit, T., & Auchère, F. (2007). Inferring temperature from morphology in solar EUV images. *Astronomy & Astrophysics, 466*, 347–355. doi:10.1051/0004-6361:20066764

Dudok de Wit, T., Lilensten, J., Aboudarham, J., Amblard, P.-O., & Kretzschmar, M. (2005). Retrieving the solar EUV spectrum from a reduced set of spectral lines. *Annales Geophysicae, 23*, 3055–3069. doi:10.5194/angeo-23-3055-2005

Dudok de Wit, T. (2006). Fast segmentation of solar extreme ultraviolet images. *Solar Physics, 239*, 519–530. doi:10.1007/s11207-006-0140-3

Eberhart, R. C., & Kennedy, J. (1995). A new optimizer using particle swarm theory. *Proceedings of the Sixth International Symposium on Micromachine and Human Science*, (pp. 39-43). Japan: Nagoya.

Ebrahimi, A., & Kabir, E. (2008). A pictorial dictionary for printed Farsi subwords. *Pattern Recognition Letters, 29*, 656–663. doi:10.1016/j.patrec.2007.11.008

Eddy, J. A. (1976). The Maunder minimum. *Science, 192*, 1189–1202. doi:10.1126/science.192.4245.1189

Edward, G. E., Taylor, C. J., & Cootes, T. F. (1998). *Interpreting face images using active appearance models*. Paper presented at the International Conference on Automatic Face and Gesture Recognition.

Egges, A., Kshirsagar, S., & Magnenat-Thalmann, N. (2003). *A model for personality and emotion simulation*. Paper presented at the Knowledge-Based Intelligent Information & Engineering Systems (KES 2003).

Ekman, P. (Ed.). (1973). *Darwin and facial expression-a century of research in review*. New York, London: Academic Press.

Ekman, P., & Friesen, W. V. (1976). Measuring facial movement. *Environmental Psychology and Nonverbal Behavior, 1*(1), 56–75. doi:10.1007/BF01115465

Ekman, P., & Friesen, W. V. (1978). *Facial action coding system: A technique for the measurement of facial movement*. Consulting Psychologies Press.

El Abed, H., & Margner, V. (2008). *Arabic text recognition systems-state of the art and future trends*. International Conference Innovations in Information Technology (IIT 2008), (pp. 692-696). Al-Ain, Qatar.

Elarian, Y. S. (2006). *A lexicon of connected components for Arabic optical text recognition*. Amman, Jordan: Jordan University of Science and Technology.

Elgammal, A. M., & Ismail, M. A. (2001). *A graph-based segmentation and feature extraction framework for Arabic text recognition*. The 6th International Conference on Document Analysis and Recognition (ICDAR01), (pp. 622-626).

El-Mahallawy, M. (2008). *A large scale HMM-based omni font-written OCR system for cursive scripts*. Cairo University, Faculty of Engineering.

Endres, T. J., Anderson, B. D. O., Johnson, C. R., & Green, M. (1999). Robustness to fractionally spaced equalizer length using the constant modulus criterion. *IEEE Transactions on Signal Processing, 47*(2), 544549. doi:10.1109/78.740141

Epema, D. H. J., & de Jongh, F. C. M. (1999). Proportional share-scheduling in single-server and multiple-server computing systems. *Performance Evaluation Review, 27*(3), 7–10. doi:10.1145/340242.340295

European Commission. (2008). *Broadband access in the EU: Situation at 1 July 2008*. Document COCOM08-41 FINAL.

Falch, M., & Henten, A. (2009). Achieving universal access to broadband. *Informatica Economică, 13*(2), 166–174.

Falconer, D. A., Moore, R. L., & Gary, G. A. (2008). Magnetogram measures of total nonpotentiality for prediction of solar coronal mass ejections from active regions of any degree of magnetic complexity. *The Astrophysical Journal, 689*, 1433–1442. doi:10.1086/591045

Farah, N., Souici, L., & Sellami, M. (2006). Classifiers combination and syntax analysis for Arabic literal amount recognition. *Engineering Applications of Artificial Intelligence, 29*–39. doi:10.1016/j.engappai.2005.05.005

Fasel, B., & Luettin, J. (2003). Automatic facial expression analysis: A survey. *Pattern Recognition, 36*, 259–275. doi:10.1016/S0031-3203(02)00052-3

Feldman, U., & Landi, E. (2008). The temperature stricture of solar coronal plasmas. *Physics of Plasmas, 15*, 056501. doi:10.1063/1.2837044

Fetterolf, P. C., & Anandalingam, G. (1991). Optimizing interconnections of local area networks: An approach using simulated annealing. *ORSA Journal on Computing, 3*(4), 275–287.

Fijalkow, I., Touzni, A., & Treichler, J. R. (1997). Fractionally spaced equalization using CMA: Robustness to channel noise and lack of disparity. *IEEE Transactions on Signal Processing, 45*(1), 5666. doi:10.1109/78.552205

Fletcher, R. (2000). *Practical methods of optimization.* John Wiley & Sons Ltd.

Fleute, M., Lavallee, S., & Julliard, R. (1999). Incorporating a statistically based shape model into a system for computer-assisted anterior cruciate ligament surgery. *Medical Image Analysis, 3*(3), 209–222. doi:10.1016/S1361-8415(99)80020-6

Fleute, M., & Lavallee, S. (1998). *Building a complete surface model from sparse data using statistical shape models: Application to computer assisted knee surgery system.* Paper presented at the 1st International Conference of Medical Image Computing and Computer-Assisted Intervention.

Fleute, M., & Lavallee, S. (1999). *Nonrigid 3D/2D registration of images using statistical models.* Paper presented at the 2nd International Conference on Medical Image Computing and Computer-Assisted Intervention.

Forney, C. D. Jr. (1972). Maximum-likelihood sequence estimation of digital sequences in the presence of inter-symbol interference. *IEEE Transactions on Information Theory, 18*(3), 363–378. doi:10.1109/TIT.1972.1054829

Foster, K. R., Schepps, J. L., Stoy, R. D., & Schwan, H. P. (1979). Dielectric properties of brain tissue between 0.01 and 10 GHz. *Physics in Medicine and Biology, 24*(6), 1187–1197. doi:10.1088/0031-9155/24/6/008

Fuller, N., Aboudarham, J., & Bentley, R. D. (2005). Filament recognition and image cleaning on Meudon H-alpha spectroheliograms. *Solar Physics, 227*, 61–73. doi:10.1007/s11207-005-8364-1

Gabriel, C. (1996). *Compilation of the dielectric properties of body tissues at RF and microwave frequencies.* (Brooks Air Force Technical Report AL/OE-TR-1996-0037).

Gagne, C., & Parizeau, M. (2006). Genetic engineering of hierarchical fuzzy regional representations for handwritten character recognition. [IJDAR]. *International Journal on Document Analysis and Recognition, 8*, 223–231. doi:10.1007/s10032-005-0005-6

Gallagher, P. T., Moon, Y. J., & Wang, H. (2002). Active-region monitoring and flare forecasting I. Data processing and first results. *Solar Physics, 209*, 171–183. doi:10.1023/A:1020950221179

Gallagher, P. T., Young, C. A., Byrne, J. P., & McAteer, R. T. J. (2010). (in press). Coronal mass ejections detection using wavelets, curvelets and ridgelets: Applications for space weather monitoring. *Advances in Space Research.* doi:10.1016/j.asr.2010.03.028

Garcia-Zambrana, A., & Puerta-Notario, A. (2003). Novel approach for increasing the peak-to-average optical power ratio in rate-adaptive optical wireless communication systems. *IEEE Proceedings in Optoelectronics, 150*(5), 439–444. doi:10.1049/ip-opt:20030526

Gardner, W. A. (1991). Exploitation of spectral redundancy in cyclostationary signals. *IEEE Signal Processing Magazine, 8*(2), 14–36. doi:10.1109/79.81007

Gazzah, S., & Ben Amara, N. (2008). Neural networks and support vector machines classifiers for writer identification using Arabic script. *The International Arab Journal of Information Technology, 5*(1), 92–101.

Gentle, J. E. (2004). *Random number generation and Monte Carlo methods* (2nd ed.). New York: Springer.

Georgoulis, M. K. (2005). Turbulence in the solar atmosphere: Manifestations and diagnostics via solar image processing. *Solar Physics, 228*, 5–27. doi:10.1007/s11207-005-2513-4

Geusebroek, J. M., Smeulders, A. W. M., & Geerts, H. (2001). A minimum cost approach for segmenting networks of lines. *International Journal of Computer Vision, 43*(2), 99–lll. doi:10.1023/A:1011118718821

Gfeller, F. R., & Bapst, U. (1979). Wireless in-house data communication via diffuse infrared radiation. *Proceedings of the IEEE, 67*(11), 1474–1486. doi:10.1109/PROC.1979.11508

Giannakis, G. B., & Halford, S. D. (1997). Blind fractionally spaced equalization of noisy FIR channels: Direct and adaptive solutions. *IEEE Transactions on Signal Processing, 45*(9), 2277–2292. doi:10.1109/78.622950

Gissot, S. F., & Hochedez, J.-F. (2007). Multiscale optical flow probing of dynamics in solar EUV images. Algorithm, calibration, and first results. *Astronomy & Astrophysics, 464*, 1107–1118. doi:10.1051/0004-6361:20065553

Gitlin, R. D., & Weinstein, S. B. (1981). Fractionally spaced equalisation: An improved digital transversal equalizer. *The Bell System Technical Journal, 60*(2), 275–296.

Godard, D. N. (1980). Selfrecovering equalization and carrier tracking in twodimensional data communication systems. *IEEE Transactions on Communications, 28*(11), 1876–1875. doi:10.1109/TCOM.1980.1094608

Golub, G. H., & van Loan, C. F. (1996). *Matrix computations*. Baltimore: The Johns Hopkins University Press.

Gonzalez, R. (2005). OFDM over indoor wireless optical channel. *IEEE Proceedings in Optoelectronics, 152*, 199–204. doi:10.1049/ip-opt:20045065

Gonzalez, R. C., & Woods, R. E. (1992). *Digital image processing*. Reading, MA: Addison Wesley.

González, O., Militello, C., Rodriguez, S., Pérez-Jiménez, R., & Ayala, A. (2002). Error estimation of the impulse response on diffuse wireless infrared indoor channels using a Monte Carlo ray-tracing algorithm. *IEEE Proceedings in Optoelectronics, 149*(5-6), 222–227. doi:10.1049/ip-opt:20020545

González-Nuevo, J., Argüeso, F., López-Caniego, M., Toffolatti, L., Sanz, J. L., & Vielva, P. (2006). The Mexican hat wavelet family: Application to point-source detection in cosmic microwave background maps. *Monthly Notices of the Royal Astronomical Society, 369*, 1603–1610. doi:10.1111/j.1365-2966.2006.10442.x

Goral, C. M., Torrance, K. E., Greenberg, D. P., & Battaile, B. (1984). Modeling the interaction of light between diffuse surfaces. *SIGGRAPH '84: Proceedings of the 11th annual conference on Computer graphics and interactive techniques.* (pp. 213–222). New York: ACM Press.

Graps, A. L. (1995). An introduction to wavelets. *IEEE Computational Science & Engineering, 2*(2), 50–61. doi:10.1109/99.388960

Green, R. J., Joshi, H., Higgins, M. D., & Leeson, M. S. (2008). Recent developments in indoor optical wireless systems. *IET Communications, 2*(1), 3–10. doi:10.1049/iet-com:20060475

Green, R. J., & McNeill, M. G. (1989). The bootstrap transimpedance amplifier-a new configuration. *IEE Proceedings on Circuits & Systems, 136*(2), 57–61. doi:10.1049/ip-g-2.1989.0009

Green, R. J., Sweet, C., & Idrus, S. (2005). Optical wireless links with enhanced linearity and selectivity. *Journal of Optical Networking, 4*(10), 671–684. doi:10.1364/JON.4.000671

Green, R. J., Joshi, H., Higgins, M. D., & Leeson, M. S. (2008). Bandwidth extension in optical wireless receiver-amplifiers. In *International Conference on International Transparent Optical Networks, ICTON 2008.* (pp. 201–204.)

Gu, M., & Tong, L. (1999). Geometrical characterizations of constant modulus receivers. *IEEE Transactions on Signal Processing, 47*(10), 27452756.

Gunst, A. W., & Bentum, M. J. (2007). *Signal processing aspects of the LOFAR*. Paper presented at the IEEE International Conference on Signal Processing and Communications, Dubai, United Arab Emirates.

Gunst, A. W., & Bentum, M. J. (2008). *The current design of the LOFAR instrument*. Paper presented at the URSI General Assembly, Chicago, IL.

Guo, Y., & Han, Y. Zhou, Q. & Duo, Z. (2006). Decision circle based dual-mode constant modulus blind equalization algorithm. *Proceedings of 8th International Conference on Signal Processing*, Beijing.

Gupta, A., Nagendraprasad, M. V., Liu, A. P., Wang, P., & Ayyadurai, S. (1993). An integrated architecture for recognition of totally unconstrained handwritten numerals. [IJDAR]. *International Journal on Document Analysis and Recognition, 7*, 753–773.

H.264/AVC Reference Software. (2009). *JM 15.1*. Retrieved from http://iphome.hhi.de/suehring/tml

Hadjar, K., & Ingold, R. (2003). Arabic newspaper page segmentation. *Seventh International Conference on Document Analysis and Recognition (ICDAR 2003), 2*, 895-899.

Haiguang, C., & Li, A. (1994). A fast filtering algorithm for image enhancement. *IEEE Transactions on Medical Imaging, 13*(3), 557–564. doi:10.1109/42.310887

Halang, W. A., Gumzej, R., Colnaric, M., & Druzovec, M. (2000). Measuring the performance of real-time systems. *The International Journal of Time-Critical Computing Systems, 18*, 59–68.

Hamami, L., & Berkani, D. (2002). Recognition system for printed multi-font and multi-size Arabic characters. *The Arabian Journal for Science and Engineering, 27*, 57–72.

Hameed, S. & Vaidya, N.H. (1999). Efficient algorithms for scheduling data broadcast. *ACM/Baltzer Journal of Wireless Network, 5*(3), 183-193.

Hameed, S., & Vaidya, N. H. (1997). Log-time algorithms for scheduling single and multiple channel data broadcast. In *Proceedings of the 3rd ACM MOBICOM*, (pp. 90-99).

Hamid, A., & Haraty, R. (2001). *A neuro-heuristic approach for segmenting handwritten Arabic text*. ACS/IEEE International Conference on Computer Systems and Applications (AICCSA'01), (pp. 110-113). Beirut, Lebanon.

Han, X. (2007). Sonic infrared imaging: A novel NDE technology for detection of cracks/ delaminations/ disbands in materials and structures. In Chen, C. H. (Ed.), *Ultrasonic and advanced methods for nondestructive testing and material characterization* (pp. 369–384). Dartmouth, RI: World Scientific. doi:10.1142/9789812770943_0016

Haralick, R. M., Shanmugam, K., & Dinstein, I. (1973). Textural features for image classification. *IEEE Transactions on Systems, Man, and Cybernetics, 3*(6), 610–621. doi:10.1109/TSMC.1973.4309314

Haratcherev, I., Taal, J., Langendoen, K., Lagendijk, R., & Sips, H. (2005). Automatic IEEE802.11 rate control for streaming applications. *Wireless Communications in Mobile Computing, 5*(4), 421–437. doi:10.1002/wcm.301

Haraty, R. A., & Ghaddar, C. (2004). Arabic text recognition. *International Arab Journal of Information Technology, 1*, 156–163.

Harvey, K. L., & Recely, F. (2002). Polar coronal holes during cycles 22 and 23. *Solar Physics, 211*, 31–52. doi:10.1023/A:1022469023581

Hashemi, H., Yun, G., Kavehrad, M., Behbahani, F., & Galko, P. A. (1994). Indoor propagation measurements at infrared frequencies for wireless local area networks applications. *IEEE Transactions on Vehicular Technology, 43*(3), 562–576. doi:10.1109/25.312790

Haupt, R. L. (1995). An introduction to genetic algorithms for electromagnetics. *IEEE Antennas and Propagation Magazine, 37*(2), 7–15. doi:10.1109/74.382334

Hayes, M. H. (1996). *Statistical digital signal processing and modeling*. John Wiley & Sons Inc.

Haykin, S. (2001). *Adaptive filter theory* (4th ed.). Englewood Cliffs, NJ: Prentice-Hall.

Haykin, S., & Chen, Z. (2005). The cocktail party problem. *Neural Computation, 17*, 1875–1902. doi:10.1162/0899766054322964

He, Z., Liang, Y., Chen, L., Ahmad, I., & Wu, D. (2005). Power-rate-distortion analysis for wireless video communication under energy constraints. *IEEE Transactions on Circuits and Systems for Video Technology, 15*(5), 645–658. doi:10.1109/TCSVT.2005.846433

Henkel, W., & Kessler, T. (1999). *A simplified impulse-noise model for the xDSL test environment*. Paper presented at the Broadcast Access Conference (BAC'99), Cracow, Poland.

Henney, C. J., & Harvey, J. W. (2005). Automated coronal hole detection using He 1083 nm spectroheliograms and photospheric magnetograms. In K. Sankarasubramanian, M. Penn, & A. Pevtsov, (Eds.), *Large-scale structures and their role in solar activity*. (p. 261).

Hiasat, A. A., Al-Ibrahim, M. M., & Gharaibeh, K. M. (1999). Design and implementation of a new efficient median filtering algorithm. *IEEE Proceedings on Vision. Image and Signal Processing, 146*(5), 273–278. doi:10.1049/ip-vis:19990444

Higgins, M. D., Green, R. J., & Leeson, M. S. (2009). A genetic algorithm method for optical wireless channel control. *Journal of Lightwave Technology, 27*(6), 760–772. doi:10.1109/JLT.2008.928395

Higgins, M. D., Green, R. J., & Leeson, M. S. (2009a). Receiver alignment dependence of a GA controlled optical wireless transmitter. *Journal of Optics. A, Pure and Applied Optics, 11*(7), 375–403. doi:10.1088/1464-4258/11/7/075403

Higgins, M. D., Green, R. J., & Leeson, M. S. (2008). *Genetic algorithm channel control for indoor optical wireless communications*. In *International Conference on International Transparent Optical Networks, ICTON 2008*. (pp. 189–192).

Higgins, P. A., Gallagher, P. T., McAteer, R. T. J., & Bloomfield, D. S. (2010). Solar magnetic feature detection and tracking for space weather monitoring. In *Advances in space research: Space weather advances* (Article in press).

Hochedez, J.-F., Jacques, L., Verwichte, E., Berghmans, D., Wauters, L., Clette, F., et al. (2002). Multiscale activity observed by EIT/SoHO. In H. Sawaya-Lacoste (Ed.), *Proceedings of the second solar cycle and space weather euroconference*. (pp. 115-118). Noordwijk, The Netherlands.

Hou, J., & O'Brien, D. C. (2006). Vertical handover-decision-making algorithm using fuzzy logic for the integrated radio-and-ow system. *IEEE Transactions on Wireless Communications, 5*(1), 176–185. doi:10.1109/TWC.2006.1576541

Hou, J., & O'Brien, D. C. (2005). Adaptive inter-system handover for heterogeneous rf and ir networks. In *Proceedings of 19th IEEE International Parallel and Distributed Processing Symposium, 2005*. (pp. 125a–125a).

Howard, R. A., Moses, J. D., Vourlidas, A., & Newmark, J. S. (2008). Sun earth connection coronal and heliospheric investigation (SECCHI). *Space Science Reviews, 136*, 67–115. doi:10.1007/s11214-008-9341-4

Howson, D. P., & Smith, R. B. (1970). *Parametric amplifiers*. McGraw-Hill.

Hsu, R., Abdel-Mottaleb, M., & Jain, A. K. (2002). Face detection in color images. *IEEE Transactions on Pattern Analysis and Machine Intelligence, 24*(5), 696–706. doi:10.1109/34.1000242

Hu, Q., Lee, W.-C., & Lee, D. L. (2001). A hybrid index technique for power efficient data broadcast. *Distributed and Parallel Databases, 9*, 151–177. doi:10.1023/A:1018944523033

Hu, Q., Lee, W. C., & Lee, D. L. (1999). Indexing techniques for wireless data broadcast under data clustering and scheduling. In *Proceedings of the 8th ACM International Conference on Information and Knowledge Management*, (pp. 351-358).

Hua, Y. (1996). Fast maximum likelihood for blind identification of multiple FIR channels. *IEEE Transactions on Signal Processing, 44*(3), 661–672. doi:10.1109/78.489039

Huang, E. W., & Wornell, G. W. (2007). Peak-to-average power reduction for low-power OFDM systems. In *Proceedings of, ICC*, 2924–2929.

Huang, J.-L., & Chen, M.-S. (2004). Dependent data broadcasting for unordered queries in a multiple channel mobile environment. *IEEE Transactions on Knowledge and Data Engineering, 16*(9), 1143–1156. doi:10.1109/TKDE.2004.39

Huang, D., & Aviyente, S. (2008). Wavelet feature selection for image classification. *IEEE Transactions on Image Processing, 17*(9), 1709–1720. doi:10.1109/TIP.2008.2001050

Huang, Y., Sistla, P., & Wolfson, O. (1994). Data replication for mobile computers. In *Proceedings of the ACM SIGMOD*, (pp. 13-24).

Hubbard, B. (1998). *The world according to wavelets: The story of a mathematical technique in the making.* Wellesley, MA: AK Peters.

Hudson, H. S., Acton, L. W., & Freeland, S. L. (1996). A long-duration solar flare with mass ejection and global consequences. *The Astrophysical Journal, 470,* 629. doi:10.1086/177894

Hult, T., & Mohammed, A. (2004). Suppression of EM fields using active control algorithms and MIMO antenna system. *Radioengineering Journal, 13*(3), 22–25.

Hult, T., & Mohammed, A. (2005). Power constrained active suppression of electromagnetic fields using MIMO antenna system. *Journal of Communications Software and Systems, 1*(1).

Hunter, H., & Moreno, J. A. (2003). New look at exploiting data parallelism in embedded systems. *Proceedings of the International Conference on Compilers, Architectures, and Synthesis for Embedded Systems,* (pp. 159–169).

Hyvarinen, A., Karhunen, J., & Oja, E. (2001). *Independent component analysis.* New York: John Wiley & Sons. doi:10.1002/0471221317

Hyvärinen, A., Hurri, J., & Hoyer, P. O. (2009). *Natural image statistics: A probabilistic approach to early computational vision.* Berlin: Springer.

Hyvärinen, A., & Oja, E. (2000). Independent Component Analysis: algorithms and applications. *Neural Networks, 13,* 411–430. doi:10.1016/S0893-6080(00)00026-5

I.T.U.T.S. Sector (2005). *Advanced video coding for generic audiovisual services.* ITU-T Recommendation H.264. (ITU-T Rec.H, 14496-14410).

Iakovidis, D. K. (2009). A pattern similarity scheme for medical image retrieval. *IEEE Transactions on Information Technology in Biomedicine, 13*(4), 442–450. doi:10.1109/TITB.2008.923144

IEC. (1989). *Draft standard-Loran receivers for ships,* International Electrotechnical Commission, IEC Technical Committee No. 80.

Ikhlef, A. & Le Guennec, D. (2007). A simplified constant modulus algorithm for blind recovery of MIMO QAM and PSK signals: A criterion with convergence analysis. *Eurasip Journal on Wireless Communications and Networking.*

ILA. (2007). *Enhanced Loran (eLoran) definition document V1.0.* International Loran Association.

Imielinski, T., Viswanathan, S., & Badrinath, B. R. (1997). Data on air: Organization and access. *IEEE Transactions on Knowledge and Data Engineering, 9*(3), 353–371. doi:10.1109/69.599926

Imielinski, T., Viswanathan, S., & Badrinath, B. R. (1994). Energy efficient indexing on air. *Proceedings of the ACM Sigmod Conference,* (pp. 25-36).

Inhester, B., Feng, L., & Wiegelmann, T. (2008). Segmentation of loops from coronal EUV images. *Solar Physics, 248,* 379–393. doi:10.1007/s11207-007-9027-1

Institute, N. E. (2010). Facts about the cornea and corneal disease. Retrieved from http://www.nei.nih.gov/health/cornealdisease/

Intel Corporation. (2008). *Intel(R) VTune(TM) Performance Analyzer 9.0.*

Ireland, J., Young, C. A., McAteer, R. T. J., Whelan, C., Hewett, R. J., & Gallagher, P. T. (2008). Multiresolution analysis of active region magnetic structure and its correlation with the Mount Wilson classification and flaring activity. *Solar Physics, 252,* 121–137. doi:10.1007/s11207-008-9233-5

IRIS. (2009). *I.R.I.S.-OCR software and document management solutions.* Retrieved from http://www.irislink.com/

ISO/IEC. I.S. (2001). *Information technology-coding of audio-visual objects.* Geneva.

Ito, K., Aoki, T., et al. (2008). *Medical image registration using phase-only correlation for distorted dental radiographs.* 19th International Conference on Pattern Recognition, 2008. ICPR 2008. Erie, J.C., McLaren, A.W. & Patel, S.V. (2009). Perspectives confocal microscopy in ophthalmology. Elsevier.

Iwayama, N., & Ishigaki, K. (2000). *Adaptive context processing in online handwritten character recognition* (pp. 469–474).

Jablon, N. K. (1989). Carrier recovery for blind equalization. *Proceedings of the International Conference on Acoustics, Speech, and Signal Processing (ICASSP-89), 2,* (pp. 1211-1214).

Jähne, B., Horst Haußecker, H., & Geißler, P. (Eds.). (1999). *Handbook of computer vision and applications, volume 2: Signal processing and pattern recognition. Ch.12: Texture analysis.* Academic Press.

Jain, A. K., Duin, R. P., & Mao, J. (2000). Statistical pattern recognition: A review. *IEEE Transactions on Pattern Analysis and Machine Intelligence, 22*, 4–37. doi:10.1109/34.824819

Jang, R., & Sun, C. (1993). Neuro-fuzzy modeling and control. *IEEE Transactions on Systems, Man, and Cybernetics, 23*(3), 378–406.

Jansky, K. G. (1932). Directional studies of atmospherics at high frequencies. *Proceedings of IRE, 20*, 1920–1932. doi:10.1109/JRPROC.1932.227477

Jelinek, H. F., Milosevic, N. T., & Ristanovic, D. (2010). The morphology of alpha ganglion cells in mammalian species: A fractal analysis study. *Control Engineering and Applied Informatics, 12*(1), 3–9.

Jester, S., & Falcke, H. (2009). Science with a lunar low-frequency array: From the dark ages of the Universe to nearby exoplanets. *New Astronomy Reviews, 53*, 1–26. doi:10.1016/j.newar.2009.02.001

Johansson, S. (2000). *Active control of propeller-induced noise in aircraft.* Unpublished doctoral dissertation, Blekinge Institute of Technology.

Johns, M., & Silverman, B. G. (2001). *How emotions and personality effect the utility of alternative decisions: A terrorist target selection case study.* Paper presented at the Tenth Conference on Computer Generated Forces and Behavioral Representation.

Johnson, R. (1992). An experimental investigation of three Eigen DF techniques. *IEEE Transactions on Aerospace and Electronic Systems, 28*(3), 852–860. doi:10.1109/7.256305

Johnson, C. R. Jr, & Anderson, B. D. O. (1995). Godard blind equalizer error surface characteristics: White, zeromean, binary case. *International Journal of Adaptive Control and Signal Processing, 9*(4), 301324.

Johnson, C. R. Jr, Schniter, P., Endres, T. J., Behm, J. D., Brown, D. R., & Casas, R. A. (1998). Blind equalization using the constant modulus criterion: A review. *Proceedings of the IEEE, 86*(10), 19271950. doi:10.1109/5.720246

Kahn, J. M., Krause, W. J., & Carruthers, J. B. (1995). Experimental characterization of non-directed indoor infrared channels. *IEEE Transactions on Communications, 43*(234), 1613–1623. doi:10.1109/26.380210

Kaiser, M. L., Kucera, T. A., & Davila, J. M. (2008). The *STEREO* mission: An introduction. *Space Science Reviews, 136*, 5–16. doi:10.1007/s11214-007-9277-0

Kanade, T., Cohn, J. F., & Tian, Y. L. (2000). *Comprehensive database for facial expression analysis.* Paper presented at the Fourth IEEE International Conference on Automatic Face and Gesture Recognition (FG '00).

Kandil, A., & El-Bialy, A. (2004). *Arabic OCR: A centerline independent segmentation technique.* The 2004 International Conference on Electrical, Electronic, and Computer Engineering (ICEEC '04), (pp. 412-415).

Kanoun, S., Ennaji, A., Lecourtier, Y., & Alimi, A. M. (2002). *Linguistic integration information in the AABATAS arabic text analysis system.* Eighth International Workshop on Frontiers in Handwriting Recognition (IWFHR'02), (pp. 389-394).

Karras, D. A., & Karkanis, S. A. Iakovidis, D.K., Maroulis, D.E. & Mertzios, B.G. (2001). Improved defect detection in manufacturing using novel multidimensional wavelet feature extraction involving vector quantization and PCA techniques. *Proceedings of the 8th Panhellenic Conference on Informatics*, (pp. 139-143).

Kasap, Z., & Magnenat-Thalmann, N. (2007). Intelligent virtual humans with autonomy and personality: State of the art. *Intelligent Decision Technologies, 1*(1-2), 3–15.

Kassim, N. E., Polisensky, E. J., Clarke, T. E., Hicks, B. C., Crane, P. C., Stewart, K. P., et al. (2005). The long wavelength array. In N. Kassim, M. Perez, W. Junor & P. Henning (Eds.), *Astronomical Society of the Pacific Conference series.* (p. 392).

Kaufman, H. E., Barron, B. A., McDonald, M. B., & Kaufman, S. C. (Eds.). (2000). *Companion handbook to the Cornea.*

Kaveh, M. (1979). High resolution spectral estimation for noisy signals. *IEEE Transactions on Acoustics, Speech, and Signal Processing, 27*, 286–287. doi:10.1109/TASSP.1979.1163243

Kawamoto, M., Ohata, M., Kohno, K., Inouye, Y., & Nandi, A. K. (2005). Robust super-exponential methods for blind equalization in the presence of Gaussian noise. *IEEE Transactions on Circuits and Wystems. II, Express Briefs, 52*(10), 651–655. doi:10.1109/TCSII.2005.852174

Kaya, K., Bilgutay, N., & Murthy, R. (1994). Flaw detection in stainless steel samples using wavelet decomposition. *Proceeding of the 1994 Ultrasonics Symposium, 2*(4), 1271-1274.

Kehtarnavaz, N. (2004). *Real-time digital signal processing based on the TMS320C6000*. Amsterdam: Elsevier.

Keller, J. M., & Chen, S. (1989). Texture description and segmentation through fractal geometry. *Computer Vision Graphics and Image Processing, 45*, 150–166. doi:10.1016/0734-189X(89)90130-8

Keller, J. M., Crownover, R. M., & Chen, R. Y. (1993). On the calculation of fractal features from images. *PAMI, 10*(15), 1087–1090.

Kesler, S. (1986). *Modern spectral analysis - II*. New York: IEEE Press.

Khan, M. A., & Khan, M. K. (2007). *Endothelial cell image enhancement using non-subsampled image pyramid* (pp. 1057–1062).

Kharma, N., Ahmed, M., & Ward, R. (1999). A new comprehensive database of handwritten Arabic words, numbers, and signatures used for OCR testing. *Proceedings of the 1999 IEEE Canadian Conference on Electrical and Computer Engineering*, (pp. 766-768). IEEE.

Khedher, M. Z., & Abandah, G. (2002). *Arabic character recognition using approximate stroke sequence*. Third International Conference on Language Resources and Evaluation (LREC2002).

Khorsheed, M. S. (2007). Offline recognition of omnifont Arabic text using the HMM ToolKit (HTK). *Pattern Recognition Letters, 28*, 1563–1571. doi:10.1016/j.patrec.2007.03.014

Khorsheed, M. S., & Clocksin, W. F. (1999). Structural features of cursive Arabic script. *10th British Machine Vision Conference, 2*, 422-431.

Kim, I. S., Shim, J. H., & Yang, J. (2003). *Face detection*. Stanford University.

Kinner, K. E. (1994). *Advances in genetic programming*. Cambridge, MA: MIT Press.

Kiranyaz, S., Ferreira, M., & Gabbouj, M. (2006). Automatic object extraction over multi-scale edge field for multimedia retrieval. *IEEE Transactions on Image Processing*, 3759–3772. doi:10.1109/TIP.2006.881966

Kirkpatrick, S., Gelatt, C. D., & Vecchi, M. P. (1983). Optimization by simulated annealing. *Science, 220*(4598), 671–680. doi:10.1126/science.220.4598.671

Ko, B. C., & Byun, H. (2005). FRIP: A region-based image retrieval tool using automatic image segmentation. *IEEE Transactions on Multimedia, 7*(1), 105–113. doi:10.1109/TMM.2004.840603

Koonen, T. (2006). Fiber to the home/fiber to the premises: What, where, and when? *Proceedings of the IEEE, 94*(5), 911–934. doi:10.1109/JPROC.2006.873435

Kosko, B. (1997). *Fuzzy engineering*. New Jersey: Prentice Hall.

Koza, J. R. (1992). *Genetic programming, on the programming of computers by means of natural selection*. Cambridge, MA: MIT Press.

Krautkramer, J., & Krautkramer, H. (1990). *Ultrasonic testing of materials* (4th ed.). New York: Springer.

Kreutz-Delgado, K., & Isukapalli, Y. (2008). Use of the Newton method for blind adaptive equalization based on the constant modulus algorithm. *IEEE Transactions on Signal Processing, 56*(8), 3983–3995.

Krista, L. D., & Gallagher, P. T. (2009). Automated coronal hole detection using local intensity thresholding techniques. *Solar Physics, 256*, 87–100. doi:10.1007/s11207-009-9357-2

Kritikos, H. & Schwan, H. (1976). Formation of hot spots in multilayered spheres. *IEEE Transactions on BME*, 168–172.

Krivova, N. A., & Solanki, S. K. (2008). Models of solar irradiance variations: Current status. *Journal of Astrophysics and Astronomy, 29*, 151–158. doi:10.1007/s12036-008-0018-x

Kshirsagar, S., Molet, T., & Magnenat-Thalmann, N. (2002). *A multilayer personality model.* Paper presented at the 2nd International Symposium on Smart Graphics.

Kuo, S. M., & Morgan, D. R. (1996). *Active noise control systems.* John Wiley & Sons Inc.

Kuruoglu, E. (2010). Bayesian source separation for cosmology. *IEEE Signal Processing Magazine, 27,* 43–54. doi:10.1109/MSP.2009.934718

Kyo, S., Okazaki, S., & Arai, T. (2005). An integrated memory array processor architecture for embedded image recognition systems. *Proceedings of the 32nd International Symposium on Computer Architecture,* (pp. 134–145).

Labrosse, N., Dalla, S., & Marshall, S. (2010). Automatic detection of limb prominences in 304 Å EUV images. *Solar Physics, 262,* 449–460. doi:10.1007/s11207-009-9492-9

Lai, H. (1998). Neurological effects of radiofrequency electromagnetic radiation. *Proceedings of the Workshop on Possible Biological and Health Effects of RF Electromagnetic Fields, Mobile Phone and Health Symposium, October 25-28.*

Lamecker, H., Wenckeback, T. H., & Hege, H. C. (2006). *Atlas-based 3D-shape reconstruction from X-ray images.* Paper presented at the Proceedings of the 18th International Conference on Pattern Recognition.

Landini, G., & Rippin, J. W. (1996). How important is tumour shape? Quantification of the epithelial-connective tissue interface in oral lesions using local connected fractal dimension analysis. *The Journal of Pathology, 179,* 210–217. doi:10.1002/(SICI)1096-9896(199606)179:2<210::AID-PATH560>3.0.CO;2-T

LaPre, C., Zhao, Y., Raphael, C., Schwartz, R., & Makhoul, J. (1996). *Multi-font recognition of printed Arabic using the BBN BYBLOS speech recognition system.* IEEE International Conference On Acoustics, Speech And Signal Processing, (pp. 2136-2139).

Last, D., Farnworth, R., & Searle, M. (1992). Effects of Skywave interference on the coverage of Loran. *IEE Proceeding-F, 139*(4), 306–314.

Latif-Amet, A., Ertüzün, A. & Erçil, A. (2000). An efficient method for texture defect detection:

Lean, J.-L., Livingston, W. C., Heath, D. F., Donnelly, R. F., Skumanich, A., & White, O. R. (1982). A three-component model of the variability of the solar ultraviolet flux 145-200 nm. *Journal of Geophysical Research, 87,* 10307–10317. doi:10.1029/JA087iA12p10307

LeBlanc, J. P., Fijalkow, I., & Johnson, C. R., Jr. (1996). Fractionally-spaced constant modulus algorithm blind equalizer error surface characterization: Effects of source distributions. *Proceedings of the IEEE International Conference on Acoustics, Speech, and Signal Processing,* Atlanta, GA, (pp. 2944-2947).

Lee, D. C. M., Kahn, J. M., & Audeh, M. D. (1997). Trellis-coded pulse-position modulation for indoor wireless infrared communications. *IEEE Transactions on Communications, 45*(9), 1080–1087. doi:10.1109/26.623072

Lee, D. K., Xu, J., Zheng, B., & Lee, W.-C. (2002). Data management in location-dependent information services. *IEEE Pervasive Computing / IEEE Computer Society [and] IEEE Communications Society, 2*(3), 65–72.

Lee, G., Chen & Lo, S-C. (2003). Broadcast data allocation for efficient access on multiple data items in mobile environments. *Mobile Networks and Applications, 8,* 365–375. doi:10.1023/A:1024579512792

Lee, B., Cioffi, J. M., Jagannathan, S., & Mohseni, M. (2007). Gigabit DSL. *IEEE Transactions on Communications, 55*(9), 1689–1692. doi:10.1109/TCOMM.2007.904374

Lee, C. K., Leong, H. V., & Si, A. (2002). Semantic data access in an asymmetric mobile environment. In *Proceedings of the 3rd Mobile Data Management,* (pp. 94-101).

Lee, W. C., Hu, Q., & Lee, D. L. (1997). Channel allocation methods for data dissemination in mobile computing environments. In *Proceedings of the 6th IEEE High Performance Distributed Computing,* (pp. 274-281).

Leeson, M. S., Green, R. J., & Higgins, M. D. (2008). Photoparametric amplifier frequency converters. *Proceedings of ICTON, 08,* 197–200.

Leka, K. D., & Barnes, G. (2003). Photospheric magnetic field properties of flaring versus flare-quiet active regions. I. Data, general approach, and sample results. *The Astrophysical Journal, 595,* 1277–1295. doi:10.1086/377511

Leshem, A., van der Veen, A. J., & Boonstra, A. J. (2000). Multichannel interference mitigation techniques in radio-astronomy. *The Astrophysical Journal, 131*, 355–373. doi:10.1086/317360

Li, S., & Qiu, T. Sh. (2009). Tracking performance analysis of fractional lower order constant modulus algorithm. *Electronics Letters, 45*(11), 545–546. doi:10.1049/el.2009.0561

Li, Y., & Ding, Z. (1995). Convergence analysis of finite length blind adaptive equalizers. *IEEE Transactions on Signal Processing, 43*(9), 21209.

Li, Y., & Ding, Z. (1996). Global convergence of fractionally spaced Godard (CMA) adaptive equalizers. *IEEE Transactions on Signal Processing, 44*(4), 818–826. doi:10.1109/78.492535

Li, J., Du, Q., & Sun, C. (2009). An improved box-counting method for image fractal dimension estimation. *Pattern Recognition, 42*(11), 2460–2469. doi:10.1016/j.patcog.2009.03.001

Li, S., Yan, L., Feng, W., Shipeng, L., & Wen, G. (2009). Complexity-constrained H.264 video encoding. *IEEE Transactions on Circuits and Systems for Video Technology, 19*(4), 477–490. doi:10.1109/TCSVT.2009.2014017

Li, X. (2007). *Enhancements & optimizations to H.264/AVC video coding.* Unpublished doctoral thesis, Loughborough University.

Lievin, J., Hamaide, J., Scholiers, W. & Lechien, J. (1981). *Measurements of ECD, time of arrival, amplitude and phase of both ground and reflected Loran pulses.* AGARD-CP-305.

Lin, K.-F., & Liu, C.-M. (2006). *Broadcasting dependent data with minimized access latency in a multi-channel environment.* IWCMC'06, July 3–6, (pp. 809-814).

Liu, H., & Xu, G. (1994). A deterministic approach to blind symbol estimation. *IEEE Signal Processing Letters, 1*(12), 205–207.

Ljung, L. (1999). *System identification: Theory for the user* (2nd ed.). Upper Saddle River, NJ: Prentice Hall.

Lo, S.-C., & Chen, A. L. P. (2000). An adaptive access method for broadcast data under an error-prone mobile environment. *IEEE Transactions on Knowledge and Data Engineering, 12*(4), 609. doi:10.1109/69.868910

Lo, S., Morris, P., & Enge, P. (2005). *Early Skywave detection network: Preliminary design and analysis.* International Loran Association, 34th Annual Convention and Technical Symposium, Paper 5-1.

Lockwood, M., Harrison, R. G., Woollings, T., & Solanki, S. H. (2010). Are cold winters in Europe associated with low solar activity? *Environmental Research Letters, 5*, 024001. doi:10.1088/1748-9326/5/2/024001

Lomba, C. R., Valadas, R. T., & de Oliveira Duarte, A. M. (2000). Efficient simulation of the impulse response of the indoor wireless optical channel. *International Journal of Communication Systems, 13*(7-8), 537–549. doi:10.1002/1099-1131(200011/12)13:7/8<537::AID-DAC455>3.0.CO;2-6

Loop, C. T. (1987). *Smooth subdivision surfaces based on triangles.* Master of Science thesis, University of Utha, Salt Lake.

López-Hernández, F. J., Pérez-Jiménez, R., & Santamará, A. (1998). Monte Carlo calculation of impulse response on diffuse ir wireless indoor channels. *Electronics Letters, 34*(12), 1260–1262. doi:10.1049/el:19980825

López-Hernández, F. J., Pérez-Jiménez, R., & Santamará, A. (1998). Modified Monte Carlo scheme for high-efficiency simulation of the impulse response on diffuse IR wireless indoor channels. *Electronics Letters, 34*(19), 1819–1820. doi:10.1049/el:19981173

López-Hernández, F. J., Pérez-Jiménez, R., & Santamará, A. (2000). Ray-tracing algorithms for fast calculation of the channel impulse response on diffuse IR wireless indoor channels. *Journal of Optical Engineering, 39*(10), 2775–2780. doi:10.1117/1.1287397

Lorigo, L., & Govindaraju, V. (2006). Off-line Arabic handwriting recognition: A survey. *IEEE Transactions on Pattern Analysis and Machine Intelligence, 28*, 712–724. doi:10.1109/TPAMI.2006.102

Lorigo, L., & Govindaraju, V. (2005). *Segmentation and pre-recognition of Arabic handwriting*. Eighth International Conference on Document Analysis and Recognition (ICDAR'05), (pp. 605-609).

Losa, G. A., Merlini, D., Nonnemacher, T. F., & Weibel, E. R. (2005). Fractals in biology and medicine. *Birkhäuser Verlag, IV,* 1–314.

Lowe, D. G. (2004). Distinctive image features from scale-invariant keypoints. *International Journal of Computer Vision, 60*(2), 91–110. doi:10.1023/B:VISI.0000029664.99615.94

Lowe, D. G. (1999). *Object recognition from local scale-invariant features*. Paper presented at the International Conference on Computer Vision.

Lowe, D.G. (2004). Distinctive image features from scale-invariant keypoints. *International Journal of Computer Vision*.

Maddouri, S. S., Amiri, H., Belaïd, A., & Choisy, C. (2002). *Combination of local and global vision modelling for Arabic handwritten words recognition*. Eighth International Workshop on Frontiers in Handwriting Recognition (IWFHR'02), (pp. 128-135). IEEE.

Mahmoud, S. (1994). Arabic character-recognition using Fourier descriptors and character contour encoding. *Pattern Recognition, 27,* 815–824. doi:10.1016/0031-3203(94)90166-X

Mahmoud, S. (2008). Recognition of writer-independent off-line handwritten Arabic (Indian) numerals using hidden Markov models. *Signal Processing, 88,* 844–857. doi:10.1016/j.sigpro.2007.10.002

Majumdar, A. (2007). Bangla basic character recognition using digital curvelet transform. *Journal of Pattern Recognition Research, 2,* 17–26.

Makhoul, J. (1975). Linear prediction: A tutorial review. *Proceedings of the IEEE, 63*(4), 561–580.

Makhoul, J., Schwartz, R., Lapre, C., & Bazzi, I. (1998). A script-independent methodology for optical character recognition. *Pattern Recognition, 31,* 1285–1294. doi:10.1016/S0031-3203(97)00152-0

Mallat, S. (2008). *A wavelet tour of signal processing: The sparse way*. New York: Academic Press.

Mandelbrot, B. B. (1982). *Fractal geometry of nature*. New York: Freeman.

Manfredi, M. L., Cha, S.-H., Yoon, S., & Tappert, C. (2005). Handwriting copybook style analysis of pseudo-online data. (pp. D2.1--D2.5).

Margner, V., & Pechwitz, M. (2001). *Synthetic data for Arabic OCR system development*. The 6th International Conference on Document Analysis and Recognition, ICDAR'01, (pp. 1159-1163).

Margner, V., Pechwitz, M., & El Abed, H. (2005). *ICDAR 2005 Arabic handwriting recognition competition*. International Conference on Document Analysis and Recognition, (pp. 70-74).

Maricic, B., Luo, Z.-Q., & Davidson, T. N. (2003). Blind constant modulus equalization via convex optimization. *IEEE Transactions on Signal Processing, 51*(3), 805–818. doi:10.1109/TSP.2002.808112

Marpe, D., Wiegand, T., & Sullivan, G. (2006). The H.264/MPEG4 advanced video coding standard and its applications. *IEEE Communications Magazine, 44*(8), 134–143. doi:10.1109/MCOM.2006.1678121

Marton, G. A., Bulbul, O., & Kanungo, T. (1999). *OmniPage vs. Sakhr: Paired model evaluation of two Arabic OCR products*. SPIE Conference on Document Recognition and Retrieval VI. San Jose, CA.

Mason, S. (1997). Heuristic reasoning strategy for automated sensor placement. *Photogrammetric Engineering and Remote Sensing, 63*(9), 1093–1102.

Mason, S., & Grun, A. (1995). Automatic sensor placement for accurate dimensional inspection. *Computer Vision and Image Understanding, 61*(3), 454–467. doi:10.1006/cviu.1995.1034

Matalas, I. (1996). *Segmentation techniques suitable for medical images*. Unpublished doctoral thesis, Imperial College, University of London, London.

Mathis, H., & Douglas, S. C. (2003). Bussgang blind deconvolution for impulsive signals. *IEEE Transactions on Signal Processing, 51*(7), 1905–1915. doi:10.1109/TSP.2003.812836

Mattfeldt, T. (1997). Nonlinear deterministic analysis of tissue texture: A stereological study on mastopathic and mammary cancer tissue using chaos theory. *Journal of Microscopy*, *185*(1), 47–66. doi:10.1046/j.1365-2818.1997.1440701.x

Matthews, I., & Baker, S. (2004). Active appearance models revisited. *International Journal of Computer Vision*, *60*(2), 135–164. doi:10.1023/B:VISI.0000029666.37597.d3

McAteer, R. T. J., Gallagher, P. T., & Ireland, J. (2005b). Statistics of active region complexity: A large-scale fractal dimension survey. *The Astrophysical Journal*, *631*, 628–635. doi:10.1086/432412

McAteer, R. T. J., Gallagher, P. T., Ireland, J., & Young, C. A. (2005a). Automated boundary-extraction and region-growing techniques applied to solar magnetograms. *Solar Physics*, *228*, 55–66. doi:10.1007/s11207-005-4075-x

MedicineNet. (2010, March 9, 2010). *Definition of cornea*. Retrieved from http://www.medterms.com/script/main/art.asp?articlekey=7248

Melhi, M. (2001). *Off-line Arabic cursive handwriting recognition using artificial neural networks*. Bradford, UK: Bradford University.

Messerschmitt, D. (1974). *Design of a finite impulse response for the Viterbi algorithm and decision-feedback equalizer*. Paper presented at IEEE International Conference on Communications, Minneapolis, MN.

Miled, H., & Ben Amara, N. E. (2001). *Planar Markov modeling for Arabic writing recognition: Advancement state*. Sixth International Conference on Document Analysis and Recognition, (pp. 69-73).

Mohammed, A., & Last, D. (1995). *Loran Skywave delay detection using rational modelling techniques* (pp. 100–104). Bath, UK: Radio Receivers and Associated Systems.

Mohammed, A., Yi, B., & Last, D. (1994). *Rational modelling techniques for the identification of Loran Skywaves*. Wild Goose Association, 23rd Annual Convention and Technical Symposium, (pp. 184-191). Rhode Island, USA.

Morabito, F. C. (2000). Independent component analysis and feature extraction techniques for NDT data. *Materials Evaluation*, *58*(1), 85–92.

Morales, M. F., Bowman, J. D., Cappallo, R., Hewitt, J. N., & Lonsdale, C. J. (2006). Statistical EOR detection and the Mileura widefield array. *New Astronomy Reviews*, *50*, 173–178.

Moussaoui, S., Brie, D., Mohammad-Djafari, A., & Carteret, C. (2006). Separation of non-negative mixture of non-negative sources using a Bayesian approach and MCMC sampling. *IEEE Transactions on Signal Processing*, *11*, 4133–4145. doi:10.1109/TSP.2006.880310

Moussaoui, S., Hauksdóttir, H., Schmidt, F., Jutten, C., Chanussot, J., & Brie, D. (2008). On the decomposition of Mars hyperspectral data by ICA and Bayesian positive source separation. *Neurocomputing*, *71*(10-12), 2194–2208. doi:10.1016/j.neucom.2007.07.034

Myint, S. W. (2003). Fractal approaches in texture analysis and classification of remotely sensed data: Comparisons with spatial autocorrelation techniques and simple descriptive statistics. *International Journal of Remote Sensing*, *24*(9), 1925–1947. doi:10.1080/01431160210155992

National Geographic. (2004, 2). *English in decline as a first language, study says*. Retrieved December 12, 2009, from http://news.nationalgeographic.com/news/2004/02/0226_040226_language.html

Navathe, S. B., Yee, W. G., Omiecinski, E., & Jermaine, C. (2002). Efficient data allocation over multiple channels at broadcast servers. *IEEE Transactions on Computers*, *51*(10), 1231–1236. doi:10.1109/TC.2002.1039849

Nefian, A. V., & Hayes, M. H. (1998). *Face detection and recognition using hidden Markov models*. Paper presented at the IEEE International Conference on Image Processing.

Nicula, B., Berghmans, D., & Hochedez, J.-F. (2005). Poisson recoding of solar images for enhanced compression. *Solar Physics*, *228*, 253–264. doi:10.1007/s11207-005-4998-2

Nikkhou, M., & Choukri, K. (2005). *Survey on Arabic language resources and tools in the Mediterranean countries*. NEMLAR, Center for Sprogteknologi, University of Copenhagen.

Nilsson, M., Nordberg, J., & Claesson, I. (2007). *Face detection using local SMQT features and split up snow classifier*. Paper presented at the IEEE Conference on Acoustics, Speech and Signal Processing, ICASSP.

NovoDynamics. (2009). *NovoDynamics VERUS™ standard.* Retrieved from http://www.novodynamics.com/

NSGA-II. (2009). *Kanpur Genetic Algorithms Laboratory.* Retrieved March 2009, from http://www.iitk.ac.in/kangal/codes.shtml

Nuance-Communications. (2009). *OmniPage OCR software.* Retrieved from http://www.nuance.com/omnipage/

Nuzillard, D., & Bijaoui, A. (2000). Blind source separation and analysis of multispectral astronomical images. *Astronomy & Astrophysics, 147,* 129–138.

O'Brien, D. C., Katz, M., Wang, P., Kalliojarvi, K., Arnon, S., & Matsumoto, M. (2005). *Short range optical wireless communications.* Wireless World Research Forum.

Ochiai, H. (2004). A novel Trellis-shaping design with both peak and average power reduction for OFDM systems. *IEEE Transactions on Communications, 52,* 1916–1926. doi:10.1109/TCOMM.2004.836593

Ochiai, H., & Imai, H. (2002). Performance analysis of deliberately clipped OFDM signals. *IEEE Transactions on Communications, 50,* 89–101. doi:10.1109/26.975762

Olague, G., & Mohr, R. (2002). Optimal camera placement for accurate reconstruction. *Pattern Recognition, 35,* 927–944. doi:10.1016/S0031-3203(01)00076-0

Olmedo, O., Zhang, J., Wechsler, H., Poland, A., & Borne, K. (2008). Automatic detection and tracking of coronal mass ejections in coronagraph time series. *Solar Physics, 248,* 485–499. doi:10.1007/s11207-007-9104-5

Ortega, M., Rui, Y., Chakrabarti, K., Mehrotra, J., & Huang, T. S. (1997). Supporting similarity queries in MARS. *Proceedings of the 5th ACM International Multimedia Conference,* (pp. 403-413).

Ostermann, J., Bormans, J., List, P., Marpe, D., Narroschke, M., & Pereira, F. (2004). Video coding with H.264/AVC: Tools, performance and complexity. *IEEE Circuits and Systems Magazine, 4*(1), 7–28. doi:10.1109/MCAS.2004.1286980

Osuna, E., Freund, R., & Girosi, F. (1997). *Training support vector machines: An application to face detection.* Paper presented at the IEEE Computer Society Conference on Computer Vision and Pattern Recognition.

Özçelebi, T., Sunay, M. O., Tekalp, A. M., & Civanlar, M. R. (2007). Cross-layer optimized rate adaptation and scheduling for multiple-user wireless video streaming. *IEEE Journal on Selected Areas in Communications, 25*(4), 760–769. doi:10.1109/JSAC.2007.070512

Özçelebi, T., Tekalp, A. M., & Civanlar, M. R. (2007). Delay-distortion optimization for content-adaptive video streaming. *IEEE Transactions on Multimedia, 9*(4), 826–836. doi:10.1109/TMM.2007.895670

Pakravan, M. R., & Kavehrad, M. (2001). Indoor wireless infrared channel characterization by measurements. *IEEE Transactions on Vehicular Technology, 50*(4), 1053–1073. doi:10.1109/25.938580

Paltridge, S. (2001). *The development of broadband access in OECD countries.* OECD Working Party on Telecommunication and Information Services Policies. (Document DSTI/ICCP/TISP(2001)2/FINAL).

Pantic, M., Valstar, M. F., Rademaker, R., & Maat, L. (2005). *Web-based database for facial expression analysis.* Paper presented at the IEEE International Conference on Multimedia and Expo (ICME'05), Amsterdam, The Netherlands.

Papadias, C. B., & Paulraj, A. J. (1997). A constant modulus algorithm for multi-user signal separation in presence of delay spread using antenna arrays. *IEEE Signal Processing Letters, 4*(6), 17881. doi:10.1109/97.586042

Parand, F., Faulkner, G. E., & O'Brien, D. C. (2003). Cellular tracked optical wireless demonstration link. *IEEE Proceedings in Optoelectronics, 150*(5), 490–496. doi:10.1049/ip-opt:20030961

Parker, E. N. (1958). Dynamics of the interplanetary gas and magnetic fields. *The Astrophysical Journal, 128,* 664. doi:10.1086/146579

Pat, P. B., & Ramakrishnan, A. (2008). Word level multiscript identification. *Pattern Recognition Letters, 29,* 1218–1229. doi:10.1016/j.patrec.2008.01.027

Pavlidis, T. (1986). A vectorizer and feature extractor for document recognition. *Computer Vision Graphics and Image Processing, 35*(1), 111–127. doi:10.1016/0734-189X(86)90128-3

Pechwitz, M., & Maergner, V. (2003). *HMM based approach for handwritten Arabic word recognition using the IFN/ENIT database.* The Seventh International Conference on Document Analysis and Recognition (ICDAR 2003), (pp. 890-894).

Pechwitz, M., & Margner, V. (2002). *Baseline estimation for Arabic handwritten words.* Eighth International Workshop on Frontiers in Handwriting Recognition (IWFHR'02), (pp. 479-484).

Pechwitz, M., Maddouri, S.S., Märgner, V., Ellouze, N. & Amiri, H. (2002). IFN/ENIT - database of handwritten Arabic words. *CIFED*, 129-136.

Pentland, A., Picard, R. W., & Sclaroff, S. (1994). Photobook: Content-based manipulation of image databases. *Proceedings of SPIE Storage and Retrieval Image and Video Databases, II*, 34–47.

Pentland, A., & Scalroff, S. (1991). Closed-form soultions for physically based shape modeling and recognition. *IEEE Transactions on Pattern Analysis and Machine Intelligence, 13*, 715–729. doi:10.1109/34.85660

Pérez-Suárez, D., Higgins, P.A., McAteer, R.T.J., Bloomfield, D.S. & Gallagher, P.T. (2010). *Solar feature detection.*

Perley, R. A., Schwab, F. R., & Bridle, A. H. (1994). *Synthesis imaging in radio astronomy. Proceedings of the Third NRAO Synthesis Imaging Summer School, Astronomical Society of the Pacific Conference Series, 6.*

Perrin, J., Torrésani, B., & Fuchs, P. (1999). A localized correlation function for stereoscopic image matching. *Traitement du Signal, 16*, 3–14.

Petajan, E. (2005). MPEG-4 face and body animation coding applied to HCI. In B. Kisacanin, V. Pavlovic & T.S. Huang (Eds.), *Real-time vision for human-computer interaction.* (pp. 249-268). United States of America: Springer Science + Business Media, Inc.

Peterson, B. & Dewalt, D. (1992). *Loran and the effects of terrestrial propagation.* USCG Academy, Technical Report.

Peterson, J. B., Pen, U., & Wu, X. (2005). Searching for early ionization with the primeval structure telescope. In N. Kassim, M. Perez, W. Junor & P. Henning (Eds.), *Astronomical Society of the Pacific Conference series.* (p. 441).

Pito, R. (1999). A solution to the next best view problem for automated surface acquisition. *IEEE Transactions on Pattern Analysis and Machine Intelligence, 21*(10), 1016–1030. doi:10.1109/34.799908

Podladchikova, O., & Berghmans, D. (2005). Automated detection Of Eit waves and dimmings. *Solar Physics, 228*, 265–284. doi:10.1007/s11207-005-5373-z

Poli, R. (2005). Tournament selection, iterated coupon-collection problem, and backward-chaining evolutionary algorithms. In *Foundations of Genetic Algorithms*, (pp. 132–155).

Popescu, D., & Dobrescu, R. (2008b). Carriage road pursuit based on statistical and fractal analysis of the texture. *International Journal of Education and Information Technologies, 2*(11), 62–70.

Popescu, D. (1990). Industrial image processing. *Buletinul IPB. Seria Inginerie Electrica, 52*(2), 91–96.

Popescu, D. & Dobrescu, R. (2008c). Plastic surface similarity measurement based on textural and fractal features. *Revista de materiale plastice, 45*(2), 158-163.

Popescu, D., Dobrescu, R., & Angelescu, N. (2008a). Color textures discrimination of land images by fractal techniques. *Proceedings of the 4th WSEAS International Conference on REMOTE SENSING -REMOTE'08*, (pp. 51-56). Venice, Italy.

Popescu, D., Dobrescu, R., Avram, V., & Mocanu, S. (2006). Local processors for on line processing with applications in intelligent vehicle technologies. *Proceedings of the 10th WSEAS International Conference on SYSTEMS*, (pp. 579-584). Athens, Greece.

Popescu, D., Patarniche, D., Dobrescu, R., Nicolae, M., & Dobrescu, M. (2009). Real time mobile object tracking based on chromatic information. *Proceedings of the 5th WSEAS International Conference on Remote Sensing -REMOTE'09*, (pp. 13-18). Genova, Italy.

Popovics, S., Bilgutay, N., Karaoguz, M., & Akgul, T. (2000). High-frequency ultrasound technique for testing concrete. *ACI Materials, 97*(1), 58–65.

Porat, B., & Friedlander, B. (1991). Blind equalisation of digital communication channels using high-order moments. *IEEE Transactions on Signal Processing, 39*(2), 522–526. doi:10.1109/78.80846

Potts, H. E., Barrett, R. K., & Diver, D. A. (2004). Balltracking: An highly efficient method for tracking flow fields. *Astronomy & Astrophysics, 424*, 253–262. doi:10.1051/0004-6361:20035891

Prabhakara, K., Hua, K. A., & Oh, J. H. (2000). A new broadcasting technique for an adaptive hybrid data delivery in wireless environment. In *Proceedings of 19th IEEE International Performance, Computing and Communications Conference,* (pp. 361-367).

Prakoonwit, S., & Benjamin, R. (2007). Optimal 3D surface reconstruction from a small number of conventional 2D X-ray images. *Journal of X-Ray Science and Technology, 15*(4), 197–222.

Prangé, R., Pallier, L., Hansen, K. C., Howard, R., Vourlidas, A., & Courtin, R. (2004). An interplanetary shock traced by planetary auroral storms from the Sun to Saturn. *Nature, 432*, 78–81. doi:10.1038/nature02986

Proakis, J., & Manolakis, D. (1988). *Introduction to digital signal processing.* New York: MacMillan Publishing Company.

Proakis, J. G., & Salehi, M. (2007). *Digital communications* (5th ed.). New York: Mc-Graw-Hill.

Pu, W., Lu, Y., & Wu, F. (2006). Joint power-distortion optimization on devices with MPEG-4 AVC/H.264 codec. *IEEE International Conference on Communications, 1,* 441-446.

Puri, A., & Eleftheriadis, A. (2008). MPEG-4: An object-based multimedia coding standard. *Mobile Networks and Applications, 3*(1), 5–32. doi:10.1023/A:1019160312366

Qureshi, S. U. H. (1985). Adaptive equalizers. *Proceedings of the IEEE, 73*(9), 1349–1387. doi:10.1109/PROC.1985.13298

Qureshi, S. U. H., & Forney, G. D., Jr. (1977). Performance and properties of a T/2 equalizer. *Proceedings of the National Telecommunications Conference,* Los Angeles, CA, (pp. 1-14).

Rajamani, K. T., Styner, M. A., Talib, H., Zheng, G., Nolte, L. P., & Ballester, M. A. G. (2007). Statistical deformable bone models for robust 3D surface extrapolation from sparse data. *Medical Image Analysis, 11*, 99–109. doi:10.1016/j.media.2006.05.001

Rajamani, K. T., Ballester, M. A. G., Nolte, L. P., & Styner, M. (2005). *A novel and stable approach to anatomical structure morphing for enhanced intraoperative 3D visualization.* Paper presented at the SPIE Medical Imaging: Visualization, Image-guided Procedures, and Display.

Rajamani, K. T., Styner, M., & Joshi, S. C. (2004). *Bone model morphing for enhanced surgical visualization.* Paper presented at the 2004 IEEE Int. Symp. Biomedical Imaging: From Nano to Macro.

Rakhee, V. S., & Savita, K. (2008). Two level signature model for multiple broadcast channel using unified index hub (UIH). *Proceeding of International Conference on Soft Computing (ICSC – 2008),* November 8-10, 2008 at IET, Alwar (Rajasthan) – India, (pp. 422-431).

Rallabandi, V. P. (2008). Enhancement of ultrasound images using stochastic resonance-based wavelet transform. *Computerized Medical Imaging and Graphics Journal, 32*(1), 316–320. doi:10.1016/j.compmedimag.2008.02.001

Ramirez-Iniguez, R., & Green, R. J. (2005). Optical antenna design for indoor optical wireless communication systems. *International Journal of Communication Systems, 18*(3), 229–245. doi:10.1002/dac.701

Ramirez-Iniguez, R., & Green, R. J. (2000). *Optical antennae for infrared mobile communications.* (Patent GB01/03812).

Randen, T., & Husøy, J. (1999). Filtering for texture classification: A comparative study. *IEEE Transactions on Pattern Analysis and Machine Intelligence, 21*, 291–310. doi:10.1109/34.761261

Regalia, P., & Mboup, M. (1999). Undermodeled equalization: A characterization of stationary points for a family of blind criteria. *IEEE Transactions on Signal Processing, 47*(3), 760770.

Reinstein, D. Z., & Archer, T. J. (2008). Epithelial thickness in the normal cornea: Three-dimensional display with very high frequency ultrasound. *NIH Public Access, 24*, 571.

Rice University. (2000). Signal processing information base. Retrieved on December 8, 2009, from http://spib.rice.edu/

Rieutord, M., Roudier, T., Roques, S., & Ducottet, C. (2007). Tracking granules on the Sun's surface and reconstructing velocity fields. I. The CST algorithm. *Astronomy & Astrophysics, 471*, 687–694. doi:10.1051/0004-6361:20066491

Ritter, D., Orman, J., Schmidgunst, C., & Graumann, R. (2007). 3D soft tissue imaging with a mobile C-arm. *Computerized Medical Imaging and Graphics, 31*, 91–102. doi:10.1016/j.compmedimag.2006.11.003

Roa, B., & Arun, K. (1992). Model based processing of signals: A state space approach. *Proceedings of the IEEE, 80*(2), 283–309. doi:10.1109/5.123298

Robbrecht, E., & Berghmans, D. (2005). Entering the era of automated CME recognition: A review of existing tools. *Solar Physics, 228*, 239–251. doi:10.1007/s11207-005-5004-8

Robini, M., Magnin, I., Benoit, H., & Baskurt, A. (1997). Two-dimensional ultrasonic flaw detection based on the wavelet packet transform. *IEEE Transactions on Ultrasonics, Ferroelectrics, and Frequency Control, 44*(6), 1382–1394. doi:10.1109/58.656642

Roth, D., Yang, M., & Ahuja, N. (2000). A snow-based face detector. *Advances in Neural Information Processing Systems, 12*, 855–861.

Rothlauf, F. (2002). *Representations for genetic and evolutionary algorithms*. Physica-Verlag.

Rowley, H., Balaju, S., & Kanade, T. (1998). Neural betwork based face detection. *IEEE Pattern Analysis and Machine Intelligence, 20*, 22–38. doi:10.1109/34.655647

RTC. (1977). *Minimum Performance Standards (MPS)-Marine Loran receiving equipment*. Radio Technical Commission for Marine Services, U.S. Federal Communication Commission. Report of Special Committee No. 70.

Saeed, K., & Albakoor, M. (2009). Region growing based segmentation algorithm for typewritten and handwritten text recognition. *Applied Soft Computing, 9*, 608–617. doi:10.1016/j.asoc.2008.08.006

Sakhr. (2009). *Sakhr software*. Retrieved from http://www.sakhr.com/

Sanchez, F., Collados, M., & Vazquez, M. (1992). *Solar observations: Techniques and interpretation (first Canary Islands winter school of astrophysics)*. Cambridge Universtiy Press.

Sangwan, R., Ludwig, R., Laplante, P., & Neill, C. (2005). Performance tuning of imaging applications through pattern-based code transformation. *Proceedings of SPIE-IS&T Electronic Imaging Conference on Real-Time Imaging*, (pp. 1–7).

Santini, S., Gupta, A., & Jain, R. (2001). Emergent semantics through interaction in image databases. *IEEE Transactions on Knowledge and Data Engineering, 13*(3), 337–351. doi:10.1109/69.929893

Sarfraz, M., & Shahab, S. A. (2005). *An efficient scheme for tilt correction in Arabic OCR system*. International Conference on Computer Graphics, Imaging and Vision: New Trends, (pp. 379–384).

Sarfraz, M., Nawaz, S. N., & Al-Khuraidly, A. (2003). *Offline Arabic text recognition system*. International Conference on Geometric Modeling and Graphics (GMAG'03), (pp. 30-36). London.

Sari, T., & Sellami, M. (2005). Cursive Arabic script segmentation and recognition system. *International Journal of Computers and Applications, 27*, 161–168. doi:10.2316/Journal.202.2005.3.202-1518

Sari, T. & Sellami, M. (2002). Morpho-lexical analysis for correcting OCR-generated Arabic words (MOLEX). *Frontiers in Handwriting Recognition*, 461-466.

Sarker, N., & Chaudhuri, B. B. (1994). An efficient differential box-counting approach to compute fractal dimension of image. *IEEE Transactions on Systems, Man, and Cybernetics, 24*, 115–120. doi:10.1109/21.259692

Saroukhani, S., & Vafadari, R. (2009). *Application of wavelet analysis in crack identification of beams*. Non-Destructive Testing 2009 conference. Blackpool, UK: BINDT.

Sato, Y. (1975). A method of self-recovering equalisation for multilevel amplitude modulation. *IEEE Transactions on Communications, 23*(6), 679–682. doi:10.1109/TCOM.1975.1092854

Savita, K. (2008). Dynamic broadcast scheduling at Unified Index Hub (UIH). *International Journal of Intelligent Information Processing, 2*(2), 243–250.

Scarpa, F., Fiorin, D., et al. (2007). *In vivo three-dimensional reconstruction of the cornea from confocal microscopy images.* 29th Annual International Conference of the IEEE Engineering in Medicine and Biology Society, 2007. EMBS 2007.

Scherrer, P. H. (1995). The solar oscillations investigation-the Michelson Doppler imager. *Solar Physics, 162*, 129–188. doi:10.1007/BF00733429

Schilizzi, R. T. (2004). The square kilometer array. In J.M. Oschmann, Jr. (Ed.), *Ground-based telescopes. Proceedings of the SPIE, 5489*, (pp. 62-71).

Schneiderman, H., & Kanade, T. (1998). *Probabilistic modeling of local appearance and spatial relationship for object recognition.* Paper presented at the IEEE International Conference Computer Vision and Pattern Recognition.

Schniter, P., & Johnson, C. R. Jr. (2000a). Bounds for the MSE performance of constant modulus estimators. *IEEE Transactions on Signal Processing, 48*(10), 2785–2796.

Schniter, P., & Johnson, C. R. Jr. (2000b). Bounds for the MSE performance of constant modulus estimators. *IEEE Transactions on Information Theory, 46*(7), 2544–2560. doi:10.1109/18.887862

Schrijver, C. J. (2007). A characteristic magnetic field pattern associated with all major solar flares and its use in flare forecasting. *The Astrophysical Journal, 655*, L117–L120. doi:10.1086/511857

Schrijver, K., Hurlburt, N. E., Cheung, M. C., Title, A. M., Delouille, V., Hochedez, J.-F., et al. (2007). *Helioinformatics: Preparing for the future of heliophysics research.* Paper presented at the 210th American Astronomical Society Meeting, Honolulu, Hawaii. Retrieved January 30, 2009, from http://www.lmsal.com/helioinformatics/Welcome.html

Schwan, H. P. (1957). Electrical properties of tissue and cell suspension. *Advances in Biological and Medical Physics, 5.*

Seinstra, F. J., Koelma, D., & Geusebroek, J. M. (2002). A software architecture for user transparent parallel image processing. *Parallel Computing, 28*(7-8), 967–993. doi:10.1016/S0167-8191(02)00103-5

Semwogerere, D.W. (2005). *Confocal microscopy.*

Sermadevi, Y., & Hemami, S. S. (2003). Linear programming optimization for video coding under multiple constraints. In *Proceedings of the IEEE Data Compression Conference,* (pp. 53–62).

Sezer, O.G., Ercil, A., & Ertuzun, A. (2007). Using perceptual relation of regularity in the texture with

Shaaban, Z. (2008). *A new recognition scheme for machine-printed Arabic texts based on neural networks* (pp. 707–710).

Shalvi, O., & Weinstein, E. (1990). New criteria for blind deconvolution of nonminimum phase systems (Channels). *IEEE Transactions on Information Theory, 36*(2), 312–321. doi:10.1109/18.52478

Shapiro, L., & Stockman, G. (2001). *Computer vision.* Prentice Hall.

Shekhar, C., Shitole, N., Zahran, O., & Al-Nuaimy, W. (2006). *Combining fuzzy logic and neural networks in classification of weld defects using ultrasonic time-of-flight diffraction.* Non-Destructive Testing 2006 conference. Derby, UK: BINDT.

Shen, G., & Apostolos, S. (2008). Application of texture analysis in land cover classification of high resolution image. *2008 Fifth International Conference on Fuzzy Systems and Knowledge Discovery,* vol. 3, (pp. 513-517).

Shih, F. Y., & Kowalski, A. J. (2003). Automatic extraction of filaments in H solar images. *Solar Physics, 218*, 99–122. doi:10.1023/B:SOLA.0000013052.34180.58

Shin, M. C., Chang, K. I., & Tsap, L. V. (2002). *Does colorspace transformation make any difference on skin detection?* Paper presented at the Sixth IEEE Workshop on Applications of Computer Vision. Retrieved from http://marathon.csee.usf.edu/~tsap/papers/wacv02.pdf

Shirali-Shahreza, M. H., & Shirali-Shahreza, S. (2006). *Persian/Arabic text font estimation using dots.* Sixth IEEE International Symposium on Signal Processing and Information Technology, (pp. 420-425).

Shivkumar, N., & Venkatasubramanian, S. (1996). Energy efficient indexing for information dissemination in wireless systems. *MONET, 1*(4), 433–446.

Sienkiewicz, Z. J. & Kowalczuk, C. I. (2005). *A summary of recent reports on mobile phones and Health* (2000-2004). (National Radiological Protection Board Report, January).

Simone, G., & Morabito, F. C. (2000). ICA-NN based data fusion approach in ECT signal restoration. *Proceedings of the International Joint Conference on Neural Networks (IJCNN2000), 5*(1), 59-64.

Slimane, F., Ingold, R., Kanoun, S., Alimi, A. M., & Hennebert, J. (2009). *A new Arabic printed text image database and evaluation protocols* (pp. 946–950).

Slock, D. T. M., & Papadias, C. (1995). Further results on blind identification and equalization of multiple FIR channels. *Proceedings of the IEEE International Conference on Acoustics, Speech, and Signal Processing*, Detroit, MI, (pp. 1964–1967).

Smeulders, A. W. M., Worring, M., Santini, S., Gupta, A., & Jain, R. (2000). Content-based image retrieval at the end of the early years. *IEEE Transanctions on PAMI, 22*, 1349–1380.

Soille, P., & Talbot, H. (2001). Directional morphological filtering. *IEEE Transactions on Pattern Analysis and Machine Intelligence, 23*(11), 1313–1329. doi:10.1109/34.969120

Soleymani, M., & Razzazi, F. (2003). *An efficient front-end system for isolated Persian/Arabic character recognition of handwritten data-entry forms. WSEAS Multiconference* (p. 6). WSEAS.

Sonka, M., Hlavac, V., & Boyle, R. (1999). *Image processing analysis and machine vision*. Cole Publishing Company.

Souček, J., Dudok de Wit, T., Décréau, P., & Dunlop, M. (2004). Local wavelet correlation: Application of timing analysis to multi-satellite CLUSTER data. *Annales Geophysicae, 22*, 4185–4196. doi:10.5194/angeo-22-4185-2004

Soviany, C. (2003). *Embedding data and task parallelism in image processing applications*. Unpublished doctoral dissertation, Delft University of Technology.

SPSS Inc. (2007). *SPSS Categories™ 16.0 manual guide.*

Staib, L. H., & Duncan, J. S. (1992). Boundary finding with parametrically deformable models. *IEEE Transactions on Pattern Analysis and Machine Intelligence, 17*, 1061–1075. doi:10.1109/34.166621

Starck, J.-L., & Murtagh, F. (2006). *Astronomical image and data analysis*. Berlin: Springer.

Stasak, J. (2004). *A contribution to image semantic analysis*. 10th Conference on Professional Information Resources.

Stegmann, M. B., Ersboll, B. K., & Larsen, B. (2003). FAME-a flexible appearance modelling environment. *IEEE Transactions on Medical Imaging, 22*(10), 1319–1331. doi:10.1109/TMI.2003.817780

Stenborg, G., & Cobelli, P. J. (2003). A wavelet packets equalization technique to reveal the multiple spatial-scale nature of coronal structures. *Astronomy & Astrophysics, 398*, 1185–1193. doi:10.1051/0004-6361:20021687

Stewart, D. B. (2006). Measuring execution time and real–time performance. *Proceedings of Embedded System Conference*.

Stix, M. (2004). *The Sun: An introduction* (2nd ed.). Springer-Verlag.

Storring, M., Andersen, H. J., & Granum, E. (1999). *Skin colour detection under changing lighting conditions*. Paper presented at the 7th Symposium on Intelligent Robotics Systems, Coimbra, Portugal.

Subband domain co-occurrence matrices. *Image and Vision Computing, 18*, 543–553.

Subsol, G., Thirion, J. P., & Ayache, N. (1996). *Application of an automatically built 3D morphometric brain atlas: Study of cerebral ventricle shape*. Paper presented at the Visualization in Biomedical Computing.

Sun, W., Xu, G., Gong, P., & Liang, S. (2006). Fractal analysis of remotely sensed images: A review of methods and applications. *International Journal of Remote Sensing, 27*(22), 4963–4990. doi:10.1080/01431160600676695

Sun, Q.-S., Zeng, S.-G., Liu, Y., Heng, P.-A., & Xia, D.-S. (2005). A new method of feature fusion and its application in image recognition. *Pattern Recognition*, 2437–2448. doi:10.1016/j.patcog.2004.12.013

Sun, B., & Huang, D. (2004). Texture classification based on support vector machine and Wavelet transform. *Proceedings of the 5th World Congress on Intelligent Control and Automation*. Hangzhou, China, (pp. 1862-1864).

Sung, K. K., & Poggio, T. (1998). Example-based learning for view-based human face detection. *IEEE Transactions on Pattern Analysis and Machine Intelligence*, *20*(1), 39–51. doi:10.1109/34.655648

Swarup, G., Kapahi, V. K., Velusamy, T., Ananthakrishnan, S., Balasubramanian, V., & Pramesh-Rao, A. (1991). Twenty-five years of radio astronomy at Tata Institute for Fundamental Research. *Current Science*, *60*(2).

Syiam, M., Nazmy, T. M., Fahmy, A. E., Fathi, H., & Ali, K. (2006). *Histogram clustering and hybrid classifier for handwritten Arabic characters recognition. The 24th IASTED international conference on Signal processing, pattern recognition, and applications* (pp. 44–49). ACTA Press.

Szelely, R., Lelemen, A., Brechbuler, C., & Gerig, G. (1996). Segmentation of 2D and 3D objects from MRI volume data using constrained elastic deformations of flexible Fourier surface models. *Medical Image Analysis*, *1*, 19–34.

Tan, M., & Bar-Ness, Y. (2003). OFDM peak-to-average power ratio reduction by combined symbol rotation and inversion with limited complexity. In. *Proceedings of IEEE GLOBECOM*, *2003*, 605–610.

Taplin, S.H., Ichikawa, L.E. & Kerlikowske, K. (2002). Concordance of breast imaging reporting

Telzhensky, N., & Leviatan, Y. (2006). Novel method of UWB antenna optimization for specified input signal forms by means of genetic algorithm. *IEEE Transactions on Antennas and Propagation*, *54*(8), 2216–2225. doi:10.1109/TAP.2006.879201

Thomas, J., Quatieri, F., & Oppenheim, A. V. (1981). Iterative techniques for minimum phase signal reconstruction from phase magnitude. *IEEE Transactions on Acoustics, Speech, and Signal Processing*, *29*, 1187–1193. doi:10.1109/TASSP.1981.1163714

Thompson, B. J., Plunkett, S. P., Gurman, J. B., Newmark, J. S., St. Cyr, O. C., & Michels, D. J. (1998). *SoHO*/EIT observations of an Earth-directed coronal mass ejection on May 12, 1997. *Geophysical Research Letters*, *25*, 2465–2468. doi:10.1029/98GL50429

Tian, Y. L., Kanade, T., & Cohn, J. F. (2001). Recognizing action units for facial expression analysis. *IEEE Transactions on Pattern Analysis and Machine Intelligence*, *23*(2), 97–115. doi:10.1109/34.908962

Tian, Y. L., Cohn, J. F., & Kanade, T. (2005). Facial expressions analysis. In Li, S. Z., & Jain, A. K. (Eds.), *Handbook of face recognition* (pp. 247–276). New York: Springer. doi:10.1007/0-387-27257-7_12

Tian, Z., Bell, K. L., & Van Trees, H. L. (1998). A recursive least squares implementation for adaptive beamforming under quadratic constraint. *Proceedings of 9th IEEE Signal Processing Workshop on Statistical Signal and Array Processing, Portland OR USA*, (pp. 9-12).

Tinena, F., Sobrevilla, P., et al. (2009). *On quality assessment of corneal endothelium and its possibility to be used for surgical corneal transplantation*. IEEE International Conference on Fuzzy Systems, 2009. FUZZ-IEEE 2009.

Tong, L., Xu, G., & Kailath, T. (1994). Blind identification and equalization based on secondorder statistics: A timedomain Approach. *IEEE Transactions on Information Theory*, *40*(2), 340–349. doi:10.1109/18.312157

Tong, L. (1995). Blind sequence estimation. *IEEE Transactions on Communications*, *43*(12), 2986–2994. doi:10.1109/26.477501

Tong, L., Xu, G., & Kailath, T. (1995). Blind identification and equalization based on secondorder statistics: A frequencydomain approach. *IEEE Transactions on Information Theory*, *41*(1), 329–334.

Tong, L., & Zeng, H. H. (1997). Channel surfing reinitialization for the constant modulus algorithm. *IEEE Signal Processing Letters*, *4*(3), 85–87.

Touj, S., Ben Amara, N., & Amiri, H. (2005). Arabic handwritten words recognition based on a planar hidden Markov model. *International Arab Journal of Information Technology*, *2*, 318–325.

Touj, S., Ben, N. E., & Amiri, H. (2003). *Generalized Hough transform for Arabic optical character recognition*. International Conference on Document Analysis and Recognition (ICDAR), (pp. 1242-1246).

Tran, D. A., Hua, K. A., & Jiang, N. (2001). A generalized design for broadcasting on multiple physical channel air-cache. In *Proceedings of the ACM SIGAPP Symposium on Applied Computing* (SAC'01), (pp. 387-392).

Treichler, J. R., & Agee, M. G. (1983). A new approach to multipath correction of constant modulus signals. *IEEE Transactions on Acoustics, Speech, and Signal Processing, 31*(2), 459–472. doi:10.1109/TASSP.1983.1164062

Treichler, J. R., & Larimore, M. G. (1985). New processing techniques based on the constant modulus adaptive algorithm. *IEEE Transactions on Acoustics, Speech, and Signal Processing, 33*(2), 420–431.

Treichler, J. R., Larimore, M., & Harp, J. C. (1998). Practical blind demodulators for high-order QAM signals. *Proceedings of the IEEE, 86*(1), 1907–1926. doi:10.1109/5,720245

Trenkle, J., Erlandson, E., Gillies, A., & Schlosser, S. (1995). *Arabic character recognition*. Symposium on Document Image, (pp. 191-195). Bowie, Maryland.

Trenkle, J., Gillies, A., Erlandson, E., Schlosser, S., & Cavin, S. (2001). *Advances In Arabic text recognition*. Symposium on Document Image Understanding Technology (SDIUT 2001), (pp. 159-168). Columbia, Maryland.

Tugnait, J. K. (1995). On blind identifiability of multipath channels using fractional sampling and second-order cyclostationary statistics. *IEEE Transactions on Information Theory, 41*(1), 308–311.

Tugnait, J. K., Tong, L., & Ding, Z. (2000). Single-user channel estimation and equalization. *IEEE Signal Processing Magazine, 17*(3), 17–28. doi:10.1109/MSP.2000.841720

Turmon, M., Jone, H. P., Malanushenko, O. V., & Pap, J. M. (2010). Statistical feature recognition for multidimensional solar imagery. *Solar Physics, 262*, 277–298. doi:10.1007/s11207-009-9490-y

Turmon, M., Pap, J. M., & Mukhtar, S. (2002). Statistical pattern recognition for labeling solar active regions: Application to *SoHO*/MDI imagery. *The Astrophysical Journal, 568*, 396–407. doi:10.1086/338681

United Nations. (2006). *United Nations Arabic Language Programme*. Retrieved December 14, 2009, from http://www.un.org/depts/OHRM/sds/lcp/Arabic/

Uno, H., Kumatani, K., Okuhata, H., Shirakawa, I., & Chiba, T. (1997). ASK digital demodulation scheme for noise immune infrared data communication. *Wireless Networks, 3*(2), 121–129. doi:10.1023/A:1019188729979

Unser, M. (1986). Sum and difference histograms for texture analysis. *IEEE Transactions on Pattern Analysis and Machine Intelligence, 8*, 118–125. doi:10.1109/TPAMI.1986.4767760

USCG. (1985). *US Coast Guard Academy: Loran Engineering Course*. USCG.

Vafadari, R., & Saroukhani, S. (2009). *Application of wavelet analysis in crack identification of frame structures*. Non-Destructive Testing 2009 conference. Blackpool, UK: BINDT.

Van der Schaar, M., & Andreopoulos, Y. (2005). Rate-distortion-complexity modeling for network and receiver aware adaptation. *IEEE Transactions on Multimedia, 7*(3), 471–479. doi:10.1109/TMM.2005.846790

Van der Veen, A. J., Leshem, A., & Boonstra, A. J. (2004). Array signal processing for radio astronomy. In Hall, P. J. (Ed.), *The square kilometre array: An engineering perspective* (pp. 231–249). Dordrecht: Springer.

van Willigen, D. (1989). *Eurofix: Differential hybridized integrated navigation*. Wild Goose Association, 18th Annual Convention and Technical Symposium, Hyannis, MA, November 1989.

Vanam, R., Riskin, E. A., Hemami, S. S., & Ladner, R. E. (2007). Distortion-complexity optimization of the h.264/mpeg-4 avc encoder using the gbfos algorithm. *Proceedings of the 2007 Data Compression Conference*, Washington, DC, USA, (pp. 303–312).

Vanam, R., Riskin, E. A., Hemami, S. S., & Ladner, R. E. (2009). H.264/MPEG-4 AVC encoder parameter selection algorithms for complexity distortion tradeoff. *Proceedings of the Data Compression Conference*.

Vapnik, V. N. (1995). *The nature of statistical learning theory*. New York: Springer.

Varma, M., & Garg, R. (2007). Locally invariant fractal features for statistical texture classification. *Proceedings of the IEEE International Conference on Computer Vision*.

Vasilescu, C., Herlea, V., Talos, F., Ivanov, B., & Dobrescu, R. (2004). Differences between intestinal and diffuse gastric carcinoma: A fractal analysis. In Dobrescu, R., & Vasilescu, C. (Eds.), *Interdisciplinary applications of fractal and chaos theory* (pp. 144–149). Bucuresti: Editura Academiei Romane.

Verma, S., et al. (2008). Two level signature model for multiple broadcast channel using Unified Index Hub (UIH). *Proceeding of 2nd International Conference on Soft Computing ICSC - 2008, IET, Alwar* (Rajasthan) – India, (pp. 422-431).

Verma, S., Rakhee, S., & Sheoran, K. (2008). Signature model for heterogeneous multiple broadcast channel using Unified Index Hub (UIH). *Proceeding of NCRAET, Govt. Engg. College, Ajmer* (Rajasthan) – India, (pp. 45-46).

Vgenis, A., Petrou, C. S., Papadias, C. B., Roudas, I., & Raptis, L. (2010). Nonsingular constant modulus equalizer for PDM-QPSK coherent optical receivers. *IEEE Photonics Technology Letters, 22*(1), 45–47. doi:10.1109/LPT.2009.2035820

Viola, P., & Jones, M. J. (2001). *Rapid object detection using a boosted cascade of simple features*. Paper presented at the CVPR.

Volpe. (2001). *Vulnerability assessment of the transportation infrastructure relying on the global positioning system*. Volpe National Transportation Systems Center.

Vukadinovic, D., & Pantic, M. (2005). *Fully automatic facial feature point detection using Gabor feature based boosted classifiers*. Paper presented at the IEEE International Conference on Systems, Man and Cybernatics.

Walkoe, W., & Starr, T. J. J. (1991). High bit rate digital subscriber line: A copper bridge to the network of the future. *IEEE Journal on Selected Areas in Communications, 9*(6), 765–768. doi:10.1109/49.93087

Wang, J., Li, J., & Wiederhold, G. (2002). SIMPLicity: Semantics-sensitive integrated matching for picture libraries. *IEEE Transactions on PAMI, 23*, 947–963.

Wang, T. C., & Karayiannis, N. B. (1998). Detection of microcalcifications in digital mammograms using wavelets. *IEEE Transactions on Medical Imaging, 17*(4), 498–509. doi:10.1109/42.730395

Wang, D., & Wang, D. (2009). Generalized derivation of neural network constant modulus algorithm for blind equalization. *Proceedings of the 5th International Conference on Wireless Communications, Networking and Mobile Computing, (WiCOM 2009)*, Beijing.

Wang, L.-L., Li, Y.-X., Ding, J.-L., & Ying, W.-Q. (2005). *ARG-based segmentation of overlapping objects in multi-energy X-ray image of passenger accompanied baggage*. Paper presented at the MIPPR 2005: Image Analysis Techniques.

Weil, C. (1975). Absorption characteristic of multilayered sphere models exposed to UHF / Microwave radiation. *IEEE Transactions on BME, 22*(6), 468–476. doi:10.1109/TBME.1975.324467

Weiler, K. (2000). The promise of long wavelength radio astronomy. In Stone, R. G. (Eds.), *Radio astronomy at long wavelengths* (pp. 243–256). American Geophysical Union.

Weismann, M., & Schiper, A. (2005). Comparison of database replication techniques on total order broadcast. *IEEE Transactions on Knowledge and Data Engineering, 17*(4), 551–566. doi:10.1109/TKDE.2005.54

Wen, M., Yao, J., Wong, D. W. K., & Chen, G. C. K. (2005). Holographic diffuser design using a modified genetic algorithm. *Optical Engineering (Redondo Beach, Calif.), 44*(8), 85801–85808. doi:10.1117/1.2031268

Westerveld, T. (2000). Image retrieval: Content versus context. *Content-Based Multimedia Information Access*, 276-284

Widrow, B., & Stearns, S. D. (1985). *Adaptive signal processing*. Prentice-Hall.

Wiegand, T., Sullivan, G., Bjntegaard, G., & Luthra, A. (2003). Overview of the H.264/AVC video coding standard. *IEEE Transactions on Circuits and Systems for Video Technology, 13*, 560–576. doi:10.1109/TCSVT.2003.815165

Wijnholds, S. J., van der Tol, S., Nijboer, R., & van der Veen, A. J. (2010). Calibration challenges for the next generation of radio telescopes. *IEEE Signal Processing Magazine, 27*(1), 32–42.

Wilson, R., & Spann, M. (1988). *Image segmentation and uncertainty*. New York: John Wiley and Sons Inc.

Wnuk, M. (2008). Remarks on hardware implementation of image processing algorithms. *International Journal of Applied Mathematics and Computer Science, 18*(1), 105–110. doi:10.2478/v10006-008-0010-2

Wong, D. W. K., Chen, G., & Yao, J. (2005). Optimization of spot pattern in indoor diffuse optical wireless local area networks. *Optics Express, 13*(8), 3000–3014. doi:10.1364/OPEX.13.003000

Wong, J. W. (1988). Broadcast delivery. *Proceedings of the IEEE, 76*(12), 1566–1577. doi:10.1109/5.16350

Wootten, A., & Emerson, D. (2005). *ALMA: Imaging at the outer limits of radio astronomy*. Paper presented at the IEEE International Conference on Acoustics, Speech, and Signal Processing (ICASSP '05), Philadelphia, USA.

Xu, J., Lee, W. C., & Tang, X. (2004). Exponential index: A parameterized distributed indexing scheme for data on air. [Boston.]. *MobiSYS, 04*(June), 6–9.

Xu, J., Lee, W.-C., Tang, X., Gao, Q., & Li, S. (2006). An error-resilient and tunable distributed indexing scheme for wireless data broadcast. *IEEE Transactions on Knowledge and Data Engineering, 18*(3), 392–404. doi:10.1109/TKDE.2006.37

Xu, Y., Ji, H., & Fermuller, C. (2009). Viewpoint invariant texture description using fractal analysis. *International Journal of Computer Vision, 83*(1), 85–100. doi:10.1007/s11263-009-0220-6

Xu, T. (2008). *Performance benchmarking of FreeRTOS and its hardware abstraction*. Unpublished doctoral thesis, Technical University of Eindhoven.

Xujia, Q., Shishuang, L., et al. (2008). *Medical image enhancement method based on 2D empirical mode decomposition*. The 2nd International Conference on Bioinformatics and Biomedical Engineering, 2008. ICBBE 2008.

Yang, H., & Lu, C. (2000). Infrared wireless LAN using multiple optical sources. *IEE Proceedings. Optoelectronics, 147*(4), 301–307. doi:10.1049/ip-opt:20000610

Yang, Y., & Su, Z. (2010). Medical image enhancement algorithm based on wavelet transform. *Electronics Letters, 46*(2), 120–121. doi:10.1049/el.2010.2063

Yao, M., Yi, W., Shen, B. & Dai, H. (2003). An image retrieval system based on fractal dimension. *Journal of Zhejiang University - Science A, 4*(4), 421-425.

Yao, W. (2008). *AAM library*. Retrieved December 15, 2009, from http://code.google.com/p/aam-library/

Yener, B., & Boult, T. E. (1994). A study of upper and lower bounds for minimum congestion routing in lightwave networks. In *13th Proceedings IEEE INFOCOM '94: Networking for Global Communications*, (pp. 138–147).

Yi, B., & Last, D. (1992). Novel techniques for the identification of Loran Skywaves. *Proceeding of the 21st Annual Technical Symposium, Wild Goose Association*, Birmingham, UK, (pp. 239-246).

Young, C. A., & Gallagher, P. T. (2008). Multiscale edge detection in the corona. *Solar Physics, 248*, 457–469. doi:10.1007/s11207-008-9177-9

Zadeh, L. A. (1996). *Fuzzy sets, fuzzy logic, and fuzzy systems*. New Jersey: World Scientific Publishing.

Zahour, A., Taconet, B., Mercy, P., & Ramdane, S. (2001). *Arabic hand-written text-line extraction*. Sixth International Conference on Document Analysis and Recognition (ICDAR '01), (pp. 281-285). IEEE.

Zahran, O. & Al-Nuaimy, W. (2005). Automatic data processing and defect detection in time-of-flight diffraction images using statistical techniques. *Insight, the Journal of the British Institute of Non-Destructive Testing, 47*(9), 538-542.

Zahran, O. (2006). Automatic detection, sizing and characterisation of weld defects using ultrasonic time-of-flight diffraction. Unpublished doctoral dissertation, University of Liverpool, Liverpool.

Zaouche, A., Dayoub, I. Rouvaen, J.M. & Tatkeu1, C. (2008). Blind channel equalization using constrained generalized pattern search optimization and reinitialization strategy. *EURASIP Journal on Advances in Signal Processing*.

Zbijewski, W., & Stayman, J. W. (2007). *xCAT: A mobile, flat-panel volumetric X-ray CT for head and neck imaging*. Paper presented at the IEEE Nuclear Science Symposium Conference Record.

Zeng, H. H., Tong, L., & Johnson, C. R. Jr. (1998). Relationships between the constant modulus and Wiener receivers. *IEEE Transactions on Information Theory, 44*(4), 152338. doi:10.1109/18.681326

Zeng, H. H., Tong, L., & Johnson, C. R. Jr. (1999). An analysis of constant modulus receivers. *IEEE Transactions on Signal Processing, 47*(11), 29902999. doi:10.1109/78.796434

Zhang, D. D. (2000). *Automated biometrics-technologies and systems.* Kluwer Academic Publishers.

Zhang, Y., Gao, W., Lu, Y., Huang, O., & Zhao, D. (2007). Joint source-channel rate-distortion optimization for H.264 video coding over error-prone networks. *IEEE Transactions on Multimedia, 9*(3), 445–454. doi:10.1109/TMM.2006.887989

Zhang, J., He, W., Yang, S., & Zhong, Y. (2003). Performance and complexity joint optimization for H.264 video coding. *Proceedings of the 2003 International Symposium on Circuits and Systems, 2,* II-888-II-891.

Zhang, J., Wang, Y., & Robinette, S. (2009). *Toward creating a comprehensive digital active region catalog.* In AAS/Solar Physics Division Meeting.

Zhang, Y., Joyner, V., Yun, R., & Sonkusale, S. (2008). A 700mbit/s cmos capacitive feedback front-end amplifier with automatic gain control for broadband optical wireless links. In *IEEE International Symposium on Circuits and Systems, 2008. ISCAS 2008.* (pp. 185–188).

Zharkova, V. V., Ipson, S., Benkhalil, A., & Zharkov, S. (2005). Feature recognition in solar images. *Artificial Intelligence Review, 23,* 209–266. doi:10.1007/s10462-004-4104-4

Zharkova, V. V., Aboudarham, J., Zharkov, S., Ipson, S. S., Benkhalil, A. K., & Fuller, N. (2005). Solar feature catalogues in EGSO. *Solar Physics, 228,* 361–375. doi:10.1007/s11207-005-5623-0

Zheng, G., Dong, X., Rajamani, K. T., Zhang, X., & Styner, M. (2007). Accurate and robust reconstruction of a surface model of the proximal femur from sparse-point data and dense-point distribution model for surgical navigation. *IEEE Transactions on Bio-Medical Engineering, 54*(12), 2109–2122. doi:10.1109/TBME.2007.895736

Zheng, L., Hassin, A., & Tang, X. (2004). A new algorithm for machine printed Arabic character segmentation. *Pattern Recognition Letters, 25,* 1723–1729. doi:10.1016/j.patrec.2004.06.015

Zheng, G., Dong, X., & Nolte, L. P. (2006). *Robust and accurate reconstruction of patient-specific 3D surface models from sparse point sets: a sequential three-stage trimmed optimization approach.* Paper presented at the 3rd Int. Workshop Medical Imaging and Augmented Reality.

Zhivov, A., Stachs, O. S., Stave, J., & Guthoff, R. F. (2009). In vivo three-dimensional confocal laser scanning microscopy of corneal surface and epithelium. *The British Journal of Ophthalmology, 93,* 667–672. doi:10.1136/bjo.2008.137430

Zhu, J., Cao, X.-R., & Liu, R.-W. (1999). A blind fractionally spaced equalizer using higher order statistics. *IEEE Transactions on Circuits and Systems II, 46*(6), 755–764. doi:10.1109/82.769783

Zhu, W., Rose, J., Barshinger, J., & Agarwala, V. (1998). Ultrasonic guided wave NDT for hidden corrosion detection. *Research in Nondestructive Evaluation, 10*(4), 205–225.

Zidouri, A. (2004). *ORAN: A basis for an Arabic OCR system.* International Symposium on Intelligent Multimedia, Video and Speech Processing, (pp. 703-706). Hong Kong.

Zidouri, A., Sarfraz, M., Shahab, S. A., & Jafri, S. M. (2005). *Adaptive dissection based subword segmentation of printed Arabic text.* Ninth International Conference on Information Visualization (IV'05), (pp. 239-243). IEEE Computer Society.

Zitzler, E. (1999). *Evolutionary algorithms for multiobjective optimization: Methods and applications.* Master's thesis, Swiss Federal Institute of technology (ETH), Zurich, Switzerland.

About the Contributors

Rami Qahwaji is a Reader in Visual Computing at the School of Computing, Informatics and Media at Bradford University (UK). Dr. Qahwaji received a first class BSc honours degree in Electrical Engineering, followed by an MSc in Control and Computer Engineering, and a PhD in Computer Vision Systems in 2002 from the University of Bradford. He is the principal investigator 2 major EPSRC grants (EP/F022948/1) and (GR/T17588/01) and also involved in 2 small EPSRC grants. He has done consultancy work for Hines Engineering Ltd, which was funded by Yorkshire Forward. He is also involved in research activities funded by the EU COST action ES0803, where he is the work package (WP) manager for WP142, and by the International Space Science Institute (ISSI) (Bern, Switzerland). He has around 85 refereed publications including 2 edited books, 8 book chapters and 25 journal papers. In addition, he has around 32 invited talks and conference presentations and 11 PhD completions. He is Chartered Engineer (CEng) and a member of various professional organisations, and has refereed research proposals for different national and international grant awarding bodies. He is also a reviewer for several international journals and conferences and he is the conference Co-Chair for the International Conference on Cyberworlds 2009 (UK) and also (CSAA09) (Egypt).

Roger Green received a BSc in Electronics from UMIST, Manchester University, and a PhD in Video Communications from Bradford University, and then a DSc in Photonic Communications, Systems and Devices from the University of Warwick. His research interests are very wide, including photonic systems, communications, signal processing, and optoelectronics. He is a Fellow of the IET, a Fellow of the Institute of Physics, and a Senior Member of the IEEE. He currently holds the Chair in Electronic Communication Systems at the University of Warwick, and is a member of the EPSRC College.

Evor Hines joined the School at the University of Warwick in 1984. He obtained his DSc (Warwick) in 2007. He is FHEA, CEng, and FIET. His main research interest is concerned with intelligent systems (also known by other names such as computational intelligence, soft computing, etc.) and their applications. Most of the work has focused on artificial neural networks, evolutionary algorithms, fuzzy Logic, neuro-fuzzy systems, and so on. Typical application areas include intelligent sensors (e.g. electronic nose); medicine; non-destructive testing of, for example, composite materials; computer vision; telecommunications; amongst others. He has co-authored some 260 articles, and he currently leads the School's Intelligent Systems Engineering Laboratory and Information and Communication Technologies Research Group.

* * *

Taha Y. Ahmed qualified from the medical school, Baghdad in 1997; he became a member of the Royal College of Ophthalmologists, London in 2006. Now, training at one of the nation's most prestigious rotations at the Tennent Institute of Ophthalmology in Glasgow. An author of number of peer-reviewed publications in general ophthalmology and currently developing his surgical and academic experience in cornea and anterior segment surgery.

Fatma Al-Abri received the B.Sc degree in Computer Science from Sultan Qaboos University, Muscat, Sultanate of Oman, in 2003, and the MSc degree in Multimedia and Internet Computing from Loughborough University, Loughborough, UK, in 2006. She is currently a PhD student in the department of Computer Science, Loughborough University. Mrs Al-Abri research interest includes H.264 video coding and multi-objective optimization of video coding and transcoding.

Ali Al-Ataby received BSc degree in Electronic and Communications Engineering from Saddam University in Baghdad, Iraq, in 1997, and subsequently, an MSc in Electronic Circuits and Systems Engineering from the same University in 1999. His main research interests include digital signal and image processing, with special interest in ultrasonic NDT techniques. Other research interests include computer aided design of large circuits and systems and information security and data hiding (Steganography). He is currently a member of the Signal Processing Group at the Department of Electrical Engineering and Electronics, University of Liverpool, UK, and is studying towards his PhD in the subject of automatic detection, sizing and characterisation of flaws using ultrasonic time-of-flight diffraction NDT.

Husni A. Al-Muhtaseb was born in 1961. He received his BSc in electrical engineering, computer option from Yarmouk University, Irbid, Jordan in 1984 and his MSc in computer science and engineering from King Fahd University of Petroleum & Minerals (KFUPM), Dhahran, Saudi Arabia in 1988. He obtained his PhD in Optical text recognition in 2010 from Bradford University (UK). He is currently an Instructor with the Department of Information & Computer Science of KFUPM. He worked as a technical consultant for the dean of admissions and registration at KFUPM for 10 years. His research interests include software development, Arabic Computing, computer Arabization, Arabic OCR, e-learning & online tutoring, and natural Arabic understanding. Mr. Al-Muhtaseb has participated in several industrial projects and worked as a consultant with different institutes/ organisations. Mr. Al-Muhtaseb has more than 60 research publications.

Waleed Al-Nuaimy received a degree in electronic engineering and telecommunications from Saddam University in Baghdad, Iraq, in 1995, and subsequently, a PhD in electrical engineering from the University of Liverpool in the UK in 1998. He joined the Department of Electrical Engineering and Electronics at the University of Liverpool as a lecturer in signal processing in 1999 after a brief period working in industry. His current research is centred on techniques to intelligently process and interpret non-destructive testing data, with a particular interest in geophysical data such as ground-penetrating radar and magnetometric data, and also ultrasonic data. Other research interests include the development of intelligent agent-based implantable bio-devices for the treatment of medical conditions such as hydrocephalus.

Mark J. Bentum was born in Smilde, The Netherlands, in 1967. He received the MSc degree in electrical engineering (with honors) from the University of Twente, Enschede, The Netherlands, in

August 1991. In December 1995, he received the PhD degree for his thesis "Interactive Visualization of Volume Data" also from the University of Twente. From December 1995 to June 1996, he was a research assistant at the University of Twente in the field of signal processing for mobile telecommunications and medical data processing. In June 1996, he joined the Netherlands Foundation for Research in Astronomy (ASTRON). He was in various positions at ASTRON. In 2005, he was involved in the eSMA project in Hawaii to correlate the Dutch JCMT mm-telescope with the Submillimeter Array (SMA) of Harvard University. From 2005 to 2008, he was responsible for the construction of the first software radio telescope in the world, LOFAR (Low Frequency Array). In 2008, he became an Associate Professor in the Telecommunication Engineering Group at the University of Twente. He is now involved with research and education in mobile radio communications. His current research interests are short-range radio communications, novel receiver technologies (for instance in the field of radio astronomy), and sensor networks. Dr. Bentum is a Senior Member of the IEEE, the Dutch Electronics and Radio Society NERG, the Dutch Royal Institute of Engineers KIVI NIRIA, and the Dutch Pattern Recognition Society, and has acted as a reviewer for various conferences and journals.

Albert-Jan Boonstra was born in The Netherlands in 1961. He received the BSc and MSc degrees in applied physics from Groningen University, Groningen, the Netherlands, in 1984 and 1987, respectively. In 2005, he received the PhD degree for his thesis "Radio frequency interference mitigation in radio astronomy" from the Delft University of Technology, Delft, The Netherlands. He was with the Laboratory for Space Research, Groningen, from 1987 to 1991, where he was involved in developing the short wavelength spectrometer (SWS) for the infrared space observatory satellite (ISO). In 1992, he joined ASTRON, the Netherlands Foundation for Research in Astronomy, initially at the Radio Observatory Westerbork, The Netherlands. He is currently with the ASTRON R&D Department, Dwingeloo, The Netherlands, where he heads the DSP group. His research interests lie in the area of signal processing, specifically RFI mitigation by digital filtering.

Min Chen received his BSc degree in Computer Science from Fudan University in 1982 and his PhD degree from University of Wales in 1991. He is currently a professor in Department of Computer Science, University of Wales Swansea. In 1990, he took up a lectureship in Swansea. He became a senior lecturer in 1998, and was awarded a personal chair (professorship) in 2001. His main research interests include visualization, video processing, aging simulation, computer graphics, and multimedia communications. He is a fellow of British Computer Society, a member of Eurographics and ACM SIGGRAPH.

Nicholas Costen received a BA in Experimental Psychology from the University of Oxford and PhD degree in Mathematics and Psychology from the University of Aberdeen. He has undertaken research at the Advanced Telecommunications Research Laboratory, Kyoto and at the Division of Imaging Science and Biomedical Engineering, University of Manchester. He is a Senior Lecturer at Manchester Metropolitan University where his interests include face recognition, human motion analysis, and muscular ultrasound interpretation.

Radu Nicolae Dobrescu MS in Automatic Control from the Faculty of Control and Computers of the Polytechnic Institute of Bucharest, in 1968, PhD degree in Electrical Engineering from the Polytechnic Institute of Bucharest, Romania, in 1976. Currently Professor, Chief of Department of Automation and Industrial Informatics, POLITECHNICA University of Bucharest, he is PhD adviser in the field of Automatic Control. Several scientific works in three main domains: Modern structures for the numerical control of machine tools and manufacturing flexible systems; Data acquisition, processing, and transmission; Local area networks and industrial communication. New researches offer pertinent results in modeling complex biological systems using nonstandard techniques, based on fractal analysis, chaos theory, and scale-free networks. Prof. Dobrescu is a pioneer in Romania in fractal theory applications for medical image processing and biological systems modeling and simulation. He is SRAIT and IEEE member since 1991 (Senior member 2005).

Thierry Dudok de Wit is currently professor at the University of Orléans and researcher at the Laboratoire de Physique et Chimie de l'Environnement et de l'Espace (LPC2E, Orléans). His research domains are solar-terrestrial physics and space weather, with a keen interest in "multis": multivariate statistical, multispectral and multiscale methods applied to multipoint measurements in space. He's also involved in instrumental developments for various satellite missions that are devoted to the observation of the solar-terrestrial system.

Eran Edirisinghe completed his BScEng (hons) degree in Electronic & Telecommunication Engineering at Moratuwa University, Sri Lanka in 1994. He obtained his MSc degree in Digital Communication Systems and PhD degree from Loughborough University, UK in 1996 and 1999. He was appointed a lecturer of the Department of Computer Science, Loughborough University in 2000 and promoted to a Senior Lecturer in 2004. In 2008 he was awarded the title, Reader in Digital Imaging at the same institution. He is currently the Director of Research of the Department of Computer Science and heads the Digital Imaging Research Group. His research interests include compression and coding standards, computer and machine vision, pattern recognition, and mobile technologies. His research has been funded by the EPSRC, TSB, and industry. He has published more than 100 scholarly articles in international conferences and journals and has acted as an expert reviewer for many journals, grant awarding bodies, and conferences.

Abdulhakim Elbita received a BSc Degree in computer science from Faculty of Science, University of Garyounis, Libya in 1989 and an MSc in software engineering from University of Bradford, UK in 2004. Since 2005, he has been employed at the University of Misurata, Libya, where Mr. A. Elbita has taught modules: Programming with C++, Pascal programming language, Fortran programming language, Software Engineering and System Analysis and Design. He has also supervised projects of four groups of BSc students. Currently, he is a PhD research student in the Informatics Research Institute, Digital Imaging Division, School of Computing, Informatics and Media. His research is funded by Libyan higher education.

Hui Fang received his BSc from Science and Technology University, Beijing, China in 2000 and his PhD from Bradford University, U.K. in 2006. Then he worked as a Post-doc research associate in Manchester Metropolitan University. He is now a research officer in computer science department, Swansea University, Wales since 2009. His research interests include facial recognition and analysis, image registration, and image modeling.

Reza Ghaffari is a current PhD student at School of Engineering, University of Warwick, United Kingdom. He has graduated with a BSc Software Engineering degree in 2007 and received his MRes degree in Electronic Systems from Nottingham Trent University in 2008. Reza joined Intelligent Systems Lab at Warwick University in 2008. His major research interests are artificial neural networks, machine learning, genetic algorithms and software architecture. He is currently working on a plant pest and disease diagnosis system for commercial greenhouses using electronic nose and artificial intelligence.

Phil Grant is Head of the Computer Science Department at the Swansea University. He obtained his BSc and Diploma in Mathematics from the University of Manchester and DPhil in Mathematical Logic from the University of Oxford. His research interests include interactive computing, multimedia communication, distributed computation, facial ageing, and biologically inspired models of computation. He is a member of the ACM and IEEE Computer Society.

Christos Grecos is a Professor in Visual Communication standards, Head of School of Computing and director of the Audio-Visual Communications and Networks Research Group (AVCN) in the University of West of Scotland. His main research interests are image/video processing, analysis, compression, and transmission, with emphasis on standard compliant algorithms and implementations. He is a Senior Member of IEEE and SPIE and has published in excess of 80 papers in the areas of his expertise. He has been an invited speaker and session chair in many international conferences and he is associate editor for the Journal of Real Time Image Processing, Springer. He has also been involved in a significant amount of government and industry funded projects.

André W. Gunst was born in Havelte, The Netherlands, in 1972. He received the MSc degree in electrical engineering from the University of Twente, Enschede, The Netherlands, in June 1999. Since then he works for the R&D department at ASTRON as a digital system engineer. Since 2004, he has worked on the development of the station systems in LOFAR. In 2006, he became responsible for the development of the overall LOFAR system for the astronomical applications. His research interests include (digital) system design and digital signal processing.

Matthew D. Higgins received his MEng and PhD degrees from the University of Warwick in 2005 and 2009 respectively. He then progressed to become a Research Fellow, conducting research in the areas of optical wireless communications and channel modelling. Matthew was then seconded for a short time to a leading telecommunications company to advise on novel coding and encryption techniques before returning to the University of Warwick as a Senior Teaching Fellow where he teaches both undergraduate and postgraduate level general communications and optical communications modules. Matthew's primary research interests are in the area of optical wireless communication channel modelling, photonic systems, coding and encryption wherein he has published numerous papers. Matthew is also member of the IEEE and IET.

Tommy Hult received his MSc degree in Electrical Engineering with emphasize on signal processing from Blekinge Institute of Technology, Sweden in 2002 and his Ph.D. in 2008. Since 2009, he is working as a research fellow in radio systems at the communications group of the department of Electrical and Information Technology, Lund University, Sweden. He is also the representative and management committee member of Sweden in five EU COST actions, COST280, COST296, COST297, IC0802 and IC0902. He is the author of more than 60 conference and journal papers in the area of telecommunication. His research interests are mainly in the areas of smart antenna systems, wave propagation, channel modelling, channel measurements, wireless sensor networks and cognitive radio.

Doina Daciana Iliescu graduated in 1991 from the Polytechnic Institute of Bucharest, Romania, Faculty of Electronics and Telecommunications, specialisation in Telecommunications and Data Networks. She received her PhD in Engineering in 1998 from the University of Warwick, UK, in the field of Optical Engineering. Currently she is Associate Professor in the School of Engineering, University of Warwick and a research member of the Systems, Measurement and Modelling Research Group and associate member of the Information and Communication Technologies Group.

Stanley S. Ipson is a Senior Lecturer in EIMC who originally trained as a physicist and has a first class BSc (Hons) in Applied Physics and a PhD in Theoretical Nuclear Physics. Since 1986, his main research interests have related to imaging and have included: extracting 3D information from uncalibrated images, deblurring infrared and millimetre wave images, optical control of industrial machinery, video camera surveillance, texture analysis, digital image watermarking, and a variety of pattern recognition applications. He has been principal investigator for a number of imaging-related projects, receiving funding from EPSRC (£243,791), DERA (£31,623), plus several industrial research contracts totalling £93,250. He was co-investigator on the Solar Feature Recognition part of the European project EGSO (European Grid of Solar Observations) and on the current EPSRC grant (EP/F022948/1), which is entitled "Image Processing, Machine Learning and Geometric Modelling for the 3D Representation of Solar Features". He has supervised 20 completed PhD projects and held two patents. His publications include invited chapters for four books, around 70 refereed journal papers and conference proceedings, and around 35 conference presentations.

Eugene Iwu completed his BEng in Electrical and Electronic Engineering at the University of Benin, Nigeria in 2000. He received an MSc with Distinction in Advanced Photonics and Communications in 2003. He started his carrier in with the global mail and logistics company Deutsche Post DHL in the same year, beginning in warehouse operations. He then moved into Information Technology as the Systems Analyst, before moving into systems implementation. He is now IS Implementation Manager where he is responsible for the implementation and testing of Information Systems during contract start-ups, renewals, and transitions. Eugene is a member of the UK Institution of Engineering and Technology.

Harita Joshi received the degree of Bachelor of Engineering (B.E.) in Electrical Engineering (with first class distinction) from M.S. University of Baroda, India in 2003 and the degree of M.Sc. in Optical Communications (with first class honours) from The University of Warwick, UK in 2005. Currently, she is working towards the degree of PhD in the area of Optical Wireless Communications within the Communications and Signal Processing Group at The University of Warwick, UK.

Savita Kumari did her MCA degree from Kurukshetra University, Kurukshetra, India in 2004. She has teaching experience of about 5 years. Presently, she has been working Faculty, Department of Computer Sciences, Zawia Engineering College, University of Seventh April, Zawia – Libya. Prior to this she was with Banasthali University, Banasthali. She has been pursuing PhD and has research interests in mobile computing, computer architecture, e-learning, m-learning and e-government transactions. Savita

Kumari is life member of Indian Society for Technical Education (ISTE) and has 6 international and national publications in her credit and has attended 12 international and national conferences / workshops.

David Last is a radionavigation expert and consultant, Past President of the Royal Institute of Navigation, and Professor Emeritus at the University of Bangor, Wales. Before his retirement in 2005, he headed the university's Radionavigation Group. He is a Past President of the International Loran Association, a fellow of the Institute of Engineering and Technology (IET), and a Chartered Engineer (CEng). Professor Last has published widely on navigation systems, including GPS, Loran-C, eLoran, Galileo and other global navigation satellite systems, maritime differential GPS, Argos, Decca Navigator, and Omega. He holds a BSc (Eng) from University of Bristol, a PhD from University of Sheffield, and a DSc from University of Wales. He lives and works in Conwy, UK.

Mark S. Leeson received the degrees of BSc and BEng with First Class Honours in Electrical and Electronic Engineering from the University of Nottingham in 1986. He then obtained a PhD in Engineering from the University of Cambridge in 1990. From 1990 to 1992 he worked as a Network Analyst for National Westminster Bank in London. After holding academic posts in London and Manchester, he joined the School of Engineering at Warwick, where he is now an Associate Professor. His major research interests are coding and modulation, ad hoc networking, optical communication systems and evolutionary optimization. To date, Mark has over 180 publications, is a Senior Member of the IEEE, a Chartered Member of the UK Institute of Physics, and a Fellow of the UK Higher Education Academy.

Abbas Mohammed is a Professor of Telecommunications Theory and Director of the Telecommunications and Radio Navigation Research Group at Blekinge Institute of Technology, Sweden. He was awarded the PhD from Liverpool University, UK, in 1992 and the Swedish "Docent degree" in Radio Communications and Navigation from Blekinge Institute of Technology in 2001. He was the recipient of the Blekinge Research Foundation Award *"Researcher of the Year Award and Prize for 2006."* He is a Fellow of The Institution of Engineering and Technology (IET) and the UK's Royal Institute of Navigation (RIN). He is an Associate Editor of the International Journal of Navigation and Observation, a Board Member of the IEEE Signal Processing Swedish Chapter, an Editorial Advisory Board Member of the Mediterranean Journal of Electronics and Communications, and former Editorial Board Member of the Radio Engineering Journal. He has also been a guest editor for several special issues of international journals. He is the author of 15 book chapters and over 200 papers in international journals and conference proceedings in the fields of telecommunications, signal processing, and radio navigation systems. He has also developed techniques for measuring Skywave delays in Loran-C receivers and received a Best Paper Award from the International Loran Association, USA, in connection to this work. He is the Swedish Representative and Management Committee Member to several EU Projects including COST 280, 296, 297, IC0802 and IC0902 Actions. His research interests are in space-time signal processing and MIMO systems, channel modelling, antennas and propagation, cognitive radio, satellite and High Altitude Platforms, and radio navigation systems.

Richard Napier is a plant biologist. Graduating in 1980 from the University of Reading, UK, and obtaining a PhD at the University of Leicester in 1984, he has worked in both animal and plant biochemistry research. Most of his research career has been associated with increasing understanding about how plant hormones are perceived by their protein receptors. These receptors are natural biosensors, although their output is delayed by developmental processing. More recently, Prof. Napier has been developing synthetic biosensors to report on endogenous hormone signals in real-time. An extension of these activities is an interest in applying sensor technologies to biological and crop science problems. He serves on the editorial boards of two plant biology journals.

Dan Popescu MS in Automatic Control from the Faculty of Control and Computers of the Polytechnic Institute of Bucharest, in 1974, MS in mathematics from Faculty of Mathematics of the University of Bucharest, in 1980, PhD degree in Electrical Engineering from the Polytechnic Institute of Bucharest, Romania, in 1987. Currently Professor in the Department of Automation and Industrial Informatics of the Faculty of Control and Computers, POLITECHNICA University of Bucharest, he is PhD adviser in the field of Automatic Control. The competence domains are: data and signal processing, image acquisition and processing, pattern recognition, sensors for robots, and digital circuit design. Current scientific areas are: equipments for complex measurements, data acquisition and remote control, wireless sensor networks, pattern recognitions and complex image processing, and interdisciplinary approaches. He is author of 15 books and more than 120 papers. He is IEEE and SRAIT member.

Simant Prakoonwit received BEng in electronic engineering from Chulalongkorn University, Thailand, MSc in communications and signal processing and PhD in 3D computer vision applied to medicine from Imperial College London, UK. After gaining his PhD, Dr. Prakoonwit worked as a Postdoctoral Research Assistant in the Department of Biological and Medical Systems (now Department of Bioengineering) also at Imperial College. His postdoctoral research was the development of optimum and accurate computerised 3D reconstruction methods from X-ray images. His work can be applied to both security applications (e.g. airport weapon and bomb scanning) and medical applications (e.g. optimum 3D human organ reconstruction). He is currently a lecturer with the School of Systems Engineering, University of Reading, UK. His research interests are medical imaging, computer vision, and biomedical engineering. He was the recipient of the *IEEE Innovation and Creativity Prize Paper Award* in 2005.

Rakhee Kulshrestha received the MSc and PhD degree from Dr. B.R. Ambedkar University, Agra, India, in 1999 and 2003 respectively. From 2004 to 2008 she worked at Banasthali University. In 2008, she joined Birla Institute of Technology and Science, Pilani, India. Her research interests are in the area of analysis of communication systems, especially cellular mobile systems, handoff management, mobile computing and queuing theory.

Hassan Ugail is the director for the Centre for Visual Computing at Bradford. He has a first class BSc Honours degree in Mathematics from King's College London and a PhD in the field of geometric design from the School of Mathematics at University of Leeds. Prof Ugail's research interests include geometric modelling, computer animation, functional design, numerical methods and design optimisation, applications of geometric modelling to real time interactive and parametric design, and applications of geometric modelling to general engineering problems. Prof Ugail has 3 patents on novel techniques relating to geometry modelling, animation, and 3D data exchange. He is a reviewer for various international journals, conferences and grant awarding bodies. His recent innovations have led the formation a university spin-out company, Tangentix Ltd, with investments from venture capitalists. He has recently won the vice-chancellor's award for knowledge transfer for his outstanding contribution to research and knowledge transfer activities.

Seema Verma is a researcher working as an Associate Professor in Department of Electronics, Banasthali University, Banasthali India. Her areas of interest include communication, error control coding, turbo codes, OFDM, MIMO, channel modeling, VLSI design of communication systems, indexing techniques, cryptography, cognitive radio, et cetera. She is the investigator of the research projects funded by AICTE & ISRO. She is a research guide announced by UGC. She has many papers to her credit in many national & international conferences.

Moi-Hoon Yap received her PhD in Computer Science from Loughborough University in 2009. She received her BSc(hons) in Statistics from Universiti Putra Malaysia (UPM) in 1999, and MSc(IT) in 2001. In 2002 and 2005, she served as a lecturer at the Faculty of Information Technology, Multimedia University (MMU), Malaysia. Currently she is a Postdoctoral Research Assistant in the Centre for Visual Computing, University of Bradford, UK. She is actively involving in computer vision research. Her research of interest is facial analysis, medical image analysis, human perception, and image and video processing.

Fu Zhang is currently a PhD student in the School of Engineering, University of Warwick, U.K. He received his first degree in Computer Science at the University of Cambridge and MSc degree in Electrical Engineering at the University of Warwick. His major research interests are the development and applications of artificial neural networks, genetic algorithms, fuzzy logic and Grey system theory, which is a relatively new theory. His recent research work is closely connected with horticultural production and environmental study.

Index